Advanced Textbooks in Control and Signal Processing

Springer
London
Berlin
Heidelberg
New York
Hong Kong
Milan
Paris
Tokyo

Series Editors

Professor Michael J. Grimble, Professor of Industrial Systems and Director
Professor Michael A. Johnson, Professor of Control Systems and Deputy Director
Industrial Control Centre, Department of Electronic and Electrical Engineering,
University of Strathclyde, Graham Hills Building, 50 George Street, Glasgow G1 1QE, U.K.

Other titles published in this series:

Genetic Algorithms
K.F. Man, K.S. Tang and S. Kwong

Neural Networks for Modelling and Control of Dynamic Systems
M. Nørgaard, O. Ravn, L.K. Hansen and N.K. Poulsen

Fault Detection and Diagnosis in Industrial Systems
L.H. Chiang, E.L. Russell and R.D. Braatz

Soft Computing
L. Fortuna, G. Rizzotto, M. Lavorgna, G. Nunnari, M.G. Xibilia and R. Caponetto

Statistical Signal Processing
T. Chonavel

Discrete-time Stochastic Processes (2nd Edition)
T. Söderström

Parallel Computing for Real-time Signal Processing and Control
M.O. Tokhi, M.A. Hossain and M.H. Shaheed

Multivariable Control Systems
P. Albertos and A. Sala

Control Systems with Input and Output Constraints
A.H. Glattfelder and W. Schaufelberger

Analysis and Control of Non-linear Process Systems
K. Hangos, J. Bokor and G. Szederkényi

Model Predictive Control (2nd Edition)
E.F. Camacho and C. Bordons

Principles of Adaptive Filters and Self-learning Systems
A. Zaknich
Publication due June 2005

L. Sciavicco and B. Siciliano

Modelling and Control of Robot Manipulators

Second Edition

With 190 Figures

Springer

Professor L. Sciavicco
Dipartimento di Informatica e Automazione, Università degli Studi di Roma Tre,
Via della Vasca Navale 79, 00146 Rome, Italy

Professor B. Siciliano
Dipartimento di Informatica e Sistemistica, Università degli Studi di Napoli Federico II, Via Claudio 21, 80125 Naples, Italy

British Library Cataloguing in Publication Data
Sciavicco, L.
Modelling and control of robot manipulators. - 2nd ed. -
(Advanced textbooks in control and signal processing)
1.Robots - Control systems 2.Robots - Motion 3.Manipulators
(Mechanism)
I.Title II.Siciliano, Bruno, 1959-
629.8'92
ISBN 1852332212

Library of Congress Cataloging-in-Publication Data
Sciavicco, L.. (Lorenzo)
Modelling and control of robot manipulators / L. Sciavicco and B. Siciliano.
p. cm. -- (Advanced textbooks in control and signal processing)
Includes bibliographical references and index.
ISBN 1-85233-221-2 (alk. paper)
1. Robots--Control systems. 2. Manipulators (Mechanism) I. Siciliano, Bruno, 1959-
II. Title. III. Series.
TJ211.35.S43 2000
629.8'92--dc21 99-462018

ISBN 1-85233-221-2 Springer-Verlag London Berlin Heidelberg
Springer-Verlag is a part of Springer Science+Business Media
springeronline.com

1st edition published by McGraw Hill Inc., 1996

Printed in Great Britain
5th printing 2004

Typesetting: Camera ready by authors
Printed and bound by Athenæum Press Ltd., Gateshead, Tyne & Wear
69/3830-54 Printed on acid-free paper SPIN 10988381

To our families

Series Editors' Foreword

The topics of control engineering and signal processing continue to flourish and develop. In common with general scientific investigation, new ideas, concepts and interpretations emerge quite spontaneously and these are then discussed, used, discarded or subsumed into the prevailing subject paradigm. Sometimes these innovative concepts coalesce into a new sub-discipline within the broad subject tapestry of control and signal processing. This preliminary battle between old and new usually takes place at conferences, through the Internet and in the journals of the discipline. After a little more maturity has been acquired by the new concepts then archival publication as a scientific or engineering monograph may occur.

A new concept in control and signal processing is known to have arrived when sufficient material has developed for the topic to be taught as a specialised tutorial workshop or as a course to undergraduates, graduates or industrial engineers. The *Advanced Textbooks in Control and Signal Processing Series* is designed as a vehicle for the systematic presentation of course material for both popular and innovative topics in the discipline. It is hoped that prospective authors will welcome the opportunity to publish a structured presentation of either existing subject areas or some of the newer emerging control and signal processing technologies.

The authors Lorenzo Sciavicco and Bruno Siciliano declare that robotics is more than fifteen years old and is a young subject! Yet, this textbook shows that a well-established paradigm of classical robotics exists and the book provides an invaluable presentation of the subject. The Series is fortunate in being able to welcome this text as a second edition. Thus it is an updated text which has benefited from the authors' teaching practice and an awareness of very recent developments in the field. Notable in this sense is the inclusion of material on vision sensors and trajectory planning.

As a course textbook, the authors have explained how various chapters may be drawn together to form a course. Further, the book is supported by a Solutions Manual. Last, but not least we ought to mention three very substantial Appendices giving useful supplementary material on the necessary mathematics, rigid body dynamics and feedback control. A fine new addition to the Series!

M.J. Grimble and M.A. Johnson
Industrial Control Centre
Glasgow, Scotland, U.K.
December 1999

About the Authors

Lorenzo Sciavicco was born in Rome, Italy, on December 8, 1938. He received the *Laurea* degree in Electronic Engineering from the University of Rome in 1963. From 1968 to 1995 he was with the University of Naples, where he served as Head of the Department of Computer and Systems Engineering from 1992 to 1995. He is currently *Professor* in the Department of Computer Engineering and Automation of the Third University of Rome. His research interests include automatic control theory and applications, inverse kinematics, redundant manipulator control, force/motion control and cooperative robots. He is co-author of more than 80 journal and conference papers, and he is co-author of the book *Modeling and Control of Robot Manipulators* (McGraw-Hill 1996). Professor Sciavicco has been one of the pioneers of robot control research in Italy, and has been awarded numerous grants for his robotics group. He has served as a referee for industrial and academic research projects on robotics and automation in Italy.
sciavicco@unina.it http://www.dia.uniroma3.it/autom/Sciavicco

Bruno Siciliano was born in Naples, Italy, on October 27, 1959. He received the *Laurea* degree and the *Research Doctorate* degree in Electronic Engineering from the University of Naples in 1982 and 1987, respectively. From 1983 to 2000 he was with the Department of Computer and Systems Engineering of the University of Naples. Since November 2000 he is *Professor* in the Department of Information and Electrical Engineering of the University of Salerno. From September 1985 to June 1986 he was a Visiting Scholar at the School of Mechanical Engineering of the Georgia Institute of Technology. His research interests include inverse kinematics, redundant manipulator control, modelling and control of flexible arms, force/motion and vision-based control, and cooperative robots. He is co-author of more than 180 journal and conference papers, and he is co-author of the books: *Modeling and Control of Robot Manipulators* with *Solutions Manual* (McGraw-Hill 1996), *Theory of Robot Control* (Springer 1996), *Robot Force Control* (Kluwer 1999). He is co-editor of 4 journal special issues, and he is co-editor of the books: *Control Problems in Robotics and Automation* (Springer 1998), *RAMSETE* (Springer 2001), *ISER'02* (Springer 2002). He has delivered more than 70 invited seminars abroad. Professor Siciliano has served as an *Associate Editor* of the *IEEE Transactions on Robotics and Automation* from 1991 to 1994, and of the *ASME Journal of Dynamic Systems, Measurement, and Control* from 1994 to 1998. He is co-editor of the *Advanced Robotics Series* (Springer), and he is on the Editorial Boards of *Robotica*, the *Journal of Robotic Systems* and the *JSME International Journal.*

He is an *IEEE Fellow* and an ASME Member. He has held representative positions within the *IEEE Robotics and Automation Society*: *Administrative Committee Member* from 1996 to 1999, *Vice-President for Publications* in 1999, and *Vice-President for Technical Activities* since 2000. From 1996 to 1999 he has been *Chair* of the *Technical Committee on Manufacturing and Automation Robotic Control* of the *IEEE Control Systems Society*. He is *Co-Chair for Conferences and Publications* of the undergoing *European Robotics Research Network*. He has served as chair or co-chair for numerous international conferences. He has been awarded numerous grants for his robotics group.
siciliano@unina.it http://cds.unina.it/~sicilian

Preface to the Second Edition

The subject matter of this textbook is to be considered well assessed in the classical robotics literature, in spite of the fact that robotics is generally regarded as a young science.

A key feature of the First Edition was recognized to be the blend of technological and innovative aspects with the foundations of modelling and control of robot manipulators. The purpose of this Second Edition with the new Publisher is to add some material that was not covered before as well as to streamline and improve some of the previous material.

The major additions regard Chapter 2 on kinematics; namely, the use of the unit quaternion to describe manipulator's end-effector orientation as an effective alternative to Euler angles or angle and axis representations (Section 2.6), and the adoption of a closed chain in the design of manipulator structures (Sections 2.8.3 and 2.9.2). Not only are these topics analyzed in the framework of kinematics, but also their impact on differential kinematics, statics, dynamics and control is illustrated. In particular, different types of orientation error are discussed for inverse kinematics algorithms (Section 3.7.3), and the concept of kineto-statics duality is extended to manipulators having a closed chain (Section 3.8.3). Yet, the dynamic model of a parallelogram arm (Section 4.3.3) clearly shows the potential of such design over the kinematically equivalent two-link planar arm. Further, the problem of planning a trajectory in the operational space is expanded to encompass the different descriptions of end-effector orientation (Section 5.3.3), and the implications for operational space control are briefly discussed (Section 6.6.3).

Another addition regards the presentation of the main features of vision sensors (Section 8.3.4) which have lately been receiving quite a deal of attention not only in research but also in the industrial community.

Finally, the bibliography has been updated with more reference texts in the introduction (Chapter 1) as well as with those references that have been used in the preparation of the new material (Chapters 2 to 8). New problems have been proposed and the Solutions Manual accompanying the book has been integrated accordingly.

Naples, December 1999 *Lorenzo Sciavicco and Bruno Siciliano*◊

◊ The authors have contributed equally to the work, and thus they are merely listed in alphabetical order.

The **Solutions Manual for Modelling and Control of Robot Manipulators, Second Edition** (ISBN 1-85233-221-2S) by Bruno Siciliano and Luigi Villani can be requested by textbook adopters from

Springer-Verlag London Ltd
Sweetapple House, Catteshall Road
Godalming, Surrey GU7 3DJ
UK
Tel: +44 (1483) 414113
Fax: +44 (1483) 415144
E-mail: postmaster@svl.co.uk
URL: www.springer.co.uk

Preface to the First Edition

In the last fifteen years, the field of robotics has stimulated an increasing interest in a wide number of scholars, and thus literature has been conspicuous both in terms of textbooks and monographs and in terms of specialized journals dedicated to robotics. This strong interest is also to be attributed to the interdisciplinary character of robotics, which is a science having roots in different areas. Cybernetics, mechanics, bioengineering, electronics, information science, and automatic control science—to mention the most important ones—are all cultural domains which undoubtedly have boosted the development of robotics. This science, however, is to be considered quite young yet.

Nowadays, writing a robotics book brings up a number of issues concerning the choice of topics and style of presentation. Current literature features many texts which can be grouped in scientific monographs on research themes, application-oriented handbooks, and textbooks. As for the last, there are wide-ranging textbooks covering a variety of topics with unavoidably limited depth and textbooks instead covering in detail a reduced number of topics believed to be basic for robotics study. Among these, mechanics and control are recognized to play a fundamental role, since these disciplines regard the preliminary know-how required to realize robot manipulators for industrial applications, *i.e.*, the only domain so far where robotics has expressed its level of a mature technology.

The goal of this work is to present the foundations of modelling and control of robot manipulators where the fundamental, technological and innovative aspects are merged on a uniform track in respect of a rigorous formalism.

Fundamental aspects are covered which regard kinematics, statics and dynamics of manipulators, trajectory planning and motion control in free space. Technological aspects include actuators, proprioceptive sensors, hardware/software control architectures and industrial robot control algorithms. Established research results with a potential for application are presented, such as kinematic redundancy and singularities, dynamic parameter identification, robust and adaptive control and interaction control. These last aspects are not systematically developed in other textbooks, even though they are recognized to be useful for applications. In the choice of the topics treated and the relative weight between them, the authors hope not to have been biased by their own research interests.

The book contents are organized into 9 chapters and 3 appendices.

In Chapter 1, the problems concerning the use of *industrial robots* are focused in the general framework of *robotics*. The most common manipulation mechanical structures are presented. *Modelling and control* topics are also introduced which are developed in the subsequent chapters.

In Chapter 2 *kinematics* is presented with a systematic and general approach which refers to Denavit-Hartenberg convention. The *direct kinematics equation* is formulated which relates joint space variables to operational space variables. This equation is utilized to find manipulator workspace as well as to derive a kinematic calibration technique. The *inverse kinematics problem* is also analyzed and closed-form solutions are found for typical manipulation structures.

Differential kinematics is presented in Chapter 3. The relationship between joint velocities and end-effector linear and angular velocities is described by the geometric *Jacobian*. The difference between geometric Jacobian and analytical Jacobian is pointed out. The Jacobian constitutes a fundamental tool to characterize a manipulator, since it allows finding singular configurations, analyzing redundancy and expressing the relationship between forces and moments applied to the end effector and the resulting joint torques at equilibrium configurations (*statics*). Moreover, the Jacobian allows formulating inverse kinematics algorithms that solve the inverse kinematics problem even for manipulators not having a closed-form solution.

Chapter 4 deals with derivation of manipulator *dynamics*, which plays a fundamental role for motion simulation, manipulation structure analysis and control algorithm synthesis. The dynamic model is obtained by explicitly taking into account the presence of actuators. Two approaches are considered; namely, one based on *Lagrange* formulation, and the other based on *Newton-Euler* formulation. The former is conceptually simpler and systematic, whereas the latter allows computation of dynamic model in a recursive form. Notable properties of the dynamic model are presented, including linearity in the parameters which is utilized to develop a model identification technique. Finally, the transformations needed to express the dynamic model in the operational space are illustrated.

As a premise to the motion control problem, in Chapter 5, *trajectory planning* techniques are illustrated which regard the computation of interpolating polynomials through a sequence of desired points. Both the case of *point-to-point motion* and that of *path motion* are treated. Techniques are developed for generating trajectories both in the joint and in the operational space, with a special concern to orientation for the latter. Finally, a trajectory dynamic scaling technique is presented to keep the joint torques within the maximum available limits at the actuators.

In Chapter 6 the problem of *motion control* in free space is treated. The distinction between joint space *decentralized* and *centralized* control strategies is pointed out. With reference to the former, the independent joint control technique is presented which is typically used for industrial robot control. As a premise to centralized control, the computed torque feedforward control technique is introduced. Advanced schemes are then introduced including PD control with gravity compensation, inverse dynamics control, robust control, and adaptive control. Centralized techniques are extended to operational space control.

Interaction control of a manipulator in contact with the working environment is tackled in Chapter 7. The concepts of mechanical *compliance* and *impedance* are defined as a natural extension of operational space control schemes to the constrained motion case. *Force control* schemes are then presented which are obtained by the addition of an outer force feedback loop to a motion control scheme. The hybrid force/position control strategy is finally presented with reference to the formulation of natural and artificial constraints describing an interaction task.

Chapter 8 is devoted to the presentation of *actuators* and *sensors*. After an illustration of the general features of an actuating system, methods to control electric and hydraulic servomotors are presented. A few proprioceptive sensors are then described, including encoders, resolvers, tachometers, and force sensors.

In Chapter 9, the functional *architecture* of a robot *control* system is illustrated. The characteristics of programming environments are presented with an emphasis on teaching-by-showing and robot-oriented programming. A general model for the hardware architecture of an industrial robot control system is finally discussed.

Appendix A is devoted to *linear algebra* and presents the fundamental notions on matrices, vectors and related operations.

Appendix B recalls those basic concepts of *rigid body mechanics* which are preliminary to the study of manipulator kinematics, statics and dynamics.

Finally, Appendix C illustrates the principles of *feedback control* of linear systems and presents a general method based on Lyapunov theory for control of nonlinear systems.

The book is the evolution of the lecture notes prepared for the course "Industrial Robotics" taught by the first author in 1990 and 1991 and by the second author since 1992 at the University of Naples. The course is offered to Computer, Electronic and Mechanical Engineering graduate students and is developed with a teaching commitment of about 90 hours.

By a proper selection of topics, the book may be utilized to teach a course on robotics fundamentals at a senior undergraduate level. The advised selection foresees coverage of the following parts*: Chapter 1, Chapter 2, Chapter 4 (Sections 4.1 and 4.3), Chapter 5, Chapter 6 (Sections 6.1, 6.2, 6.3, and 6.4), Chapter 8, and Chapter 9. The teaching commitment is of about 50 hours. In this case, the availability of an industrial robot in the laboratory is strongly recommended to accompany class work with training work.

From a pedagogical viewpoint, the various topics are presented in an instrumental manner and are developed with a gradually increasing level of difficulty. Problems are raised and proper tools are established to find engineering-oriented solutions. Each chapter is introduced by a brief preamble providing the rationale and the objectives of the subject matter. The topics needed for a proficient study of the text are presented into three considerable appendices, whose purpose is to provide students of different extraction with a homogeneous background. Mechanical Engineering students will benefit from reading of the appendices on linear algebra and feedback control, whereas Computer and Electronic Engineering students are advised to study the appendix on rigid body mechanics.

The book contains more than 170 illustrations and more than 50 worked-out examples and case studies throughout the text with frequent resort to simulation.

---

* Those parts that shall be covered only at a graduate level are marked with an asterisk in the table of contents.

The results of computer implementations of inverse kinematics algorithms, inverse dynamics computation, trajectory planning techniques, motion and interaction control algorithms are presented in much detail in order to facilitate the comprehension of the theoretical development as well as to increase sensitivity to application in practical problems. More than 80 problems are proposed and the book is accompanied by a Solutions Manual that comes with a toolbox created in MATLAB® with Simulink® to solve those problems requiring computer simulation. Special care has been devoted to the selection of bibliographical references (more than 200) which are collected at the end of each chapter.

Naples, July 1995 *LS & BS*

® MATLAB and Simulink are registered trademarks of The MathWorks Inc

Acknowledgements

The authors wish to acknowledge all those who have been helpful in the preparation of this book.

Particular thanks go to Pasquale Chiacchio and Stefano Chiaverini, with whom the authors have been collaborating on robotics research activities for several years; the discussions and exchange of viewpoints with them in the planning stage of the text have been stimulating. Significant have been Pasquale Chiacchio's contribution to the writing of Chapter 5 on trajectory planning and Stefano Chiaverini's contribution to the writing of Chapter 7 on interaction control.

The valuable engagement of Luigi Villani, first as a graduate student of the course and then as a doctorate student, is acknowledged. He has substantially contributed to the writing of Appendix B on rigid body mechanics as well as of the final version of Chapter 4 on dynamics. His careful reading of the entire manuscript has allowed for an improvement of a few topics. He has provided relevant support in the development of those examples requiring computer simulation. The educational potential of the text is certainly increased by the availability of a solutions manual which features Luigi Villani as a co-author.

A special note of thanks goes to the colleagues Wayne Book, Alessandro De Luca, Gianantonio Magnani, Claudio Melchiorri and Deirdre Meldrum for having provided constructive criticisms on the contents of the book, which they have adopted in the form of lecture notes for university courses they have taught. A number of useful suggestions have also come from the colleagues Olav Egeland, Mark Spong and Antonio Tornambè, to whom the authors' sincere thanks are to be presented.

Students' participation in the refinement of the various versions of the lecture notes has been active with their requests of clarification and pointing out of numerous imprecisions.

Fabrizio Caccavale and Ciro Natale have been precious in the revision of the material for the second edition of the book. The latter is to be acknowledged also for his contribution to the writing of Section 8.3.4 on vision sensors. The second edition has also benefited from the comments on the first edition by the colleagues Thomas Alberts, Jon Kieffer, George Lee, Carlos Lück, Norberto Pires and Juris Vagners.

Finally, the authors wish to thank Nicholas Pinfield, Engineering Editor of Springer-Verlag, London, for his great enthusiasm in the project and for bringing the second edition of the book to fruition. His assistant, Oliver Jackson, also deserves a warm note of mention for his precious collaboration and patience during the preparation of the manuscript.

Table of Contents

1. Introduction

Robotics is concerned with the study of those machines that can replace human beings in the execution of a task, as regards both physical activity and decision making. The goal of the introductory chapter is to point out the problems related to the use of *robots* in *industrial* applications, with reference to the general framework of robotics. A classification of the most common *manipulator* mechanical structures is presented. Topics of *modelling* and *control* are introduced which will be examined in the following chapters. The chapter ends with a list of references dealing with subjects both of specific interest and of related interest to those covered by this textbook.

1.1 Robotics

Robotics has profound cultural roots. In the course of centuries, human beings have constantly attempted to seek substitutes that would be able to mimic their behaviour in the various instances of interaction with the surrounding environment. Several motivations have inspired this continuous search referring to philosophical, economic, social and scientific principles.

One of human beings' greatest ambitions has been to give life to their artefacts. The legend of the titan Prometheus, who molded humankind from clay, as well as that of the giant Talus, the bronze slave forged by Hephaestus, testify how Greek mythology was influenced by that ambition, which has been revisited in the tale of Frankenstein in modern times.

So as the giant Talus was entrusted with the task of protecting the island of Crete from invaders, in the Industrial Age a mechanical creature (*automaton*) has been entrusted with the task of substituting a human being in subordinate labor duties. This concept was introduced by the Czech playwright Karel Čapek who wrote the play *Rossum's Universal Robots (R.U.R.)* in 1921. On that occasion he coined the term *robot*—derived from the Slav *robota* that means executive labor—to denote the automaton built by Rossum who ends up by rising against humankind in the science fiction tale.

In the subsequent years, in view of the development of science fiction, the behaviour conceived for the robot has often been conditioned by feelings. This has contributed to render the robot more and more similar to its creator.

It was in the forties when the Russian Isaac Asimov, the well-known science fiction writer, conceived the robot as an automaton of human appearance but devoid of feelings. Its behaviour was dictated by a "positronic" brain programmed by a human being in such a way as to satisfy certain rules of ethical conduct. The term *robotics* was then introduced by Asimov as the symbol of the science devoted to the study of robots which was based on the *three fundamental laws*:

1. A robot may not injure a human being or, through inaction, allow a human being to come to harm.
2. A robot must obey the orders given by human beings, except when such orders would conflict with the first law.
3. A robot must protect its own existence, as long as such protection does not conflict with the first or second law.

These laws established rules of behaviour to consider as specifications for the design of a robot, which since then has attained the connotation of an industrial product designed by engineers or specialized technicians.

Science fiction has influenced common people that continue to imagine the robot as a humanoid who can speak, walk, see, and hear, with an exterior very much like that presented by the robots of the movie *Star Wars*.

According to a scientific interpretation of the science-fiction scenario, the robot is seen as a machine that, independently of its exterior, is able to modify the environment in which it operates. This is accomplished by carrying out actions that are conditioned by certain rules of behaviour intrinsic in the machine as well as by some data the robot acquires on its status and on the environment. In fact, *robotics* has recently been defined as the science studying the *intelligent connection of perception to action*.

The robot's capacity for action is provided by a *mechanical system* which is in general constituted by a locomotion apparatus to move in the environment and by a manipulation apparatus to operate on the objects present in the environment. The realization of such a system refers to a scientific framework concerning the design of articulated mechanical systems, choice of materials, and type of actuators that ensure mobility to the structure.

The robot's capacity for perception is provided by a *sensory system* which can acquire data on the internal status of the mechanical system (proprioceptive sensors) as well as on the external status of the environment (exteroceptive sensors). The realization of such a system refers to a scientific framework concerning materials science, signal conditioning, data processing, and information retrieval.

The robot's capacity for connecting action to perception in an intelligent fashion is provided by a *control system* which can decide the execution of the action in respect of the constraints imposed by the mechanical system and the environment. The realization of such a system refers to the scientific framework of cybernetics, concerning artificial intelligence and expert systems, programming environments, computational architectures, and motion control.

Therefore, it can be recognized that robotics is an interdisciplinary subject concerning the cultural areas of *mechanics*, *electronics*, *information theory*, and *automation theory*.

The above considerations point out both the conceptual and technological complexity that influences development of robots endowed with good characteristics of *autonomy*. This is needed for the execution of missions in unstructured or scarcely structured environments, *i.e.*, when geometrical or physical description of the environment is not completely known a priori.

The expression *advanced robotics* usually refers to the science studying robots with marked characteristics of autonomy, whose applications are conceived to solve problems of operation in hostile environments (space, underwater, nuclear, military, *etc.*) or to execute service missions (domestic applications, medical aids, assistance to the disabled, agriculture, *etc.*).

Nowadays, advanced robotics is still in its infancy. It has indeed featured the realization of prototypes only, because the associated technology is not yet mature. The motivations urging an advance of knowledge in this field are multiple; they range from the need for automata whenever human operators are not available or are not safe (e.g., applications in hostile environments) to the opportunity of developing products for potentially wide markets which are aimed at improving quality of life (e.g., service robotics).

If a robot is assumed to operate in a strongly structured environment, the degree of autonomy required for the automaton is radically decreased. The industrial environment, at least for a conspicuous number of significant applications, presents the above characteristic. *Industrial robotics* is the discipline concerning robot design, control and applications in industry, and its products are by now reaching the level of a mature technology.

Industrial robots have gained a wide popularity as essential components for the realization of automated manufacturing systems. Reduction of manufacturing costs, increase of productivity, improvement of product quality standards and, last but not least, the possibility of eliminating harmful or alienating tasks for the human operator in a manufacturing system, represent the main factors that have determined spreading of robotics technology in a wider and wider range of applications in manufacturing industry.

In view of the above, it should be clear how an important chapter of robotics science is constituted by industrial robotics, whose fundamentals are treated in this textbook.

1.2 Industrial Robot

By its usual meaning, the term *automation* denotes a technology aimed at replacing human beings with machines in a manufacturing process, as regards not only the execution of physical operations but also the intelligent processing of information on the status of the process. Automation is then the synthesis of industrial technologies typical of the manufacturing process and computer technology allowing information management. The three levels of automation one may refer to are: rigid automation, programmable automation, and flexible automation.

Rigid automation deals with a factory context oriented to the mass manufacturing of products of the same type. The need to manufacture large numbers of parts with high productivity and quality standards demands the use of fixed operational sequences to be executed on the workpiece by special purpose machines.

Programmable automation deals with a factory context oriented to the manufacturing of low-to-medium batches of products of different types. A programmable automated system allows easily changing the sequence of operations to be executed on the workpieces in order to vary the range of products. The machines employed are more versatile and are capable to manufacture different objects belonging to the same group technology. The majority of the products available on the market today are manufactured by programmable automated systems.

Flexible automation represents the evolution of programmable automation. Its goal is to allow manufacturing of variable batches of different products by minimizing the time lost for reprogramming the sequence of operations and the machines employed to pass from one batch to the next. The realization of a flexible manufacturing system demands a strong integration of computer technology with industrial technology.

The *industrial robot* is a machine with significant characteristics of versatility and flexibility. According to the widely accepted definition of the Robot Institute of America, *a robot is a reprogrammable multifunctional manipulator designed to move materials, parts, tools or specialized devices through variable programmed motions for the performance of a variety of tasks*. Such a definition, dating to 1980, reflects the current status of robotics technology.

By virtue of its programmability, the industrial robot is a typical component of programmable automated systems. Nonetheless, robots can be entrusted with tasks both in rigid automated systems and in flexible automated systems. An industrial robot is constituted by:

- A mechanical structure or *manipulator* that consists of a sequence of rigid bodies (*links*) connected by means of articulations (*joints*); a manipulator is characterized by an *arm* that ensures mobility, a *wrist* that confers dexterity, and an *end effector* that performs the task required of the robot.
- *Actuators* that set the manipulator in motion through actuation of the joints; the motors employed are typically electric and hydraulic, and occasionally pneumatic.
- *Sensors* that measure the status of the manipulator (proprioceptive sensors) and, if necessary, the status of the environment (exteroceptive sensors).
- A *control system* (computer) that enables control and supervision of manipulator motion.

The essential feature that differentiates an industrial robot from a numerically controlled machine tool is its enhanced versatility; this is endowed by the manipulator's end effector, which can be many a tool of different type, as well as by the large workspace compared to manipulator encumbrance.

Industrial robots present three fundamental capacities that make them useful for a manufacturing process: *material handling*, *manipulation*, and *measurement*.

In a manufacturing process, each object has to be transferred from one location of the factory to another in order to be stored, manufactured, assembled, and packed. During transfer, the physical characteristics of the object do not undergo any alteration. The robot's capacity to pick up an object, move it in space on predefined paths and release it makes the robot itself an ideal candidate for material handling operations. Typical applications include:

- palletizing (placing objects on a pallet in an ordered way),
- warehouse loading and unloading,
- mill and machine tool tending,
- part sorting,
- packaging.

Manufacturing consists of transforming objects from raw material into finished products; during this process, the part either changes its own physical characteristics as a result of machining or loses its identity as a result of an assembly of more parts. The robot's capacity to manipulate both objects and tools make it suitable to be employed for manufacturing. Typical applications include:

- arc and spot welding,
- painting and coating,
- gluing and sealing,
- laser and water jet cutting,
- milling and drilling,
- casting and die spraying,
- deburring and grinding,
- screwing, wiring and fastening,
- assembly of mechanical and electrical groups,
- assembly of electronic boards.

Besides material handling and manipulation, in a manufacturing process it is necessary to perform measurements to test product quality. The robot's capacity to explore the three-dimensional space together with the availability of measurements on the manipulator's status allow using a robot as a measuring device. Typical applications include:

- object inspection,
- contour finding,
- detection of manufacturing imperfections.

The listed applications describe the current employment of robots as components of industrial automation systems. They all refer to strongly structured working environments and thus they do not exhaust all the possible utilizations of robots for industrial applications. The fall-outs of advanced robotics products may be of concern for industrial robotics whenever one attempts to solve problems regarding the adaptation of the robot to a changeable working environment.

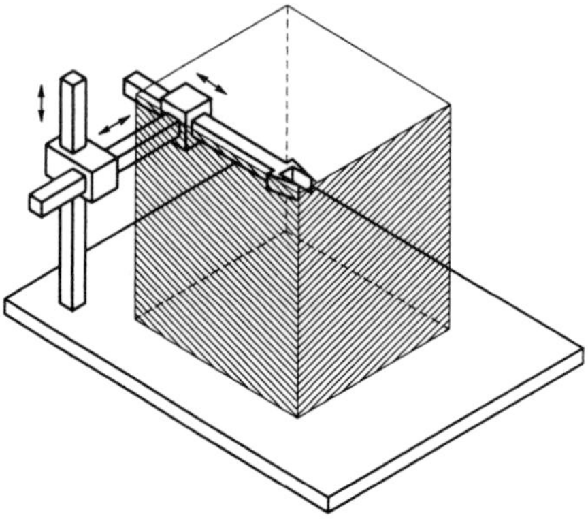

Figure 1.1 Cartesian manipulator and its workspace.

1.3 Manipulator Structures

The fundamental structure of a manipulator is the *open kinematic chain*. From a topological viewpoint, a kinematic chain is termed open when there is only one sequence of links connecting the two ends of the chain. Alternatively, a manipulator contains a *closed kinematic chain* when a sequence of links forms a loop.

Manipulator's mobility is ensured by the presence of joints. The articulation between two consecutive links can be realized by means of either a *prismatic* or a *revolute* joint. In an open kinematic chain, each prismatic or revolute joint provides the structure with a single degree of mobility. A prismatic joint realizes a relative translational motion between the two links, whereas a revolute joint realizes a relative rotational motion between the two links. Revolute joints are usually preferred to prismatic joints in view of their compactness and reliability. On the other hand, in a closed kinematic chain, the number of degrees of mobility is less than the number of joints in view of the constraints imposed by the loop.

The *degrees of mobility* shall be properly distributed along the mechanical structure in order to provide the *degrees of freedom* required for the execution of a given task. Typically, each joint providing a degree of mobility is actuated. In the most general case of a task consisting of arbitrarily positioning and orienting an object in the three-dimensional space, *six* are the required degrees of freedom, three for positioning a point on the object and three for orienting the object with respect to a reference coordinate frame. If more degrees of mobility than degrees of freedom are available, the manipulator is said to be *kinematically redundant*.

The *workspace* represents that portion of the environment the manipulator's end effector can access. Its shape and volume depend on the manipulator structure as well as on the presence of mechanical joint limits.

The task required of the arm is to position the wrist which then is required to orient the end effector; at least three degrees of mobility are then necessary in the three-dimensional workspace. The type and sequence of the arm's degrees of mobility,

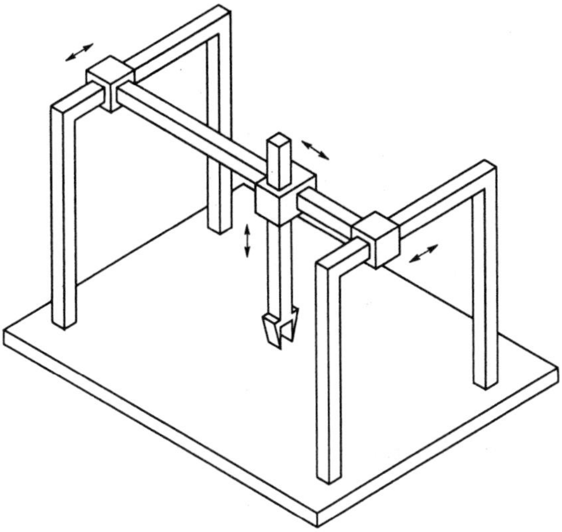

Figure 1.2 Gantry manipulator.

starting from the base joint, allows classifying manipulators as: *Cartesian*, *cylindrical*, *spherical*, *SCARA*, and *anthropomorphic*.

The *Cartesian* geometry is realized by three prismatic joints whose axes typically are mutually orthogonal (Figure 1.1). In view of the simple geometry, each degree of mobility corresponds to a degree of freedom in the Cartesian space and thus it is natural to perform straight motions in space. The Cartesian structure offers very good mechanical stiffness. Wrist positioning accuracy is constant everywhere in the workspace. This is the volume enclosed by a rectangular parallelepiped (Figure 1.1). As opposed to high accuracy, the structure has low dexterity since all the joints are prismatic. The approach to manipulate an object is sideways. On the other hand, if it is desired to approach an object from the top, the Cartesian manipulator can be realized by a *gantry* structure as illustrated in Figure 1.2. Such a structure allows obtaining a large volume workspace and manipulating objects of gross dimensions and weight. Cartesian manipulators are employed for material handling and assembly. The motors actuating the joints of a Cartesian manipulator are typically electric and occasionally pneumatic.

The *cylindrical* geometry differs from the Cartesian one in that the first prismatic joint is replaced with a revolute joint (Figure 1.3). If the task is described in cylindrical coordinates, also in this case each degree of mobility corresponds to a degree of freedom. The cylindrical structure offers good mechanical stiffness. Wrist positioning accuracy decreases as the horizontal stroke increases. The workspace is a portion of a hollow cylinder (Figure 1.3). The horizontal prismatic joint makes the wrist of a cylindrical manipulator suitable to access horizontal cavities. Cylindrical manipulators are mainly employed for carrying objects even of gross dimensions; in such a case the use of hydraulic motors is to be preferred to that of electric motors.

The *spherical* manipulator differs from the cylindrical one in that the second pris-

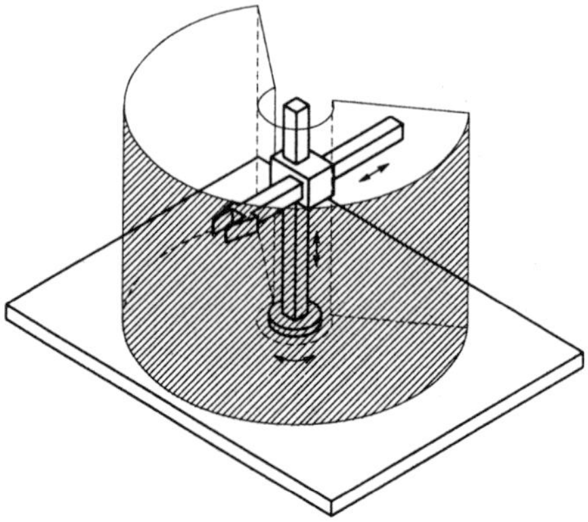

Figure 1.3 Cylindrical manipulator and its workspace.

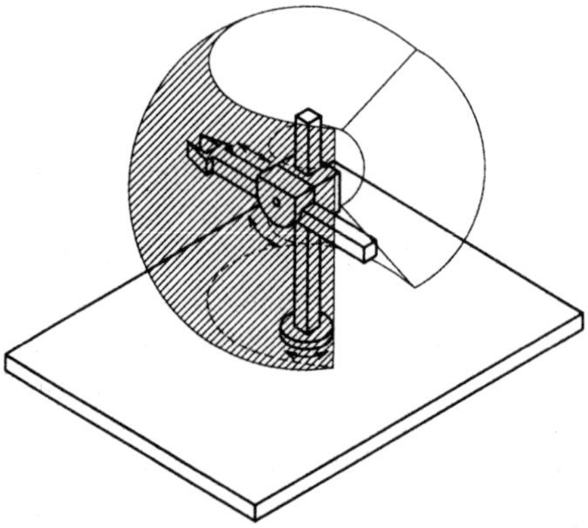

Figure 1.4 Spherical manipulator and its workspace.

matic joint is replaced with a revolute joint (Figure 1.4). Each degree of mobility corresponds to a degree of freedom only if the task is described in spherical coordinates. Mechanical stiffness is lower than the above two geometries and mechanical construction is more complex. Wrist positioning accuracy decreases as the radial stroke increases. The workspace is a portion of a hollow sphere (Figure 1.4); it can include also the supporting base of the manipulator and thus it can allow manipulation of objects on the floor. Spherical manipulators are mainly employed for machining. Electric motors are typically used to actuate the joints.

A special geometry is the *SCARA* geometry that can be realized by disposing two revolute joints and one prismatic joint in such a way that all the axes of motion are parallel (Figure 1.5). The acronym SCARA stands for *Selective Compliance Assembly Robot Arm* and characterizes the mechanical features of a structure offering high stiffness to vertical loads and compliance to horizontal loads. As such, the SCARA

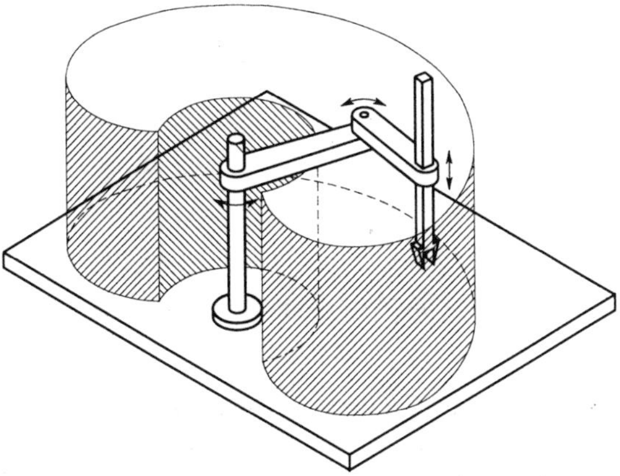

Figure 1.5 SCARA manipulator and its workspace.

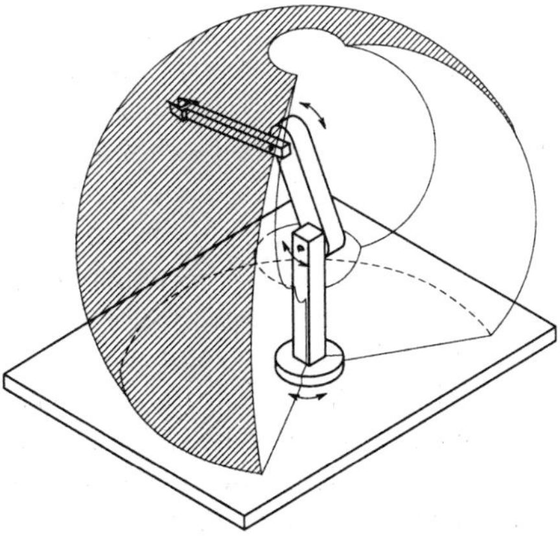

Figure 1.6 Anthropomorphic manipulator and its workspace.

structure is congenial to vertical assembly tasks. The correspondence between the degrees of mobility and the degrees of freedom is maintained only for the vertical component of a task described in Cartesian coordinates. Wrist positioning accuracy decreases as the distance of the wrist from the first joint axis increases. The typical workspace is illustrated in Figure 1.5. The SCARA manipulator is suitable for manipulation of small objects; joints are actuated by electric motors.

The *anthropomorphic* geometry is realized by three revolute joints; the revolute axis of the first joint is orthogonal to the axes of the other two which are parallel (Figure 1.6). By virtue of its similarity with the human arm, the second joint is called the shoulder joint and the third joint is called the elbow joint since it connects the "arm" with the "forearm." The anthropomorphic structure is the most dexterous one,

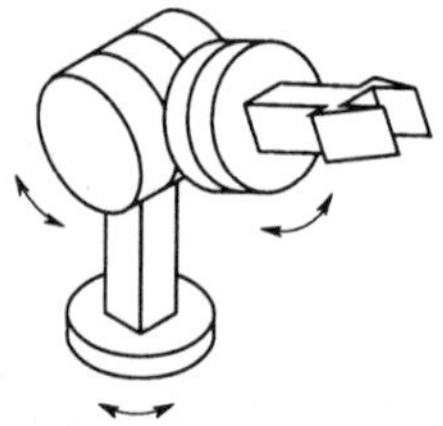

Figure 1.7 Spherical wrist.

since all the joints are revolute. On the other hand, the correspondence between the degrees of mobility and the degrees of freedom is lost and wrist positioning accuracy varies inside the workspace. This is approximately a portion of a sphere (Figure 1.6) and its volume is large compared to manipulator encumbrance. Joints are typically actuated by electric motors. The range of industrial applications of anthropomorphic manipulators is wide.

All the above manipulators have an open kinematic chain structure. Whenever larger payloads are to be manipulated, the mechanical structure shall have higher stiffness to guarantee comparable positioning accuracy. In this case, one has to resort to structures having closed kinematic chains. For instance, sometimes a parallelogram design is adopted between the shoulder and elbow joints of the arm for an anthropomorphic structure, thus creating a closed kinematic chain; nonetheless, a substantial kinematic equivalence with the open chain can be shown.

The manipulator structures presented above are required to position the wrist which then is required to orient the manipulator's end effector. If arbitrary orientation in the three-dimensional space is desired, the wrist must possess at least three degrees of mobility provided by revolute joints. Since the wrist constitutes the terminal part of the manipulator, it has to be compact; this often complicates its mechanical design. Without entering into construction details, the realization endowing the wrist with the highest dexterity is one where the three revolute axes intersect at a single point. In such a case, the wrist is called a *spherical wrist*, as represented in Figure 1.7. The key feature of a spherical wrist is the decoupling between position and orientation of the end effector; the arm is entrusted with the task of positioning the above point of intersection, whereas the wrist determines the end-effector orientation. Those realizations where the wrist is not spherical are simpler from a mechanical viewpoint, but position and orientation are coupled, and this complicates the coordination between the motion of the arm and that of the wrist to perform a given task.

The *end effector* is specified according to the task the robot shall execute. For material handling tasks, the end effector is constituted by a gripper of proper shape and dimensions determined by the object to grasp. For machining and assembly tasks, the end effector is a tool or a specialized device, e.g., a welding torch, a spray gun, a mill, a drill, or a screwdriver.

The versatility and flexibility of a robot manipulator shall not induce the conviction that all mechanical structures are equivalent for the execution of a given task. The choice of a robot is indeed conditioned by the application which sets constraints on the

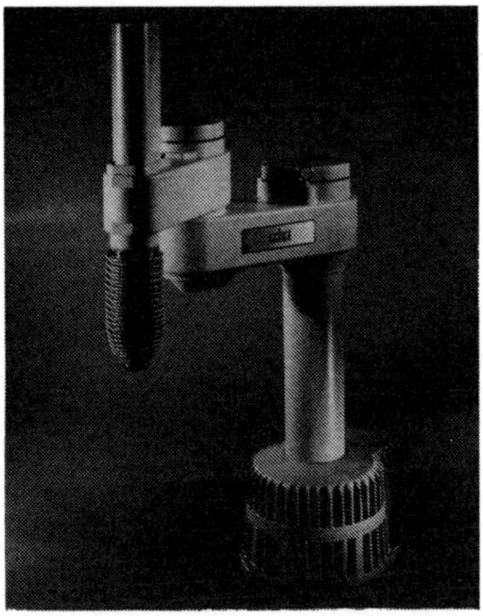

Figure 1.8 The AdeptOne XL robot (courtesy of Adept Technology Inc).

workspace dimensions and shape, the maximum payload, positioning accuracy, and dynamic performance of the manipulator.

The photographs of a few industrial robots are illustrated in Figures 1.8 to 1.13.

The AdeptOne XL robot in Figure 1.8 has a four-joint SCARA structure. Direct drive motors are employed. The maximum reach is 800 mm, with a repeatability[1] of 0.025 mm horizontally and 0.038 mm vertically. Maximum speeds are 1200 mm/s for the prismatic joint, while range from to 650 to 3300 deg/s for the three revolute joints. The maximum payload is 12 kg. Typical industrial applications include small-parts material handling, assembly and packaging.

The Comau SMART S2 robot in Figure 1.9 has a six-joint anthropomorphic structure with nonspherical wrist. The maximum reach is 1458 mm horizontally and 2208 mm vertically, with a repeatability of 0.1 mm. Maximum speeds range from 115 to 200 deg/s for the inner three joints, and from 300 to 430 deg/s for the outer three joints. The maximum payload is 16 kg. Both floor and ceiling mounting positions are allowed. Typical industrial applications include arc welding, light handling, assembly and technological processes.

The ABB IRB 4400 robot in Figure 1.10 has also a six-joint anthropomorphic structure, but differently from the previous open-chain structure, it possesses a closed chain of parallelogram type between the shoulder and elbow joints. The maximum reach ranges from 1960 to 2740 mm for the various versions, with a repeatability from 0.07 to 0.1 mm. The maximum speed at the end effector is 2200 mm/s. The maximum payload is 60 kg. Floor or shelf mounting is available. Typical industrial applications include material handling, machine tending, grinding, gluing, casting, die

[1] Repeatability is a classical parameter found in industrial robot data sheets. It gives a measure of the manipulator's ability to return to a previously reached position.

Figure 1.9 The Comau SMART S2 robot (courtesy of Comau SpA Robotica).

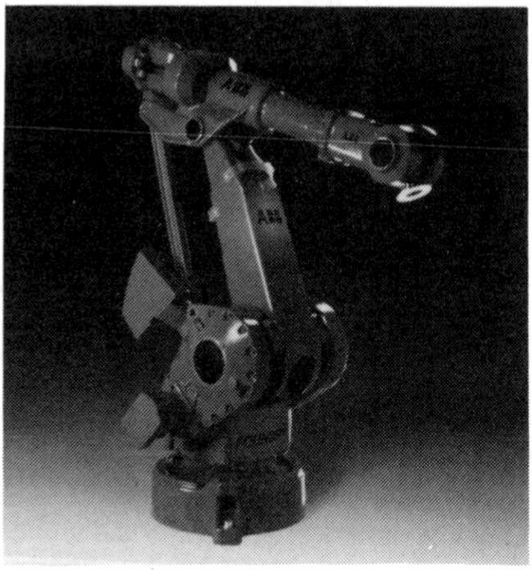

Figure 1.10 The ABB IRB 4400 robot (courtesy of ABB Flexible Automation AB).

spraying and assembly.

The Kuka KL 250 linear unit with KR 15/2 robot in Figure 1.11 is composed by a six-joint anthropomorphic structure with spherical wrist, which is mounted on a sliding track with a gantry type installation; the upright installation is also available. The maximum payload of the linear unit is 250 kg with a stroke of 6200 mm, a maximum speed of 1310 mm/s and a repeatability of 0.2 mm. On the other hand, the robot is characterized by a maximum payload of 25 kg, a maximum reach of 1570 mm and a repeatability of 0.1 mm. Maximum speeds are 152 deg/s for the inner three joints, while range from 284 to 604 deg/s for the outer three joints. Since motion control of the linear

Figure 1.11 The Kuka KL 250 linear unit with KR 15/2 robot (courtesy of Kuka Roboter GmbH).

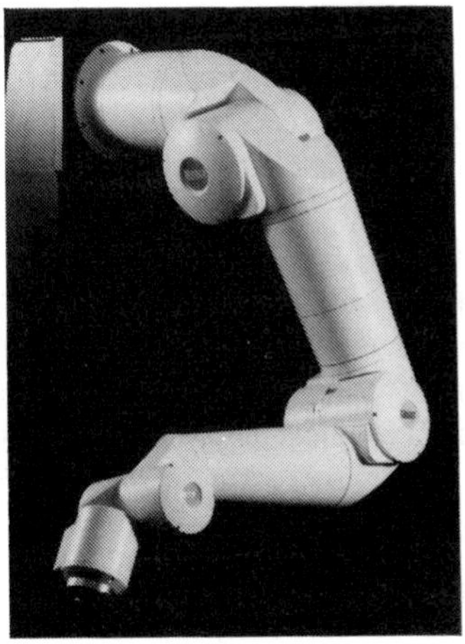

Figure 1.12 The Robotics Research K-1207i robot (courtesy of Robotics Research Corporation).

unit is integrated in the robot control as a seventh joint axis, kinematic redundancy with respect to six-degree-of-freedom tasks is achieved and in turn enhanced mobility throughout the workspace. Typical industrial applications include machine tending, arc welding, deburring, coating, sealing, plasma and waterjet cutting.

The next two structures are to be considered less conventional than the previous four, as long as industrial applications are concerned. The Robotics Research K-1207i robot in Figure 1.12 has also a seven-joint structure, but the additional joint is of revolute type and is integrated into the articulated robot. Enhanced dexterity

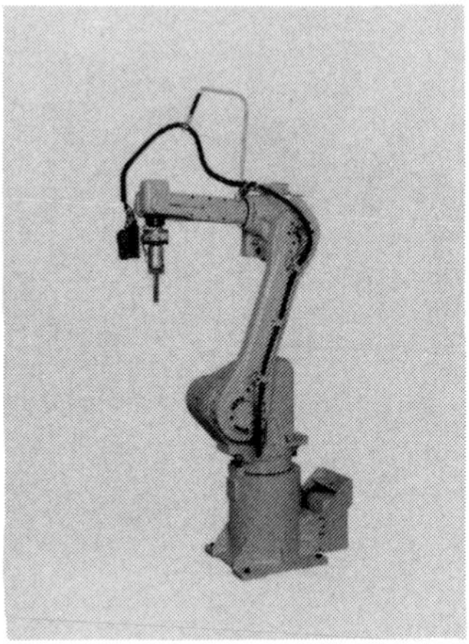

Figure 1.13 The Fanuc I-21i robot (courtesy of Fanuc Ltd).

is achieved allowing the robot to fold compactly, a feature exploitable in operations requiring manipulation through risers or small portholes, as well as minimizing the storage requirements. A modular arm construction concept is adopted so that units with seven joints in series up to seventeen axes in branching topologies can be assembled. The robot is used for manufacturing operations, especially in the aerospace industry and research field.

The Fanuc I-21i robot in Figure 1.13 has a six-joint anthropomorphic structure with a nonspherical wrist. The novelty for an industrial product is represented by a sensor-based control unit including 3D vision guidance and six-axis force sensor. It is used for handling of randomly positioned objects, e.g., workpieces scattered on a tray, as well as for sophisticated mechanical parts assembly, e.g., bolt fastening.

1.4 Modelling and Control of Robot Manipulators

In all industrial robot applications, completion of a generic task requires the execution of a specific motion prescribed to the manipulator's end effector. The motion can be either unconstrained, if there is no physical interaction between the end effector and the environment, or constrained if contact forces arise between the end effector and the environment.

The correct execution of the end-effector motion is entrusted to the control system which shall provide the joint actuators of the manipulator with the commands consistent with the desired motion trajectory. Control of end-effector motion demands an accurate analysis of the characteristics of the mechanical structure, actuators, and sensors. The goal of such analysis is derivation of mathematical models of robot components. Modelling a robot manipulator is therefore a necessary premise to finding motion control strategies.

In the remainder, those significant topics in the study of modelling and control of robot manipulators are illustrated which constitute the subjects of the subsequent chapters.

1.4.1 Modelling

Kinematic analysis of a manipulator structure concerns the description of the manipulator motion with respect to a fixed reference Cartesian frame by ignoring the forces and moments that cause motion of the structure. It is meaningful to distinguish between kinematics and differential kinematics. *Kinematics* describes the analytical relationship between the joint positions and the end-effector position and orientation. *Differential kinematics* describes the analytical relationship between the joint motion and the end-effector motion in terms of velocities.

The formulation of the kinematics relationship allows studying two key problems of robotics; namely, the direct kinematics problem and the inverse kinematics problem. The former concerns the determination of a systematic, general method to describe the end-effector motion as a function of the joint motion by means of linear algebra tools. The latter concerns the inverse problem; its solution is of fundamental importance to transform the desired motion naturally prescribed to the end effector in the workspace into the corresponding joint motion.

The availability of a manipulator's kinematic model is useful also to determine the relationship between the forces and torques applied to the joints and the forces and moments applied to the end effector in *static* equilibrium configurations.

Chapter 2 is dedicated to the study of kinematics; Chapter 3 is dedicated to the study of differential kinematics and statics; whereas Appendix A provides a useful brush-up on *linear algebra*.

Kinematics of a manipulator represents the basis of a systematic, general derivation of its *dynamics*, *i.e.*, the equations of motion of the manipulator as a function of the forces and moments acting on it. The availability of the dynamic model is very useful for mechanical design of the structure, choice of actuators, determination of control strategies, and computer simulation of manipulator motion. Chapter 4 is dedicated to the study of dynamics; whereas Appendix B recalls some fundamentals on *rigid body mechanics*.

1.4.2 Control

With reference to the tasks assigned to a manipulator, the issue is whether to specify the motion at the joints or directly at the end effector. In material handling tasks, it is sufficient to assign only the pick-up and release locations of an object (point-to-point motion), whereas, in machining tasks, the end effector has to follow a desired trajectory (path motion). The goal of *trajectory planning* is to generate the time laws for the relevant variables (joint or end-effector) starting from a concise description of the desired motion. Chapter 5 is dedicated to trajectory planning.

The trajectories generated constitute the reference inputs to the *motion control* system of the mechanical structure. The problem of manipulator control is to find the time behaviour of the forces and torques to be delivered by the joint actuators so as to ensure the execution of the reference trajectories. This problem is quite complex, since a manipulator is an articulated system and, as such, the motion of one link influences the motion of the others. Manipulator equations of motion indeed reveal the presence of coupling dynamic effects among the joints, except in the case of a Cartesian structure with mutually orthogonal axes. The synthesis of the joint forces and torques cannot be made on the basis of the sole knowledge of the dynamic model, since this does not completely describe the real structure. Therefore, manipulator control is entrusted to the closure of feedback loops; by computing the deviation between the reference inputs and the data provided by the proprioceptive sensors, a feedback control system is capable to satisfy accuracy requirements on the execution of the prescribed trajectories.

Chapter 6 is dedicated to the presentation of motion control techniques; whereas Appendix C illustrates the basic principles of *feedback control*.

If a manipulation task requires interaction between the end effector and the environment, the control problem is further complicated by observing that besides the (constrained) motion, also the contact forces have to be controlled. Chapter 7 is dedicated to the presentation of *interaction control* techniques.

Realization of the motion specified by the control law requires the employment of *actuators* and *sensors*. The functional characteristics of the most commonly used actuators and sensors for industrial robots are described in Chapter 8.

Finally, Chapter 9 is concerned with the hardware/software *architecture* of a robot's *control system* which is in charge of implementation of control laws as well as of interface with the operator.

1.5 Bibliographical Reference Texts

In the last twenty years, the robotics field has stimulated the interest of an increasing number of scholars. A truly respectable international research community has been established. Literature production has been conspicuous, both in terms of textbooks and scientific monographs and in terms of journals dedicated to robotics. Therefore, it seems appropriate to close this introduction by offering a selection of *bibliographical reference texts* to those readers who wish to make a thorough study of robotics.

Besides indicating those textbooks sharing an affinity of contents with this one, the following lists include general books and specialized texts on related subjects, collections of contributions on the state of the art of research, scientific journals, and series of international conferences.

Textbooks on Modelling and Control of Robot Manipulators

Asada H., Slotine J.-J.E. (1986) *Robot Analysis and Control.* Wiley, New York.

Craig J.J. (1989) *Introduction to Robotics: Mechanics and Control.* 2nd ed., Addison-Wesley, Reading, Mass.

Khalil W., Dombre E. (1999) *Modélisation Identification et Commande des Robots.* 2ème éd., Hermès, Paris.

Koivo A.J. (1989) *Fundamentals for Control of Robotic Manipulators*. Wiley, New York.

Lewis F.L., Abdallah C.T., Dawson D.M. (1993) *Control of Robot Manipulators*. Macmillan, New York.

Paul R.P. (1981) *Robot Manipulators: Mathematics, Programming, and Control*. MIT Press, Cambridge, Mass.

Schilling R.J. (1990) *Fundamentals of Robotics: Analysis and Control*. Prentice-Hall, Englewood Cliffs, N.J.

Spong M.W., Vidyasagar M. (1989) *Robot Dynamics and Control*. Wiley, New York.

Yoshikawa T. (1990) *Foundations of Robotics*. MIT Press, Cambridge, Mass.

General Books

Critchlow A.J. (1985) *Introduction to Robotics*. Macmillan, New York.

Dorf R.C. (1988) *International Encyclopedia of Robotics*. Wiley, New York.

Engelberger J.F. (1980) *Robotics in Practice*. Amacom, New York.

Engelberger J.F. (1989) *Robotics in Service*. MIT Press, Cambridge, Mass.

Fu K.S., Gonzalez R.C., Lee C.S.G. (1987) *Robotics: Control, Sensing, Vision, and Intelligence*. McGraw-Hill, New York.

Hunt V.D. (1983) *Industrial Robotics Handbook*. Industrial Press, New York.

Koren Y. (1985) *Robotics for Engineers*. McGraw-Hill, New York.

McKerrow P.J. (1991) *Introduction to Robotics*. Addison-Wesley, Sydney.

Snyder W.E. (1985) *Industrial Robots: Computer Interfacing and Control*. Prentice-Hall, Englewood Cliffs, N.J.

Vukobratović M. (1989) *Introduction to Robotics*. Springer-Verlag, Berlin.

Specialized Texts

Topics of related interest to modelling and control of robot manipulators are:

- manipulator mechanical design,
- manipulation tools,
- manipulators with elastic members,
- parallel robots,
- locomotion apparatus,
- motion planning of mobile robots,
- force control,
- robot vision,
- multisensory data fusion.

The following texts are dedicated to these topics:

Angeles J. (1997) *Fundamentals of Robotic Mechanical Systems: Theory, Methods, and Algorithms*. Springer-Verlag, New York.

Canny J.F. (1988) *The Complexity of Robot Motion Planning*. MIT Press, Cambridge, Mass.

Canudas de Wit C., Siciliano B., Bastin G. (Eds.) (1996) *Theory of Robot Control*. Springer-Verlag, London.

Corke, P.I. (1996) *Visual Control of Robots*. Research Studies Press, Taunton, England.

Cutkosky M.R. (1985) *Robotic Grasping and Fine Manipulation*. Kluwer Academic Publishers, Boston, Mass.

Durrant-Whyte H.F. (1988) *Integration, Coordination and Control of Multi-Sensor Robot Systems*. Kluwer Academic Publishers, Boston, Mass.

Fraser A.R., Daniel R.W. (1991) *Perturbation Techniques for Flexible Manipulators*. Kluwer Academic Publishers, Boston, Mass.

Hirose, S. (1993) *Biologically Inspired Robots*. Oxford University Press, Oxford, England.

Horn B.K.P. (1986) *Robot Vision*. McGraw-Hill, New York.

Latombe J.-C. (1991) *Robot Motion Planning*. Kluwer Academic Publishers, Boston, Mass.

Mason M.T., Salisbury J.K. (1985) *Robot Hands and the Mechanics of Manipulation*. MIT Press, Cambridge, Mass.

Merlet J.-P. (2000) *Parallel Robots*. Kluwer Academic Publishers, Dordrecht, The Netherlands.

Murray R.M., Li Z., Sastry S.S. (1994) *A Mathematical Introduction to Robotic Manipulation*. CRC Press, Boca Raton, Fla.

Raibert M. (1985) *Legged Robots that Balance*. MIT Press, Cambridge, Mass.

Rivin E.I. (1987) *Mechanical Design of Robots*. McGraw-Hill, New York.

Siciliano B., Villani L. (1999) *Robot Force Control*. Kluwer Academic Publishers, Boston, Mass.

Todd D.J. (1985) *Walking Machines, an Introduction to Legged Robots*. Chapman Hall, London.

Tsai L.-W. (1999) *Robot Analysis: The Mechanics of Serial and Parallel Manipulators*. Wiley, New York.

Collections of Contributions on the State of the Art of Research

Brady M. (1989) *Robotics Science*. MIT Press, Cambridge, Mass.

Brady M., Hollerbach J.M., Johnson T.L., Lozano-Pérez T., Mason M.T. (1982) *Robot Motion: Planning and Control*. MIT Press, Cambridge, Mass.

Khatib O., Craig J.J., Lozano-Pérez T. (1989) *The Robotics Review 1*. MIT Press, Cambridge, Mass.

Khatib O., Craig J.J., Lozano-Pérez T. (1992) *The Robotics Review 2*. MIT Press, Cambridge, Mass.

Lee C.S.G., Gonzalez R.C., Fu K.S. (1986) *Tutorial on Robotics*. 2nd ed., IEEE Computer Society Press, Silver Spring, Md.

Spong M.W., Lewis F.L., Abdallah C.T. (1993) *Robot Control: Dynamics, Motion Planning, and Analysis*. IEEE Press, New York.

Scientific Journals on Robotics

Advanced Robotics

IEEE Robotics and Automation Magazine

IEEE Transactions on Robotics and Automation

International Journal of Robotics and Intelligent Systems

International Journal of Robotics Research

Journal of Robotic Systems

Robotica

Robotics and Autonomous Systems

Series of Scientific International Conferences on Robotics

IEEE International Conference on Robotics and Automation

IEEE/RSJ International Conference on Intelligent Robots and Systems

IFAC Symposium on Robot Control

International Conference on Advanced Robotics

International Symposium of Robotics Research

International Symposium on Experimental Robotics

Several prestigious journals and conferences give substantial space to robotics subjects. Such references are not cited here because they are not purely dedicated to robotics.

2. Kinematics

A *manipulator* can be schematically represented from a mechanical viewpoint as a kinematic chain of rigid bodies (*links*) connected by means of revolute or prismatic *joints*. One end of the chain is constrained to a base, while an *end effector* is mounted to the other end. The resulting motion of the structure is obtained by composition of the elementary motions of each link with respect to the previous one. Therefore, in order to manipulate an object in space, it is necessary to describe the end-effector position and orientation. This chapter is dedicated to the derivation of the *direct kinematics equation* through a systematic, general approach based on linear algebra. This allows the end-effector position and orientation to be expressed as a function of the joint variables of the mechanical structure with respect to a reference frame. Both open-chain and closed-chain kinematic structures are considered. With reference to a *minimal representation of orientation*, the concept of *operational space* is introduced and its relationship with the *joint space* is established. Furthermore, a *calibration* technique of the manipulator kinematic parameters is presented. The chapter ends with the derivation of solutions to the *inverse kinematics problem*, which consists of the determination of the joint variables corresponding to a given end-effector configuration.

2.1 Position and Orientation of a Rigid Body

A *rigid body* is completely described in space by its *position* and *orientation* with respect to a reference frame. As shown in Figure 2.1, let $O–xyz$ be the orthonormal reference frame and $\boldsymbol{x}$, $\boldsymbol{y}$, $\boldsymbol{z}$ be the unit vectors of the frame axes.

The position of a point O' on the rigid body with respect to the coordinate frame $O–xyz$ is expressed by the relation

$$\boldsymbol{o}' = o'_x \boldsymbol{x} + o'_y \boldsymbol{y} + o'_z \boldsymbol{z},$$

where o'_x, o'_y, o'_z denote the components of the vector $\boldsymbol{o}'$ along the frame axes; the position of O' can be compactly written as the (3×1) vector

$$\boldsymbol{o}' = \begin{bmatrix} o'_x \\ o'_y \\ o'_z \end{bmatrix}. \tag{2.1}$$

Vector $\boldsymbol{o}'$ is a bound vector since its line of application and point of application are both prescribed, in addition to its direction and norm.

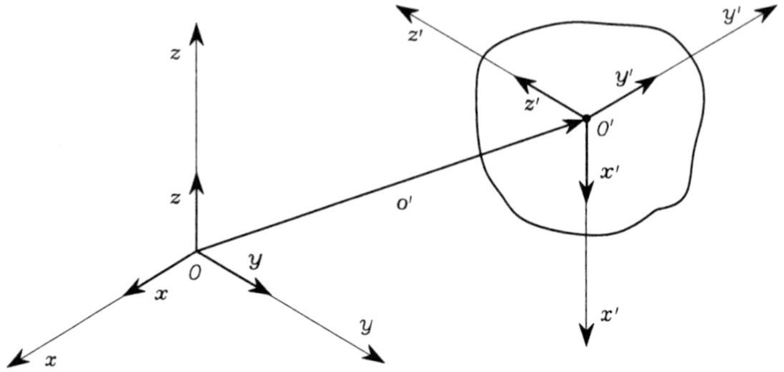

Figure 2.1 Position and orientation of a rigid body.

In order to describe the rigid body orientation, it is convenient to consider an orthonormal frame attached to the body and express its unit vectors with respect to the reference frame. Let then O'–$x'y'z'$ be such frame with origin in O' and $\boldsymbol{x}'$, $\boldsymbol{y}'$, $\boldsymbol{z}'$ be the unit vectors of the frame axes. These vectors are expressed with respect to the reference frame O–xyz by the equations:

$$\begin{aligned}\boldsymbol{x}' &= x'_x\boldsymbol{x} + x'_y\boldsymbol{y} + x'_z\boldsymbol{z}\\ \boldsymbol{y}' &= y'_x\boldsymbol{x} + y'_y\boldsymbol{y} + y'_z\boldsymbol{z}\\ \boldsymbol{z}' &= z'_x\boldsymbol{x} + z'_y\boldsymbol{y} + z'_z\boldsymbol{z}.\end{aligned} \tag{2.2}$$

The components of each unit vector are the direction cosines of the axes of frame O'–$x'y'z'$ with respect to the reference frame O–xyz.

2.2 Rotation Matrix

By adopting a compact notation, the three unit vectors in (2.2) describing the body orientation with respect to the reference frame can be combined in the (3×3) matrix

$$\boldsymbol{R} = \begin{bmatrix} \boldsymbol{x}' & \boldsymbol{y}' & \boldsymbol{z}' \end{bmatrix} = \begin{bmatrix} x'_x & y'_x & z'_x \\ x'_y & y'_y & z'_y \\ x'_z & y'_z & z'_z \end{bmatrix} = \begin{bmatrix} \boldsymbol{x}'^T\boldsymbol{x} & \boldsymbol{y}'^T\boldsymbol{x} & \boldsymbol{z}'^T\boldsymbol{x} \\ \boldsymbol{x}'^T\boldsymbol{y} & \boldsymbol{y}'^T\boldsymbol{y} & \boldsymbol{z}'^T\boldsymbol{y} \\ \boldsymbol{x}'^T\boldsymbol{z} & \boldsymbol{y}'^T\boldsymbol{z} & \boldsymbol{z}'^T\boldsymbol{z} \end{bmatrix}, \tag{2.3}$$

which is termed *rotation matrix*.

It is worth noting that the column vectors of matrix $\boldsymbol{R}$ are mutually orthogonal since they represent the unit vectors of an orthonormal frame, *i.e.*,

$$\boldsymbol{x}'^T\boldsymbol{y}' = 0 \qquad \boldsymbol{y}'^T\boldsymbol{z}' = 0 \qquad \boldsymbol{z}'^T\boldsymbol{x}' = 0.$$

Also, they have unit norm

$$\boldsymbol{x}'^T\boldsymbol{x}' = 1 \qquad \boldsymbol{y}'^T\boldsymbol{y}' = 1 \qquad \boldsymbol{z}'^T\boldsymbol{z}' = 1.$$

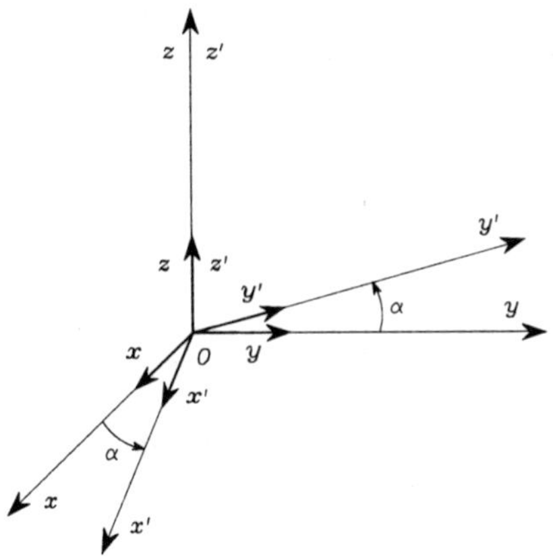

Figure 2.2 Rotation of the frame O–xyz by an angle α about axis z.

As a consequence, $\boldsymbol{R}$ is an *orthogonal* matrix meaning that

$$\boldsymbol{R}^T\boldsymbol{R} = \boldsymbol{I} \tag{2.4}$$

where $\boldsymbol{I}$ denotes the (3×3) identity matrix.

If both sides of (2.4) are postmultiplied by the inverse matrix $\boldsymbol{R}^{-1}$, the useful result is obtained:

$$\boldsymbol{R}^T = \boldsymbol{R}^{-1}, \tag{2.5}$$

that is, the transpose of the rotation matrix is equal to its inverse. Further, observe that $\det(\boldsymbol{R}) = 1$ if the frame is right-handed, while $\det(\boldsymbol{R}) = -1$ if the frame is left-handed.

2.2.1 Elementary Rotations

Consider the frames that can be obtained via *elementary rotations* of the reference frame about one of the coordinate axes. These rotations are positive if they are made counter-clockwise about the relative axis.

Suppose that the reference frame O–xyz is rotated by an angle α about axis z (Figure 2.2), and let O–$x'y'z'$ be the rotated frame. The unit vectors of the new frame can be described in terms of their components with respect to the reference frame, *i.e.*,

$$\boldsymbol{x}' = \begin{bmatrix} \cos\alpha \\ \sin\alpha \\ 0 \end{bmatrix} \qquad \boldsymbol{y}' = \begin{bmatrix} -\sin\alpha \\ \cos\alpha \\ 0 \end{bmatrix} \qquad \boldsymbol{z}' = \begin{bmatrix} 0 \\ 0 \\ 1 \end{bmatrix}.$$

Hence, the rotation matrix of frame O–$x'y'z'$ with respect to frame O–xyz is

$$\boldsymbol{R}_z(\alpha) = \begin{bmatrix} \cos\alpha & -\sin\alpha & 0 \\ \sin\alpha & \cos\alpha & 0 \\ 0 & 0 & 1 \end{bmatrix}. \tag{2.6}$$

In a similar manner, it can be shown that the rotations by an angle β about axis y and by an angle γ about axis x are respectively given by:

$$\boldsymbol{R}_y(\beta) = \begin{bmatrix} \cos\beta & 0 & \sin\beta \\ 0 & 1 & 0 \\ -\sin\beta & 0 & \cos\beta \end{bmatrix} \tag{2.7}$$

$$\boldsymbol{R}_x(\gamma) = \begin{bmatrix} 1 & 0 & 0 \\ 0 & \cos\gamma & -\sin\gamma \\ 0 & \sin\gamma & \cos\gamma \end{bmatrix}. \tag{2.8}$$

These matrices will be useful to describe rotations about an arbitrary axis in space.

It is easy to verify that for the elementary rotation matrices in (2.6)–(2.8) the following property holds:

$$\boldsymbol{R}_k(-\vartheta) = \boldsymbol{R}_k^T(\vartheta) \qquad k = x, y, z. \tag{2.9}$$

In view of (2.6)–(2.8), the rotation matrix can be attributed a geometrical meaning; namely, the matrix $\boldsymbol{R}$ describes the rotation about an axis in space needed to align the axes of the reference frame with the corresponding axes of the body frame.

2.2.2 Representation of a Vector

In order to understand a further geometrical meaning of a rotation matrix, consider the case when the origin of the body frame coincides with the origin of the reference frame (Figure 2.3); it follows that $\boldsymbol{o}' = \boldsymbol{0}$, where $\boldsymbol{0}$ denotes the (3×1) null vector. A point P in space can be represented either as

$$\boldsymbol{p} = \begin{bmatrix} p_x \\ p_y \\ p_z \end{bmatrix}$$

with respect to frame O–xyz, or as

$$\boldsymbol{p}' = \begin{bmatrix} p'_x \\ p'_y \\ p'_z \end{bmatrix}$$

with respect to frame O–$x'y'z'$.

Since $\boldsymbol{p}$ and $\boldsymbol{p}'$ are representations of the same point P, it is

$$\boldsymbol{p} = p'_x\boldsymbol{x}' + p'_y\boldsymbol{y}' + p'_z\boldsymbol{z}' = \begin{bmatrix} \boldsymbol{x}' & \boldsymbol{y}' & \boldsymbol{z}' \end{bmatrix} \boldsymbol{p}'$$

and, accounting for (2.3), it is

$$\boldsymbol{p} = \boldsymbol{R}\boldsymbol{p}'. \tag{2.10}$$

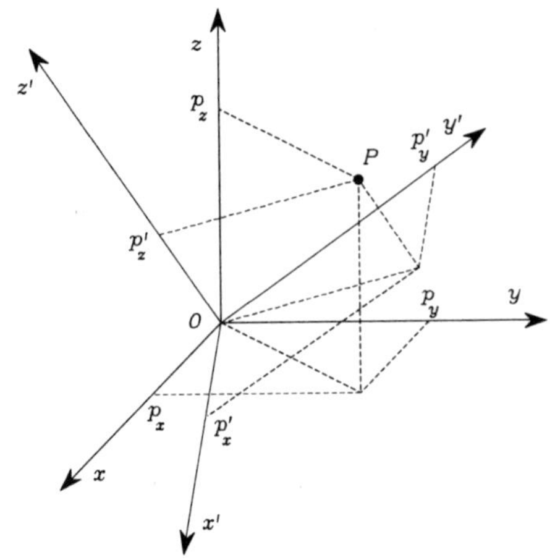

Figure 2.3 Representation of a point P in two different coordinate frames.

The rotation matrix $\boldsymbol{R}$ represents the *transformation matrix* of the vector coordinates in frame $O\text{–}x'y'z'$ into the coordinates of the same vector in frame $O\text{–}xyz$. In view of the orthogonality property (2.4), the inverse transformation is simply given by

$$\boldsymbol{p}' = \boldsymbol{R}^T\boldsymbol{p}. \tag{2.11}$$

Example 2.1

Consider two frames with common origin mutually rotated by an angle α about the axis z. Let $\boldsymbol{p}$ and $\boldsymbol{p}'$ be the vectors of the coordinates of a point P, expressed in the frames $O\text{–}xyz$ and $O\text{–}x'y'z'$, respectively (Figure 2.4). On the basis of simple geometry, the relationship between the coordinates of P in the two frames is:

$$\begin{aligned} p_x &= p'_x \cos\alpha - p'_y \sin\alpha \\ p_y &= p'_x \sin\alpha + p'_y \cos\alpha \\ p_z &= p'_z. \end{aligned}$$

Therefore, the matrix (2.6) represents not only the orientation of a frame with respect to another frame, but it also describes the transformation of a vector from a frame to another frame with the same origin.

2.2.3 Rotation of a Vector

A rotation matrix can be also interpreted as the matrix operator allowing rotation of a vector by a given angle about an arbitrary axis in space. In fact, let $\boldsymbol{p}'$ be a vector in the reference frame $O\text{–}xyz$; in view of orthogonality of the matrix $\boldsymbol{R}$, the product $\boldsymbol{R}\boldsymbol{p}'$ yields a vector $\boldsymbol{p}$ with the same norm as that of $\boldsymbol{p}'$ but rotated with respect to

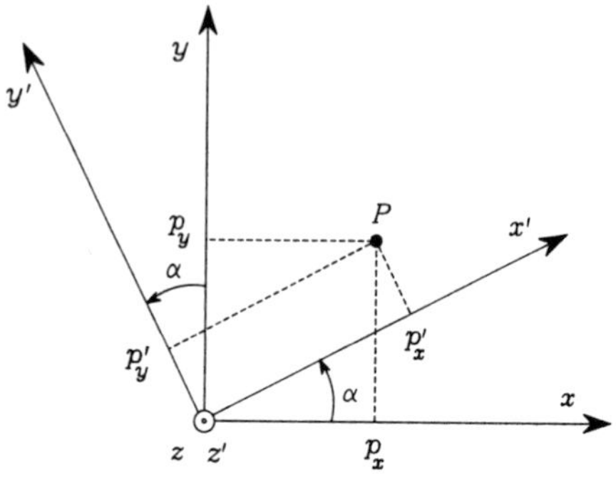

Figure 2.4 Representation of a point P in rotated frames.

$\boldsymbol{p}'$ according to the matrix $\boldsymbol{R}$. The norm equality can be proved by observing that $\boldsymbol{p}^T\boldsymbol{p} = \boldsymbol{p}'^T\boldsymbol{R}^T\boldsymbol{R}\boldsymbol{p}'$ and applying (2.4). This interpretation of the rotation matrix will be revisited later.

Example 2.2

Consider the vector $\boldsymbol{p}$ which is obtained by rotating a vector $\boldsymbol{p}'$ in the plane xy by an angle α about axis z of the reference frame (Figure 2.5). Let (p'_x, p'_y, p'_z) be the coordinates of the vector $\boldsymbol{p}'$. The vector $\boldsymbol{p}$ has components

$$\begin{aligned} p_x &= p'_x \cos\alpha - p'_y \sin\alpha \\ p_y &= p'_x \sin\alpha + p'_y \cos\alpha \\ p_z &= p'_z. \end{aligned}$$

It is easy to recognize that $\boldsymbol{p}$ can be expressed as

$$\boldsymbol{p} = \boldsymbol{R}_z(\alpha)\boldsymbol{p}',$$

where $\boldsymbol{R}_z(\alpha)$ is the same rotation matrix as in (2.6).

In sum, a rotation matrix attains three *equivalent geometrical meanings*:

- It describes the mutual orientation between two coordinate frames; its column vectors are the direction cosines of the axes of the rotated frame with respect to the original frame.
- It represents the coordinate transformation between the coordinates of a point expressed in two different frames (with common origin).
- It is the operator that allows rotating a vector in the same coordinate frame.

2.3 Composition of Rotation Matrices

In order to derive composition rules of rotation matrices, it is useful to consider the

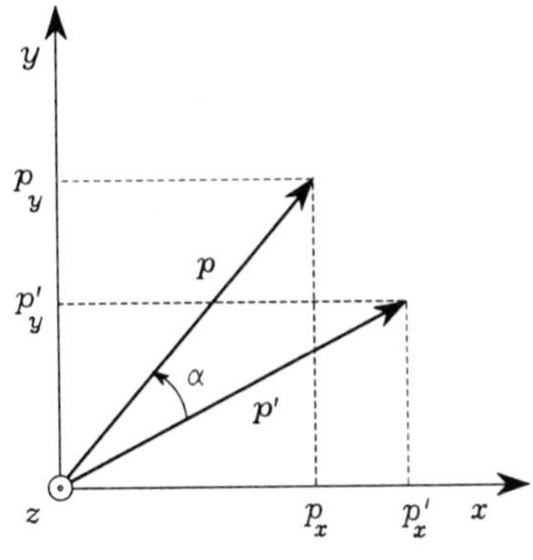

Figure 2.5 Rotation of a vector.

expression of a vector in two different reference frames. Let then O–$x_0y_0z_0$, O–$x_1y_1z_1$, O–$x_2y_2z_2$ be three frames with common origin O. The vector $\boldsymbol{p}$ describing the position of a generic point in space can be expressed in each of the above frames; let $\boldsymbol{p}^0$, $\boldsymbol{p}^1$, $\boldsymbol{p}^2$ denote the expressions of $\boldsymbol{p}$ in the three frames[1].

At first, consider the relationship between the expression $\boldsymbol{p}^2$ of the vector $\boldsymbol{p}$ in Frame 2 and the expression $\boldsymbol{p}^1$ of the same vector in Frame 1. If $\boldsymbol{R}_i^j$ denotes the rotation matrix of Frame i with respect to Frame j, it is

$$\boldsymbol{p}^1 = \boldsymbol{R}_2^1\boldsymbol{p}^2. \tag{2.12}$$

Similarly, it turns out that

$$\boldsymbol{p}^0 = \boldsymbol{R}_1^0\boldsymbol{p}^1 \tag{2.13}$$

$$\boldsymbol{p}^0 = \boldsymbol{R}_2^0\boldsymbol{p}^2. \tag{2.14}$$

On the other hand, substituting (2.12) in (2.13) and using (2.14) gives

$$\boldsymbol{R}_2^0 = \boldsymbol{R}_1^0\boldsymbol{R}_2^1. \tag{2.15}$$

The relationship in (2.15) can be interpreted as the composition of successive rotations. Consider a frame initially aligned with the frame O–$x_0y_0z_0$. The rotation expressed by matrix $\boldsymbol{R}_2^0$ can be regarded as obtained in two steps:

- first rotate the given frame according to $\boldsymbol{R}_1^0$, so as to align it with frame O–$x_1y_1z_1$;
- then rotate the frame, now aligned with frame O–$x_1y_1z_1$, according to $\boldsymbol{R}_2^1$, so as to align it with frame O–$x_2y_2z_2$.

[1] Hereafter, the superscript of a vector or a matrix denotes the frame in which its components are expressed.

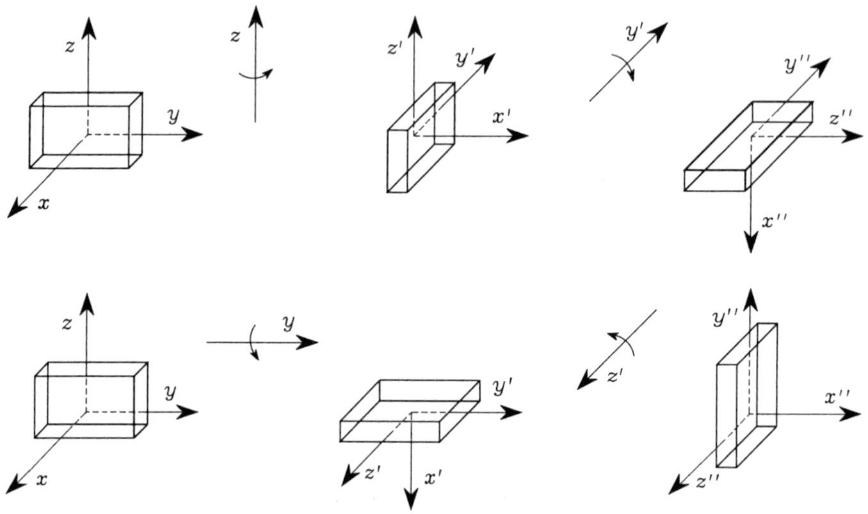

Figure 2.6 Successive rotations of an object about axes of current frame.

Notice that the overall rotation can be expressed as a sequence of partial rotations; each rotation is defined with respect to the preceding one. The frame with respect to which the rotation occurs is termed *current frame*. Composition of successive rotations is then obtained by postmultiplication of the rotation matrices following the given order of rotations, as in (2.15). With the adopted notation, in view of (2.5), it is

$$\boldsymbol{R}_i^j = (\boldsymbol{R}_j^i)^{-1} = (\boldsymbol{R}_j^i)^T. \tag{2.16}$$

Successive rotations can be also specified by constantly referring them to the initial frame; in this case, the rotations are made with respect to a *fixed frame*. Let $\boldsymbol{R}_1^0$ be the rotation matrix of frame O–$x_1y_1z_1$ with respect to the fixed frame O–$x_0y_0z_0$. Let then $\bar{\boldsymbol{R}}_2^0$ denote the matrix characterizing frame O–$x_2y_2z_2$ with respect to Frame 0, which is obtained as a rotation of Frame 1 according to the matrix $\bar{\boldsymbol{R}}_2^1$. Since (2.15) gives a composition rule of successive rotations about the axes of the current frame, the overall rotation can be regarded as obtained in the following steps:

- first realign Frame 1 with Frame 0 by means of rotation $\boldsymbol{R}_0^1$;
- then make the rotation expressed by $\bar{\boldsymbol{R}}_2^1$ with respect to the current frame;
- finally compensate for the rotation made for the realignment by means of the inverse rotation $\boldsymbol{R}_1^0$.

Since the above rotations are described with respect to the current frame, application of the composition rule (2.15) yields

$$\bar{\boldsymbol{R}}_2^0 = \boldsymbol{R}_1^0 \boldsymbol{R}_0^1 \bar{\boldsymbol{R}}_2^1 \boldsymbol{R}_1^0.$$

In view of (2.16), it is

$$\bar{\boldsymbol{R}}_2^0 = \bar{\boldsymbol{R}}_2^1 \boldsymbol{R}_1^0 \tag{2.17}$$

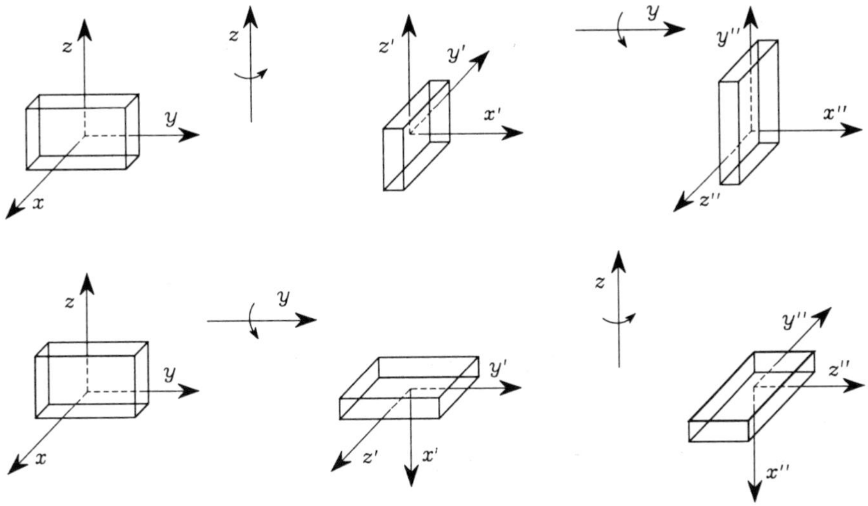

Figure 2.7 Successive rotations of an object about axes of fixed frame.

where the resulting $\bar{\boldsymbol{R}}_2^0$ is different from the matrix $\boldsymbol{R}_2^0$ in (2.15). Hence, it can be stated that composition of successive rotations with respect to a fixed frame is obtained by premultiplication of the single rotation matrices in the order of the given sequence of rotations.

By recalling the meaning of a rotation matrix in terms of the orientation of a current frame with respect to a fixed frame, it can be recognized that its columns are the direction cosines of the axes of the current frame with respect to the fixed frame, while its rows (columns of its transpose and inverse) are the direction cosines of the axes of the fixed frame with respect to the current frame.

An important issue of composition of rotations is that the matrix product is not commutative. In view of this, it can be concluded that two rotations in general do not commute and its composition depends on the order of the single rotations.

Example 2.3

Consider an object and a frame attached to it. Figure 2.6 shows the effects of two successive rotations of the object with respect to the current frame by changing the order of rotations. It is evident that the final object orientation is different in the two cases. Also in the case of rotations made with respect to the current frame, the final orientations differ (Figure 2.7). It is interesting to note that the effects of the sequence of rotations with respect to the fixed frame are interchanged with the effects of the sequence of rotations with respect to the current frame. This can be explained by observing that the order of rotations in the fixed frame commutes with respect to the order of rotations in the current frame.

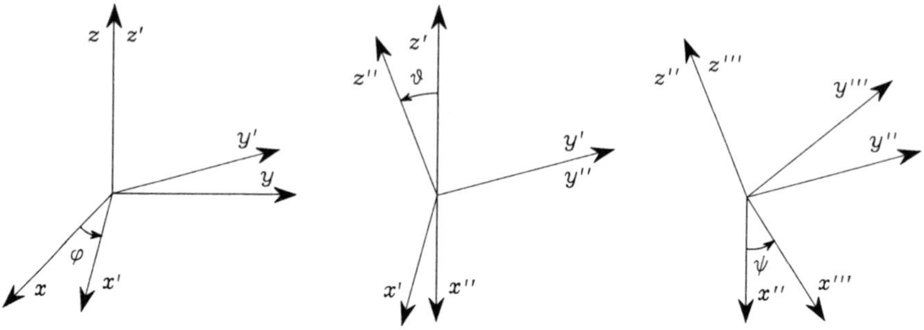

Figure 2.8 Representation of Euler angles ZYZ.

2.4 Euler Angles

Rotation matrices give a redundant description of frame orientation; in fact, they are characterized by nine elements which are not independent but related by six constraints due to the orthogonality conditions given in (2.4). This implies that *three parameters* are sufficient to describe orientation of a rigid body in space. A representation of orientation in terms of three independent parameters constitutes a *minimal representation*.

A minimal representation of orientation can be obtained by using a set of three angles $\boldsymbol{\phi} = [\varphi \quad \vartheta \quad \psi]^T$. Consider the rotation matrix expressing the elementary rotation about one of the coordinate axes as a function of a single angle. Then, a generic rotation matrix can be obtained by composing a suitable sequence of three elementary rotations while guaranteeing that two successive rotations are not made about parallel axes. This implies that 12 distinct sets of angles are allowed out of all 27 possible combinations; each set represents a triplet of *Euler angles*. In the following, two sets of Euler angles are analyzed; namely, the ZYZ angles and the ZYX (or Roll–Pitch–Yaw) angles.

2.4.1 ZYZ Angles

The rotation described by *ZYZ angles* is obtained as composition of the following elementary rotations (Figure 2.8):

- Rotate the reference frame by the angle φ about axis z; this rotation is described by the matrix $\boldsymbol{R}_z(\varphi)$ which is formally defined in (2.6).
- Rotate the current frame by the angle ϑ about axis y'; this rotation is described by the matrix $\boldsymbol{R}_{y'}(\vartheta)$ which is formally defined in (2.7).
- Rotate the current frame by the angle ψ about axis z''; this rotation is described by the matrix $\boldsymbol{R}_{z''}(\psi)$ which is again formally defined in (2.6).

The resulting frame orientation is obtained by composition of rotations with respect to *current frames*, and then it can be computed via postmultiplication of the matrices of elementary rotation, *i.e.*,[2]

$$
\begin{aligned}
\boldsymbol{R}(\boldsymbol{\phi}) &= \boldsymbol{R}_z(\varphi)\boldsymbol{R}_{y'}(\vartheta)\boldsymbol{R}_{z''}(\psi) \\
&= \begin{bmatrix} c_\varphi c_\vartheta c_\psi - s_\varphi s_\psi & -c_\varphi c_\vartheta s_\psi - s_\varphi c_\psi & c_\varphi s_\vartheta \\ s_\varphi c_\vartheta c_\psi + c_\varphi s_\psi & -s_\varphi c_\vartheta s_\psi + c_\varphi c_\psi & s_\varphi s_\vartheta \\ -s_\vartheta c_\psi & s_\vartheta s_\psi & c_\vartheta \end{bmatrix}.
\end{aligned} \tag{2.18}
$$

It is useful to solve the *inverse problem*, that is to determine the set of Euler angles corresponding to a given rotation matrix

$$
\boldsymbol{R} = \begin{bmatrix} r_{11} & r_{12} & r_{13} \\ r_{21} & r_{22} & r_{23} \\ r_{31} & r_{32} & r_{33} \end{bmatrix}.
$$

Compare this expression with that of $\boldsymbol{R}(\boldsymbol{\phi})$ in (2.18). By considering the elements $[1,3]$ and $[2,3]$, on the assumption that $r_{13} \neq 0$ and $r_{23} \neq 0$, it follows that

$$
\varphi = \text{Atan2}(r_{23}, r_{13}),
$$

where $\text{Atan2}(y, x)$ is the arctangent function of two arguments[3]. Then, squaring and summing the elements $[1,3]$ and $[2,3]$ and using the element $[3,3]$ yields

$$
\vartheta = \text{Atan2}\left(\sqrt{r_{13}^2 + r_{23}^2}, r_{33}\right).
$$

The choice of the positive sign for the term $\sqrt{r_{13}^2 + r_{23}^2}$ limits the range of feasible values of ϑ to $(0, \pi)$. On this assumption, considering the elements $[3,1]$ and $[3,2]$ gives

$$
\psi = \text{Atan2}(r_{32}, -r_{31}).
$$

In sum, the requested solution is

$$
\begin{aligned}
\varphi &= \text{Atan2}(r_{23}, r_{13}) \\
\vartheta &= \text{Atan2}\left(\sqrt{r_{13}^2 + r_{23}^2}, r_{33}\right) \\
\psi &= \text{Atan2}(r_{32}, -r_{31}).
\end{aligned} \tag{2.19}
$$

---

[2] The notations c_ϕ and s_ϕ are the abbreviations for $\cos\phi$ and $\sin\phi$, respectively; short-hand notations of this kind will be adopted often throughout the text.

[3] The function $\text{Atan2}(y, x)$ computes the arctangent of the ratio y/x but utilizes the sign of each argument to determine which quadrant the resulting angle belongs to; this allows the correct determination of an angle in a range of 2π.

It is possible to derive another solution which produces the same effects as solution (2.19). Choosing ϑ in the range $(-\pi, 0)$ leads to

$$\begin{aligned}\varphi &= \text{Atan2}(-r_{23}, -r_{13})\\ \vartheta &= \text{Atan2}\left(-\sqrt{r_{13}^2 + r_{23}^2}, r_{33}\right)\\ \psi &= \text{Atan2}(-r_{32}, r_{31}).\end{aligned} \tag{2.19'}$$

Solutions (2.19) and (2.19′) degenerate when $s_\vartheta = 0$; in this case, it is possible to determine only the sum or difference of φ and ψ. In fact, if $\vartheta = 0, \pi$, the successive rotations of φ and ψ are made about axes of current frames which are parallel, thus giving equivalent contributions to the rotation.[4]

2.4.2 Roll–Pitch–Yaw Angles

Another set of Euler angles originates from a representation of orientation in the (aero)nautical field. These are the ZYX angles, also called *Roll–Pitch–Yaw angles*, to denote the typical motions of an (air)craft. In this case, the angles $\boldsymbol{\phi} = [\varphi \quad \vartheta \quad \psi]^T$ represent rotations defined with respect to a fixed frame attached to the centre of mass of the craft (Figure 2.9).

The rotation resulting from Roll–Pitch–Yaw angles can be obtained as follows:

- Rotate the reference frame by the angle ψ about axis x (yaw); this rotation is described by the matrix $\boldsymbol{R}_x(\psi)$ which is formally defined in (2.8).
- Rotate the reference frame by the angle ϑ about axis y (pitch); this rotation is described by the matrix $\boldsymbol{R}_y(\vartheta)$ which is formally defined in (2.7).
- Rotate the reference frame by the angle φ about axis z (roll); this rotation is described by the matrix $\boldsymbol{R}_z(\varphi)$ which is formally defined in (2.6).

The resulting frame orientation is obtained by composition of rotations with respect to the *fixed frame*, and then it can be computed via premultiplication of the matrices of elementary rotation, *i.e.*,[5]

$$\begin{aligned}\boldsymbol{R}(\boldsymbol{\phi}) &= \boldsymbol{R}_z(\varphi)\boldsymbol{R}_y(\vartheta)\boldsymbol{R}_x(\psi)\\ &= \begin{bmatrix} c_\varphi c_\vartheta & c_\varphi s_\vartheta s_\psi - s_\varphi c_\psi & c_\varphi s_\vartheta c_\psi + s_\varphi s_\psi \\ s_\varphi c_\vartheta & s_\varphi s_\vartheta s_\psi + c_\varphi c_\psi & s_\varphi s_\vartheta c_\psi - c_\varphi s_\psi \\ -s_\vartheta & c_\vartheta s_\psi & c_\vartheta c_\psi \end{bmatrix}.\end{aligned} \tag{2.20}$$

---

[4] In the following chapter, it will be seen that these configurations characterize the so-called representation *singularities* of the Euler angles.

[5] The ordered sequence of rotations XYZ about axes of the fixed frame is equivalent to the sequence ZYX about axes of the current frame.

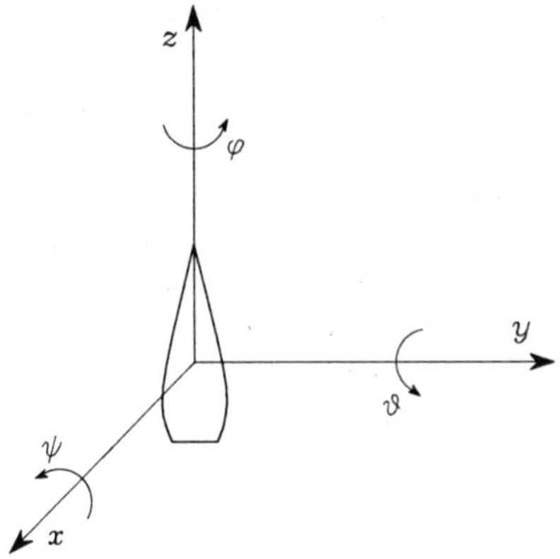

Figure 2.9 Representation of Roll–Pitch–Yaw angles.

As for the Euler angles ZYZ, the *inverse solution* to a given rotation matrix

$$\boldsymbol{R} = \begin{bmatrix} r_{11} & r_{12} & r_{13} \\ r_{21} & r_{22} & r_{23} \\ r_{31} & r_{32} & r_{33} \end{bmatrix}$$

can be obtained by comparing it with the expression of $\boldsymbol{R}(\boldsymbol{\phi})$ in (2.20). The solution for ϑ in the range $(-\pi/2, \pi/2)$ is

$$\begin{aligned} \varphi &= \text{Atan2}(r_{21}, r_{11}) \\ \vartheta &= \text{Atan2}\left(-r_{31}, \sqrt{r_{32}^2 + r_{33}^2}\right) \\ \psi &= \text{Atan2}(r_{32}, r_{33}), \end{aligned} \tag{2.21}$$

whereas the other equivalent solution for ϑ in the range $(\pi/2, 3\pi/2)$ is

$$\begin{aligned} \varphi &= \text{Atan2}(-r_{21}, -r_{11}) \\ \vartheta &= \text{Atan2}\left(-r_{31}, -\sqrt{r_{32}^2 + r_{33}^2}\right) \\ \psi &= \text{Atan2}(-r_{32}, -r_{33}). \end{aligned} \tag{2.21'}$$

Solutions (2.21) and (2.21′) degenerate when $c_\vartheta = 0$; in this case, it is possible to determine only the sum or difference of φ and ψ.

2.5 Angle and Axis

A nonminimal representation of orientation can be obtained by resorting to *four parameters* expressing a rotation of a given angle about an axis in space. This can be advantageous in the problem of trajectory planning for a manipulator's end-effector orientation.

Let $\boldsymbol{r} = [\,r_x \quad r_y \quad r_z\,]^T$ be the unit vector of a rotation axis with respect to the reference frame O–xyz. In order to derive the rotation matrix $\boldsymbol{R}(\vartheta, \boldsymbol{r})$ expressing the rotation of an *angle* ϑ about *axis* $\boldsymbol{r}$, it is convenient to compose elementary rotations about the coordinate axes of the reference frame. The angle is taken to be positive if the rotation is made counter-clockwise about axis $\boldsymbol{r}$.

As shown in Figure 2.10, a possible solution is to rotate first $\boldsymbol{r}$ by the angles necessary to align it with axis z, then to rotate by ϑ about z and finally to rotate by the angles necessary to align the unit vector with the initial direction. In detail, the sequence of rotations, to be made always with respect to axes of fixed frame, is the following:

- align $\boldsymbol{r}$ with z, which is obtained as the sequence of a rotation by $-\alpha$ about z and a rotation by $-\beta$ about y;
- rotate by ϑ about z;
- realign with the initial direction of $\boldsymbol{r}$, which is obtained as the sequence of a rotation by β about y and a rotation by α about z.

In sum, the resulting rotation matrix is

$$\boldsymbol{R}(\vartheta, \boldsymbol{r}) = \boldsymbol{R}_z(\alpha)\boldsymbol{R}_y(\beta)\boldsymbol{R}_z(\vartheta)\boldsymbol{R}_y(-\beta)\boldsymbol{R}_z(-\alpha). \tag{2.22}$$

From the components of the unit vector $\boldsymbol{r}$ it is possible to extract the transcendental functions needed to compute the rotation matrix in (2.22), so as to eliminate the dependence from α and β; in fact, it is

$$\sin\alpha = \frac{r_y}{\sqrt{r_x^2 + r_y^2}} \qquad \cos\alpha = \frac{r_x}{\sqrt{r_x^2 + r_y^2}}$$

$$\sin\beta = \sqrt{r_x^2 + r_y^2} \qquad \cos\beta = r_z.$$

Then, it can be found that the rotation matrix corresponding to a given angle and axis is

$$\boldsymbol{R}(\vartheta, \boldsymbol{r}) = \begin{bmatrix} r_x^2(1 - c_\vartheta) + c_\vartheta & r_x r_y(1 - c_\vartheta) - r_z s_\vartheta & r_x r_z(1 - c_\vartheta) + r_y s_\vartheta \\ r_x r_y(1 - c_\vartheta) + r_z s_\vartheta & r_y^2(1 - c_\vartheta) + c_\vartheta & r_y r_z(1 - c_\vartheta) - r_x s_\vartheta \\ r_x r_z(1 - c_\vartheta) - r_y s_\vartheta & r_y r_z(1 - c_\vartheta) + r_x s_\vartheta & r_z^2(1 - c_\vartheta) + c_\vartheta \end{bmatrix}. \tag{2.23}$$

For this matrix, the following property holds:

$$\boldsymbol{R}(-\vartheta, -\boldsymbol{r}) = \boldsymbol{R}(\vartheta, \boldsymbol{r}), \tag{2.24}$$

i.e., a rotation by $-\vartheta$ about $-\boldsymbol{r}$ cannot be distinguished from a rotation by ϑ about $\boldsymbol{r}$; hence, such representation is not unique.

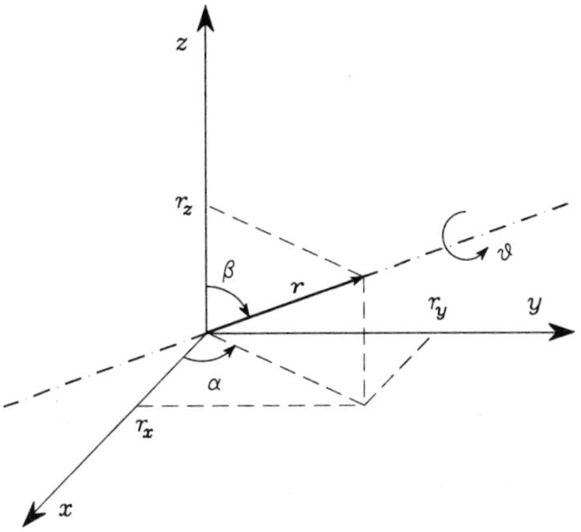

Figure 2.10 Rotation of an angle about an axis.

If it is desired to solve the *inverse problem* to compute the axis and angle corresponding to a given rotation matrix

$$\boldsymbol{R} = \begin{bmatrix} r_{11} & r_{12} & r_{13} \\ r_{21} & r_{22} & r_{23} \\ r_{31} & r_{32} & r_{33} \end{bmatrix},$$

the following result is useful:

$$\begin{aligned} \vartheta &= \cos^{-1}\left(\frac{r_{11} + r_{22} + r_{33} - 1}{2}\right) \\ \boldsymbol{r} &= \frac{1}{2\sin\vartheta}\begin{bmatrix} r_{32} - r_{23} \\ r_{13} - r_{31} \\ r_{21} - r_{12} \end{bmatrix}, \end{aligned} \tag{2.25}$$

for $\sin\vartheta \neq 0$. Notice that (2.25) expresses the rotation in terms of four parameters; namely, the angle and the three components of the axis unit vector. However, it can be observed that the three components of $\boldsymbol{r}$ are not independent but are constrained by the condition

$$r_x^2 + r_y^2 + r_z^2 = 1. \tag{2.26}$$

If $\sin\vartheta = 0$, (2.25) becomes meaningless. To solve the inverse problem, it is necessary to directly refer to the particular expressions attained by the rotation matrix $\boldsymbol{R}$ and find the solving formulæ in the two cases $\vartheta = 0$ and $\vartheta = \pi$. Notice that, when $\vartheta = 0$ (null rotation), the unit vector $\boldsymbol{r}$ is arbitrary (singularity).

2.6 Unit Quaternion

The drawbacks of the angle/axis representation can be overcome by a different four-parameter representation; namely, the unit *quaternion*, viz. Euler parameters, defined as $\mathcal{Q} = \{\eta, \boldsymbol{\epsilon}\}$ where:

$$\begin{aligned}\eta &= \cos\frac{\vartheta}{2}\\ \boldsymbol{\epsilon} &= \sin\frac{\vartheta}{2}\boldsymbol{r};\end{aligned} \tag{2.27}$$

η is called the scalar part of the quaternion while $\boldsymbol{\epsilon} = [\,\epsilon_x \quad \epsilon_y \quad \epsilon_z\,]^T$ is called the vector part of the quaternion. They are constrained by the condition

$$\eta^2 + \epsilon_x^2 + \epsilon_y^2 + \epsilon_z^2 = 1, \tag{2.28}$$

hence, the name *unit* quaternion. It is worth remarking that, differently from the angle/axis representation, a rotation by $-\vartheta$ about $-\boldsymbol{r}$ gives the same quaternion as that associated with a rotation by ϑ about $\boldsymbol{r}$; this solves the above nonuniqueness problem. In view of (2.23), (2.27) and (2.28), the rotation matrix corresponding to a given quaternion takes on the form

$$\boldsymbol{R}(\eta, \boldsymbol{\epsilon}) = \begin{bmatrix} 2(\eta^2+\epsilon_x^2)-1 & 2(\epsilon_x\epsilon_y - \eta\epsilon_z) & 2(\epsilon_x\epsilon_z + \eta\epsilon_y) \\ 2(\epsilon_x\epsilon_y + \eta\epsilon_z) & 2(\eta^2+\epsilon_y^2)-1 & 2(\epsilon_y\epsilon_z - \eta\epsilon_x) \\ 2(\epsilon_x\epsilon_z - \eta\epsilon_y) & 2(\epsilon_y\epsilon_z + \eta\epsilon_x) & 2(\eta^2+\epsilon_z^2)-1 \end{bmatrix}. \tag{2.29}$$

If it is desired to solve the *inverse problem* to compute the quaternion corresponding to a given rotation matrix

$$\boldsymbol{R} = \begin{bmatrix} r_{11} & r_{12} & r_{13} \\ r_{21} & r_{22} & r_{23} \\ r_{31} & r_{32} & r_{33} \end{bmatrix},$$

the following result is useful:

$$\begin{aligned}\eta &= \frac{1}{2}\sqrt{r_{11}+r_{22}+r_{33}+1}\\ \boldsymbol{\epsilon} &= \frac{1}{2}\begin{bmatrix} \operatorname{sgn}(r_{32}-r_{23})\sqrt{r_{11}-r_{22}-r_{33}+1} \\ \operatorname{sgn}(r_{13}-r_{31})\sqrt{r_{22}-r_{33}-r_{11}+1} \\ \operatorname{sgn}(r_{21}-r_{12})\sqrt{r_{33}-r_{11}-r_{22}+1} \end{bmatrix},\end{aligned} \tag{2.30}$$

where conventionally $\operatorname{sgn}(x) = 1$ for $x \geq 0$ and $\operatorname{sgn}(x) = -1$ for $x < 0$. Notice that in (2.30) it has been implicitly assumed $\eta \geq 0$; this corresponds to an angle $\vartheta \in [-\pi, \pi]$, and thus any rotation can be described. Also, compared to the inverse solution in (2.25) for the angle and axis representation, no singularity occurs for (2.30).

The quaternion extracted from $\boldsymbol{R}^{-1} = \boldsymbol{R}^T$ is denoted as $\mathcal{Q}^{-1}$, and can be computed as

$$\mathcal{Q}^{-1} = \{\eta, -\boldsymbol{\epsilon}\}. \tag{2.31}$$

Let $\mathcal{Q}_1 = \{\eta_1, \boldsymbol{\epsilon}_1\}$ and $\mathcal{Q}_2 = \{\eta_2, \boldsymbol{\epsilon}_2\}$ denote the quaternions corresponding to the rotation matrices $\boldsymbol{R}_1$ and $\boldsymbol{R}_2$, respectively. The quaternion corresponding to the product $\boldsymbol{R}_1\boldsymbol{R}_2$ is given by

$$\mathcal{Q}_1 * \mathcal{Q}_2 = \{\eta_1\eta_2 - \boldsymbol{\epsilon}_1^T\boldsymbol{\epsilon}_2, \eta_1\boldsymbol{\epsilon}_2 + \eta_2\boldsymbol{\epsilon}_1 + \boldsymbol{\epsilon}_1 \times \boldsymbol{\epsilon}_2\} \tag{2.32}$$

where the quaternion product operator "$*$" has been formally introduced. It is easy to see that if $\mathcal{Q}_2 = \mathcal{Q}_1^{-1}$ then the quaternion $\{1, \mathbf{0}\}$ is obtained from (2.32) which is the identity element for the product.

2.7 Homogeneous Transformations

As illustrated at the beginning of the chapter, the position of a rigid body in space is expressed in terms of the position of a suitable point on the body with respect to a reference frame (translation), while its orientation is expressed in terms of the components of the unit vectors of a frame attached to the body—with origin in the above point—with respect to the same reference frame (rotation).

As shown in Figure 2.11, consider an arbitrary point P in space. Let $\boldsymbol{p}^0$ be the vector of coordinates of P with respect to the reference frame O_0–$x_0y_0z_0$. Consider then another frame in space O_1–$x_1y_1z_1$. Let $\boldsymbol{o}_1^0$ be the vector describing the origin of Frame 1 with respect to Frame 0, and $\boldsymbol{R}_1^0$ be the rotation matrix of Frame 1 with respect to Frame 0. Let also $\boldsymbol{p}^1$ be the vector of coordinates of P with respect to Frame 1. On the basis of simple geometry, the position of point P with respect to the reference frame can be expressed as

$$\boldsymbol{p}^0 = \boldsymbol{o}_1^0 + \boldsymbol{R}_1^0\boldsymbol{p}^1. \tag{2.33}$$

Hence, (2.33) represents the *coordinate transformation* (*translation* + *rotation*) of a bound vector between two frames.

The inverse transformation can be obtained by premultiplying both sides of (2.33) by $\boldsymbol{R}_1^{0T}$; in view of (2.4), it follows that

$$\boldsymbol{p}^1 = -\boldsymbol{R}_1^{0T}\boldsymbol{o}_1^0 + \boldsymbol{R}_1^{0T}\boldsymbol{p}^0 \tag{2.34}$$

which, via (2.16), can be written as

$$\boldsymbol{p}^1 = -\boldsymbol{R}_0^1\boldsymbol{o}_1^0 + \boldsymbol{R}_0^1\boldsymbol{p}^0. \tag{2.35}$$

In order to achieve a compact representation of the relationship between the coordinates of the same point in two different frames, the *homogeneous representation* of a generic vector $\boldsymbol{p}$ can be introduced as the vector $\tilde{\boldsymbol{p}}$ formed by adding a fourth unit component, *i.e.*,

$$\tilde{\boldsymbol{p}} = \begin{bmatrix} \boldsymbol{p} \\ 1 \end{bmatrix}. \tag{2.36}$$

By adopting this representation for the vectors $\boldsymbol{p}^0$ and $\boldsymbol{p}^1$ in (2.33), the coordinate transformation can be written in terms of the (4×4) matrix

$$\boldsymbol{A}_1^0 = \begin{bmatrix} \boldsymbol{R}_1^0 & \boldsymbol{o}_1^0 \\ \boldsymbol{0}^T & 1 \end{bmatrix} \tag{2.37}$$

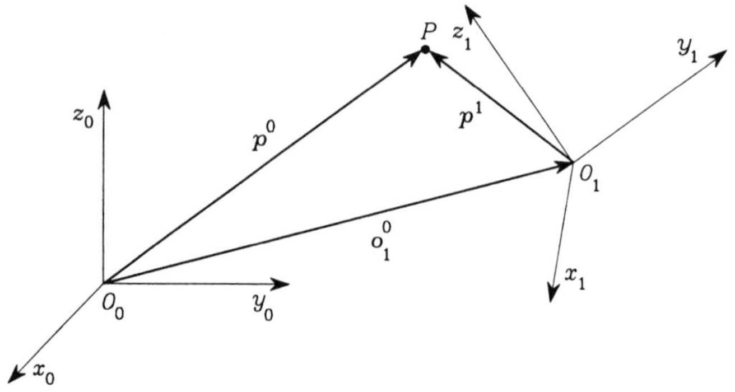

Figure 2.11 Representation of a point P in different coordinate frames.

which, according to (2.36), is termed *homogeneous transformation matrix*. As can be seen from (2.37), the transformation of a vector from Frame 1 to Frame 0 is expressed by a single matrix containing the rotation matrix of Frame 1 with respect to Frame 0 and the translation vector from the origin of Frame 0 to the origin of Frame 1[6]. Therefore, the coordinate transformation (2.33) can be compactly rewritten as

$$\tilde{\boldsymbol{p}}^0 = \boldsymbol{A}_1^0 \tilde{\boldsymbol{p}}^1. \tag{2.38}$$

The coordinate transformation between Frame 0 and Frame 1 is described by the homogeneous transformation matrix $\boldsymbol{A}_0^1$ which satisfies the equation

$$\tilde{\boldsymbol{p}}^1 = \boldsymbol{A}_0^1 \tilde{\boldsymbol{p}}^0 = \left(\boldsymbol{A}_1^0\right)^{-1} \tilde{\boldsymbol{p}}^0. \tag{2.39}$$

This matrix is expressed in a block-partitioned form as

$$\boldsymbol{A}_0^1 = \begin{bmatrix} \boldsymbol{R}_1^{0T} & -\boldsymbol{R}_1^{0T}\boldsymbol{o}_1^0 \\ \boldsymbol{0}^T & 1 \end{bmatrix} = \begin{bmatrix} \boldsymbol{R}_0^1 & -\boldsymbol{R}_0^1\boldsymbol{o}_1^0 \\ \boldsymbol{0}^T & 1 \end{bmatrix}, \tag{2.40}$$

which gives the homogeneous representation form of the result already established by (2.34) and (2.35).

Notice that for the homogeneous transformation matrix the orthogonality property does not hold; hence, in general,

$$\boldsymbol{A}^{-1} \neq \boldsymbol{A}^T. \tag{2.41}$$

---

[6] It can be shown that in (2.37) nonnull values of the first three elements of the fourth row of $\boldsymbol{A}$ produce a perspective effect, while values other than unity for the fourth element give a scaling effect.

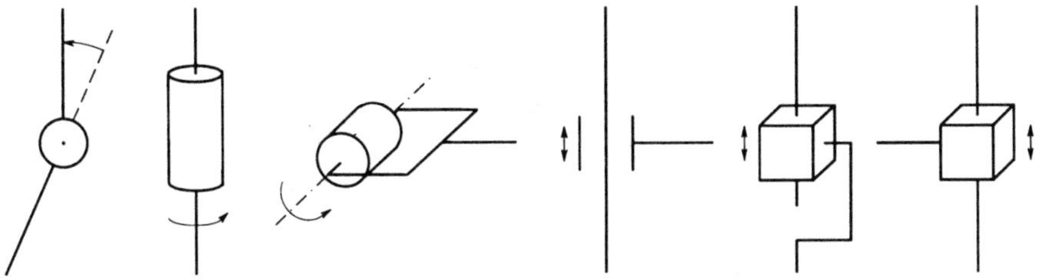

Figure 2.12 Conventional representations of joints.

In sum, a homogeneous transformation matrix expresses the coordinate transformation between two frames in a compact form. If the frames have the same origin, it reduces to the rotation matrix previously defined. Instead, if the frames have distinct origins, it allows keeping the notation with superscripts and subscripts that directly characterize the current frame and the fixed frame.

Analogously to what presented for the rotation matrices, it is easy to verify that a sequence of coordinate transformations can be composed by the product

$$\tilde{\boldsymbol{p}}^0 = \boldsymbol{A}_1^0 \boldsymbol{A}_2^1 \dots \boldsymbol{A}_n^{n-1} \tilde{\boldsymbol{p}}^n \tag{2.42}$$

where $\boldsymbol{A}_i^{i-1}$ denotes the homogeneous transformation relating the description of a point in Frame i to the description of the same point in Frame $i-1$.

2.8 Direct Kinematics

A manipulator consists of a series of rigid bodies (*links*) connected by means of kinematic pairs or *joints*. Joints can be essentially of two types: *revolute* and *prismatic*; conventional representations of the two types of joints are sketched in Figure 2.12. The whole structure forms a *kinematic chain*. One end of the chain is constrained to a base. An *end effector* (gripper, tool) is connected to the other end allowing manipulation of objects in space.

From a topological viewpoint, the kinematic chain is termed *open* when there is only one sequence of links connecting the two ends of the chain. Alternatively, a manipulator contains a *closed* kinematic chain when a sequence of links forms a loop.

The mechanical structure of a manipulator is characterized by a number of degrees of mobility which uniquely determine its configuration. Each degree of mobility is typically associated with a joint articulation and constitutes a *joint variable*. The aim of *direct kinematics* is to compute the position and orientation of the end effector as a function of the joint variables.

It was previously illustrated that the position and orientation of a body with respect to a reference frame are described by the position vector of the origin and the unit

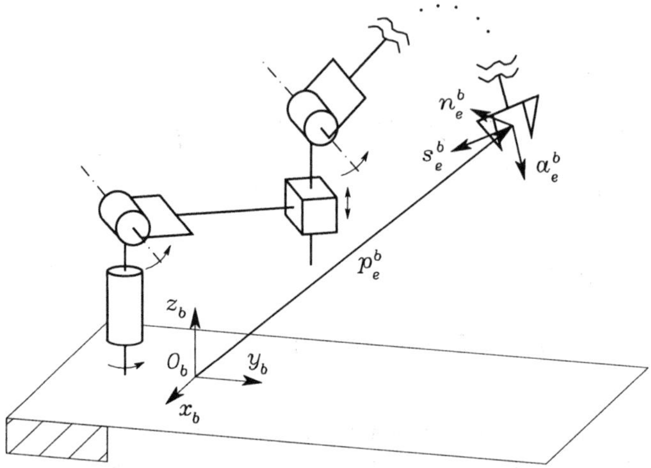

Figure 2.13 Description of the position and orientation of the end-effector frame.

vectors of a frame attached to the body. Hence, with respect to a reference frame O_b–$x_b y_b z_b$, the direct kinematics function is expressed by the homogeneous transformation matrix

$$\boldsymbol{T}_e^b(\boldsymbol{q}) = \begin{bmatrix} \boldsymbol{n}_e^b(\boldsymbol{q}) & \boldsymbol{s}_e^b(\boldsymbol{q}) & \boldsymbol{a}_e^b(\boldsymbol{q}) & \boldsymbol{p}_e^b(\boldsymbol{q}) \\ 0 & 0 & 0 & 1 \end{bmatrix}, \tag{2.43}$$

where $\boldsymbol{q}$ is the $(n \times 1)$ vector of joint variables, $\boldsymbol{n}_e$, $\boldsymbol{s}_e$, $\boldsymbol{a}_e$ are the unit vectors of a frame attached to the end effector, and $\boldsymbol{p}_e$ is the position vector of the origin of such frame with respect to the origin of the base frame (Figure 2.13). Note that $\boldsymbol{n}_e$, $\boldsymbol{s}_e$, $\boldsymbol{a}_e$ and $\boldsymbol{p}_e$ are a function of $\boldsymbol{q}$.

The frame O_b–$x_b y_b z_b$ is termed *base frame*. The frame attached to the end effector is termed *end-effector frame* and is conveniently chosen according to the particular task geometry. If the end effector is a gripper, the origin of the end-effector frame is located at the centre of the gripper, the unit vector $\boldsymbol{a}_e$ is chosen in the *approach* direction to the object, the unit vector $\boldsymbol{s}_e$ is chosen normal to $\boldsymbol{a}_e$ in the *sliding* plane of the jaws, and the unit vector $\boldsymbol{n}_e$ is chosen *normal* to the other two so that the frame $(\boldsymbol{n}_e, \boldsymbol{s}_e, \boldsymbol{a}_e)$ is right-handed.

A first way to compute direct kinematics is offered by a geometric analysis of the structure of the given manipulator.

Example 2.4

Consider the two-link planar arm in Figure 2.14. On the basis of simple trigonometry,

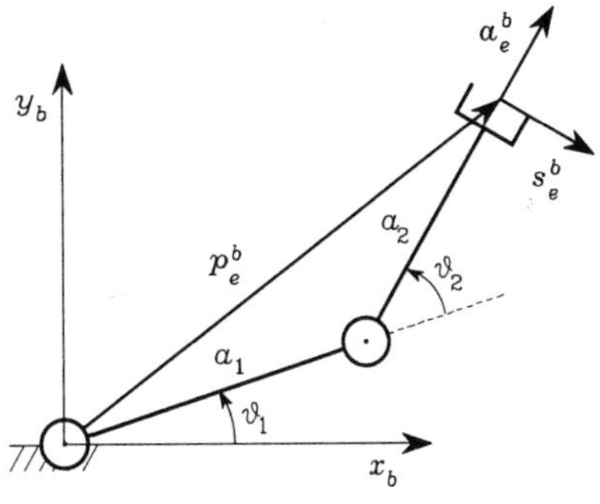

Figure 2.14 Two-link planar arm.

the choice of the joint variables, the base frame, and the end-effector frame leads to[7]

$$T_e^b(q) = \begin{bmatrix} n_e^b & s_e^b & a_e^b & p_e^b \\ 0 & 0 & 0 & 1 \end{bmatrix} = \begin{bmatrix} 0 & s_{12} & c_{12} & a_1 c_1 + a_2 c_{12} \\ 0 & -c_{12} & s_{12} & a_1 s_1 + a_2 s_{12} \\ 1 & 0 & 0 & 0 \\ 0 & 0 & 0 & 1 \end{bmatrix}. \tag{2.44}$$

It is not difficult to infer that the effectiveness of a geometric approach to the direct kinematics problem is based first on a convenient choice of the relevant quantities and then on the ability and geometric intuition of the problem solver. Whenever the manipulator structure is complex and the number of joints increases, it is preferable to adopt a less direct solution, which, though, is based on a systematic, general procedure. The problem becomes even more complex when the manipulator contains one or more closed kinematic chains. In such a case, as it will be discussed later, there is no guarantee to obtain an analytical expression for the direct kinematics function in (2.43).

2.8.1 Open Chain

Consider an *open-chain* manipulator constituted by $n+1$ links connected by n joints, where Link 0 is conventionally fixed to the ground. It is assumed that each joint provides the mechanical structure with a single degree of mobility, corresponding to the joint variable.

The construction of an operating procedure for the computation of direct kinematics is naturally derived from the typical open kinematic chain of the manipulator structure. In fact, since each joint connects two consecutive links, it is reasonable to consider first the description of kinematic relationship between consecutive links and then to obtain the overall description of manipulator kinematics in a recursive fashion. To this

[7] The notations $s_{i\dots j}$, $c_{i\dots j}$ denote respectively $\sin(q_i + \ldots + q_j)$, $\cos(q_i + \ldots + q_j)$.

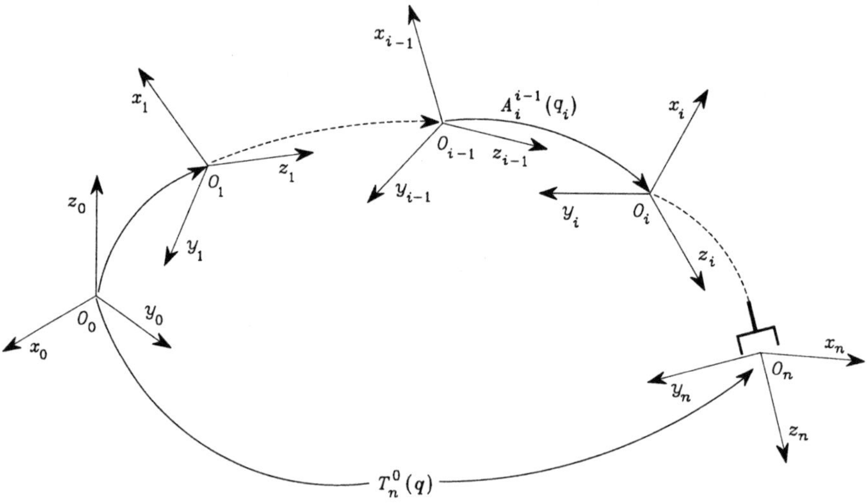

Figure 2.15 Coordinate transformations in an open kinematic chain.

purpose, it is worth defining a coordinate frame attached to each link, from Link 0 to Link n. Then, the coordinate transformation describing the position and orientation of Frame n with respect to Frame 0 (Figure 2.15) is given by

$$\boldsymbol{T}_n^0(\boldsymbol{q}) = \boldsymbol{A}_1^0(q_1)\boldsymbol{A}_2^1(q_2)\dots\boldsymbol{A}_n^{n-1}(q_n). \tag{2.45}$$

As requested, the computation of the direct kinematics function is recursive and is obtained in a systematic manner by simple products of the homogeneous transformation matrices $\boldsymbol{A}_i^{i-1}(q_i)$ (for $i = 1,\dots,n$), each of which is a function of a single joint variable.

With reference to the direct kinematics equation in (2.44), the actual coordinate transformation describing the position and orientation of the end-effector frame with respect to the base frame can be obtained as

$$\boldsymbol{T}_e^b(\boldsymbol{q}) = \boldsymbol{T}_0^b\boldsymbol{T}_n^0(\boldsymbol{q})\boldsymbol{T}_e^n \tag{2.46}$$

where $\boldsymbol{T}_0^b$ and $\boldsymbol{T}_e^n$ are two (typically) constant homogeneous transformations describing the position and orientation of Frame 0 with respect to the base frame, and of the end-effector frame with respect to Frame n, respectively. Hereafter, the subscript e is dropped whenever it is referred to $\boldsymbol{p}_e$ and $\boldsymbol{R}_e = [\,\boldsymbol{n}_e \quad \boldsymbol{s}_e \quad \boldsymbol{a}_e\,]$, *i.e.*, for brevity, $\boldsymbol{p}$ and $\boldsymbol{R} = [\,\boldsymbol{n} \quad \boldsymbol{s} \quad \boldsymbol{a}\,]$ respectively denote the position and orientation of the end effector.

2.8.2 Denavit-Hartenberg Convention

In order to compute the direct kinematics equation for an open-chain manipulator according to the recursive expression in (2.45), a systematic, general method is to be

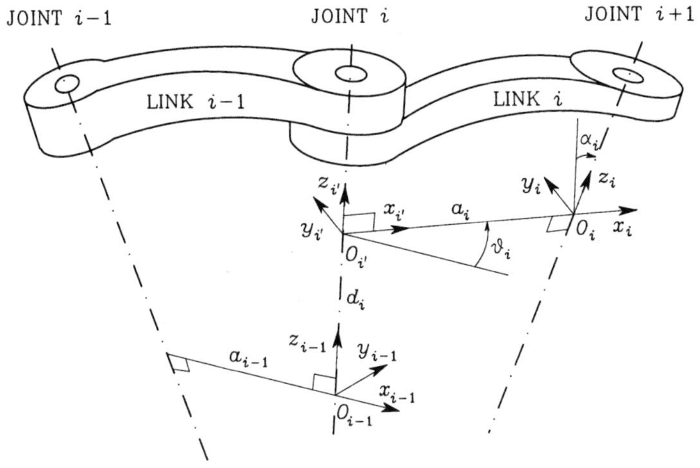

Figure 2.16 Denavit-Hartenberg kinematic parameters.

derived to define the relative position and orientation of two consecutive links; the problem is that to determine two frames attached to the two links and compute the coordinate transformations between them. In general, the frames can be arbitrarily chosen as long as they are attached to the link they are referred to. Nevertheless, it is convenient to set some rules also for the definition of the link frames.

With reference to Figure 2.16, let Axis i denote the axis of the joint connecting Link $i-1$ to Link i; the so-called *Denavit-Hartenberg convention* is adopted to define link Frame i:

- Choose axis z_i along the axis of Joint $i+1$.
- Locate the origin O_i at the intersection of axis z_i with the common normal[8] to axes z_{i-1} and z_i. Also, locate $O_{i'}$ at the intersection of the common normal with axis z_{i-1}.
- Choose axis x_i along the common normal to axes z_{i-1} and z_i with direction from Joint i to Joint $i+1$.
- Choose axis y_i so as to complete a right-handed frame.

The Denavit-Hartenberg convention gives a nonunique definition of the link frame in the following cases:

- For Frame 0, only the direction of axis z_0 is specified; then O_0 and x_0 can be arbitrarily chosen.
- For Frame n, since there is no Joint $n+1$, z_n is not uniquely defined while x_n

[8] The common normal between two lines is the line containing the minimum distance segment between the two lines.

has to be normal to axis z_{n-1}. Typically, Joint n is revolute, and thus z_n is to be aligned with the direction of z_{n-1}.

- When two consecutive axes are parallel, the common normal between them is not uniquely defined.
- When two consecutive axes intersect, the direction of x_i is arbitrary.
- When Joint i is prismatic, the direction of z_{i-1} is arbitrary.

In all such cases, the indeterminacy can be exploited to simplify the procedure; for instance, the axes of consecutive frames can be made parallel.

Once the link frames have been established, the position and orientation of Frame i with respect to Frame $i-1$ are completely specified by the following *parameters*:

a_i distance between O_i and $O_{i'}$,

d_i coordinate of $O_{i'}$ along z_{i-1},

α_i angle between axes z_{i-1} and z_i about axis x_i to be taken positive when rotation is made counter-clockwise,

ϑ_i angle between axes x_{i-1} and x_i about axis z_{i-1} to be taken positive when rotation is made counter-clockwise.

Two of the four parameters (a_i and α_i) are always constant and depend only on the geometry of connection between consecutive joints established by Link i. Of the remaining two parameters, only one is variable depending on the type of joint that connects Link $i-1$ to Link i. In particular:

- if Joint i is *revolute* the variable is ϑ_i,
- if Joint i is *prismatic* the variable is d_i.

At this point, it is possible to express the coordinate transformation between Frame i and Frame $i-1$ according to the following steps:

- Choose a frame aligned with Frame $i-1$.
- Translate the chosen frame by d_i along axis z_{i-1} and rotate it by ϑ_i about axis z_{i-1}; this sequence aligns the current frame with Frame i' and is described by the homogeneous transformation matrix

$$\boldsymbol{A}_{i'}^{i-1} = \begin{bmatrix} c_{\vartheta_i} & -s_{\vartheta_i} & 0 & 0 \\ s_{\vartheta_i} & c_{\vartheta_i} & 0 & 0 \\ 0 & 0 & 1 & d_i \\ 0 & 0 & 0 & 1 \end{bmatrix}.$$

- Translate the frame aligned with Frame i' by a_i along axis $x_{i'}$ and rotate it by α_i about axis $x_{i'}$; this sequence aligns the current frame with Frame i and is described by the homogeneous transformation matrix

$$\boldsymbol{A}_{i}^{i'} = \begin{bmatrix} 1 & 0 & 0 & a_i \\ 0 & c_{\alpha_i} & -s_{\alpha_i} & 0 \\ 0 & s_{\alpha_i} & c_{\alpha_i} & 0 \\ 0 & 0 & 0 & 1 \end{bmatrix}.$$

- The resulting coordinate transformation is obtained by postmultiplication of the single transformations as

$$A_i^{i-1}(q_i) = A_{i'}^{i-1} A_i^{i'} = \begin{bmatrix} c_{\vartheta_i} & -s_{\vartheta_i} c_{\alpha_i} & s_{\vartheta_i} s_{\alpha_i} & a_i c_{\vartheta_i} \\ s_{\vartheta_i} & c_{\vartheta_i} c_{\alpha_i} & -c_{\vartheta_i} s_{\alpha_i} & a_i s_{\vartheta_i} \\ 0 & s_{\alpha_i} & c_{\alpha_i} & d_i \\ 0 & 0 & 0 & 1 \end{bmatrix}. \qquad (2.47)$$

Notice that the transformation matrix from Frame i to Frame $i-1$ is a function only of the joint variable q_i, that is, ϑ_i for a revolute joint or d_i for a prismatic joint.

To summarize, the Denavit-Hartenberg convention allows constructing the direct kinematics function by composition of the individual coordinate transformations expressed by (2.47) into one homogeneous transformation matrix as in (2.45). The procedure can be applied to any open kinematic chain and can be easily rewritten in an operating form as follows.

1. Find and number consecutively the joint axes; set the directions of axes $z_0, \ldots, z_{n-1}$.
2. Choose Frame 0 by locating the origin on axis z_0; axes x_0 and y_0 are chosen so as to obtain a right-handed frame. If feasible, it is worth choosing Frame 0 to coincide with the base frame.

Execute steps from **3** to **5** for $i = 1, \ldots, n-1$:

3. Locate the origin O_i at the intersection of z_i with the common normal to axes z_{i-1} and z_i. If axes z_{i-1} and z_i are parallel and Joint i is revolute, then locate O_i so that $d_i = 0$; if Joint i is prismatic, locate O_i at a reference position for the joint range, e.g., a mechanical limit.
4. Choose axis x_i along the common normal to axes z_{i-1} and z_i with direction from Joint i to Joint $i+1$.
5. Choose axis y_i so as to obtain a right-handed frame.

To complete:

6. Choose Frame n; if Joint n is revolute, then align z_n with z_{n-1}, otherwise, if Joint n is prismatic, then choose z_n arbitrarily. Axis x_n is set according to step **4**.
7. For $i = 1, \ldots, n$, form the table of parameters $a_i, d_i, \alpha_i, \vartheta_i$.
8. On the basis of the parameters in **7**, compute the homogeneous transformation matrices $A_i^{i-1}(q_i)$ for $i = 1, \ldots, n$.
9. Compute the homogeneous transformation $T_n^0(q) = A_1^0 \ldots A_n^{n-1}$ that yields the position and orientation of Frame n with respect to Frame 0.
10. Given T_0^b and T_e^n, compute the direct kinematics function as $T_e^b(q) = T_0^b T_n^0 T_e^n$ that yields the position and orientation of the end-effector frame with respect to the base frame.

For what concerns the computational aspects of direct kinematics, it can be recognized that the heaviest load derives from the evaluation of transcendental functions. On the

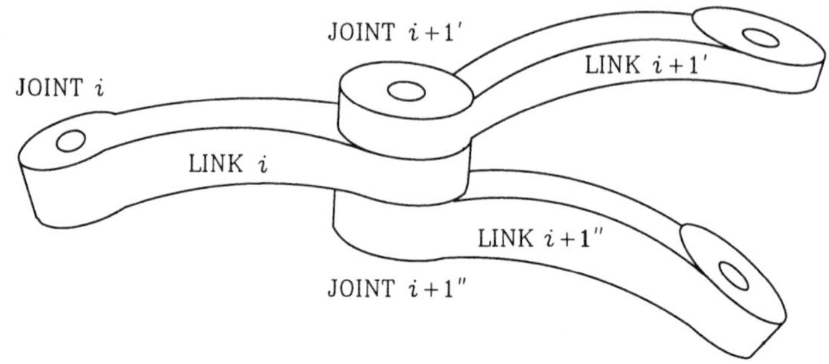

Figure 2.17 Connection of a single link in the chain with two links.

other hand, by suitably factorizing the transformation equations and introducing local variables, the number of flops (additions + multiplications) can be reduced. Finally, for computation of orientation it is convenient to evaluate the two unit vectors of the end-effector frame of simplest expression and derive the third one by vector product of the first two.

2.8.3 Closed Chain

The above direct kinematics method based on the Denavit-Hartenberg convention exploits the inherently recursive feature of an open-chain manipulator. Nevertheless, the method can be extended to the case of manipulators containing closed kinematic chains according to the technique illustrated below.

Consider a *closed-chain* manipulator constituted by $n + 1$ links. Because of the presence of a loop, the number of joints l must be greater than n; in particular, it can be understood that the number of closed loops is equal to $l - n$.

With reference to Figure 2.17, Links 0 through i are connected successively through the first i joints as in an open kinematic chain. Then, Joint $i + 1'$ connects Link i with Link $i + 1'$ while Joint $i + 1''$ connects Link i with Link $i + 1''$; the axes of Joints $i + 1'$ and $i + 1''$ are assumed to be aligned. Although not represented in the figure, Links $i + 1'$ and $i + 1''$ are members of the closed kinematic chain. In particular, Link $i + 1'$ is further connected to Link $i + 2'$ via Joint $i + 2'$ and so forth, until Link j via Joint j. Likewise, Link $i + 1''$ is further connected to Link $i + 2''$ via Joint $i + 2''$ and so forth, until Link k via Joint k. Finally, Links j and k are connected together at Joint $j + 1$ to form a closed chain. In general, $j \neq k$.

In order to attach frames to the various links and apply Denavit-Hartenberg convention, one closed kinematic chain is taken into account. The closed chain can be virtually cut open at Joint $j + 1$, *i.e.*, the joint between Link j and Link k. An equivalent tree-structured open kinematic chain is obtained, and thus link frames can be defined as in Figure 2.18. Since Links 0 through i occur before the two branches of the tree, they are left out of the analysis. For the same reason, Links $j + 1$ through n are left out as well. Notice that Frame i is to be chosen with axis z_i aligned with the axes of

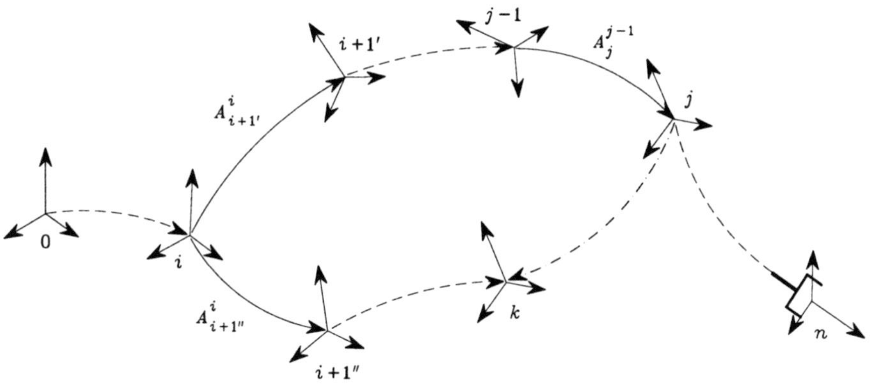

Figure 2.18 Coordinate transformations in a closed kinematic chain.

Joints $i + 1'$ and $i + 1''$.

It follows that the position and orientation of Frame j with respect to Frame i can be expressed by composing the homogeneous transformations as

$$\boldsymbol{A}_j^i(\boldsymbol{q}') = \boldsymbol{A}_{i+1'}^i(q_{i+1'}) \dots \boldsymbol{A}_j^{j-1}(q_j) \tag{2.48}$$

where $\boldsymbol{q}' = [\, q_{i+1'} \quad \dots \quad q_j \,]^T$. Likewise, the position and orientation of Frame k with respect to Frame i is given by

$$\boldsymbol{A}_k^i(\boldsymbol{q}'') = \boldsymbol{A}_{i+1''}^i(q_{i+1''}) \dots \boldsymbol{A}_k^{k-1}(q_k) \tag{2.49}$$

where $\boldsymbol{q}'' = [\, q_{i+1''} \quad \dots \quad q_k \,]^T$.

Since Links j and k are connected to each other through Joint $j + 1$, it is worth analyzing the mutual position and orientation between Frames j and k, as illustrated in Figure 2.19. Notice that, since Links j and k are connected to form a closed chain, axes z_j and z_k are aligned. Therefore, the following orientation constraint has to be imposed between Frames j and k:

$$\boldsymbol{z}_j^i(\boldsymbol{q}') = \boldsymbol{z}_k^i(\boldsymbol{q}'') \tag{2.50}$$

where the unit vectors of the two axes have been conveniently referred to Frame i.

Moreover, if Joint $j + 1$ is prismatic, the angle ϑ_{jk} between axes x_j and x_k is fixed; hence, in addition to (2.50), the following constraint is obtained:

$$\boldsymbol{x}_j^{iT}(\boldsymbol{q}')\boldsymbol{x}_k^i(\boldsymbol{q}'') = \cos\vartheta_{jk}. \tag{2.51}$$

Obviously, there is no need to impose a similar constraint on axes y_j and y_k since that would be redundant.

Regarding the position constraint between Frames j and k, let $\boldsymbol{p}_j^i$ and $\boldsymbol{p}_k^i$ respectively denote the positions of the origins of Frames j and k, when referred to Frame i.

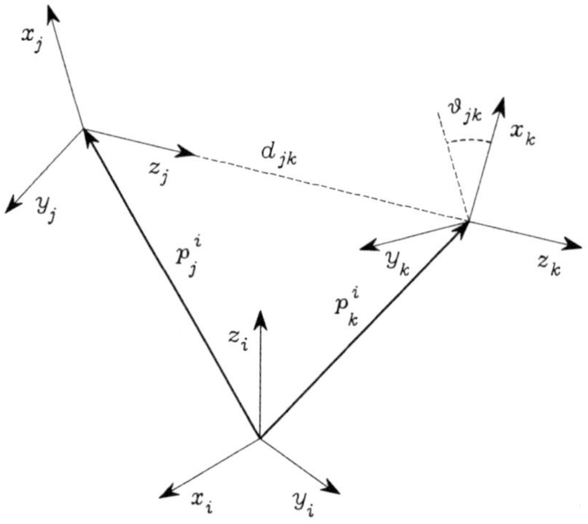

Figure 2.19 Coordinate transformation at the cut joint.

By projecting on Frame j the distance vector of the origin of Frame k from Frame j, the following constraint has to be imposed:

$$\boldsymbol{R}_i^j(\boldsymbol{q}')\left(\boldsymbol{p}_j^i(\boldsymbol{q}') - \boldsymbol{p}_k^i(\boldsymbol{q}'')\right) = [\,0 \quad 0 \quad d_{jk}\,]^T \tag{2.52}$$

where $\boldsymbol{R}_i^j = \boldsymbol{R}_j^{iT}$ denotes the orientation of Frame i with respect to Frame j. At this point, if Joint $j+1$ is revolute, then d_{jk} is a fixed offset along axis z_j; hence, the three equalities of (2.52) fully describe the position constraint. If, however, Joint $j+1$ is prismatic, then d_{jk} varies. Consequently, only the first two equalities of (2.52) describe the position constraint, *i.e.*,

$$\begin{bmatrix} \boldsymbol{x}_j^{iT}(\boldsymbol{q}') \\ \boldsymbol{y}_j^{iT}(\boldsymbol{q}') \end{bmatrix} \left(\boldsymbol{p}_j^i(\boldsymbol{q}') - \boldsymbol{p}_k^i(\boldsymbol{q}'')\right) = \begin{bmatrix} 0 \\ 0 \end{bmatrix} \tag{2.53}$$

where $\boldsymbol{R}_j^i = [\,\boldsymbol{x}_j^i \quad \boldsymbol{y}_j^i \quad \boldsymbol{z}_j^i\,]$.

In summary, if Joint $j+1$ is *revolute* the constraints are

$$\begin{cases} \boldsymbol{R}_i^j(\boldsymbol{q}')\left(\boldsymbol{p}_j^i(\boldsymbol{q}') - \boldsymbol{p}_k^i(\boldsymbol{q}'')\right) = [\,0 \quad 0 \quad d_{jk}\,]^T \\ \boldsymbol{z}_j^i(\boldsymbol{q}') = \boldsymbol{z}_k^i(\boldsymbol{q}''), \end{cases} \tag{2.54}$$

whereas if Joint $j+1$ is *prismatic* the constraints are

$$\begin{cases} \begin{bmatrix} \boldsymbol{x}_j^{iT}(\boldsymbol{q}') \\ \boldsymbol{y}_j^{iT}(\boldsymbol{q}') \end{bmatrix} \left(\boldsymbol{p}_j^i(\boldsymbol{q}') - \boldsymbol{p}_k^i(\boldsymbol{q}'')\right) = \begin{bmatrix} 0 \\ 0 \end{bmatrix} \\ \boldsymbol{z}_j^i(\boldsymbol{q}') = \boldsymbol{z}_k^i(\boldsymbol{q}'') \\ \boldsymbol{x}_j^{iT}(\boldsymbol{q}')\boldsymbol{x}_k^i(\boldsymbol{q}'') = \cos\vartheta_{jk}. \end{cases} \tag{2.55}$$

In either case, there are six equalities that must be satisfied. Those should be solved for a reduced number of independent joint variables to be keenly chosen among the components of $\boldsymbol{q}'$ and $\boldsymbol{q}''$ which characterize the degrees of mobility of the closed chain. These are the natural candidates to be the actuated joints, while the other joints in the chain (including the cut joint) are typically not actuated. Such independent variables, together with the remaining joint variables not involved in the above analysis, constitute the joint vector $\boldsymbol{q}$ that allows the direct kinematics equation to be computed as

$$\boldsymbol{T}_n^0(\boldsymbol{q}) = \boldsymbol{A}_i^0 \boldsymbol{A}_j^i \boldsymbol{A}_n^j, \tag{2.56}$$

where the sequence of successive transformations after the closure of the chain has been conventionally resumed from Frame j.

In general, there is no guarantee to solve the constraints in closed form unless the manipulator has a simple kinematic structure. In other words, for a given manipulator with a specific geometry, e.g., a planar structure, some of the above equalities may become dependent. Hence, the number of independent equalities is less than six and it should likely be easier to solve them.

To conclude, it is worth sketching the operating form of the procedure to compute the direct kinematics function for a closed-chain manipulator using the Denavit-Hartenberg convention.

1. In the closed chain, select one joint that is not actuated. Assume that the joint is cut open so as to obtain an open chain in a tree structure.
2. Compute the homogeneous transformations according to Denavit-Hartenberg convention.
3. Find the equality constraints for the two frames connected by the cut joint.
4. Solve the constraints for a reduced number of joint variables.
5. Express the homogeneous transformations in terms of the above joint variables and compute the direct kinematics function by composing the various transformations from the base frame to the end-effector frame.

2.9 Kinematics of Typical Manipulator Structures

This section contains several examples of computation of the direct kinematics function for typical manipulator structures that are often encountered in industrial robots.

2.9.1 Three-link Planar Arm

Consider the three-link planar arm in Figure 2.20, where the link frames have been illustrated. Since the revolute axes are all parallel, the simplest choice was made for all axes x_i along the direction of the relative links (the direction of x_0 is arbitrary) and all lying in the plane (x_0, y_0). In this way, all the parameters d_i are null and the angles between the axes x_i directly provide the joint variables. The Denavit-Hartenberg parameters are specified in Table 2.1.

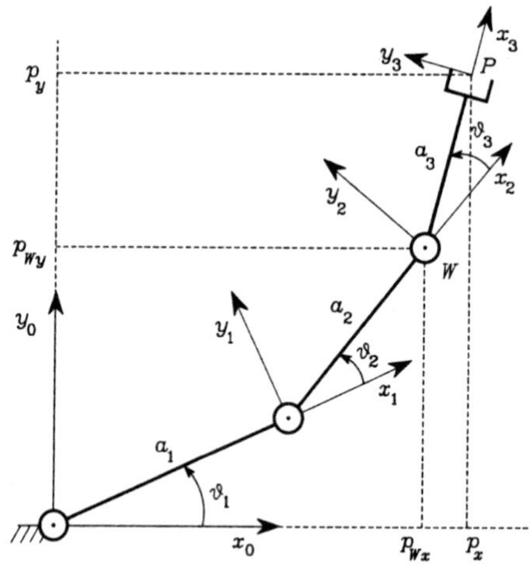

Figure 2.20 Three-link planar arm.

Table 2.1 Denavit-Hartenberg parameters for the three-link planar arm.

Link	a_i	α_i	d_i	ϑ_i
1	a_1	0	0	ϑ_1
2	a_2	0	0	ϑ_2
3	a_3	0	0	ϑ_3

Since all joints are revolute, the homogeneous transformation matrix defined in (2.47) has the same structure for each joint, *i.e.*,

$$\boldsymbol{A}_i^{i-1}(\vartheta_i) = \begin{bmatrix} c_i & -s_i & 0 & a_i c_i \\ s_i & c_i & 0 & a_i s_i \\ 0 & 0 & 1 & 0 \\ 0 & 0 & 0 & 1 \end{bmatrix} \qquad i = 1, 2, 3. \tag{2.57}$$

Computation of the direct kinematics function as in (2.45) yields

$$\boldsymbol{T}_3^0(\boldsymbol{q}) = \boldsymbol{A}_1^0 \boldsymbol{A}_2^1 \boldsymbol{A}_3^2 = \begin{bmatrix} c_{123} & -s_{123} & 0 & a_1 c_1 + a_2 c_{12} + a_3 c_{123} \\ s_{123} & c_{123} & 0 & a_1 s_1 + a_2 s_{12} + a_3 s_{123} \\ 0 & 0 & 1 & 0 \\ 0 & 0 & 0 & 1 \end{bmatrix} \tag{2.58}$$

where $\boldsymbol{q} = [\,\vartheta_1 \quad \vartheta_2 \quad \vartheta_3\,]^T$. Notice that the unit vector $\boldsymbol{z}_3^0$ of Frame 3 is aligned with $\boldsymbol{z}_0 = [\,0 \quad 0 \quad 1\,]^T$, in view of the fact that all revolute joints are parallel to axis z_0. Obviously, $p_z = 0$ and all three joints concur to determine the end-effector position in the plane of the structure.

It is worth pointing out that Frame 3 does not coincide with the end-effector frame as in Figure 2.13, since the resulting approach unit vector is aligned with $\boldsymbol{x}_3^0$ and

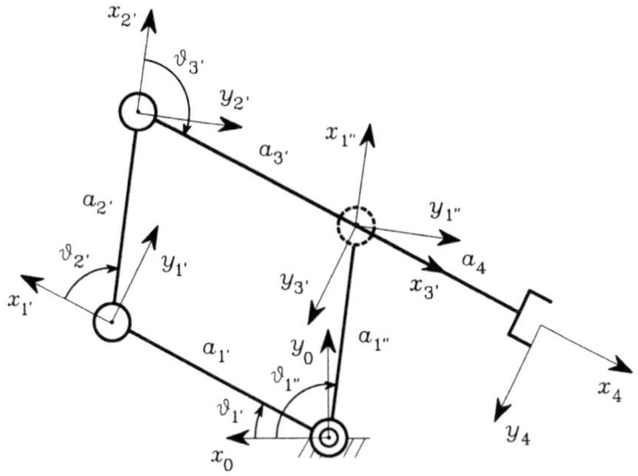

Figure 2.21 Parallelogram arm.

not with z_3^0. Thus, assuming that the two frames have the same origin, the constant transformation

$$\boldsymbol{T}_e^3 = \begin{bmatrix} 0 & 0 & 1 & 0 \\ 0 & -1 & 0 & 0 \\ 1 & 0 & 0 & 0 \\ 0 & 0 & 0 & 1 \end{bmatrix}$$

is needed, having taken $\boldsymbol{n}$ aligned with $\boldsymbol{z}_0$.

2.9.2 Parallelogram Arm

Consider the parallelogram arm in Figure 2.21. A closed chain occurs where the first two joints connect Link $1'$ and Link $1''$ to Link 0, respectively. Joint 4 was selected as the cut joint, and the link frames have been established accordingly. The Denavit-Hartenberg parameters are specified in Table 2.2, where $a_{1'} = a_{3'}$ and $a_{2'} = a_{1''}$ in view of the parallelogram structure.

Table 2.2 Denavit-Hartenberg parameters for the parallelogram arm.

Link	a_i	α_i	d_i	ϑ_i
$1'$	$a_{1'}$	0	0	$\vartheta_{1'}$
$2'$	$a_{2'}$	0	0	$\vartheta_{2'}$
$3'$	$a_{3'}$	0	0	$\vartheta_{3'}$
$1''$	$a_{1''}$	0	0	$\vartheta_{1''}$
4	a_4	0	0	0

Notice that the parameters for Link 4 are all constant. Since the joints are revolute, the homogeneous transformation matrix defined in (2.47) has the same structure for each joint, *i.e.*, as in (2.57) for Joints $1'$, $2'$, $3'$ and $1''$. Therefore, the coordinate transformations for the two branches of the tree are respectively:

$$\boldsymbol{A}_{3'}^{0}(\boldsymbol{q}') = \boldsymbol{A}_{1'}^{0}\boldsymbol{A}_{2'}^{1'}\boldsymbol{A}_{3'}^{2'} = \begin{bmatrix} c_{1'2'3'} & -s_{1'2'3'} & 0 & a_{1'}c_{1'} + a_{2'}c_{1'2'} + a_{3'}c_{1'2'3'} \\ s_{1'2'3'} & c_{1'2'3'} & 0 & a_{1'}s_{1'} + a_{2'}s_{1'2'} + a_{3'}s_{1'2'3'} \\ 0 & 0 & 1 & 0 \\ 0 & 0 & 0 & 1 \end{bmatrix}$$

where $\boldsymbol{q}' = [\,\vartheta_{1'} \quad \vartheta_{2'} \quad \vartheta_{3'}\,]^T$, and

$$\boldsymbol{A}_{1''}^{0}(q'') = \begin{bmatrix} c_{1''} & -s_{1''} & 0 & a_{1''}c_{1''} \\ s_{1''} & c_{1''} & 0 & a_{1''}s_{1''} \\ 0 & 0 & 1 & 0 \\ 0 & 0 & 0 & 1 \end{bmatrix}$$

where $q'' = \vartheta_{1''}$. To complete, the constant homogeneous transformation for the last link is

$$\boldsymbol{A}_{4}^{3'} = \begin{bmatrix} 1 & 0 & 0 & a_4 \\ 0 & 1 & 0 & 0 \\ 0 & 0 & 1 & 0 \\ 0 & 0 & 0 & 1 \end{bmatrix}.$$

With reference to (2.54), the position constraints are ($d_{3'1''} = 0$)

$$\boldsymbol{R}_{0}^{3'}(\boldsymbol{q}')\left(\boldsymbol{p}_{3'}^{0}(\boldsymbol{q}') - \boldsymbol{p}_{1''}^{0}(q'')\right) = \begin{bmatrix} 0 \\ 0 \\ 0 \end{bmatrix}$$

while the orientation constraints are satisfied independently of $\boldsymbol{q}'$ and q''. Since $a_{1'} = a_{3'}$ and $a_{2'} = a_{1''}$, two independent constraints can be extracted, *i.e.*,

$$a_{1'}(c_{1'} + c_{1'2'3'}) + a_{1''}(c_{1'2'} - c_{1''}) = 0$$
$$a_{1'}(s_{1'} + s_{1'2'3'}) + a_{1''}(s_{1'2'} - s_{1''}) = 0.$$

In order to satisfy them for any choice of $a_{1'}$ and $a_{1''}$, it must be

$$\begin{aligned} \vartheta_{2'} &= \vartheta_{1''} - \vartheta_{1'} \\ \vartheta_{3'} &= \pi - \vartheta_{2'} = \pi - \vartheta_{1''} + \vartheta_{1'} \end{aligned} \qquad (2.59)$$

Therefore, the vector of joint variables is $\boldsymbol{q} = [\,\vartheta_{1'} \quad \vartheta_{1''}\,]^T$. These joints are natural candidates to be the actuated joints.[9] Substituting the expressions of $\vartheta_{2'}$ and $\vartheta_{3'}$ into

[9] Notice that it is not possible to solve (2.59) for $\vartheta_{2'}$ and $\vartheta_{3'}$ since they are constrained by the condition $\vartheta_{2'} + \vartheta_{3'} = \pi$.

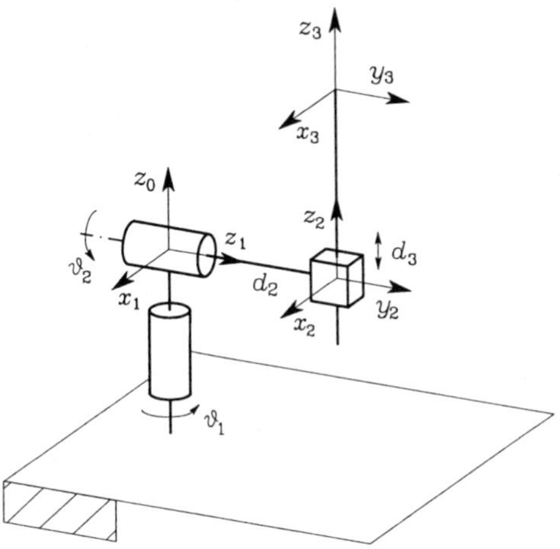

Figure 2.22 Spherical arm.

the homogeneous transformation $A^0_{3'}$ and computing the direct kinematics function as in (2.56) yields

$$T^0_4(q) = A^0_{3'}(q)A^{3'}_4 = \begin{bmatrix} -c_{1'} & s_{1'} & 0 & a_{1''}c_{1''} - a_4 c_{1'} \\ -s_{1'} & -c_{1'} & 0 & a_{1''}s_{1''} - a_4 s_{1'} \\ 0 & 0 & 1 & 0 \\ 0 & 0 & 0 & 1 \end{bmatrix}. \tag{2.60}$$

A comparison between (2.60) and (2.44) reveals that the parallelogram arm is kinematically equivalent to a two-link planar arm. The noticeable difference, though, is that the two actuated joints—providing the degrees of mobility of the structure—are located at the base. This will greatly simplify the dynamic model of the structure, as shall be seen in Section 4.3.3.

2.9.3 Spherical Arm

Consider the spherical arm in Figure 2.22, where the link frames have been illustrated. Notice that the origin of Frame 0 was located at the intersection of z_0 with z_1 so that $d_1 = 0$; analogously, the origin of Frame 2 was located at the intersection between z_1 and z_2. The Denavit-Hartenberg parameters are specified in Table 2.3.

Table 2.3 Denavit-Hartenberg parameters for the spherical arm.

Link	a_i	α_i	d_i	ϑ_i
1	0	$-\pi/2$	0	ϑ_1
2	0	$\pi/2$	d_2	ϑ_2
3	0	0	d_3	0

The homogeneous transformation matrices defined in (2.47) are for the single joints:

$$A_1^0(\vartheta_1) = \begin{bmatrix} c_1 & 0 & -s_1 & 0 \\ s_1 & 0 & c_1 & 0 \\ 0 & -1 & 0 & 0 \\ 0 & 0 & 0 & 1 \end{bmatrix} \qquad A_2^1(\vartheta_2) = \begin{bmatrix} c_2 & 0 & s_2 & 0 \\ s_2 & 0 & -c_2 & 0 \\ 0 & 1 & 0 & d_2 \\ 0 & 0 & 0 & 1 \end{bmatrix}$$

$$A_3^2(d_3) = \begin{bmatrix} 1 & 0 & 0 & 0 \\ 0 & 1 & 0 & 0 \\ 0 & 0 & 1 & d_3 \\ 0 & 0 & 0 & 1 \end{bmatrix}.$$

Computation of the direct kinematics function as in (2.45) yields

$$T_3^0(q) = A_1^0 A_2^1 A_3^2 = \begin{bmatrix} c_1c_2 & -s_1 & c_1s_2 & c_1s_2d_3 - s_1d_2 \\ s_1c_2 & c_1 & s_1s_2 & s_1s_2d_3 + c_1d_2 \\ -s_2 & 0 & c_2 & c_2d_3 \\ 0 & 0 & 0 & 1 \end{bmatrix} \tag{2.61}$$

where $q = [\,\vartheta_1 \quad \vartheta_2 \quad d_3\,]^T$. Notice that the third joint does not obviously influence the rotation matrix. Further, the orientation of the unit vector y_3^0 is uniquely determined by the first joint, since the revolute axis of the second joint z_1 is parallel to axis y_3. Differently from the previous structures, in this case Frame 3 can represent an end-effector frame of unit vectors (n, s, a), *i.e.*, $T_e^3 = I$.

2.9.4 Anthropomorphic Arm

Consider the anthropomorphic arm in Figure 2.23. Notice how this arm corresponds to a two-link planar arm with an additional rotation about an axis of the plane. In this respect, the parallelogram arm could be used in lieu of the two-link planar arm, as found in some industrial robots with an anthropomorphic structure.

The link frames have been illustrated in the figure. As for the previous structure, the origin of Frame 0 was chosen at the intersection of z_0 with z_1 ($d_1 = 0$); further, z_1 and z_2 are parallel and the choice of axes x_1 and x_2 was made as for the two-link planar arm. The Denavit-Hartenberg parameters are specified in Table 2.4.

Table 2.4 Denavit-Hartenberg parameters for the anthropomorphic arm.

Link	a_i	α_i	d_i	ϑ_i
1	0	$\pi/2$	0	ϑ_1
2	a_2	0	0	ϑ_2
3	a_3	0	0	ϑ_3

The homogeneous transformation matrices defined in (2.47) are for the single joints:

$$A_1^0(\vartheta_1) = \begin{bmatrix} c_1 & 0 & s_1 & 0 \\ s_1 & 0 & -c_1 & 0 \\ 0 & 1 & 0 & 0 \\ 0 & 0 & 0 & 1 \end{bmatrix}$$

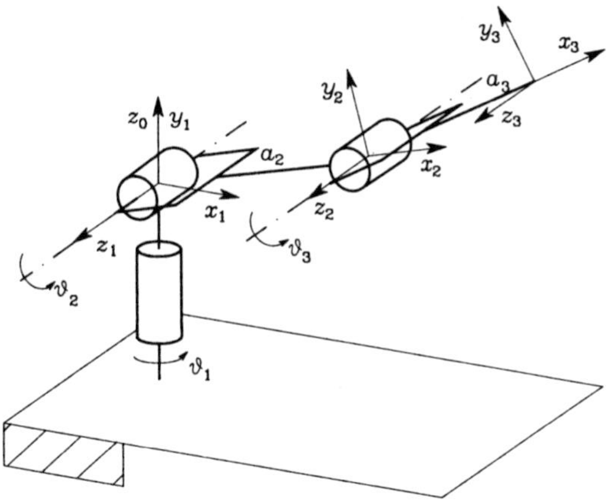

Figure 2.23 Anthropomorphic arm.

$$A_i^{i-1}(\vartheta_i) = \begin{bmatrix} c_i & -s_i & 0 & a_i c_i \\ s_i & c_i & 0 & a_i s_i \\ 0 & 0 & 1 & 0 \\ 0 & 0 & 0 & 1 \end{bmatrix} \qquad i = 2, 3.$$

Computation of the direct kinematics function as in (2.45) yields

$$T_3^0(q) = A_1^0 A_2^1 A_3^2 = \begin{bmatrix} c_1 c_{23} & -c_1 s_{23} & s_1 & c_1(a_2 c_2 + a_3 c_{23}) \\ s_1 c_{23} & -s_1 s_{23} & -c_1 & s_1(a_2 c_2 + a_3 c_{23}) \\ s_{23} & c_{23} & 0 & a_2 s_2 + a_3 s_{23} \\ 0 & 0 & 0 & 1 \end{bmatrix} \tag{2.62}$$

where $q = [\,\vartheta_1 \quad \vartheta_2 \quad \vartheta_3\,]^T$. Since z_3 is aligned with z_2, Frame 3 does not coincide with a possible end-effector frame as in Figure 2.13, and a proper constant transformation would be needed.

2.9.5 Spherical Wrist

Consider a particular type of structure consisting just of the wrist of Figure 2.24. Joint variables were numbered progressively starting from 4, since such a wrist is typically thought of as mounted on a three-degree-of-mobility arm of a six-degree-of-mobility manipulator. It is worth noticing that the wrist is spherical since all revolute axes intersect at a single point. Once z_3, z_4, z_5 have been established, and x_3 has been chosen, there is an indeterminacy on the directions of x_4 and x_5. With reference to the frames indicated in Figure 2.24, the Denavit-Hartenberg parameters are specified in Table 2.5.

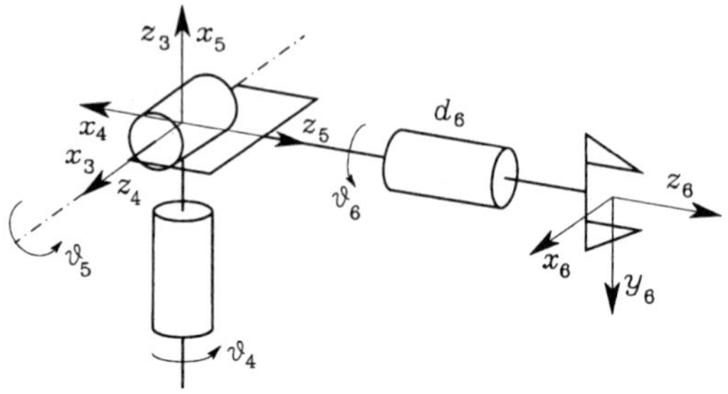

Figure 2.24 Spherical wrist.

Table 2.5 Denavit-Hartenberg parameters for the spherical wrist.

Link	a_i	α_i	d_i	ϑ_i
4	0	$-\pi/2$	0	ϑ_4
5	0	$\pi/2$	0	ϑ_5
6	0	0	d_6	ϑ_6

The homogeneous transformation matrices defined in (2.47) are for the single joints:

$$\boldsymbol{A}_4^3(\vartheta_4) = \begin{bmatrix} c_4 & 0 & -s_4 & 0 \\ s_4 & 0 & c_4 & 0 \\ 0 & -1 & 0 & 0 \\ 0 & 0 & 0 & 1 \end{bmatrix} \qquad \boldsymbol{A}_5^4(\vartheta_5) = \begin{bmatrix} c_5 & 0 & s_5 & 0 \\ s_5 & 0 & -c_5 & 0 \\ 0 & 1 & 0 & 0 \\ 0 & 0 & 0 & 1 \end{bmatrix}$$

$$\boldsymbol{A}_6^5(\vartheta_6) = \begin{bmatrix} c_6 & -s_6 & 0 & 0 \\ s_6 & c_6 & 0 & 0 \\ 0 & 0 & 1 & d_6 \\ 0 & 0 & 0 & 1 \end{bmatrix}.$$

Computation of the direct kinematics function as in (2.45) yields

$$\boldsymbol{T}_6^3(\boldsymbol{q}) = \boldsymbol{A}_4^3\boldsymbol{A}_5^4\boldsymbol{A}_6^5 = \begin{bmatrix} c_4c_5c_6 - s_4s_6 & -c_4c_5s_6 - s_4c_6 & c_4s_5 & c_4s_5d_6 \\ s_4c_5c_6 + c_4s_6 & -s_4c_5s_6 + c_4c_6 & s_4s_5 & s_4s_5d_6 \\ -s_5c_6 & s_5s_6 & c_5 & c_5d_6 \\ 0 & 0 & 0 & 1 \end{bmatrix} \tag{2.63}$$

where $\boldsymbol{q} = [\,\vartheta_4 \quad \vartheta_5 \quad \vartheta_6\,]^T$. Notice that, as a consequence of the choice made for the coordinate frames, the block matrix $\boldsymbol{R}_6^3$ that can be extracted from $\boldsymbol{T}_6^3$ coincides with the rotation matrix of Euler angles (2.18) previously derived, that is, ϑ_4, ϑ_5, ϑ_6 constitute the set of ZYZ angles with respect to the reference frame O_3–$x_3y_3z_3$. Moreover, the unit vectors of Frame 6 coincide with the unit vectors of a possible end-effector frame according to Figure 2.13.

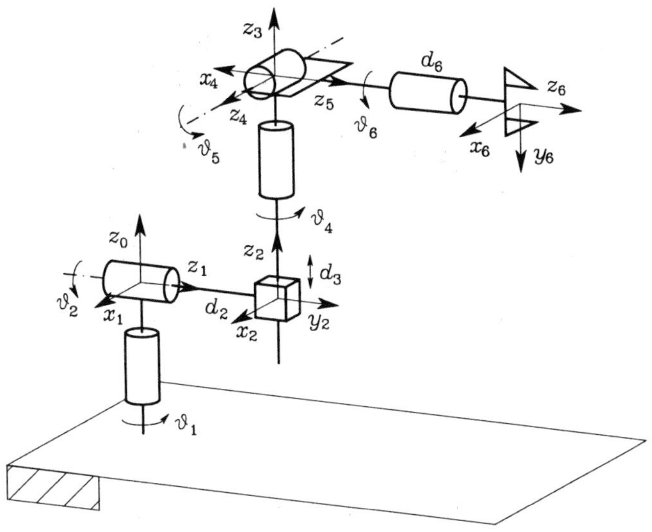

Figure 2.25 Stanford manipulator.

2.9.6 Stanford Manipulator

The so-called Stanford manipulator is composed by a spherical arm and a spherical wrist (Figure 2.25). Since Frame 3 of the spherical arm coincides with Frame 3 of the spherical wrist, the direct kinematics function can be obtained via simple composition of the transformation matrices (2.61) and (2.63) of the previous examples, *i.e.*,

$$\boldsymbol{T}_6^0 = \boldsymbol{T}_3^0 \boldsymbol{T}_6^3 = \begin{bmatrix} \boldsymbol{n}^0 & \boldsymbol{s}^0 & \boldsymbol{a}^0 & \boldsymbol{p}^0 \\ 0 & 0 & 0 & 1 \end{bmatrix}.$$

Carrying out the products yields

$$\boldsymbol{p}^0 = \begin{bmatrix} c_1 s_2 d_3 - s_1 d_2 + \big(c_1(c_2 c_4 s_5 + s_2 c_5) - s_1 s_4 s_5\big) d_6 \\ s_1 s_2 d_3 + c_1 d_2 + \big(s_1(c_2 c_4 s_5 + s_2 c_5) + c_1 s_4 s_5\big) d_6 \\ c_2 d_3 + (-s_2 c_4 s_5 + c_2 c_5) d_6 \end{bmatrix} \tag{2.64}$$

for the end-effector position, and

$$\boldsymbol{n}^0 = \begin{bmatrix} c_1\big(c_2(c_4 c_5 c_6 - s_4 s_6) - s_2 s_5 c_6\big) - s_1(s_4 c_5 c_6 + c_4 s_6) \\ s_1\big(c_2(c_4 c_5 c_6 - s_4 s_6) - s_2 s_5 c_6\big) + c_1(s_4 c_5 c_6 + c_4 s_6) \\ -s_2(c_4 c_5 c_6 - s_4 s_6) - c_2 s_5 c_6 \end{bmatrix}$$

$$\boldsymbol{s}^0 = \begin{bmatrix} c_1\big(-c_2(c_4 c_5 s_6 + s_4 c_6) + s_2 s_5 s_6\big) - s_1(-s_4 c_5 s_6 + c_4 c_6) \\ s_1\big(-c_2(c_4 c_5 s_6 + s_4 c_6) + s_2 s_5 s_6\big) + c_1(-s_4 c_5 s_6 + c_4 c_6) \\ s_2(c_4 c_5 s_6 + s_4 c_6) + c_2 s_5 s_6 \end{bmatrix} \tag{2.65}$$

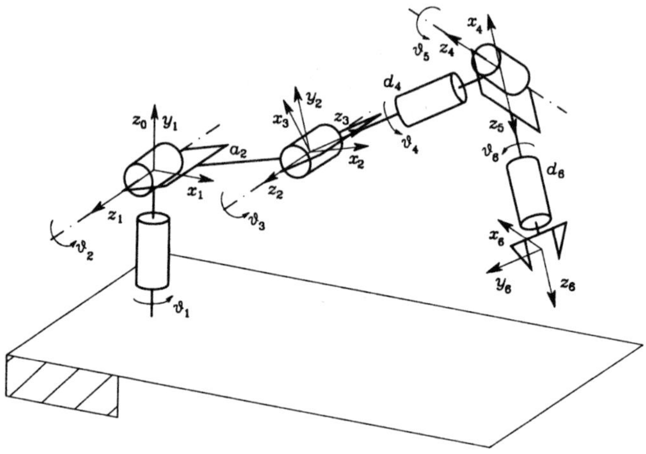

Figure 2.26 Anthropomorphic arm with spherical wrist.

$$a^0 = \begin{bmatrix} c_1(c_2c_4s_5 + s_2c_5) - s_1s_4s_5 \\ s_1(c_2c_4s_5 + s_2c_5) + c_1s_4s_5 \\ -s_2c_4s_5 + c_2c_5 \end{bmatrix}$$

for the end-effector orientation.

A comparison of the vector $\boldsymbol{p}$ in (2.64) with the vector $\boldsymbol{p}_3^0$ relative to the sole spherical arm (2.61) reveals the presence of additional contributions due to the choice of the origin of the end-effector frame at a distance d_6 from the origin of Frame 3 along the direction of $\boldsymbol{a}$. In other words, if it were $d_6 = 0$, the position vector would be the same. This feature is of fundamental importance for the solution of the inverse kinematics for this manipulator, as will be seen later.

2.9.7 Anthropomorphic Arm with Spherical Wrist

A comparison between Figure 2.23 and Figure 2.24 reveals that the direct kinematics function cannot be obtained by multiplying the transformation matrices $\boldsymbol{T}_3^0$ and $\boldsymbol{T}_6^3$, since Frame 3 of the anthropomorphic arm cannot coincide with Frame 3 of the spherical wrist.

Direct kinematics of the entire structure can be obtained in two ways. One consists of interposing a constant transformation matrix between $\boldsymbol{T}_3^0$ and $\boldsymbol{T}_6^3$ which allows aligning the two frames. The other refers to the Denavit-Hartenberg operating procedure with the frame assignment for the entire structure illustrated in Figure 2.26. The Denavit-Hartenberg parameters are specified in Table 2.6.

Table 2.6 Denavit-Hartenberg parameters for the anthropomorphic arm with spherical wrist.

Link	a_i	α_i	d_i	ϑ_i
1	0	$\pi/2$	0	ϑ_1
2	a_2	0	0	ϑ_2
3	0	$\pi/2$	0	ϑ_3
4	0	$-\pi/2$	d_4	ϑ_4
5	0	$\pi/2$	0	ϑ_5
6	0	0	d_6	ϑ_6

Since rows 3 and 4 differ from the corresponding rows of the tables for the two single structures, the relative homogeneous transformation matrices $\boldsymbol{A}_3^2$ and $\boldsymbol{A}_4^3$ have to be modified into

$$\boldsymbol{A}_3^2(\vartheta_3) = \begin{bmatrix} c_3 & 0 & s_3 & 0 \\ s_3 & 0 & -c_3 & 0 \\ 0 & 1 & 0 & 0 \\ 0 & 0 & 0 & 1 \end{bmatrix} \qquad \boldsymbol{A}_4^3(\vartheta_4) = \begin{bmatrix} c_4 & 0 & -s_4 & 0 \\ s_4 & 0 & c_4 & 0 \\ 0 & -1 & 0 & d_4 \\ 0 & 0 & 0 & 1 \end{bmatrix}$$

while the other transformation matrices remain the same. Computation of the direct kinematics function leads to expressing the position and orientation of the end-effector frame as:

$$\boldsymbol{p}^0 = \begin{bmatrix} a_2c_1c_2 + d_4c_1s_{23} + d_6\big(c_1(c_{23}c_4s_5 + s_{23}c_5) + s_1s_4s_5\big) \\ a_2s_1c_2 + d_4s_1s_{23} + d_6\big(s_1(c_{23}c_4s_5 + s_{23}c_5) - c_1s_4s_5\big) \\ a_2s_2 - d_4c_{23} + d_6(s_{23}c_4s_5 - c_{23}c_5) \end{bmatrix} \tag{2.66}$$

and

$$\boldsymbol{n}^0 = \begin{bmatrix} c_1\big(c_{23}(c_4c_5c_6 - s_4s_6) - s_{23}s_5c_6\big) + s_1(s_4c_5c_6 + c_4s_6) \\ s_1\big(c_{23}(c_4c_5c_6 - s_4s_6) - s_{23}s_5c_6\big) - c_1(s_4c_5c_6 + c_4s_6) \\ s_{23}(c_4c_5c_6 - s_4s_6) + c_{23}s_5c_6 \end{bmatrix}$$

$$\boldsymbol{s}^0 = \begin{bmatrix} c_1\big(-c_{23}(c_4c_5s_6 + s_4c_6) + s_{23}s_5s_6\big) + s_1(-s_4c_5s_6 + c_4c_6) \\ s_1\big(-c_{23}(c_4c_5s_6 + s_4c_6) + s_{23}s_5s_6\big) - c_1(-s_4c_5s_6 + c_4c_6) \\ -s_{23}(c_4c_5s_6 + s_4c_6) - c_{23}s_5s_6 \end{bmatrix} \tag{2.67}$$

$$\boldsymbol{a}^0 = \begin{bmatrix} c_1(c_{23}c_4s_5 + s_{23}c_5) + s_1s_4s_5 \\ s_1(c_{23}c_4s_5 + s_{23}c_5) - c_1s_4s_5 \\ s_{23}c_4s_5 - c_{23}c_5 \end{bmatrix}.$$

By setting $d_6 = 0$, the position of the wrist axes intersection is obtained. In that case, the vector $\boldsymbol{p}^0$ in (2.66) corresponds to the vector $\boldsymbol{p}_3^0$ for the sole anthropomorphic arm in (2.62), because d_4 gives the length of the forearm (a_3) and axis x_3 in Figure 2.26 is rotated by $\pi/2$ with respect to axis x_3 in Figure 2.23.

2.10 Joint Space and Operational Space

As described in the previous sections, the direct kinematics equation of a manipulator allows the position and orientation of the end-effector frame to be expressed as a function of the joint variables with respect to the base frame.

If a task is to be specified for the end effector, it is necessary to assign the end-effector position and orientation, eventually as a function of time (trajectory). This is quite easy for the position. On the other hand, specifying the orientation through the unit vector triplet $(\boldsymbol{n}, \boldsymbol{s}, \boldsymbol{a})$ is quite difficult, since their nine components must be guaranteed to satisfy the orthonormality constraints imposed by (2.4) at each time instant. This problem will be resumed in Chapter 5.

The problem of describing end-effector orientation admits a natural solution if one of the above minimal representations is adopted. In this case, indeed, a motion trajectory can be assigned to the set of angles chosen to represent orientation.

Therefore, position can be given by a minimal number of coordinates with regard to the geometry of the structure, and orientation can be specified in terms of a minimal representation (Euler angles) describing the rotation of the end-effector frame with respect to the base frame. In this way, it is possible to describe a manipulator posture by means of the $(m \times 1)$ vector, with $m \leq n$,

$$\boldsymbol{x} = \begin{bmatrix} \boldsymbol{p} \\ \boldsymbol{\phi} \end{bmatrix} \tag{2.68}$$

where $\boldsymbol{p}$ describes the end-effector position and $\boldsymbol{\phi}$ its orientation.

This representation of position and orientation allows the description of an end-effector task in terms of a number of inherently independent parameters. The vector $\boldsymbol{x}$ is defined in the space in which the manipulator task is specified; hence, this space is typically called *operational space*.

On the other hand, the *joint space* (configuration space) denotes the space in which the $(n \times 1)$ vector of joint variables

$$\boldsymbol{q} = \begin{bmatrix} q_1 \\ \vdots \\ q_n \end{bmatrix}, \tag{2.69}$$

is defined; it is $q_i = \vartheta_i$ for a revolute joint and $q_i = d_i$ for a prismatic joint.

Accounting for the dependence of position and orientation from the joint variables, the direct kinematics equation can be written in a form other than (2.45), *i.e.*,

$$\boldsymbol{x} = \boldsymbol{k}(\boldsymbol{q}). \tag{2.70}$$

The $(m \times 1)$ vector function $\boldsymbol{k}(\cdot)$—nonlinear in general—allows computation of the operational space variables from the knowledge of the joint space variables.

It is worth noticing that the dependence of the orientation components of the function $\boldsymbol{k}(\boldsymbol{q})$ in (2.70) on the joint variables is not easy to express except for simple cases. In fact, in the most general case of a six-dimensional operational space ($m = 6$), the computation of the three components of the function $\boldsymbol{\phi}(\boldsymbol{q})$ cannot be performed in closed form but goes through the computation of the elements of the rotation matrix, *i.e.*, $\boldsymbol{n}(\boldsymbol{q}), \boldsymbol{s}(\boldsymbol{q}), \boldsymbol{a}(\boldsymbol{q})$. The equations that allow determining the Euler angles from the triplet of unit vectors $\boldsymbol{n}$, $\boldsymbol{s}$, $\boldsymbol{a}$ were given in Section 2.4.

Example 2.5

Consider again the three-link planar arm in Figure 2.20. The geometry of the structure suggests that the end-effector position is determined by the two coordinates p_x and p_y, while its orientation is determined by the angle ϕ formed by the end effector with the axis x_0. Expressing these operational variables as a function of the joint variables, the two position coordinates are given by the first two elements of the fourth column of the homogeneous transformation matrix (2.58), while the orientation angle is simply given by the sum of joint variables. In sum, the direct kinematics equation can be written in the form

$$\boldsymbol{x} = \begin{bmatrix} p_x \\ p_y \\ \phi \end{bmatrix} = \boldsymbol{k}(\boldsymbol{q}) = \begin{bmatrix} a_1 c_1 + a_2 c_{12} + a_3 c_{123} \\ a_1 s_1 + a_2 s_{12} + a_3 s_{123} \\ \vartheta_1 + \vartheta_2 + \vartheta_3 \end{bmatrix}. \tag{2.71}$$

This expression shows that three joint space variables allow specification of at most three independent operational space variables. On the other hand, if orientation is of no concern, it is $\boldsymbol{x} = [\,p_x \quad p_y\,]^T$ and there is *kinematic redundancy* of degrees of mobility with respect to a pure positioning end-effector task; this concept will be widely treated later.

2.10.1 Workspace

With reference to the operational space, an index of robot performance is the so-called *workspace*; this is the region described by the origin of the end-effector frame when all the manipulator joints execute all possible motions. It is often customary to distinguish between *reachable* workspace and *dexterous* workspace. The latter is the region that the origin of the end-effector frame can describe while attaining different orientations, while the former is the region that the origin of the end-effector frame can reach with at least one orientation. Obviously, the dexterous workspace is a subspace of the reachable workspace. A manipulator with less than six degrees of mobility cannot take any arbitrary position and orientation in space.

The workspace is characterized by the manipulator geometry and the mechanical joint limits. For an n-degree-of-mobility manipulator the reachable workspace is the geometric locus of the points that can be achieved by considering the direct kinematics equation for the sole position part, *i.e.*,

$$\boldsymbol{p} = \boldsymbol{p}(\boldsymbol{q}) \qquad q_{im} \le q_i \le q_{iM} \quad i = 1, \dots, n,$$

where q_{im} (q_{iM}) denotes the minimum (maximum) limit at Joint i. This volume is finite, closed, connected—$\boldsymbol{p}(\boldsymbol{q})$ is a continuous function—and thus is defined by its bordering surface. Since the joints are revolute or prismatic, it is easy to recognize that this surface is constituted by surface elements of planar, spherical, toroidal and cylindrical type. The manipulator workspace (without end effector) is reported in the data sheet given by the robot manufacturer in terms of a top view and a side view. It represents a basic element to evaluate robot performance for a desired application.

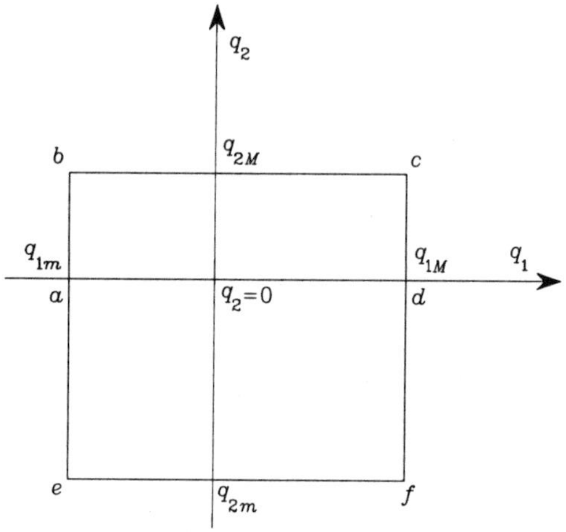

Figure 2.27 Region of admissible configurations for a two-link arm.

Example 2.6

Consider the simple two-link planar arm. If the mechanical joint limits are known, the arm can attain all the joint space configurations corresponding to the points in the rectangle in Figure 2.27.

The reachable workspace can be derived via a graphical construction of the image of the rectangle perimeter in the plane of the arm. To this purpose, it is worth considering the images of the segments ab, bc, cd, da, ae, ef, fd. Along the segments ab, bc, cd, ae, ef, fd a loss of mobility occurs due to a joint limit; a loss of mobility occurs also along the segment ad because the arm and forearm are aligned[10]. Further, a change of the arm posture occurs at points a and d: for $q_2 > 0$ the *elbow-down* posture is obtained, while for $q_2 < 0$ the arm is in the *elbow-up* posture.

In the plane of the arm, start drawing the arm in configuration A corresponding to q_{1m} and $q_2 = 0$ (a); then, the segment ab describing motion from $q_2 = 0$ to q_{2M} generates the arc AB; the subsequent arcs BC, CD, DA, AE, EF, FD are generated in a similar way (Figure 2.28). The external contour of the area $CDAEFHC$ delimits the requested workspace. Further, the area $BCDAB$ is relative to elbow-down postures while the area $DAEFD$ is relative to elbow-up postures; hence, the points in the area $BADHB$ are reachable by the end effector with both postures.

In a real manipulator, for a given set of joint variables, the actual values of the operational space variables deviate from those computed via direct kinematics. The direct kinematics equation has indeed a dependence from the Denavit-Hartenberg parameters

---

[10] In the following chapter, it will be seen that this configuration characterizes a kinematic *singularity* of the arm.

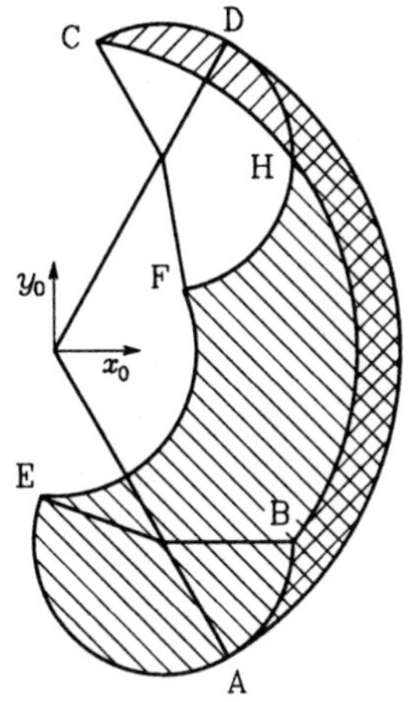

Figure 2.28 Workspace of a planar two-link arm.

which is not explicit in (2.70). If the mechanical dimensions of the structure differ from the corresponding parameter of the table because of mechanical tolerances, a deviation arises between the position reached in the assigned posture and the position computed via direct kinematics. Such a deviation is defined *accuracy*; this parameter attains typical values below one millimeter and depends on the structure as well as on manipulator dimensions. Accuracy varies with the end-effector position in the workspace and it is a relevant parameter when robot programming oriented environments are adopted, as will be seen in the last chapter.

Another parameter that is usually listed in the performance data sheet of an industrial robot is *repeatability* which gives a measure of the manipulator's ability to return to a previously reached position; this parameter is relevant for programming an industrial robot by the teaching-by-showing technique which will be presented in the last chapter. Repeatability depends not only on the characteristics of the mechanical structure but also on the transducers and controller; it is expressed in metric units and is typically smaller than accuracy. For instance, for a manipulator with a maximum reach of 1.5 m, accuracy varies from 0.2 to 1 mm in the workspace, while repeatability varies from 0.02 to 0.2 mm.

2.10.2 Kinematic Redundancy

A manipulator is termed *kinematically redundant* when it has a number of degrees of mobility which is greater than the number of variables that are necessary to describe a given task. With reference to the above-defined spaces, a manipulator is intrinsically redundant when the dimension of the operational space is smaller than the dimension of the joint space ($m < n$). Redundancy is, anyhow, a concept *relative* to the task assigned to the manipulator; a manipulator can be redundant with respect to a task and nonredundant with respect to another. Even in the case of $m = n$, a manipulator can be *functionally* redundant when only a number of r components of operational space are of concern for the specific task, with $r < m$.

Consider again the three-degree-of-mobility planar arm of Section 2.9.1. If only the end-effector position (in the plane) is specified, that structure presents a functional redundancy ($n = m = 3, r = 2$); this is lost when also the end-effector orientation in the plane is specified ($n = m = r = 3$). On the other hand, a four-degree-of-mobility planar arm is intrinsically redundant ($n = 4, m = 3$).

Yet, take the typical industrial robot with six degrees of mobility; such manipulator is not intrinsically redundant ($n = m = 6$), but it can become functionally redundant with regard to the task to execute. Thus, for instance, in a laser-cutting task a functional redundancy will occur since the end-effector rotation about the approach direction is irrelevant to completion of the task ($r = 5$).

At this point, a question should arise spontaneously: Why to intentionally utilize a redundant manipulator? The answer is to recognize that redundancy can provide the manipulator with dexterity and versatility in its motion. The typical example is constituted by the human arm that has *seven* degrees of mobility: three in the shoulder, one in the elbow and three in the wrist, without considering the degrees of mobility in the fingers. This manipulator is intrinsically redundant; in fact, if the base and the hand position and orientation are both fixed—requiring six degrees of freedom—the elbow can be moved, thanks to the additional available degree of mobility. Then, for instance, it is possible to avoid obstacles in the workspace. Further, if a joint of a redundant manipulator reaches its mechanical limit, there might be other joints that allow execution of the prescribed end-effector motion.

A formal treatment of redundancy will be presented in the following chapter.

2.11 Kinematic Calibration

The Denavit-Hartenberg parameters for direct kinematics need to be computed as precisely as possible in order to improve manipulator accuracy. *Kinematic calibration* techniques are devoted to finding accurate estimates of Denavit-Hartenberg parameters from a series of measurements on the manipulator's end-effector location. Hence, they do not allow direct measurement of the geometric parameters of the structure.

Consider the direct kinematics equation in (2.70) which can be rewritten by emphasizing the dependence of the operational space variables on the fixed Denavit-Hartenberg parameters, besides the joint variables. Let $\boldsymbol{a} = [\,a_1 \quad \dots \quad a_n\,]^T$, $\boldsymbol{\alpha} = [\,\alpha_1 \quad \dots \quad \alpha_n\,]^T$, $\boldsymbol{d} = [\,d_1 \quad \dots \quad d_n\,]^T$, and $\boldsymbol{\vartheta} = [\,\theta_1 \quad \dots \quad \theta_n\,]^T$ denote the vectors of Denavit-Hartenberg parameters for the whole structure; then (2.70) becomes

$$\boldsymbol{x} = \boldsymbol{k}(\boldsymbol{a}, \boldsymbol{\alpha}, \boldsymbol{d}, \boldsymbol{\vartheta}). \tag{2.72}$$

Manipulator's end-effector location shall be measured with high precision for the effectiveness of the kinematic calibration procedure. To this purpose a mechanical apparatus can be used that allows constraining the end effector at given locations with a priori known precision. Alternatively, direct measurement systems of object position and orientation in the Cartesian space can be used which employ triangulation techniques.

Let $\boldsymbol{x}_m$ be the measured location and $\boldsymbol{x}_n$ the nominal location that can be computed via (2.72) with the nominal values of the parameters $\boldsymbol{a}$, $\boldsymbol{\alpha}$, $\boldsymbol{d}$, $\boldsymbol{\vartheta}$. The nominal values of the fixed parameters are set equal to the design data of the mechanical structure, whereas the nominal values of the joint variables are set equal to the data provided by the position transducers at the given manipulator posture. The deviation $\Delta\boldsymbol{x} = \boldsymbol{x}_m - \boldsymbol{x}_n$ gives a measure of accuracy at the given posture. On the assumption of small deviations, at first approximation, it is possible to derive the following relation from (2.72):

$$\Delta\boldsymbol{x} = \frac{\partial\boldsymbol{k}}{\partial\boldsymbol{a}}\Delta\boldsymbol{a} + \frac{\partial\boldsymbol{k}}{\partial\boldsymbol{\alpha}}\Delta\boldsymbol{\alpha} + \frac{\partial\boldsymbol{k}}{\partial\boldsymbol{d}}\Delta\boldsymbol{d} + \frac{\partial\boldsymbol{k}}{\partial\boldsymbol{\vartheta}}\Delta\boldsymbol{\vartheta} \tag{2.73}$$

where $\Delta\boldsymbol{a}$, $\Delta\boldsymbol{\alpha}$, $\Delta\boldsymbol{d}$, $\Delta\boldsymbol{\vartheta}$ denote the deviations between the values of the parameters of the real structure and the nominal ones. Moreover, $\partial\boldsymbol{k}/\partial\boldsymbol{a}$, $\partial\boldsymbol{k}/\partial\boldsymbol{\alpha}$, $\partial\boldsymbol{k}/\partial\boldsymbol{d}$, $\partial\boldsymbol{k}/\partial\boldsymbol{\vartheta}$ denote the $(m \times n)$ matrices whose elements are the partial derivatives of the components of the direct kinematics function with respect to the single parameters[11].

Group the parameters in the $(4n \times 1)$ vector $\boldsymbol{\zeta} = [\,\boldsymbol{a}^T \quad \boldsymbol{\alpha}^T \quad \boldsymbol{d}^T \quad \boldsymbol{\vartheta}^T\,]^T$. Let $\Delta\boldsymbol{\zeta} = \boldsymbol{\zeta}_m - \boldsymbol{\zeta}_n$ denote the parameter variations with respect to the nominal values, and $\boldsymbol{\Phi} = [\,\partial\boldsymbol{k}/\partial\boldsymbol{a} \quad \partial\boldsymbol{k}/\partial\boldsymbol{\alpha} \quad \partial\boldsymbol{k}/\partial\boldsymbol{d} \quad \partial\boldsymbol{k}/\partial\boldsymbol{\vartheta}\,]$ the $(m \times 4n)$ *kinematic calibration matrix* computed for the nominal values of the parameters $\boldsymbol{\zeta}_n$. Then (2.73) can be compactly rewritten as

$$\Delta\boldsymbol{x} = \boldsymbol{\Phi}(\boldsymbol{\zeta}_n)\Delta\boldsymbol{\zeta}. \tag{2.74}$$

It is desired to compute $\Delta\boldsymbol{\zeta}$ starting from the knowledge of $\boldsymbol{\zeta}_n$, $\boldsymbol{x}_n$ and the measurement of $\boldsymbol{x}_m$. Since (2.74) constitutes a system of m equations into $4n$ unknowns with $m < 4n$, a sufficient number of end-effector location measures has to be performed so as to obtain a system of at least $4n$ equations. Therefore, if measurements are made for a number of l locations, (2.74) yields

$$\Delta\bar{\boldsymbol{x}} = \begin{bmatrix} \Delta\boldsymbol{x}_1 \\ \vdots \\ \Delta\boldsymbol{x}_l \end{bmatrix} = \begin{bmatrix} \boldsymbol{\Phi}_1 \\ \vdots \\ \boldsymbol{\Phi}_l \end{bmatrix} \Delta\boldsymbol{\zeta} = \bar{\boldsymbol{\Phi}}\Delta\boldsymbol{\zeta}. \tag{2.75}$$

As regards the nominal values of the parameters needed for the computation of the matrices $\boldsymbol{\Phi}_i$, it should be observed that the geometric parameters are constant whereas the joint variables depend on the manipulator configuration at the location i.

In order to avoid ill-conditioning of matrix $\bar{\boldsymbol{\Phi}}$, it is advisable to choose l so that $lm \gg 4n$ and then solve (2.75) with a least-squares technique; in this case the solution is of the form

$$\Delta\boldsymbol{\zeta} = (\bar{\boldsymbol{\Phi}}^T\bar{\boldsymbol{\Phi}})^{-1}\bar{\boldsymbol{\Phi}}^T\Delta\bar{\boldsymbol{x}} \tag{2.76}$$

---

[11] These matrices are the Jacobians of the transformations between the parameter space and the operational space.

where $(\bar{\Phi}^T\bar{\Phi})^{-1}\bar{\Phi}^T$ is the *left pseudo-inverse* matrix of $\bar{\Phi}$. By computing $\bar{\Phi}$ with the nominal values of the parameters ζ_n, the first parameter *estimate* is given by

$$\zeta' = \zeta_n + \Delta\zeta. \tag{2.77}$$

This is a nonlinear parameter estimate problem and, as such, the procedure shall be iterated until $\Delta\zeta$ converges within a given threshold. At each iteration, the calibration matrix $\bar{\Phi}$ is to be updated with the parameter estimates ζ' obtained via (2.77) at the previous iteration. In a similar manner, the deviation $\Delta\bar{x}$ is to be computed as the difference between the measured values for the l end-effector locations and the corresponding locations computed by the direct kinematics function with the values of the parameters at the previous iteration.

As a result of the kinematic calibration procedure, more accurate estimates of the real manipulator geometric parameters as well as possible corrections to make on the joint transducers measurements are obtained.

Kinematic calibration is an operation that is performed by the robot manufacturer to guarantee the accuracy reported in the data sheet. There is another kind of calibration that is performed by the robot user which is needed for the measurement system *start-up* to guarantee that the position transducers data are consistent with the attained manipulator posture. For instance, in the case of incremental (nonabsolute) position transducers, such calibration consists of taking the mechanical structure into a given reference posture (*home*) and initializing the position transducers with the values at that posture.

2.12 Inverse Kinematics Problem

The direct kinematics equation, either in the form (2.45) or in the form (2.70), establishes the functional relationship between the joint variables and the end-effector position and orientation. The *inverse kinematics problem* consists of the determination of the joint variables corresponding to a given end-effector position and orientation. The solution to this problem is of fundamental importance in order to transform the motion specifications, assigned to the end effector in the operational space, into the corresponding joint space motions that allow execution of the desired motion.

As regards the direct kinematics equation in (2.45), the end-effector position and rotation matrix are computed in a unique manner, once the joint variables are known[12]. On the other hand, the inverse kinematics problem is much more complex for the following reasons:

- The equations to solve are in general nonlinear, and thus it is not always possible to find a *closed-form solution*.
- *Multiple solutions* may exist.

---

[12] In general, this cannot be said for (2.70) too, since the Euler angles are not uniquely defined.

- *Infinite solutions* may exist, e.g., in the case of a kinematically redundant manipulator.
- There might be no *admissible* solutions, in view of the manipulator kinematic structure.

For what concerns existence of solutions, this is guaranteed if the given end-effector position and orientation belong to the manipulator dexterous workspace.

On the other hand, the problem of multiple solutions depends not only on the number of degrees of mobility but also on the number of nonnull Denavit-Hartenberg parameters; in general, the greater is the number of nonnull parameters, the greater is the number of admissible solutions. For a six-degree-of-mobility manipulator without mechanical joint limits, there are in general up to 16 admissible solutions. Such occurrence demands some criterion to choose among admissible solutions (e.g., the elbow-up/elbow-down case of Example 2.6). The existence of mechanical joint limits may eventually reduce the number of admissible multiple solutions for the real structure.

Computation of closed-form solutions requires either *algebraic intuition* to find out those significant equations containing the unknowns or *geometric intuition* to find out those significant points on the structure with respect to which it is convenient to express position and/or orientation as a function of a reduced number of unknowns. The following examples will point out the ability required to an inverse kinematics problem solver. On the other hand, in all those cases when there are no—or it is difficult to find—closed-form solutions, it might be appropriate to resort to *numerical solution techniques*; these clearly have the advantage to be applicable to any kinematic structure, but in general they do not allow computation of all admissible solutions.

2.12.1 Solution of Three-link Planar Arm

Consider the arm shown in Figure 2.20 whose direct kinematics was given in (2.58). It is desired to find the joint variables $\vartheta_1, \vartheta_2, \vartheta_3$ corresponding to a given end-effector position and orientation.

As already pointed out, it is convenient to specify position and orientation in terms of a minimal number of parameters: the two coordinates p_x, p_y and the angle ϕ with axis x_0, in this case. Hence, it is possible to refer to the direct kinematics equation in the form (2.71).

A first *algebraic solution* technique is illustrated below. Having specified the orientation, the relation

$$\phi = \vartheta_1 + \vartheta_2 + \vartheta_3 \tag{2.78}$$

is one of the equations of the system to solve[13]. From (2.58) the following equations

---

[13] If ϕ is not specified, then the arm is redundant and there exist infinite solutions to the inverse kinematics problem.

can be obtained:

$$\begin{aligned} p_{Wx} &= p_x - a_3 c_\phi = a_1 c_1 + a_2 c_{12} \\ p_{Wy} &= p_y - a_3 s_\phi = a_1 s_1 + a_2 s_{12} \end{aligned} \tag{2.79}$$

which describe the position of point W, *i.e.*, the origin of Frame 2; this depends only on the first two angles ϑ_1 and ϑ_2. Squaring and summing the two equations in (2.79) yields

$$p_{Wx}^2 + p_{Wy}^2 = a_1^2 + a_2^2 + 2a_1 a_2 c_2$$

from which

$$c_2 = \frac{p_{Wx}^2 + p_{Wy}^2 - a_1^2 - a_2^2}{2a_1 a_2}.$$

The existence of a solution obviously imposes that $-1 \leq c_2 \leq 1$, otherwise the given point would be outside the arm reachable workspace. Then, set

$$s_2 = \pm\sqrt{1 - c_2^2},$$

where the positive sign is relative to the elbow-down posture and the negative sign to the elbow-up posture. Hence, the angle ϑ_2 can be computed as

$$\vartheta_2 = \mathrm{Atan2}(s_2, c_2).$$

Having determined ϑ_2, the angle ϑ_1 can be found as follows. Substituting ϑ_2 into (2.79) yields an algebraic system of two equations in the two unknowns s_1 and c_1, whose solution is

$$\begin{aligned} s_1 &= \frac{(a_1 + a_2 c_2) p_{Wy} - a_2 s_2 p_{Wx}}{p_{Wx}^2 + p_{Wy}^2} \\ c_1 &= \frac{(a_1 + a_2 c_2) p_{Wx} + a_2 s_2 p_{Wy}}{p_{Wx}^2 + p_{Wy}^2}. \end{aligned}$$

In analogy to the above, it is

$$\vartheta_1 = \mathrm{Atan2}(s_1, c_1).$$

Finally, the angle ϑ_3 is found from (2.78) as

$$\vartheta_3 = \phi - \vartheta_1 - \vartheta_2.$$

An alternative *geometric solution* technique is presented below. As above, the orientation angle is given as in (2.78) and the coordinates of the origin of Frame 2 are computed as in (2.79). The application of the cosine theorem to the triangle formed by links a_1, a_2 and the segment connecting points W and O gives

$$p_{Wx}^2 + p_{Wy}^2 = a_1^2 + a_2^2 - 2a_1 a_2 \cos(\pi - \vartheta_2);$$

the two admissible configurations of the triangle are shown in Figure 2.29. Observing that $\cos(\pi - \vartheta_2) = -\cos\vartheta_2$ leads to

$$c_2 = \frac{p_{Wx}^2 + p_{Wy}^2 - a_1^2 - a_2^2}{2a_1 a_2}.$$

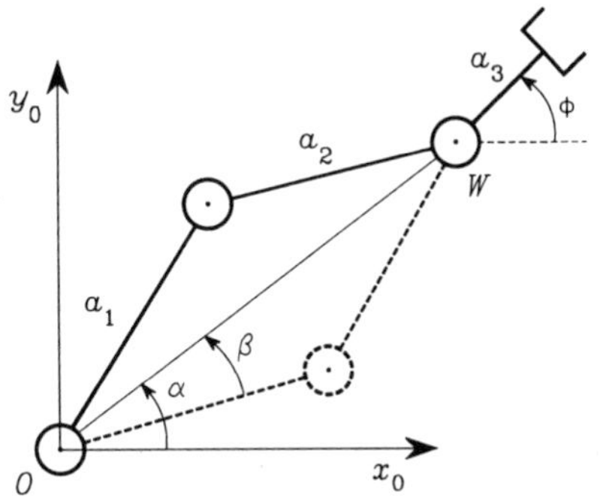

Figure 2.29 Admissible postures for a two-link planar arm.

For the existence of the triangle, it must be $\sqrt{p_{Wx}^2 + p_{Wy}^2} \le a_1 + a_2$. This condition is not satisfied when the given point is outside the arm reachable workspace. Then, on the assumption of admissible solutions, it is

$$\vartheta_2 = \cos^{-1}(c_2);$$

the elbow-up posture is obtained for $\vartheta_2 \in (-\pi, 0)$ while the elbow-down posture is obtained for $\vartheta_2 \in (0, \pi)$.

To find ϑ_1 consider the angles α and β in Figure 2.29. Notice that the determination of α depends on the sign of p_{Wx} and p_{Wy}; then, it is necessary to compute α as

$$\alpha = \mathrm{Atan2}(p_{Wy}, p_{Wx}).$$

To compute β, applying again the cosine theorem yields

$$c_\beta \sqrt{p_{Wx}^2 + p_{Wy}^2} = a_1 + a_2 c_2,$$

and resorting to the expression of c_2 given above leads to

$$\beta = \cos^{-1}\left(\frac{p_{Wx}^2 + p_{Wy}^2 + a_1^2 - a_2^2}{2a_1\sqrt{p_{Wx}^2 + p_{Wy}^2}}\right)$$

with $\beta \in (0, \pi)$ so as to preserve the existence of triangles. Then, it is

$$\vartheta_1 = \alpha \pm \beta,$$

where the positive sign holds for $\vartheta_2 < 0$ and the negative sign for $\vartheta_2 > 0$. Finally, ϑ_3 is computed from (2.78).

It is worth noticing that, in view of the substantial equivalence between the two-link planar arm and the parallelogram arm, the above techniques can be formally applied to solve the inverse kinematics of the arm in Section 2.9.2.

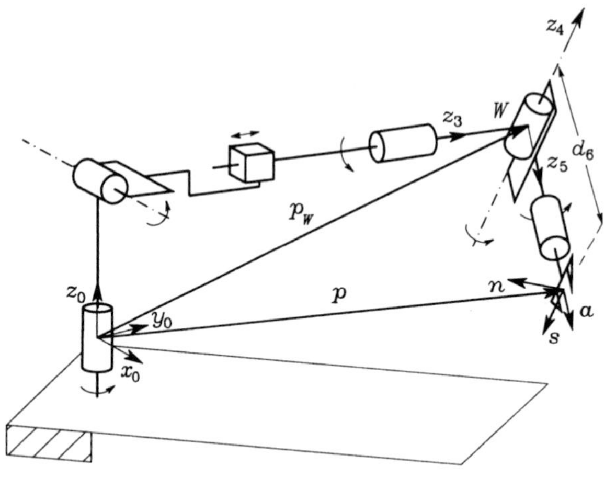

Figure 2.30 Manipulator with spherical wrist.

2.12.2 Solution of Manipulators with Spherical Wrist

Most of the existing manipulators are kinematically simple, since they are typically formed by an arm, of the kind presented above, and a spherical wrist; both the Stanford manipulator presented in Section 2.9.6 and the manipulator presented in Section 2.9.7 belong to this class of manipulators. This choice is partly motivated by the difficulty to find solutions to the inverse kinematics problem in the general case. In particular, a *six-degree-of-mobility* kinematic structure has closed-form inverse kinematics solutions if:

- three consecutive revolute joint axes intersect at a common point, like for the spherical wrist;
- three consecutive revolute joint axes are parallel.

In any case, algebraic or geometric intuition is required to obtain closed-form solutions.

Inspired by the previous solution to a three-link planar arm, a suitable point along the structure can be found whose position can be expressed both as a function of the given end-effector position and orientation and as a function of a reduced number of joint variables. This is equivalent to articulate the inverse kinematics problem into two subproblems, since the solution for the *position* is *decoupled* from that for the *orientation*.

For a manipulator with spherical wrist, the natural choice is to locate such point W at the intersection of the three terminal revolute axes (Figure 2.30). In fact, once the end-effector position and orientation are specified in terms of $\boldsymbol{p}$ and $\boldsymbol{R} = [\,\boldsymbol{n} \quad \boldsymbol{s} \quad \boldsymbol{a}\,]$, the wrist position can be found as

$$\boldsymbol{p}_W = \boldsymbol{p} - d_6\boldsymbol{a} \tag{2.80}$$

which is a function of the sole joint variables that determine the arm position[14]. Hence, in the case of a (nonredundant) three-degree-of-mobility arm, the inverse kinematics can be solved according to the following steps:

- compute the wrist position $\boldsymbol{p}_W(q_1, q_2, q_3)$ as in (2.80);
- solve inverse kinematics for (q_1, q_2, q_3);
- compute $\boldsymbol{R}_3^0(q_1, q_2, q_3)$;
- compute $\boldsymbol{R}_6^3(\vartheta_4, \vartheta_5, \vartheta_6) = \boldsymbol{R}_3^{0T}\boldsymbol{R}$;
- solve inverse kinematics for orientation $(\vartheta_4, \vartheta_5, \vartheta_6)$.

Therefore, on the basis of this kinematic decoupling, it is possible to solve the inverse kinematics for the arm separately from the inverse kinematics for the spherical wrist. Below are presented the solutions for two typical arms (spherical and anthropomorphic) as well as the solution for the spherical wrist.

2.12.3 Solution of Spherical Arm

Consider the spherical arm shown in Figure 2.22, whose direct kinematics was given in (2.61). It is desired to find the joint variables ϑ_1, ϑ_2, d_3 corresponding to a given end-effector position $\boldsymbol{p}_W$. In order to separate the variables on which $\boldsymbol{p}_W$ depends, it is convenient to express the position of $\boldsymbol{p}_W$ with respect to Frame 1; then, consider the matrix equation

$$(\boldsymbol{A}_1^0)^{-1}\boldsymbol{T}_3^0 = \boldsymbol{A}_2^1\boldsymbol{A}_3^2.$$

Equating the first three elements of the fourth columns of the matrices on both sides yields

$$\boldsymbol{p}_W^1 = \begin{bmatrix} p_{Wx}c_1 + p_{Wy}s_1 \\ -p_{Wz} \\ -p_{Wx}s_1 + p_{Wy}c_1 \end{bmatrix} = \begin{bmatrix} d_3 s_2 \\ -d_3 c_2 \\ d_2 \end{bmatrix} \tag{2.81}$$

which depends only on ϑ_2 and d_3. To solve this equation, set

$$t = \tan\frac{\vartheta_1}{2}$$

so that

$$c_1 = \frac{1-t^2}{1+t^2} \qquad s_1 = \frac{2t}{1+t^2}.$$

Substituting this equation in the third component on the left-hand side of (2.81) gives

$$(d_2 + p_{Wy})t^2 + 2p_{Wx}t + d_2 - p_{Wy} = 0,$$

---

[14] Note that the same reasoning was implicitly adopted in Section 2.12.1 for the three-link planar arm; $\boldsymbol{p}_W$ provided the one-degree-of-mobility wrist position for the two-degree-of-mobility arm obtained by considering only the first two links.

whose solution is

$$t = \frac{-p_{Wx} \pm \sqrt{p_{Wx}^2 + p_{Wy}^2 - d_2^2}}{d_2 + p_{Wy}}.$$

The two solutions correspond to two different postures. Further, if the discriminant is negative, the solution is not admissible. Hence, it is

$$\vartheta_1 = 2\mathrm{Atan2}\left(-p_{Wx} \pm \sqrt{p_{Wx}^2 + p_{Wy}^2 - d_2^2},\ d_2 + p_{Wy}\right).$$

Once ϑ_1 is known, from the first two components of (2.81) it is

$$\frac{p_{Wx}c_1 + p_{Wy}s_1}{-p_{Wz}} = \frac{d_3 s_2}{-d_3 c_2},$$

from which

$$\vartheta_2 = \mathrm{Atan2}(p_{Wx}c_1 + p_{Wy}s_1, p_{Wz}).$$

Finally, squaring and summing the first two components of (2.81) yields

$$d_3 = \sqrt{(p_{Wx}c_1 + p_{Wy}s_1)^2 + p_{Wz}^2},$$

where only the solution with $d_3 > 0$ has been considered.

2.12.4 Solution of Anthropomorphic Arm

Consider the anthropomorphic arm shown in Figure 2.23. It is desired to find the joint variables $\vartheta_1, \vartheta_2, \vartheta_3$ corresponding to a given end-effector position $\boldsymbol{p}_W$. Notice that the direct kinematics for $\boldsymbol{p}_W$ is expressed by (2.62) which can be obtained from (2.66) by setting $d_6 = 0$, $d_4 = a_3$ and replacing ϑ_3 with the angle $\vartheta_3 + \pi/2$ because of the misalignment of the Frames 3 for the structures in Figure 2.23 and in Figure 2.26, respectively.

From the particular geometry it is

$$\vartheta_1 = \mathrm{Atan2}(p_{Wy}, p_{Wx}).$$

Observe that another admissible solution is

$$\vartheta_1 = \pi + \mathrm{Atan2}(p_{Wy}, p_{Wx})$$

on condition that ϑ_2 be modified into $\pi - \vartheta_2$. Once ϑ_1 is known, the resulting structure is planar with regard to the variables ϑ_2 and ϑ_3. Hence, exploiting the previous solution of the two-link planar arm in Section 2.12.1 directly gives

$$\vartheta_3 = \mathrm{Atan2}(s_3, c_3)$$

where

$$c_3 = \frac{p_{Wx}^2 + p_{Wy}^2 + p_{Wz}^2 - a_2^2 - a_3^2}{2a_2a_3} \qquad s_3 = \pm\sqrt{1 - c_3^2}$$

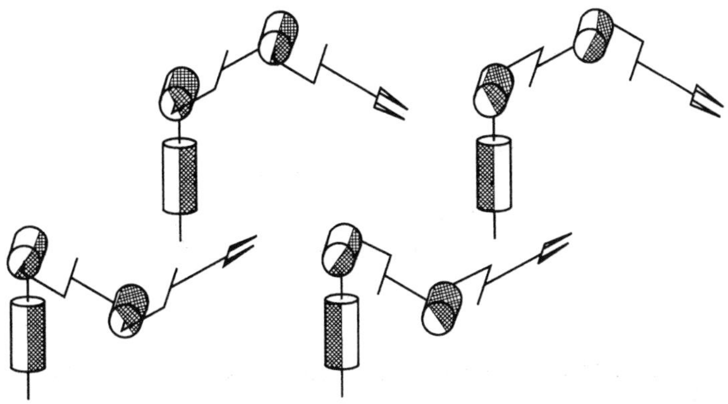

Figure 2.31 The four configurations of an anthropomorphic arm compatible with a given wrist position.

and

$$\vartheta_2 = \text{Atan2}(s_2, c_2)$$

where

$$s_2 = \frac{(a_2 + a_3c_3)p_{Wz} - a_3s_3\sqrt{p_{Wx}^2 + p_{Wy}^2}}{p_{Wx}^2 + p_{Wy}^2 + p_{Wz}^2}$$

$$c_2 = \frac{(a_2 + a_3c_3)\sqrt{p_{Wx}^2 + p_{Wy}^2} + a_3s_3p_{Wz}}{p_{Wx}^2 + p_{Wy}^2 + p_{Wz}^2}.$$

It can be recognized that four solutions exist according to the values of ϑ_1, ϑ_2, ϑ_3 (Figure 2.31): shoulder-right/elbow-up, shoulder-left/elbow-up, shoulder-right/elbow-down, shoulder-left/elbow-down; obviously, the forearm orientation is different for the two pairs of solutions. Notice also that it is possible to find the solutions only if

$$p_{Wx} \neq 0 \qquad p_{Wy} \neq 0.$$

In the case $p_{Wx} = p_{Wy} = 0$, an infinity of solutions is obtained, since it is possible to determine the joint variables ϑ_2 and ϑ_3 independently of the value of ϑ_1; in the following, it will be seen that the arm in such configuration is kinematically *singular*.

2.12.5 Solution of Spherical Wrist

Consider the spherical wrist shown in Figure 2.24, whose direct kinematics was given in (2.63). It is desired to find the joint variables ϑ_4, ϑ_5, ϑ_6 corresponding to a given end-effector orientation $\boldsymbol{R}_6^3$. As previously pointed out, these angles constitute a set of Euler angles ZYZ with respect to Frame 3. Hence, having computed the rotation matrix

$$\boldsymbol{R}_6^3 = \begin{bmatrix} n_x^3 & s_x^3 & a_x^3 \\ n_y^3 & s_y^3 & a_y^3 \\ n_z^3 & s_z^3 & a_z^3 \end{bmatrix},$$

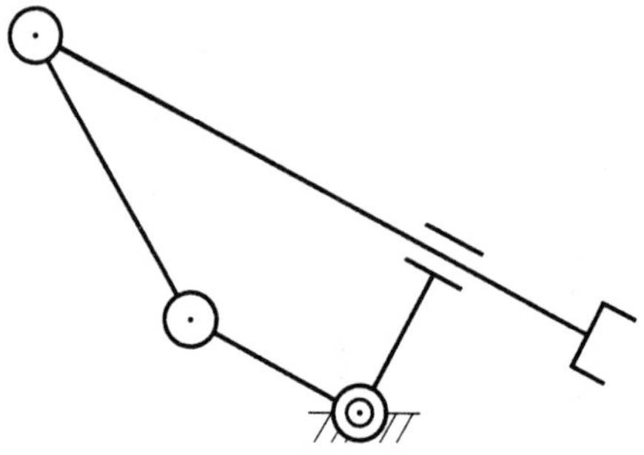

Figure 2.32 Four-link closed-chain planar arm with prismatic joint.

from its expression in terms of the joint variables in (2.63), it is possible to compute the solutions directly as in (2.19) and (2.19′), *i.e.*,

$$\begin{aligned} \vartheta_4 &= \text{Atan2}(a_y^3, a_x^3) \\ \vartheta_5 &= \text{Atan2}\left(\sqrt{(a_x^3)^2 + (a_y^3)^2}, a_z^3\right) \\ \vartheta_6 &= \text{Atan2}(s_z^3, -n_z^3) \end{aligned} \tag{2.82}$$

for $\vartheta_5 \in (0, \pi)$, and

$$\begin{aligned} \vartheta_4 &= \text{Atan2}(-a_y^3, -a_x^3) \\ \vartheta_5 &= \text{Atan2}\left(-\sqrt{(a_x^3)^2 + (a_y^3)^2}, a_z^3\right) \\ \vartheta_6 &= \text{Atan2}(-s_z^3, n_z^3) \end{aligned} \tag{2.82′}$$

for $\vartheta_5 \in (-\pi, 0)$.

Problems

2.1 Find the rotation matrix corresponding to the set of Euler angles ZXZ.

2.2 Discuss the inverse solution for the Euler angles ZYZ in the case $s_\vartheta = 0$.

2.3 Discuss the inverse solution for the Roll–Pitch–Yaw angles in the case $c_\vartheta = 0$.

2.4 Verify that the rotation matrix corresponding to the rotation by an angle about an arbitrary axis is given by (2.23).

2.5 Prove that the angle and the unit vector of the axis corresponding to a rotation matrix are given by (2.25). Find inverse formulæ in the case of $\sin\vartheta = 0$.

2.6 Verify that the rotation matrix corresponding to the unit quaternion is given by (2.29).

2.7 Prove that the unit quaternion is invariant with respect to the rotation matrix and its transpose, *i.e.*, $\boldsymbol{R}(\eta, \boldsymbol{\epsilon})\boldsymbol{\epsilon} = \boldsymbol{R}^T(\eta, \boldsymbol{\epsilon})\boldsymbol{\epsilon} = \boldsymbol{\epsilon}$.

2.8 Prove that the unit quaternion corresponding to a rotation matrix is given by (2.30).

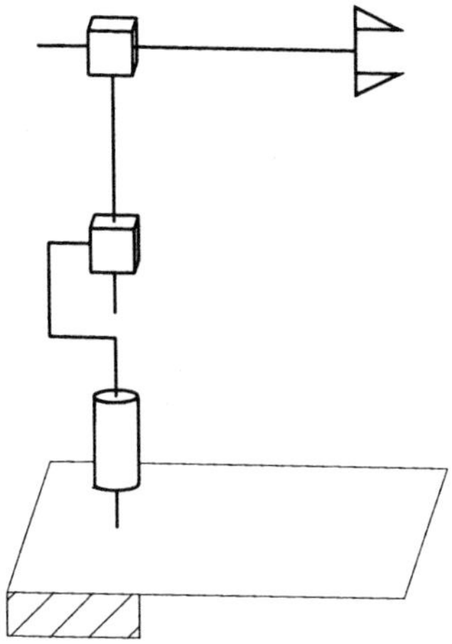

Figure 2.33 Cylindrical arm.

2.9 Prove that the quaternion product is expressed by (2.32).

2.10 By applying the rules for inverting a block-partitioned matrix, prove that matrix $\boldsymbol{A}_0^1$ is given by (2.40).

2.11 Find the direct kinematics equation of the four-link closed-chain planar arm in Figure 2.32, where the two links connected by the prismatic joint are orthogonal to each other.

2.12 Find the direct kinematics equation for the cylindrical arm in Figure 2.33.

2.13 Find the direct kinematics equation for the SCARA manipulator in Figure 2.34.

2.14 For the set of minimal representations of orientation ϕ, define the sum operation in terms of the composition of rotations. By means of an example, show that the commutative property does not hold for that operation.

2.15 Consider the elementary rotations about coordinate axes given by infinitesimal angles. Show that the rotation resulting from any two elementary rotations does not depend on the order of rotations. [Hint: for an infinitesimal angle $d\phi$, approximate $\cos(d\phi) \approx 1$ and $\sin(d\phi) \approx d\phi \ldots$].
Further, define $\boldsymbol{R}(d\phi_x, d\phi_y, d\phi_z) = \boldsymbol{R}_x(d\phi_x)\boldsymbol{R}_y(d\phi_y)\boldsymbol{R}_z(d\phi_z)$; show that

$$\boldsymbol{R}(d\phi_x, d\phi_y, d\phi_z)\boldsymbol{R}(d\phi_x', d\phi_y', d\phi_z') = \boldsymbol{R}(d\phi_x + d\phi_x', d\phi_y + d\phi_y', d\phi_z + d\phi_z').$$

2.16 Draw the workspace of the three-link planar arm in Figure 2.20 with the data:

$$a_1 = 0.5 \qquad a_2 = 0.3 \qquad a_3 = 0.2$$

$$-\pi/3 \le q_1 \le \pi/3 \qquad -2\pi/3 \le q_2 \le 2\pi/3 \qquad -\pi/2 \le q_3 \le \pi/2.$$

2.17 Solve the inverse kinematics for the cylindrical arm in Figure 2.33.

2.18 Solve the inverse kinematics for the SCARA manipulator in Figure 2.34.

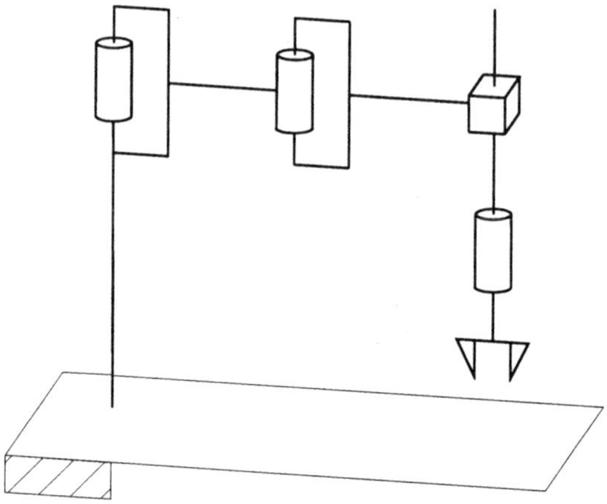

Figure 2.34 SCARA manipulator.

Bibliography

Angeles J. (1982) *Spatial Kinematic Chains: Analysis, Synthesis, Optimization*. Springer-Verlag, Berlin.

Asada H., Slotine J.-J.E. (1986) *Robot Analysis and Control*. Wiley, New York.

Bottema O., Roth B. (1979) *Theoretical Kinematics*. North Holland, Amsterdam.

Craig J.J. (1989) *Introduction to Robotics: Mechanics and Control*. 2nd ed., Addison-Wesley, Reading, Mass.

Denavit J., Hartenberg R.S. (1955) A kinematic notation for lower-pair mechanisms based on matrices. *ASME J. Applied Mechanics* 22:215–221.

Hollerbach J.M., Sahar G. (1983) Wrist-partitioned inverse kinematic accelerations and manipulator dynamics. *Int. J. Robotics Research* 2(4):61–76.

Hunt K.H. (1978) *Kinematic Geometry of Mechanisms*. Clarendon Press, Oxford.

Khalil W., Dombre E. (1999) *Modélisation Identification et Commande des Robots*. 2ème éd., Hermès, Paris.

Lee C.S.G. (1982) Robot kinematics, dynamics and control. *IEEE Computer* 15(12):62–80.

Liégeois A. (1985) *Performance and Computer-Aided Design*. Robot Technology 7, Kogan Page, London.

Luh J.Y.S., Zheng Y.-F. (1985) Computation of input generalized forces for robots with closed kinematic chain mechanisms. *IEEE J. Robotics and Automation* 1:95–103.

McCarthy J.M. (1990) *An Introduction to Theoretical Kinematics*. MIT Press, Cambridge, Mass.

Paul R.P. (1981) *Robot Manipulators: Mathematics, Programming, and Control*. MIT Press, Cambridge, Mass.

Paul R.P., Shimano B.E., Mayer G. (1981) Kinematic control equations for simple manipulators. *IEEE Trans. Systems, Man, and Cybernetics* 11:449–455.

Paul R.P., Stevenson C.N. (1983) Kinematics of robot wrists. *Int. J. Robotics Research* 2(1):31–38.

Paul R.P., Zhang H. (1986) Computationally efficient kinematics for manipulators with spherical wrists based on the homogeneous transformation representation. *Int. J. Robotics Research* 5(2):32–44.

Pieper D.L. (1968) *The Kinematics of Manipulators Under Computer Control*. Memo. AIM 72, Stanford Artificial Intelligence Laboratory.

Scheinman V.D. (1969) *Design of a Computer Controlled Manipulator*. Memo. AIM 92, Stanford Artificial Intelligence Laboratory.

Spong M.W., Vidyasagar M. (1989) *Robot Dynamics and Control*. Wiley, New York.

Whitney D.E. (1972) The mathematics of coordinated control of prosthetic arms and manipulators. *ASME J. Dynamic Systems, Measurement, and Control* 94:303–309.

Yoshikawa T. (1990) *Foundations of Robotics*. MIT Press, Boston, Mass.

3. Differential Kinematics and Statics

In the previous chapter, direct and inverse kinematics equations establishing the relationship between the joint variables and the end-effector position and orientation were derived. In this chapter, *differential kinematics* is presented which gives the relationship between the joint velocities and the corresponding end-effector linear and angular velocity. This mapping is described by a matrix, termed *geometric Jacobian*, which depends on the manipulator configuration. Alternatively, if the end-effector location is expressed with reference to a minimal representation in the operational space, it is possible to compute the Jacobian matrix via differentiation of the direct kinematics function with respect to the joint variables. The resulting Jacobian, termed *analytical Jacobian*, in general differs from the geometric one. The Jacobian constitutes one of the most important tools for manipulator characterization; in fact, it is useful for finding singular configuration, analyzing redundancy, determining inverse kinematics algorithms, describing the mapping between forces applied to the end effector and resulting torques at the joints (*statics*) and, as will be seen in the following chapters, for deriving dynamic equations of motion and designing operational space control schemes. Finally, the *kineto-static duality* concept is illustrated, which is at the basis of the definition of velocity and force *manipulability ellipsoids*.

3.1 Geometric Jacobian

Consider an n-degree-of-mobility manipulator. The direct kinematics equation can be written in the form

$$\boldsymbol{T}(\boldsymbol{q}) = \begin{bmatrix} \boldsymbol{R}(\boldsymbol{q}) & \boldsymbol{p}(\boldsymbol{q}) \\ \boldsymbol{0}^T & 1 \end{bmatrix}$$

where $\boldsymbol{q} = [\, q_1 \quad \dots \quad q_n \,]^T$ is the vector of joint variables. Both end-effector position and orientation vary as $\boldsymbol{q}$ varies.

The goal of differential kinematics is to find the relationship between the joint velocities and the end-effector linear and angular velocities. In other words, it is desired to express the end-effector linear velocity $\dot{\boldsymbol{p}}$ and angular velocity $\boldsymbol{\omega}$ as a function of the joint velocities $\dot{\boldsymbol{q}}$ by means of the following relations:

$$\dot{\boldsymbol{p}} = \boldsymbol{J}_P(\boldsymbol{q})\dot{\boldsymbol{q}} \tag{3.1}$$

$$\boldsymbol{\omega} = \boldsymbol{J}_O(\boldsymbol{q})\dot{\boldsymbol{q}}; \tag{3.2}$$

notice that $\boldsymbol{v}$ and $\boldsymbol{\omega}$ are free vectors since their directions in space are prescribed but their points of application and lines of application are not prescribed.

In (3.1) $\boldsymbol{J}_P$ is the $(3 \times n)$ matrix relative to the contribution of the joint velocities $\dot{\boldsymbol{q}}$ to the end-effector *linear* velocity $\dot{\boldsymbol{p}}$, while in (3.2) $\boldsymbol{J}_O$ is the $(3 \times n)$ matrix relative to the contribution of the joint velocities $\dot{\boldsymbol{q}}$ to the end-effector *angular* velocity $\boldsymbol{\omega}$. In compact form, (3.1) and (3.2) can be written as

$$\boldsymbol{v} = \begin{bmatrix} \dot{\boldsymbol{p}} \\ \boldsymbol{\omega} \end{bmatrix} = \boldsymbol{J}(\boldsymbol{q})\dot{\boldsymbol{q}} \tag{3.3}$$

which represents the manipulator *differential kinematics equation*. The $(6 \times n)$ matrix $\boldsymbol{J}$ is the manipulator *geometric Jacobian*

$$\boldsymbol{J} = \begin{bmatrix} \boldsymbol{J}_P \\ \boldsymbol{J}_O \end{bmatrix}, \tag{3.4}$$

which in general is a function of the joint variables.

In order to compute the geometric Jacobian, it is worth recalling a number of properties of rotation matrices and some important results of rigid body kinematics.

3.1.1 Derivative of a Rotation Matrix

The manipulator direct kinematics equation in (2.45) describes the end-effector position and orientation, as a function of the joint variables, in terms of a position vector and a rotation matrix. Since the aim is to characterize the end-effector linear and angular velocity, it is worth considering first the *derivative of a rotation matrix* with respect to time.

Consider a time-varying rotation matrix $\boldsymbol{R} = \boldsymbol{R}(t)$. In view of the orthogonality of $\boldsymbol{R}$, one has the relation

$$\boldsymbol{R}(t)\boldsymbol{R}^T(t) = \boldsymbol{I}$$

which, differentiated with respect to time, gives the identity

$$\dot{\boldsymbol{R}}(t)\boldsymbol{R}^T(t) + \boldsymbol{R}(t)\dot{\boldsymbol{R}}^T(t) = \boldsymbol{O}.$$

Set

$$\boldsymbol{S}(t) = \dot{\boldsymbol{R}}(t)\boldsymbol{R}^T(t); \tag{3.5}$$

the (3×3) matrix $\boldsymbol{S}$ is *skew-symmetric* since

$$\boldsymbol{S}(t) + \boldsymbol{S}^T(t) = \boldsymbol{O}. \tag{3.6}$$

Postmultiplying both sides of (3.5) by $\boldsymbol{R}(t)$ gives

$$\dot{\boldsymbol{R}}(t) = \boldsymbol{S}(t)\boldsymbol{R}(t) \tag{3.7}$$

that allows expressing the time derivative of $\boldsymbol{R}(t)$ as a function of $\boldsymbol{R}(t)$ itself.

The equation in (3.7) relates the rotation matrix $\boldsymbol{R}$ to its derivative by means of the skew-symmetric operator $\boldsymbol{S}$ and has a meaningful physical interpretation. Consider a constant vector $\boldsymbol{p}'$ and the vector $\boldsymbol{p}(t) = \boldsymbol{R}(t)\boldsymbol{p}'$. The time derivative of $\boldsymbol{p}(t)$ is

$$\dot{\boldsymbol{p}}(t) = \dot{\boldsymbol{R}}(t)\boldsymbol{p}'$$

which, in view of (3.7), can be written as

$$\dot{\boldsymbol{p}}(t) = \boldsymbol{S}(t)\boldsymbol{R}(t)\boldsymbol{p}'.$$

If the vector $\boldsymbol{\omega}(t)$ denotes the *angular velocity* of frame $\boldsymbol{R}(t)$ with respect to the reference frame at time t, it is known from mechanics that

$$\dot{\boldsymbol{p}}(t) = \boldsymbol{\omega}(t) \times \boldsymbol{R}(t)\boldsymbol{p}'.$$

Therefore, the matrix operator $\boldsymbol{S}(t)$ describes the vector product between the vector $\boldsymbol{\omega}$ and the vector $\boldsymbol{R}(t)\boldsymbol{p}'$. The matrix $\boldsymbol{S}(t)$ is so that its symmetric elements with respect to the main diagonal represent the components of the vector $\boldsymbol{\omega}(t) = [\,\omega_x \quad \omega_y \quad \omega_z\,]^T$ in the form

$$\boldsymbol{S} = \begin{bmatrix} 0 & -\omega_z & \omega_y \\ \omega_z & 0 & -\omega_x \\ -\omega_y & \omega_x & 0 \end{bmatrix}, \tag{3.8}$$

which justifies the expression $\boldsymbol{S}(t) = \boldsymbol{S}(\boldsymbol{\omega}(t))$.

Furthermore, if $\boldsymbol{R}$ denotes a rotation matrix, it can be shown that the following relation holds:

$$\boldsymbol{R}\boldsymbol{S}(\boldsymbol{\omega})\boldsymbol{R}^T = \boldsymbol{S}(\boldsymbol{R}\boldsymbol{\omega}) \tag{3.9}$$

which will be useful later.

Example 3.1

Consider the elementary rotation matrix about axis z given in (2.6). If α is a function of time, by computing the time derivative of $\boldsymbol{R}_z(\alpha(t))$, (3.5) becomes

$$\begin{aligned} \boldsymbol{S}(t) &= \begin{bmatrix} -\dot{\alpha}\sin\alpha & -\dot{\alpha}\cos\alpha & 0 \\ \dot{\alpha}\cos\alpha & -\dot{\alpha}\sin\alpha & 0 \\ 0 & 0 & 0 \end{bmatrix} \begin{bmatrix} \cos\alpha & \sin\alpha & 0 \\ -\sin\alpha & \cos\alpha & 0 \\ 0 & 0 & 1 \end{bmatrix} \\ &= \begin{bmatrix} 0 & -\dot{\alpha} & 0 \\ \dot{\alpha} & 0 & 0 \\ 0 & 0 & 0 \end{bmatrix} = \boldsymbol{S}(\boldsymbol{\omega}(t)). \end{aligned}$$

According to (3.8), it is

$$\boldsymbol{\omega} = [\,0 \quad 0 \quad \dot{\alpha}\,]^T$$

that expresses the angular velocity of the frame about axis z.

With reference to Figure 2.11, consider the coordinate transformation of a point P from Frame 1 to Frame 0; in view of (2.33), this is given by

$$\boldsymbol{p}^0 = \boldsymbol{o}_1^0 + \boldsymbol{R}_1^0\boldsymbol{p}^1. \tag{3.10}$$

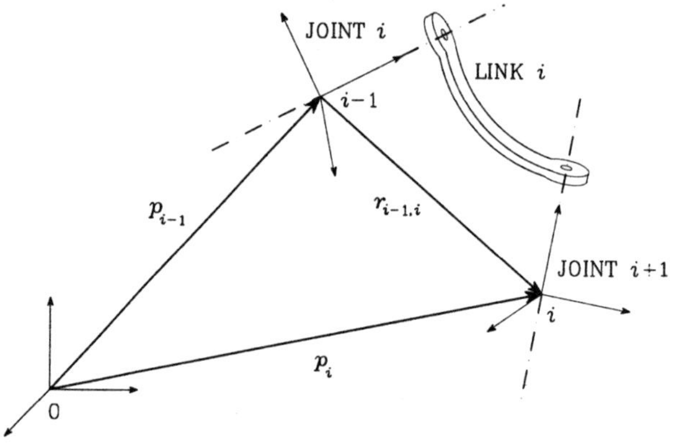

Figure 3.1 Characterization of generic Link i of a manipulator.

Differentiating (3.10) with respect to time gives

$$\dot{\boldsymbol{p}}^0 = \dot{\boldsymbol{o}}_1^0 + \boldsymbol{R}_1^0\dot{\boldsymbol{p}}^1 + \dot{\boldsymbol{R}}_1^0\boldsymbol{p}^1; \tag{3.11}$$

utilizing the expression of the derivative of a rotation matrix (3.7) and specifying the dependence on the angular velocity gives

$$\dot{\boldsymbol{p}}^0 = \dot{\boldsymbol{o}}_1^0 + \boldsymbol{R}_1^0\dot{\boldsymbol{p}}^1 + \boldsymbol{S}(\boldsymbol{\omega}_1^0)\boldsymbol{R}_1^0\boldsymbol{p}^1.$$

Further, denoting the vector $\boldsymbol{R}_1^0\boldsymbol{p}^1$ by $\boldsymbol{r}_1^0$, it is

$$\dot{\boldsymbol{p}}^0 = \dot{\boldsymbol{o}}_1^0 + \boldsymbol{R}_1^0\dot{\boldsymbol{p}}^1 + \boldsymbol{\omega}_1^0 \times \boldsymbol{r}_1^0 \tag{3.12}$$

which is the known form of the velocity composition rule.

Notice that, if $\boldsymbol{p}^1$ is *fixed* in Frame 1, it is

$$\dot{\boldsymbol{p}}^0 = \dot{\boldsymbol{o}}_1^0 + \boldsymbol{\omega}_1^0 \times \boldsymbol{r}_1^0 \tag{3.13}$$

since $\dot{\boldsymbol{p}}^1 = \boldsymbol{0}$.

3.1.2 Link Velocity

Consider the generic Link i of a manipulator with an open kinematic chain. According to the Denavit-Hartenberg convention adopted in the previous chapter, Link i connects Joints i and $i+1$; Frame i is attached to Link i and has origin along Joint $i+1$ axis, while Frame $i-1$ has origin along Joint i axis (Figure 3.1).

Let $\boldsymbol{p}_{i-1}$ and $\boldsymbol{p}_i$ be the position vectors of the origins of Frames $i-1$ and i, respectively. Also, let $\boldsymbol{r}_{i-1,i}^{i-1}$ denote the position of the origin of Frame i with respect to

Frame $i-1$ expressed in Frame $i-1$. According to the coordinate transformation (3.10), one can write[1]

$$\boldsymbol{p}_i = \boldsymbol{p}_{i-1} + \boldsymbol{R}_{i-1}\boldsymbol{r}^{i-1}_{i-1,i}.$$

Then, by virtue of (3.12), it is

$$\begin{aligned}\dot{\boldsymbol{p}}_i &= \dot{\boldsymbol{p}}_{i-1} + \boldsymbol{R}_{i-1}\dot{\boldsymbol{r}}^{i-1}_{i-1,i} + \boldsymbol{\omega}_{i-1} \times \boldsymbol{R}_{i-1}\boldsymbol{r}^{i-1}_{i-1,i} \\ &= \dot{\boldsymbol{p}}_{i-1} + \boldsymbol{v}_{i-1,i} + \boldsymbol{\omega}_{i-1} \times \boldsymbol{r}_{i-1,i}\end{aligned} \tag{3.14}$$

which gives the expression of the linear velocity of Link i as a function of the translational and rotational velocities of Link $i-1$. Note that $\boldsymbol{v}_{i-1,i}$ denotes the velocity of the origin of Frame i with respect to the origin of Frame $i-1$.

Concerning link angular velocity, it is worth starting from the rotation composition

$$\boldsymbol{R}_i = \boldsymbol{R}_{i-1}\boldsymbol{R}^{i-1}_i;$$

from (3.7), its time derivative can be written as

$$\boldsymbol{S}(\boldsymbol{\omega}_i)\boldsymbol{R}_i = \boldsymbol{S}(\boldsymbol{\omega}_{i-1})\boldsymbol{R}_i + \boldsymbol{R}_{i-1}\boldsymbol{S}(\boldsymbol{\omega}^{i-1}_{i-1,i})\boldsymbol{R}^{i-1}_i \tag{3.15}$$

where $\boldsymbol{\omega}^{i-1}_{i-1,i}$ denotes the angular velocity of Frame i with respect to Frame $i-1$ expressed in Frame $i-1$. From (2.4), the second term on the right-hand side of (3.15) can be rewritten as

$$\boldsymbol{R}_{i-1}\boldsymbol{S}(\boldsymbol{\omega}^{i-1}_{i-1,i})\boldsymbol{R}^{i-1}_i = \boldsymbol{R}_{i-1}\boldsymbol{S}(\boldsymbol{\omega}^{i-1}_{i-1,i})\boldsymbol{R}^T_{i-1}\boldsymbol{R}_{i-1}\boldsymbol{R}^{i-1}_i;$$

in view of property (3.9), it is

$$\boldsymbol{R}_{i-1}\boldsymbol{S}(\boldsymbol{\omega}^{i-1}_{i-1,i})\boldsymbol{R}^{i-1}_i = \boldsymbol{S}(\boldsymbol{R}_{i-1}\boldsymbol{\omega}^{i-1}_{i-1,i})\boldsymbol{R}_i.$$

Then, (3.15) becomes

$$\boldsymbol{S}(\boldsymbol{\omega}_i)\boldsymbol{R}_i = \boldsymbol{S}(\boldsymbol{\omega}_{i-1})\boldsymbol{R}_i + \boldsymbol{S}(\boldsymbol{R}_{i-1}\boldsymbol{\omega}^{i-1}_{i-1,i})\boldsymbol{R}_i$$

leading to the result

$$\begin{aligned}\boldsymbol{\omega}_i &= \boldsymbol{\omega}_{i-1} + \boldsymbol{R}_{i-1}\boldsymbol{\omega}^{i-1}_{i-1,i} \\ &= \boldsymbol{\omega}_{i-1} + \boldsymbol{\omega}_{i-1,i},\end{aligned} \tag{3.16}$$

which gives the expression of the angular velocity of Link i as a function of the angular velocities of Link $i-1$ and of Link i with respect to Link $i-1$.

---

[1] Hereafter, the indication of superscript '0' is omitted for quantities referred to Frame 0. Also, without loss of generality, Frame 0 and Frame n are taken as the base frame and the end-effector frame, respectively.

The relations in (3.14) and (3.16) attain different expressions depending on the type of Joint i (*prismatic* or *revolute*).

Prismatic Joint

Since orientation of Frame i with respect to Frame $i-1$ does not vary by moving Joint i, it is

$$\boldsymbol{\omega}_{i-1,i} = \mathbf{0}. \tag{3.17}$$

Further, the linear velocity is

$$\boldsymbol{v}_{i-1,i} = \dot{d}_i \boldsymbol{z}_{i-1} \tag{3.18}$$

where $\boldsymbol{z}_{i-1}$ is the unit vector of Joint i axis. Hence, the expressions of angular velocity (3.16) and linear velocity (3.14) respectively become

$$\boldsymbol{\omega}_i = \boldsymbol{\omega}_{i-1} \tag{3.19}$$

$$\dot{\boldsymbol{p}}_i = \dot{\boldsymbol{p}}_{i-1} + \dot{d}_i \boldsymbol{z}_{i-1} + \boldsymbol{\omega}_i \times \boldsymbol{r}_{i-1,i}, \tag{3.20}$$

where the relation $\boldsymbol{\omega}_i = \boldsymbol{\omega}_{i-1}$ has been exploited to derive (3.20).

Revolute Joint

For the angular velocity it is obviously

$$\boldsymbol{\omega}_{i-1,i} = \dot{\vartheta}_i \boldsymbol{z}_{i-1}, \tag{3.21}$$

while for the linear velocity it is

$$\boldsymbol{v}_{i-1,i} = \boldsymbol{\omega}_{i-1,i} \times \boldsymbol{r}_{i-1,i} \tag{3.22}$$

due to the rotation of Frame i with respect to Frame $i-1$ induced by the motion of Joint i. Hence, the expressions of angular velocity (3.16) and linear velocity (3.14) respectively become

$$\boldsymbol{\omega}_i = \boldsymbol{\omega}_{i-1} + \dot{\vartheta}_i \boldsymbol{z}_{i-1} \tag{3.23}$$

$$\dot{\boldsymbol{p}}_i = \dot{\boldsymbol{p}}_{i-1} + \boldsymbol{\omega}_i \times \boldsymbol{r}_{i-1,i}, \tag{3.24}$$

where (3.16) has been exploited to derive (3.24).

3.1.3 Jacobian Computation

Let the Jacobian in (3.4) be partitioned into the (3×1) column vectors as:

$$\boldsymbol{J} = \begin{bmatrix} \boldsymbol{\jmath}_{P1} & & \boldsymbol{\jmath}_{Pn} \\ & \dots & \\ \boldsymbol{\jmath}_{O1} & & \boldsymbol{\jmath}_{On} \end{bmatrix}. \tag{3.25}$$

The term $\dot{q}_i \boldsymbol{\jmath}_{Pi}$ represents the contribution of single Joint i to the end-effector linear velocity, while the term $\dot{q}_i \boldsymbol{\jmath}_{Oi}$ represents the contribution of single Joint i to the end-effector angular velocity. In order to compute the Jacobian it is convenient to compute the single contributions by distinguishing the case of a *prismatic* joint ($q_i = d_i$) from the case of a *revolute* joint ($q_i = \vartheta_i$).

For the contribution to the *angular velocity*:

- If Joint i is *prismatic*, from (3.17) it is

$$\dot{q}_i \boldsymbol{\jmath}_{Oi} = \mathbf{0}$$

and then

$$\boldsymbol{\jmath}_{Oi} = \mathbf{0}.$$

- If Joint i is *revolute*, from (3.21) it is

$$\dot{q}_i \boldsymbol{\jmath}_{Oi} = \dot{\vartheta}_i \boldsymbol{z}_{i-1}$$

and then

$$\boldsymbol{\jmath}_{Oi} = \boldsymbol{z}_{i-1}.$$

For the contribution to the *linear velocity*:

- If Joint i is *prismatic*, from (3.18) it is

$$\dot{q}_i \boldsymbol{\jmath}_{Pi} = \dot{d}_i \boldsymbol{z}_{i-1}$$

and then

$$\boldsymbol{\jmath}_{Pi} = \boldsymbol{z}_{i-1}.$$

- If Joint i is *revolute*, observing that the contribution to the linear velocity is to be computed with reference to the origin of the end-effector frame (Figure 3.2), it is

$$\begin{aligned} \dot{q}_i \boldsymbol{\jmath}_{Pi} &= \boldsymbol{\omega}_{i-1,i} \times \boldsymbol{r}_{i-1,n} \\ &= \dot{\vartheta}_i \boldsymbol{z}_{i-1} \times (\boldsymbol{p} - \boldsymbol{p}_{i-1}) \end{aligned}$$

and then

$$\boldsymbol{\jmath}_{Pi} = \boldsymbol{z}_{i-1} \times (\boldsymbol{p} - \boldsymbol{p}_{i-1}).$$

In sum, it is:

$$\begin{bmatrix} \boldsymbol{\jmath}_{Pi} \\ \boldsymbol{\jmath}_{Oi} \end{bmatrix} = \begin{cases} \begin{bmatrix} \boldsymbol{z}_{i-1} \\ \mathbf{0} \end{bmatrix} & \text{for a } \textit{prismatic} \text{ joint} \\ \begin{bmatrix} \boldsymbol{z}_{i-1} \times (\boldsymbol{p} - \boldsymbol{p}_{i-1}) \\ \boldsymbol{z}_{i-1} \end{bmatrix} & \text{for a } \textit{revolute} \text{ joint.} \end{cases} \tag{3.26}$$

The equation in (3.26) allow Jacobian computation in a simple, systematic way on the basis of direct kinematics relations. In fact, the vectors $\boldsymbol{z}_{i-1}$, $\boldsymbol{p}$ and $\boldsymbol{p}_{i-1}$ are all functions of the joint variables. In particular:

- $\boldsymbol{z}_{i-1}$ is given by the third column of the rotation matrix $\boldsymbol{R}_{i-1}^0$, *i.e.*,

$$\boldsymbol{z}_{i-1} = \boldsymbol{R}_1^0(q_1) \dots \boldsymbol{R}_{i-1}^{i-2}(q_{i-1}) \boldsymbol{z}_0 \tag{3.27}$$

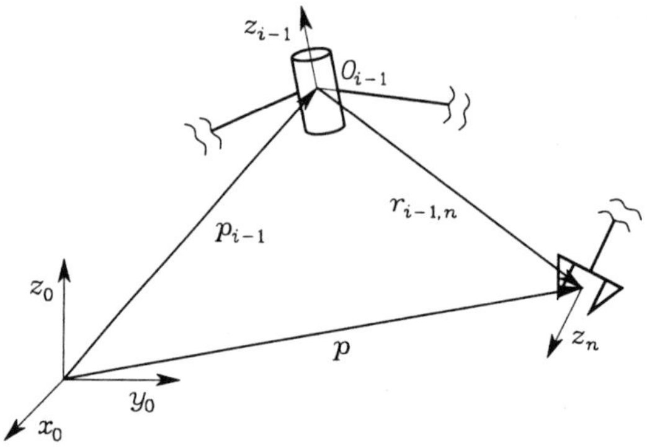

Figure 3.2 Representation of vectors needed for the computation of the velocity contribution of a revolute joint to the end-effector linear velocity.

where $z_0 = [0 \quad 0 \quad 1]^T$ allows selecting the third column.

- $\boldsymbol{p}$ is given by the first three elements of the fourth column of the transformation matrix $\boldsymbol{T}_n^0$, *i.e.*, by expressing $\tilde{\boldsymbol{p}}$ in the (4×1) homogeneous form

$$\tilde{\boldsymbol{p}} = \boldsymbol{A}_1^0(q_1)\dots\boldsymbol{A}_n^{n-1}(q_n)\tilde{\boldsymbol{p}}_0 \tag{3.28}$$

where $\tilde{\boldsymbol{p}}_0 = [0 \quad 0 \quad 0 \quad 1]^T$ allows selecting the fourth column.

- $\boldsymbol{p}_{i-1}$ is given by the first three elements of the fourth column of the transformation matrix $\boldsymbol{T}_{i-1}^0$, *i.e.*, it can be extracted from

$$\tilde{\boldsymbol{p}}_{i-1} = \boldsymbol{A}_1^0(q_1)\dots\boldsymbol{A}_{i-1}^{i-2}(q_{i-1})\tilde{\boldsymbol{p}}_0. \tag{3.29}$$

Remarkably, the above equations can be conveniently used to compute the translational and rotational velocities of any point along the manipulator structure, as long as the direct kinematics functions relative to that point are known.

Finally, notice that the Jacobian matrix depends on the frame in which the end-effector velocity is expressed. The above equations allow computation of the geometric Jacobian with respect to the base frame. If it is desired to represent the Jacobian in a different Frame u, it is sufficient to know the relative rotation matrix $\boldsymbol{R}^u$. The relationship between velocities in the two frames is

$$\begin{bmatrix} \dot{\boldsymbol{p}}^u \\ \boldsymbol{\omega}^u \end{bmatrix} = \begin{bmatrix} \boldsymbol{R}^u & \boldsymbol{O} \\ \boldsymbol{O} & \boldsymbol{R}^u \end{bmatrix} \begin{bmatrix} \dot{\boldsymbol{p}} \\ \boldsymbol{\omega} \end{bmatrix},$$

which, substituted in (3.3), gives

$$\begin{bmatrix} \dot{\boldsymbol{p}}^u \\ \boldsymbol{\omega}^u \end{bmatrix} = \begin{bmatrix} \boldsymbol{R}^u & \boldsymbol{O} \\ \boldsymbol{O} & \boldsymbol{R}^u \end{bmatrix} \boldsymbol{J}\dot{\boldsymbol{q}}.$$

On the assumption of a time-invariant Frame u, it is

$$\boldsymbol{J}^u = \begin{bmatrix} \boldsymbol{R}^u & \boldsymbol{O} \\ \boldsymbol{O} & \boldsymbol{R}^u \end{bmatrix} \boldsymbol{J}, \tag{3.30}$$

where $\boldsymbol{J}^u$ denotes the geometric Jacobian in Frame u.

3.2 Jacobian of Typical Manipulator Structures

In the following, the Jacobian is computed for some of the typical manipulator structures of the previous chapter.

3.2.1 Three-link Planar Arm

In this case, from (3.26) the Jacobian is

$$\boldsymbol{J}(\boldsymbol{q}) = \begin{bmatrix} \boldsymbol{z}_0 \times (\boldsymbol{p} - \boldsymbol{p}_0) & \boldsymbol{z}_1 \times (\boldsymbol{p} - \boldsymbol{p}_1) & \boldsymbol{z}_2 \times (\boldsymbol{p} - \boldsymbol{p}_2) \\ \boldsymbol{z}_0 & \boldsymbol{z}_1 & \boldsymbol{z}_2 \end{bmatrix}.$$

Computation of the position vectors of the various links gives

$$\boldsymbol{p}_0 = \begin{bmatrix} 0 \\ 0 \\ 0 \end{bmatrix} \quad \boldsymbol{p}_1 = \begin{bmatrix} a_1 c_1 \\ a_1 s_1 \\ 0 \end{bmatrix} \quad \boldsymbol{p}_2 = \begin{bmatrix} a_1 c_1 + a_2 c_{12} \\ a_1 s_1 + a_2 s_{12} \\ 0 \end{bmatrix}$$

$$\boldsymbol{p} = \begin{bmatrix} a_1 c_1 + a_2 c_{12} + a_3 c_{123} \\ a_1 s_1 + a_2 s_{12} + a_3 s_{123} \\ 0 \end{bmatrix},$$

while computation of the unit vectors of revolute joint axes gives

$$\boldsymbol{z}_0 = \boldsymbol{z}_1 = \boldsymbol{z}_2 = \begin{bmatrix} 0 \\ 0 \\ 1 \end{bmatrix}$$

since they are all parallel to axis z_0. From (3.25) it is

$$\boldsymbol{J} = \begin{bmatrix} -a_1 s_1 - a_2 s_{12} - a_3 s_{123} & -a_2 s_{12} - a_3 s_{123} & -a_3 s_{123} \\ a_1 c_1 + a_2 c_{12} + a_3 c_{123} & a_2 c_{12} + a_3 c_{123} & a_3 c_{123} \\ 0 & 0 & 0 \\ 0 & 0 & 0 \\ 0 & 0 & 0 \\ 1 & 1 & 1 \end{bmatrix}. \tag{3.31}$$

In the Jacobian (3.31), only the three nonnull rows are relevant (the rank of the matrix is at most 3); these refer to the two components of linear velocity along axes x_0, y_0

and the component of angular velocity about axis z_0. This result can be derived by observing that three degrees of mobility allow specification of at most three end-effector variables; v_z, ω_x, ω_y are always null for this kinematic structure. If orientation is of no concern, the (2×3) Jacobian for the positional part can be derived by considering just the first two rows, *i.e.*,

$$\boldsymbol{J}_P = \begin{bmatrix} -a_1 s_1 - a_2 s_{12} - a_3 s_{123} & -a_2 s_{12} - a_3 s_{123} & -a_3 s_{123} \\ a_1 c_1 + a_2 c_{12} + a_3 c_{123} & a_2 c_{12} + a_3 c_{123} & a_3 c_{123} \end{bmatrix}. \tag{3.32}$$

3.2.2 Anthropomorphic Arm

In this case, from (3.26) the Jacobian is

$$\boldsymbol{J} = \begin{bmatrix} \boldsymbol{z}_0 \times (\boldsymbol{p} - \boldsymbol{p}_0) & \boldsymbol{z}_1 \times (\boldsymbol{p} - \boldsymbol{p}_1) & \boldsymbol{z}_2 \times (\boldsymbol{p} - \boldsymbol{p}_2) \\ \boldsymbol{z}_0 & \boldsymbol{z}_1 & \boldsymbol{z}_2 \end{bmatrix}.$$

Computation of the position vectors of the various links gives

$$\boldsymbol{p}_0 = \boldsymbol{p}_1 = \begin{bmatrix} 0 \\ 0 \\ 0 \end{bmatrix} \qquad \boldsymbol{p}_2 = \begin{bmatrix} a_2 c_1 c_2 \\ a_2 s_1 c_2 \\ a_2 s_2 \end{bmatrix}$$

$$\boldsymbol{p} = \begin{bmatrix} c_1(a_2 c_2 + a_3 c_{23}) \\ s_1(a_2 c_2 + a_3 c_{23}) \\ a_2 s_2 + a_3 s_{23} \end{bmatrix},$$

while computation of the unit vectors of revolute joint axes gives

$$\boldsymbol{z}_0 = \begin{bmatrix} 0 \\ 0 \\ 1 \end{bmatrix} \qquad \boldsymbol{z}_1 = \boldsymbol{z}_2 = \begin{bmatrix} s_1 \\ -c_1 \\ 0 \end{bmatrix}.$$

From (3.25) it is

$$\boldsymbol{J} = \begin{bmatrix} -s_1(a_2 c_2 + a_3 c_{23}) & -c_1(a_2 s_2 + a_3 s_{23}) & -a_3 c_1 s_{23} \\ c_1(a_2 c_2 + a_3 c_{23}) & -s_1(a_2 s_2 + a_3 s_{23}) & -a_3 s_1 s_{23} \\ 0 & a_2 c_2 + a_3 c_{23} & a_3 c_{23} \\ 0 & s_1 & s_1 \\ 0 & -c_1 & -c_1 \\ 1 & 0 & 0 \end{bmatrix}. \tag{3.33}$$

Only three of the six rows of the Jacobian (3.33) are linearly independent. Having three degrees of mobility only, it is worth considering the upper (3×3) block of the Jacobian

$$\boldsymbol{J}_P = \begin{bmatrix} -s_1(a_2 c_2 + a_3 c_{23}) & -c_1(a_2 s_2 + a_3 s_{23}) & -a_3 c_1 s_{23} \\ c_1(a_2 c_2 + a_3 c_{23}) & -s_1(a_2 s_2 + a_3 s_{23}) & -a_3 s_1 s_{23} \\ 0 & a_2 c_2 + a_3 c_{23} & a_3 c_{23} \end{bmatrix} \tag{3.34}$$

that describes the relationship between the joint velocities and the end-effector linear velocity. This structure does not allow obtaining an arbitrary angular velocity ω; in fact, the two components ω_x and ω_y are not independent ($s_1\omega_y = -c_1\omega_x$).

3.2.3 Stanford Manipulator

In this case, from (3.26) it is

$$J = \begin{bmatrix} z_0 \times (p - p_0) & z_1 \times (p - p_1) & z_2 \\ z_0 & z_1 & 0 \end{bmatrix.$$
$$\begin{matrix} z_3 \times (p - p_3) & z_4 \times (p - p_4) & z_5 \times (p - p_5) \\ z_3 & z_4 & z_5 \end{matrix}\Bigg].$$

Computation of the position vectors of the various links gives

$$p_0 = p_1 = \begin{bmatrix} 0 \\ 0 \\ 0 \end{bmatrix} \qquad p_3 = p_4 = p_5 = \begin{bmatrix} c_1 s_2 d_3 - s_1 d_2 \\ s_1 s_2 d_3 + c_1 d_2 \\ c_2 d_3 \end{bmatrix}$$

$$p = \begin{bmatrix} c_1 s_2 d_3 - s_1 d_2 + \big(c_1(c_2 c_4 s_5 + s_2 c_5) - s_1 s_4 s_5\big) d_6 \\ s_1 s_2 d_3 + c_1 d_2 + \big(s_1(c_2 c_4 s_5 + s_2 c_5) + c_1 s_4 s_5\big) d_6 \\ c_2 d_3 + (-s_2 c_4 s_5 + c_2 c_5) d_6 \end{bmatrix},$$

while computation of the unit vectors of joint axes gives

$$z_0 = \begin{bmatrix} 0 \\ 0 \\ 1 \end{bmatrix} \qquad z_1 = \begin{bmatrix} -s_1 \\ c_1 \\ 0 \end{bmatrix} \qquad z_2 = z_3 = \begin{bmatrix} c_1 s_2 \\ s_1 s_2 \\ c_2 \end{bmatrix}$$

$$z_4 = \begin{bmatrix} -c_1 c_2 s_4 - s_1 c_4 \\ -s_1 c_2 s_4 + c_1 c_4 \\ s_2 s_4 \end{bmatrix} \qquad z_5 = \begin{bmatrix} c_1(c_2 c_4 s_5 + s_2 c_5) - s_1 s_4 s_5 \\ s_1(c_2 c_4 s_5 + s_2 c_5) + c_1 s_4 s_5 \\ -s_2 c_4 s_5 + c_2 c_5 \end{bmatrix}.$$

The sought Jacobian can be obtained by developing the computations as in (3.25), leading to expressing end-effector linear and angular velocity as a function of joint velocities.

3.3 Kinematic Singularities

The Jacobian in the differential kinematics equation of a manipulator defines a linear mapping

$$v = J(q)\dot{q} \tag{3.35}$$

between the vector $\dot{q}$ of joint velocities and the vector $v = [\,\dot{p}^T \quad \omega^T\,]^T$ of end-effector velocity. The Jacobian is, in general, a function of the configuration q; those configurations at which J is rank-deficient are termed *kinematic singularities*. To find the singularities of a manipulator is of great interest for the following reasons:

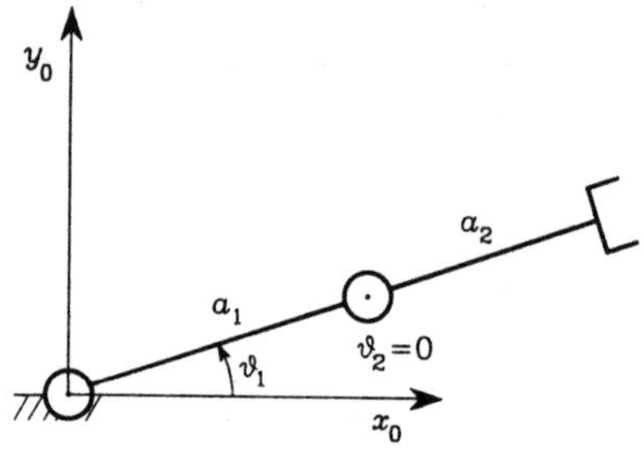

Figure 3.3 Two-link planar arm at a boundary singularity.

(a) Singularities represent configurations at which mobility of the structure is reduced, *i.e.*, it is not possible to impose an arbitrary motion to the end effector.

(b) When the structure is at a singularity, infinite solutions to the inverse kinematics problem may exist.

(c) In the neighbourhood of a singularity, small velocities in the operational space may cause large velocities in the joint space.

Singularities can be classified into:

- *Boundary* singularities that occur when the manipulator is either outstretched or retracted. It may be understood that these singularities do not represent a true drawback, since they can be avoided on condition that the manipulator is not driven to the boundaries of its reachable workspace.
- *Internal* singularities that occur inside the reachable workspace and are generally caused by the alignment of two or more axes of motion, or else by the attainment of particular end-effector configurations. Differently from the above, these singularities constitute a serious problem, as they can be encountered anywhere in the reachable workspace for a planned path in the operational space.

Example 3.2

To illustrate the behaviour of a manipulator at a singularity, consider a two-link planar arm. In this case, it is worth considering only the components $\dot{p}_x$ and $\dot{p}_y$ of the linear velocity in the plane. Thus, the Jacobian is the (2×2) matrix

$$\boldsymbol{J} = \begin{bmatrix} -a_1 s_1 - a_2 s_{12} & -a_2 s_{12} \\ a_1 c_1 + a_2 c_{12} & a_2 c_{12} \end{bmatrix}. \tag{3.36}$$

To analyze matrix rank, consider its determinant given by

$$\det(\boldsymbol{J}) = a_1 a_2 s_2. \tag{3.37}$$

For $a_1, a_2 \neq 0$, it is easy to find that the determinant in (3.37) vanishes whenever

$$\vartheta_2 = 0 \qquad \vartheta_2 = \pi,$$

ϑ_1 being irrelevant for the determination of singular configurations. These occur when the arm tip is located either on the outer ($\vartheta_2 = 0$) or on the inner ($\vartheta_2 = \pi$) boundary of the reachable workspace. Figure 3.3 illustrates the arm posture for $\vartheta_2 = 0$.

By analyzing the differential motion of the structure in such configuration, it can be observed that the two column vectors $[\,-(a_1 + a_2)s_1 \quad (a_1 + a_2)c_1\,]^T$ and $[\,-a_2 s_1 \quad a_2 c_1\,]^T$ of the Jacobian become parallel, and thus the Jacobian rank becomes one; this means that the tip velocity components are not independent (see point **(a)** above).

3.3.1 Singularity Decoupling

Computation of internal singularities via the Jacobian determinant may be tedious and of no easy solution for complex structures. For manipulators having a spherical wrist, by analogy with what already has been seen for inverse kinematics, it is possible to split the problem of singularity computation into two separate problems:

- computation of *arm singularities* resulting from the motion of the first three or more links,
- computation of *wrist singularities* resulting from the motion of the wrist joints.

For the sake of simplicity, consider the case $n = 6$; the Jacobian can be partitioned into (3×3) blocks as follows:

$$\boldsymbol{J} = \begin{bmatrix} \boldsymbol{J}_{11} & \boldsymbol{J}_{12} \\ \boldsymbol{J}_{21} & \boldsymbol{J}_{22} \end{bmatrix} \tag{3.38}$$

where, since the outer three joints are all revolute, the expressions of the two right blocks are respectively

$$\boldsymbol{J}_{12} = \begin{bmatrix} \boldsymbol{z}_3 \times (\boldsymbol{p} - \boldsymbol{p}_3) & \boldsymbol{z}_4 \times (\boldsymbol{p} - \boldsymbol{p}_4) & \boldsymbol{z}_5 \times (\boldsymbol{p} - \boldsymbol{p}_5) \end{bmatrix}$$

and

$$\boldsymbol{J}_{22} = \begin{bmatrix} \boldsymbol{z}_3 & \boldsymbol{z}_4 & \boldsymbol{z}_5 \end{bmatrix}. \tag{3.39}$$

As singularities are typical of the mechanical structure and do not depend on the frames chosen to describe kinematics, it is convenient to choose the origin of the end-effector frame at the intersection of the wrist axes (see Figure 2.30). The choice $\boldsymbol{p} = \boldsymbol{p}_W$ leads to

$$\boldsymbol{J}_{12} = \begin{bmatrix} \boldsymbol{0} & \boldsymbol{0} & \boldsymbol{0} \end{bmatrix},$$

since all vectors $\boldsymbol{p}_W - \boldsymbol{p}_i$ are parallel to the unit vectors $\boldsymbol{z}_i$, for $i = 3, 4, 5$, no matter how Frames $3, 4, 5$ are chosen according to Denavit-Hartenberg convention. In view of this choice, the overall Jacobian becomes a block lower-triangular matrix. In this case, computation of the determinant is greatly simplified, as this is given by the product of the determinants of the two blocks on the diagonal, *i.e.*,

$$\det(\boldsymbol{J}) = \det(\boldsymbol{J}_{11})\det(\boldsymbol{J}_{22}). \tag{3.40}$$

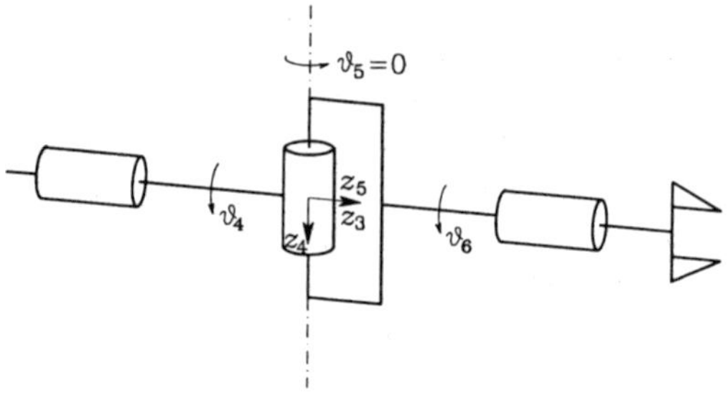

Figure 3.4 Spherical wrist at a singularity.

In turn, a true *singularity decoupling* has been achieved; the condition

$$\det(\boldsymbol{J}_{11}) = 0$$

leads to determining the *arm singularities*, while the condition

$$\det(\boldsymbol{J}_{22}) = 0$$

leads to determining the *wrist singularities*.

Notice, however, that this form of Jacobian does not provide the relationship between the joint velocities and the end-effector velocity, but it allows simplifying singularity computation. Below the two types of singularities are analyzed in detail.

3.3.2 Wrist Singularities

On the basis of the above singularity decoupling, wrist singularities can be determined by inspecting the block $\boldsymbol{J}_{22}$ in (3.39). It can be recognized that the wrist is at a singular configuration whenever the unit vectors $\boldsymbol{z}_3$, $\boldsymbol{z}_4$, $\boldsymbol{z}_5$ are linearly dependent. The wrist kinematic structure reveals that a singularity occurs when $\boldsymbol{z}_3$ and $\boldsymbol{z}_5$ are *aligned*, *i.e.*, whenever

$$\vartheta_5 = 0 \qquad \vartheta_5 = \pi.$$

Taking into consideration only the first configuration (Figure 3.4), the loss of mobility is caused by the fact that rotations of equal magnitude about opposite directions on ϑ_4 and ϑ_6 do not produce any end-effector rotation. Further, the wrist is not allowed to rotate about the axis orthogonal to $\boldsymbol{z}_4$ and $\boldsymbol{z}_3$, (see point **(a)** above). This singularity is naturally described in the joint space and can be encountered anywhere inside the manipulator reachable workspace; as a consequence, special care is to be taken in programming an end-effector motion.

3.3.3 Arm Singularities

Arm singularities are characteristic of a specific manipulator structure; to illustrate their determination, consider the anthropomorphic arm (Figure 2.23), whose Jacobian

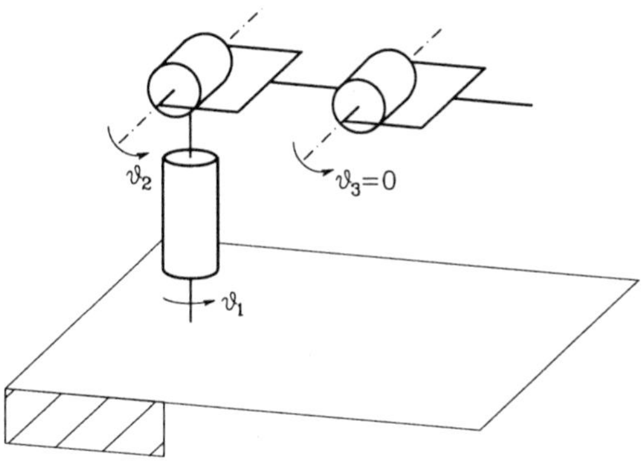

Figure 3.5 Anthropomorphic arm at an elbow singularity.

for the linear velocity part is given by (3.34). Its determinant is

$$\det(\boldsymbol{J}_P) = -a_2 a_3 s_3 (a_2 c_2 + a_3 c_{23}).$$

Like in the case of the planar arm of Example 3.3, the determinant does not depend on the first joint variable.

For $a_2, a_3 \neq 0$, the determinant vanishes if $s_3 = 0$ and/or $(a_2 c_2 + a_3 c_{23}) = 0$. The first situation occurs whenever

$$\vartheta_3 = 0 \qquad \vartheta_3 = \pi$$

meaning that the elbow is outstretched (Figure 3.5) or retracted, and is termed *elbow singularity*. Notice that this type of singularity is conceptually equivalent to the singularity found for the two-link planar arm.

By recalling the direct kinematics equation in (2.62), it can be observed that the second situation occurs when the wrist point lies on axis z_0 (Figure 3.6); it is thus characterized by

$$p_x = p_y = 0$$

and is termed *shoulder singularity*.

Notice that the whole axis z_0 describes a continuum of singular configurations; a rotation of ϑ_1 does not cause any translation of the wrist position (the first column of $\boldsymbol{J}_P$ is always null at a shoulder singularity), and then the kinematics equation admits infinite solutions; moreover, motions starting from the singular configuration that take the wrist along the z_1 direction are not allowed (see point **(b)** above).

If a spherical wrist is connected to an anthropomorphic arm (Figure 2.26), the arm direct kinematics is different. In this case the Jacobian to consider represents the block $\boldsymbol{J}_{11}$ of the Jacobian in (3.38) with $\boldsymbol{p} = \boldsymbol{p}_W$. Analyzing its determinant leads to finding the same singular configurations, which are relative to different values of the third joint variables, though—compare (2.62) and (2.66).

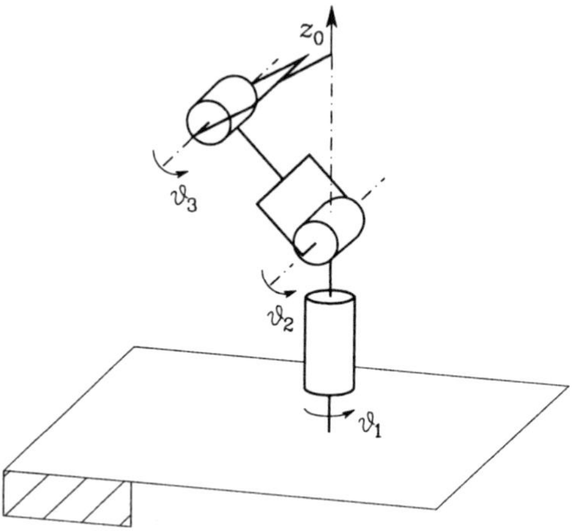

Figure 3.6 Anthropomorphic arm at a shoulder singularity.

Finally, it is important to remark that, differently from the wrist singularities, the arm singularities are well identified in the operational space, and thus they can be suitably avoided in the end-effector path planning stage.

3.4 Analysis of Redundancy

The concept of *kinematic redundancy* has been introduced in Section 2.10.2; redundancy is related to the number n of degrees of mobility of the structure, the number m of operational space variables, and the number r of operational space variables necessary to specify a given task.

In order to perform a systematic analysis of redundancy, it is worth considering differential kinematics in lieu of direct kinematics (2.70). To this purpose, (3.35) is to be interpreted as the differential kinematics mapping relating the n components of the joint velocity vector to the $r \leq m$ components of the velocity vector $\boldsymbol{v}$ of concern for the specific task. To clarify this point, consider the case of a three-link planar arm; that is not intrinsically redundant ($n = m = 3$) and its Jacobian (3.31) has three null rows accordingly. If the task does not specify ω_z ($r = 2$), the arm becomes functionally redundant and the Jacobian to consider for redundancy analysis is the one in (3.32).

A different case is that of the anthropomorphic arm for which only position variables are of concern ($n = m = 3$). The relevant Jacobian is the one in (3.34). The arm is neither intrinsically redundant nor can become functionally redundant if it is assigned a planar task; in that case, indeed, the task would set constraints on the three components of end-effector linear velocity.

Therefore, the differential kinematics equation to consider can be formally written as in (3.35), *i.e.*,

$$\boldsymbol{v} = \boldsymbol{J}(\boldsymbol{q})\dot{\boldsymbol{q}}, \tag{3.41}$$

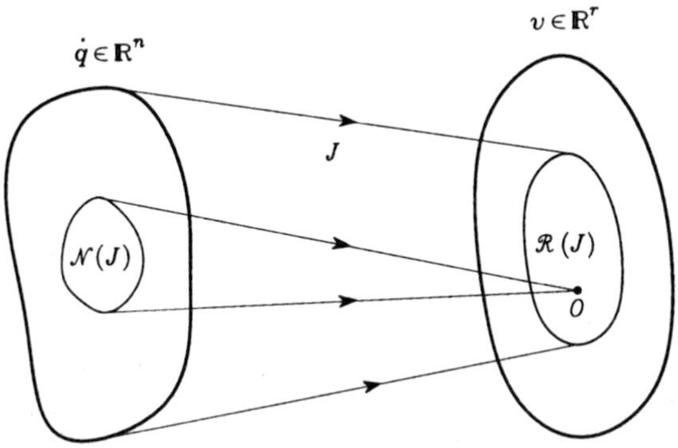

Figure 3.7 Mapping between the joint velocity space and the end-effector velocity space.

where now $\boldsymbol{v}$ is meant to be the $(r \times 1)$ vector of end-effector velocity of concern for the specific task and $\boldsymbol{J}$ is the corresponding $(r \times n)$ Jacobian matrix that can be extracted from the geometric Jacobian; $\dot{\boldsymbol{q}}$ is the $(n \times 1)$ vector of joint velocities. If $r < n$, the manipulator is kinematically redundant and there exist $(n - r)$ *redundant degrees of mobility*.

The Jacobian describes the linear mapping from the joint velocity space to the end-effector velocity space. In general, it is a function of the configuration. In the context of differential kinematics, however, the Jacobian has to be regarded as a constant matrix, since the instantaneous velocity mapping is of interest for a given posture. The mapping is schematically illustrated in Figure 3.7 with a typical notation from set theory.

The relationship in (3.41) can be characterized in terms of the *range* and *null* spaces of the mapping; specifically, one has that:

- The *range* of $\boldsymbol{J}$ is the subspace $\mathcal{R}(\boldsymbol{J})$ in $\mathbb{R}^r$ of the end-effector velocities that can be generated by the joint velocities, in the given manipulator posture.
- The *null* of $\boldsymbol{J}$ is the subspace $\mathcal{N}(\boldsymbol{J})$ in $\mathbb{R}^n$ of joint velocities that do not produce any end-effector velocity, in the given manipulator posture.

If the Jacobian has *full rank*, one has:

$$\dim\big(\mathcal{R}(\boldsymbol{J})\big) = r \qquad \dim\big(\mathcal{N}(\boldsymbol{J})\big) = n - r$$

and the range of $\boldsymbol{J}$ spans the entire space $\mathbb{R}^r$. Instead, if the Jacobian degenerates at a *singularity*, the dimension of the range space decreases while the dimension of the null space increases, since the following relation holds:

$$\dim\big(\mathcal{R}(\boldsymbol{J})\big) + \dim\big(\mathcal{N}(\boldsymbol{J})\big) = n$$

independently of the rank of the matrix $\boldsymbol{J}$.

The existence of a subspace $\mathcal{N}(\boldsymbol{J}) \neq \emptyset$ for a redundant manipulator allows determination of systematic techniques for handling redundant degrees of mobility. To this purpose, if $\dot{\boldsymbol{q}}^*$ denotes a solution to (3.41) and $\boldsymbol{P}$ is an $(n \times n)$ matrix so that

$$\mathcal{R}(\boldsymbol{P}) \equiv \mathcal{N}(\boldsymbol{J}),$$

also the joint velocity vector

$$\dot{\boldsymbol{q}} = \dot{\boldsymbol{q}}^* + \boldsymbol{P}\dot{\boldsymbol{q}}_0, \tag{3.42}$$

with arbitrary $\dot{\boldsymbol{q}}_0$, is a solution to (3.41). In fact, premultiplying both sides of (3.42) by $\boldsymbol{J}$ yields

$$\boldsymbol{J}\dot{\boldsymbol{q}} = \boldsymbol{J}\dot{\boldsymbol{q}}^* + \boldsymbol{J}\boldsymbol{P}\dot{\boldsymbol{q}}_0 = \boldsymbol{J}\dot{\boldsymbol{q}}^* = \boldsymbol{v}$$

since $\boldsymbol{J}\boldsymbol{P}\dot{\boldsymbol{q}}_0 = \boldsymbol{0}$ for any $\dot{\boldsymbol{q}}_0$. This result is of fundamental importance for redundancy resolution; a solution of the kind (3.42) points out the possibility of choosing the vector of arbitrary joint velocities $\dot{\boldsymbol{q}}_0$ so as to advantageously exploit the redundant degrees of mobility. In fact, the effect of $\dot{\boldsymbol{q}}_0$ is to generate *internal motions* of the structure that do not change the end-effector position and orientation but may allow, for instance, manipulator reconfiguration into more dexterous postures for execution of a given task.

3.5 Differential Kinematics Inversion

In Section 2.12 it was shown how the inverse kinematics problem admits closed-form solutions only for manipulators having a simple kinematic structure. Problems arise whenever the end effector attains a particular position and/or orientation in the operational space, or the structure is complex and it is not possible to relate end-effector position and orientation to different sets of joint variables, or else the manipulator is redundant. These limitations are caused by the highly nonlinear relationship between joint space variables and operational space variables.

On the other hand, the differential kinematics equation represents a linear mapping between the joint velocity space and the operational velocity space, although it varies with the current configuration. This fact suggests the possibility to utilize the differential kinematics equation to tackle the inverse kinematics problem.

Suppose that a motion trajectory is assigned to the end effector in terms of $\boldsymbol{v}$ and the initial conditions on position and orientation. The aim is to determine a feasible joint trajectory $(\boldsymbol{q}(t), \dot{\boldsymbol{q}}(t))$ that reproduces the given trajectory.

By considering (3.41) with $n = r$, the joint velocities can be obtained via simple inversion of the Jacobian matrix

$$\dot{\boldsymbol{q}} = \boldsymbol{J}^{-1}(\boldsymbol{q})\boldsymbol{v}. \tag{3.43}$$

If the initial manipulator posture $\boldsymbol{q}(0)$ is known, joint positions can be computed by integrating velocities over time, *i.e.*,

$$\boldsymbol{q}(t) = \int_0^t \dot{\boldsymbol{q}}(\varsigma)d\varsigma + \boldsymbol{q}(0).$$

The integration can be performed in discrete time by resorting to numerical techniques. The simplest technique is based on the Euler integration method; given an integration interval Δt, if the joint positions and velocities at time t_k are known, the joint positions at time $t_{k+1} = t_k + \Delta t$ can be computed as

$$\boldsymbol{q}(t_{k+1}) = \boldsymbol{q}(t_k) + \dot{\boldsymbol{q}}(t_k)\Delta t. \tag{3.44}$$

This technique for inverting kinematics is independent of the solvability of the kinematic structure. Nonetheless, it is necessary that the *Jacobian* be *square* and of *full rank*; this demands further insight into the cases of *redundant* manipulators and kinematic *singularity* occurrence.

3.5.1 Redundant Manipulators

When the manipulator is *redundant* ($r < n$), the Jacobian matrix has more columns than rows and infinite solutions exist to (3.41). A viable solution method is to formulate the problem as a constrained linear optimization problem.

In detail, once the end-effector velocity $\boldsymbol{v}$ and Jacobian $\boldsymbol{J}$ are given (for a given configuration $\boldsymbol{q}$), it is desired to find the solutions $\dot{\boldsymbol{q}}$ that satisfy the linear equation in (3.41) and *minimize* the quadratic cost functional of joint velocities

$$g(\dot{\boldsymbol{q}}) = \frac{1}{2}\dot{\boldsymbol{q}}^T \boldsymbol{W} \dot{\boldsymbol{q}}$$

where $\boldsymbol{W}$ is a suitable $(n \times n)$ symmetric positive definite weighting matrix.

This problem can be solved with the *method of Lagrangian multipliers*. Consider the modified cost functional

$$g(\dot{\boldsymbol{q}}, \boldsymbol{\lambda}) = \frac{1}{2}\dot{\boldsymbol{q}}^T \boldsymbol{W} \dot{\boldsymbol{q}} + \boldsymbol{\lambda}^T(\boldsymbol{v} - \boldsymbol{J}\dot{\boldsymbol{q}})$$

where $\boldsymbol{\lambda}$ is an $(r \times 1)$ vector of unknown multipliers that allows incorporating the constraint (3.41) in the functional to minimize. The requested solution has to satisfy the necessary conditions:

$$\left(\frac{\partial g}{\partial \dot{\boldsymbol{q}}}\right)^T = \boldsymbol{0} \qquad \left(\frac{\partial g}{\partial \boldsymbol{\lambda}}\right)^T = \boldsymbol{0}.$$

From the first one, it is $\boldsymbol{W}\dot{\boldsymbol{q}} - \boldsymbol{J}^T\boldsymbol{\lambda} = \boldsymbol{0}$ and thus

$$\dot{\boldsymbol{q}} = \boldsymbol{W}^{-1}\boldsymbol{J}^T\boldsymbol{\lambda} \tag{3.45}$$

where the inverse of $\boldsymbol{W}$ exists. Notice that the solution (3.45) is a minimum, since $\partial^2 g/\partial \dot{\boldsymbol{q}}^2 = \boldsymbol{W}$ is positive definite. From the second condition above, the constraint

$$\boldsymbol{v} = \boldsymbol{J}\dot{\boldsymbol{q}}$$

is recovered. Combining the two conditions gives

$$\boldsymbol{v} = \boldsymbol{J}\boldsymbol{W}^{-1}\boldsymbol{J}^T\boldsymbol{\lambda};$$

on the assumption that $\boldsymbol{J}$ has full rank, $\boldsymbol{J}\boldsymbol{W}^{-1}\boldsymbol{J}^T$ is an $(r \times r)$ square matrix of rank r and thus can be inverted. Solving for $\boldsymbol{\lambda}$ yields

$$\boldsymbol{\lambda} = (\boldsymbol{J}\boldsymbol{W}^{-1}\boldsymbol{J}^T)^{-1}\boldsymbol{v}$$

which, substituted into (3.45), gives the sought optimal solution

$$\dot{\boldsymbol{q}} = \boldsymbol{W}^{-1}\boldsymbol{J}^T(\boldsymbol{J}\boldsymbol{W}^{-1}\boldsymbol{J}^T)^{-1}\boldsymbol{v}. \tag{3.46}$$

Premultiplying both sides of (3.46) by $\boldsymbol{J}$, it is easy to verify that this solution satisfies the differential kinematics equation in (3.41).

A particular case occurs when the weighting matrix $\boldsymbol{W}$ is the identity matrix $\boldsymbol{I}$ and the solution simplifies into

$$\dot{\boldsymbol{q}} = \boldsymbol{J}^{\dagger}\boldsymbol{v}; \tag{3.47}$$

the matrix

$$\boldsymbol{J}^{\dagger} = \boldsymbol{J}^T(\boldsymbol{J}\boldsymbol{J}^T)^{-1} \tag{3.48}$$

is the *right pseudo-inverse* of $\boldsymbol{J}$. The obtained solution locally minimizes the norm of joint velocities.

It was pointed out above that if $\dot{\boldsymbol{q}}^*$ is a solution to (3.41), also $\dot{\boldsymbol{q}}^* + \boldsymbol{P}\dot{\boldsymbol{q}}_0$ is a solution, where $\dot{\boldsymbol{q}}_0$ is a vector of arbitrary joint velocities and $\boldsymbol{P}$ is a projector in the null space of $\boldsymbol{J}$. Therefore, thanks to the presence of redundant degrees of mobility, the solution (3.47) can be modified by the introduction of another term of the kind $\boldsymbol{P}\dot{\boldsymbol{q}}_0$. In particular, $\dot{\boldsymbol{q}}_0$ can be specified so as to satisfy an additional constraint to the problem.

In that case, it is necessary to consider a new cost functional in the form

$$g'(\dot{\boldsymbol{q}}) = \frac{1}{2}(\dot{\boldsymbol{q}} - \dot{\boldsymbol{q}}_0)^T(\dot{\boldsymbol{q}} - \dot{\boldsymbol{q}}_0);$$

this choice is aimed at minimizing the norm of vector $\dot{\boldsymbol{q}} - \dot{\boldsymbol{q}}_0$; in other words, solutions are sought which satisfy the constraint (3.41) and are as close as possible to $\dot{\boldsymbol{q}}_0$. In this way, the objective specified through $\dot{\boldsymbol{q}}_0$ becomes unavoidably a secondary objective to satisfy with respect to the primary objective specified by the constraint (3.41).

Proceeding in a way similar to the above yields

$$g'(\dot{\boldsymbol{q}}, \boldsymbol{\lambda}) = \frac{1}{2}(\dot{\boldsymbol{q}} - \dot{\boldsymbol{q}}_0)^T(\dot{\boldsymbol{q}} - \dot{\boldsymbol{q}}_0) + \boldsymbol{\lambda}^T(\boldsymbol{v} - \boldsymbol{J}\dot{\boldsymbol{q}});$$

from the first necessary condition it is

$$\dot{\boldsymbol{q}} = \boldsymbol{J}^T\boldsymbol{\lambda} + \dot{\boldsymbol{q}}_0 \tag{3.49}$$

which, substituted into (3.41), gives

$$\boldsymbol{\lambda} = (\boldsymbol{J}\boldsymbol{J}^T)^{-1}(\boldsymbol{v} - \boldsymbol{J}\dot{\boldsymbol{q}}_0).$$

Finally, substituting $\boldsymbol{\lambda}$ back in (3.49) gives

$$\dot{\boldsymbol{q}} = \boldsymbol{J}^{\dagger}\boldsymbol{v} + (\boldsymbol{I} - \boldsymbol{J}^{\dagger}\boldsymbol{J})\dot{\boldsymbol{q}}_0. \tag{3.50}$$

As can be easily recognized, the obtained solution is composed of two terms. The first one is relative to minimum norm joint velocities. The second one, termed *homogeneous solution*, attempts to satisfy the additional constraint to specify via $\dot{q}_0$[2]; the matrix $(I - J^\dagger J)$ is one of those matrices P introduced in (3.42) which allows projecting the vector $\dot{q}_0$ in the null space of J, so as not to violate the constraint (3.41). A direct consequence is that, in the case $v = 0$, is is possible to generate *internal motions* described by $(I - J^\dagger J)\dot{q}_0$ that reconfigure the manipulator structure without changing the end-effector position and orientation.

Finally, it is worth discussing the way to specify the vector $\dot{q}_0$ for a convenient utilization of redundant degrees of mobility. A typical choice is

$$\dot{q}_0 = k_0 \left(\frac{\partial w(q)}{\partial q}\right)^T \tag{3.51}$$

where $k_0 > 0$ and $w(q)$ is a (secondary) objective function of the joint variables. Since the solution moves along the direction of the gradient of the objective function, it attempts to *locally maximize* it compatible to the primary objective (kinematic constraint). Typical objective functions are:

- The *manipulability measure*, defined as

$$w(q) = \sqrt{\det\left(J(q)J^T(q)\right)} \tag{3.52}$$

  which vanishes at a singular configuration; thus, by maximizing this measure, redundancy is exploited to move away from singularities.

- The *distance from mechanical joint limits*, defined as

$$w(q) = -\frac{1}{2n}\sum_{i=1}^{n}\left(\frac{q_i - \bar{q}_i}{q_{iM} - q_{im}}\right)^2 \tag{3.53}$$

  where q_{iM} (q_{im}) denotes the maximum (minimum) joint limit and $\bar{q}_i$ the middle value of the joint range; thus, by maximizing this distance, redundancy is exploited to keep the joint variables as close as possible to the centre of their ranges.

- The *distance from an obstacle*, defined as

$$w(q) = \min_{p,o} \|p(q) - o\| \tag{3.54}$$

  where o is the position vector of a suitable point on the obstacle (its centre, for instance, if the obstacle is modeled as a sphere) and p is the position vector of a generic point along the structure; thus, by maximizing this distance, redundancy is exploited to avoid collision of the manipulator with an obstacle[3].

---

[2] It should be recalled that the additional constraint has secondary priority with respect to the primary kinematic constraint.

[3] If an obstacle occurs along the end-effector path, it is opportune to invert the order of priority between the kinematic constraint and the additional constraint; in this way the obstacle may be avoided, but one gives up tracking the desired path.

3.5.2 Kinematic Singularities

Both solutions (3.43) and (3.47) can be computed only when the Jacobian has full rank. Hence, they become meaningless when the manipulator is at a singular configuration; in such a case, the system $\boldsymbol{v} = \boldsymbol{J}\dot{\boldsymbol{q}}$ contains linearly dependent equations.

It is possible to find a solution $\dot{\boldsymbol{q}}$ by extracting all the linearly independent equations only if $\boldsymbol{v} \in \mathcal{R}(\boldsymbol{J})$. The occurrence of this situation means that the assigned path is physically executable by the manipulator, even though it is at a singular configuration. If instead $\boldsymbol{v} \notin \mathcal{R}(\boldsymbol{J})$, the system of equations has no solution; this means that the operational space path cannot be executed by the manipulator at the given posture.

It is important to underline that the inversion of the Jacobian can represent a serious inconvenience not only at a singularity but also in the neighbourhood of a singularity. For instance, for the Jacobian inverse it is well known that its computation requires the computation of the determinant; in the neighbourhood of a singularity, the determinant takes on a relatively small value which can cause large joint velocities (see point **(c)** in Section 3.3). Consider again the above example of the shoulder singularity for the anthropomorphic arm. If a path is assigned to the end effector which passes nearby the base rotation axis (geometric locus of singular configurations), the base joint is forced to make a rotation of about π in a relatively short time to allow the end effector to keep tracking the imposed trajectory.

A more rigorous analysis of the solution features in the neighbourhood of singular configurations can be developed by resorting to the singular value decomposition (SVD) of matrix $\boldsymbol{J}$.

An alternative solution overcoming the problem of inverting differential kinematics in the neighbourhood of a singularity is provided by the so-called *damped least-squares (DLS) inverse*

$$\boldsymbol{J}^{\star} = \boldsymbol{J}^T(\boldsymbol{J}\boldsymbol{J}^T + k^2\boldsymbol{I})^{-1} \tag{3.55}$$

where k is a damping factor that renders the inversion better conditioned from a numerical viewpoint. It can be shown that such a solution can be obtained by reformulating the problem in terms of the minimization of the cost functional

$$g''(\dot{\boldsymbol{q}}) = \frac{1}{2}(\boldsymbol{v} - \boldsymbol{J}\dot{\boldsymbol{q}})^T(\boldsymbol{v} - \boldsymbol{J}\dot{\boldsymbol{q}}) + \frac{1}{2}k^2\dot{\boldsymbol{q}}^T\dot{\boldsymbol{q}},$$

where the introduction of the first term allows tolerating a finite inversion error with the advantage of norm-bounded velocities. The factor k establishes the relative weight between the two objectives, and there exist techniques for selecting optimal values for the damping factor.

3.6 Analytical Jacobian

The above sections have shown the way to compute the end-effector velocity in terms of the velocity of the end-effector frame. The Jacobian is computed by following a *geometric technique* in which the contributions of each joint velocity to the components of end-effector linear and angular velocity are determined.

If the end-effector position and orientation are specified in terms of a minimal number of parameters in the operational space as in (2.68), it is natural to ask whether it is possible to compute the Jacobian via differentiation of the direct kinematics function with respect to the joint variables. To this purpose, below an *analytical technique* is presented to compute the Jacobian, and the existing relationship between the two Jacobians is found.

The translational velocity of the end-effector frame can be expressed as the time derivative of vector $\boldsymbol{p}$, representing the origin of the end-effector frame with respect to the base frame, *i.e.*,

$$\dot{\boldsymbol{p}} = \frac{\partial \boldsymbol{p}}{\partial \boldsymbol{q}}\dot{\boldsymbol{q}} = \boldsymbol{J}_P(\boldsymbol{q})\dot{\boldsymbol{q}}. \tag{3.56}$$

For what concerns the rotational velocity of the end-effector frame, the minimal representation of orientation in terms of three variables $\boldsymbol{\phi}$ can be considered. Its time derivative $\dot{\boldsymbol{\phi}}$ in general differs from the angular velocity vector defined above. In any case, once the function $\boldsymbol{\phi}(\boldsymbol{q})$ is known, it is formally correct to consider the Jacobian obtained as

$$\dot{\boldsymbol{\phi}} = \frac{\partial \boldsymbol{\phi}}{\partial \boldsymbol{q}}\dot{\boldsymbol{q}} = \boldsymbol{J}_\phi(\boldsymbol{q})\dot{\boldsymbol{q}}. \tag{3.57}$$

Computing the Jacobian $\boldsymbol{J}_\phi(\boldsymbol{q})$ as $\partial\boldsymbol{\phi}/\partial\boldsymbol{q}$ is not straightforward, since the function $\boldsymbol{\phi}(\boldsymbol{q})$ is not usually available in direct form, but requires computation of the elements of the relative rotation matrix.

Upon these premises, the differential kinematics equation can be obtained as the time derivative of the direct kinematics equation in (2.70), *i.e.*,

$$\dot{\boldsymbol{x}} = \begin{bmatrix} \dot{\boldsymbol{p}} \\ \dot{\boldsymbol{\phi}} \end{bmatrix} = \begin{bmatrix} \boldsymbol{J}_P(\boldsymbol{q}) \\ \boldsymbol{J}_\phi(\boldsymbol{q}) \end{bmatrix} \dot{\boldsymbol{q}} = \boldsymbol{J}_A(\boldsymbol{q})\dot{\boldsymbol{q}} \tag{3.58}$$

where the *analytical Jacobian*

$$\boldsymbol{J}_A(\boldsymbol{q}) = \frac{\partial \boldsymbol{k}(\boldsymbol{q})}{\partial \boldsymbol{q}} \tag{3.59}$$

is different from the geometric Jacobian $\boldsymbol{J}$, since the end-effector angular velocity $\boldsymbol{\omega}$ with respect to the base frame is not given by $\dot{\boldsymbol{\phi}}$.

It is possible to find the relationship between the angular velocity $\boldsymbol{\omega}$ and the rotational velocity $\dot{\boldsymbol{\phi}}$ for a given set of orientation angles. For instance, consider the Euler angles ZYZ defined in Section 2.4.1; in Figure 3.8, the vectors corresponding to the rotational velocities $\dot{\varphi}$, $\dot{\vartheta}$, $\dot{\psi}$ have been represented with reference to the current frame. Figure 3.9 illustrates how to compute the contributions of each rotational velocity to the components of angular velocity about the axes of the reference frame:

- as a result of $\dot{\varphi}$: $[\omega_x \quad \omega_y \quad \omega_z]^T = \dot{\varphi}\,[0 \quad 0 \quad 1]^T$
- as a result of $\dot{\vartheta}$: $[\omega_x \quad \omega_y \quad \omega_z]^T = \dot{\vartheta}\,[-s_\varphi \quad c_\varphi \quad 0]^T$
- as a result of $\dot{\psi}$: $[\omega_x \quad \omega_y \quad \omega_z]^T = \dot{\psi}\,[c_\varphi s_\vartheta \quad s_\varphi s_\vartheta \quad c_\vartheta]^T$,

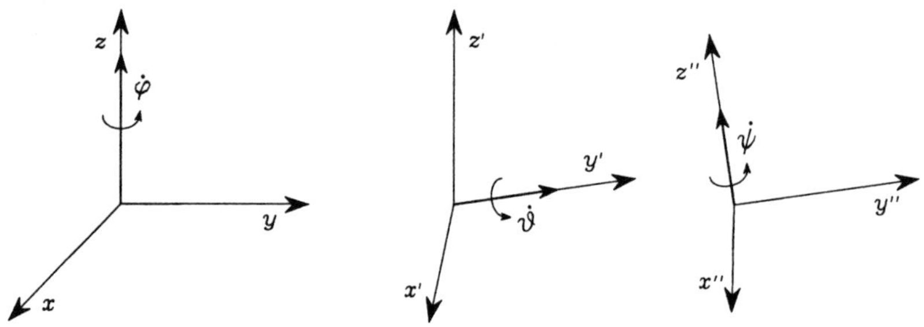

Figure 3.8 Rotational velocities of Euler angles ZYZ in current frame.

and then the equation relating the angular velocity $\boldsymbol{\omega}$ to the time derivative of the Euler angles $\dot{\boldsymbol{\phi}}$ is

$$\boldsymbol{\omega} = \begin{bmatrix} 0 & -s_\varphi & c_\varphi s_\vartheta \\ 0 & c_\varphi & s_\varphi s_\vartheta \\ 1 & 0 & c_\vartheta \end{bmatrix} \dot{\boldsymbol{\phi}} = \boldsymbol{T}(\boldsymbol{\phi})\dot{\boldsymbol{\phi}}. \tag{3.60}$$

The determinant of matrix $\boldsymbol{T}$ is $-s_\vartheta$, which implies that the relationship cannot be inverted for $\vartheta = 0, \pi$. This means that, even though all rotational velocities of the end-effector frame can be expressed by means of a suitable angular velocity vector $\boldsymbol{\omega}$, there exist angular velocities which cannot be expressed by means of $\dot{\boldsymbol{\phi}}$ when the orientation of the end-effector frame causes $s_\vartheta = 0$[4]. In fact, in this situation, the angular velocities that can be described by $\dot{\boldsymbol{\phi}}$ shall have linearly dependent components in the directions orthogonal to axis z ($\omega_x^2 + \omega_y^2 = \dot{\vartheta}^2$). An orientation for which the determinant of the transformation matrix vanishes is termed *representation singularity* of $\boldsymbol{\phi}$.

From a physical viewpoint, the meaning of $\boldsymbol{\omega}$ is more intuitive than that of $\dot{\boldsymbol{\phi}}$. The three components of $\boldsymbol{\omega}$ represent the components of angular velocity with respect to the base frame. Instead, the three elements of $\dot{\boldsymbol{\phi}}$ represent nonorthogonal components of angular velocity defined with respect to the axes of a frame that varies as the end-effector orientation varies. On the other hand, while the integral of $\dot{\boldsymbol{\phi}}$ over time gives $\boldsymbol{\phi}$, the integral of $\boldsymbol{\omega}$ does not admit a clear physical interpretation, as can be seen in the following example.

Example 3.3

Consider an object whose orientation with respect to a reference frame is known at time $t = 0$. Assign the following time profiles to $\boldsymbol{\omega}$:

$$\boldsymbol{\omega} = [\pi/2 \quad 0 \quad 0]^T \quad 0 \le t \le 1 \qquad \boldsymbol{\omega} = [0 \quad \pi/2 \quad 0]^T \quad 1 < t \le 2,$$
$$\boldsymbol{\omega} = [0 \quad \pi/2 \quad 0]^T \quad 0 \le t \le 1 \qquad \boldsymbol{\omega} = [\pi/2 \quad 0 \quad 0]^T \quad 1 < t \le 2.$$

---

[4] In Section 2.5, it was shown that for this orientation the inverse solution of the Euler angles degenerates.

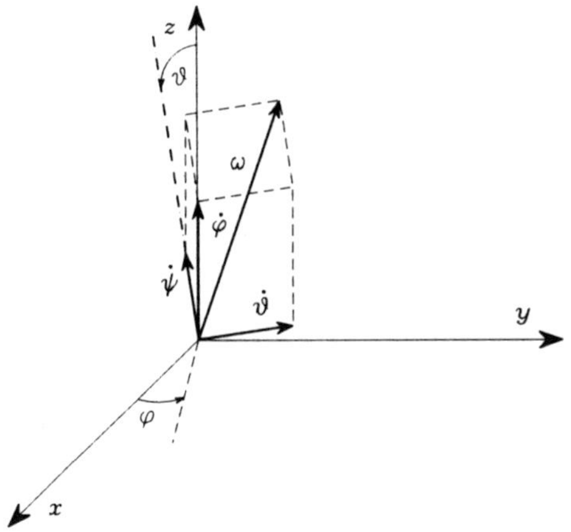

Figure 3.9 Composition of elementary rotational velocities for computing angular velocity.

The integral of ω gives the same result in the two cases

$$\int_0^2 \omega dt = [\,\pi/2 \quad \pi/2 \quad 0\,]^T$$

but the final object orientation corresponding to the second time law is clearly different from the one obtained with the first time law (Figure 3.10).

Once the transformation $\boldsymbol{T}$ between $\boldsymbol{\omega}$ and $\dot{\boldsymbol{\phi}}$ is given, the analytical Jacobian can be related to the geometric Jacobian as

$$\boldsymbol{v} = \begin{bmatrix} \boldsymbol{I} & \boldsymbol{O} \\ \boldsymbol{O} & \boldsymbol{T}(\boldsymbol{\phi}) \end{bmatrix} \dot{\boldsymbol{x}} = \boldsymbol{T}_A(\boldsymbol{\phi})\dot{\boldsymbol{x}} \tag{3.61}$$

which, in view of (3.3) and (3.58), yields

$$\boldsymbol{J} = \boldsymbol{T}_A(\boldsymbol{\phi})\boldsymbol{J}_A. \tag{3.62}$$

This relationship shows that $\boldsymbol{J}$ and $\boldsymbol{J}_A$, in general, differ. Regarding the use of either one or the other in all those problems where the influence of the Jacobian matters, it is anticipated that the geometric Jacobian will be adopted whenever it is necessary to refer to quantities of clear physical meaning, while the analytical Jacobian will be adopted whenever it is necessary to refer to differential quantities of variables defined in the operational space.

For certain manipulator geometries, it is possible to establish a substantial equivalence between $\boldsymbol{J}$ and $\boldsymbol{J}_A$. In fact, when the degrees of mobility cause rotations of the end effector all about the same fixed axis in space, the two Jacobians are essentially the same. This is the case of the above three-link planar arm. Its geometric Jacobian (3.31)

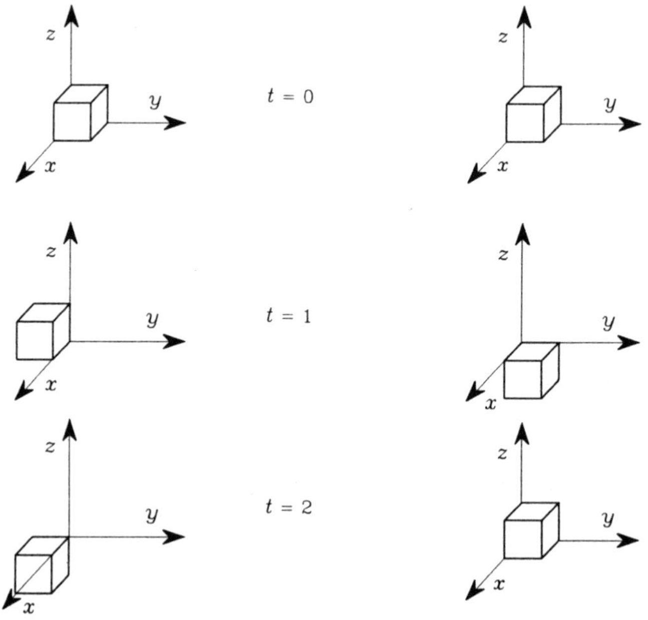

Figure 3.10 Nonuniqueness of orientation computed as the integral of angular velocity.

reveals that only rotations about axis z_0 are permitted. The (3×3) analytical Jacobian that can be derived by considering the end-effector position components in the plane of the structure and defining the end-effector orientation as $\phi = \vartheta_1 + \vartheta_2 + \vartheta_3$ coincides with the matrix that is obtained by eliminating the three null rows of the geometric Jacobian.

3.7 Inverse Kinematics Algorithms

In Section 3.5 it was shown how to invert kinematics by using the differential kinematics equation. In the numerical implementation of (3.44), computation of joint velocities is obtained by using the inverse of the Jacobian evaluated with the joint variables at the previous instant of time

$$\boldsymbol{q}(t_{k+1}) = \boldsymbol{q}(t_k) + \boldsymbol{J}^{-1}(\boldsymbol{q}(t_k))\boldsymbol{v}(t_k)\Delta t.$$

It follows that the computed joint velocities $\dot{\boldsymbol{q}}$ do not coincide with those satisfying (3.43) in the continuous time. Therefore, reconstruction of joint variables $\boldsymbol{q}$ is entrusted to a numerical integration which involves *drift* phenomena of the solution; as a consequence, the end-effector location corresponding to the computed joint variables differs from the desired one.

This inconvenience can be overcome by resorting to a solution scheme that accounts for the operational space error between the desired and the actual end-effector position and orientation. Let

$$e = x_d - x = x_d - k(q) \tag{3.63}$$

be the expression of such error, where (2.70) has been used.

Consider the time derivative of (3.63)

$$\dot{e} = \dot{x}_d - \dot{x} \tag{3.64}$$

which, according to differential kinematics (3.58), can be written as

$$\dot{e} = \dot{x}_d - J_A(q)\dot{q}. \tag{3.65}$$

It is worth noticing that the use of operational space quantities has naturally lead to using the analytical Jacobian in lieu of the geometric Jacobian in (3.65). For this equation to lead to an *inverse kinematics algorithm*, it is worth relating the computed joint velocity vector $\dot{q}$ to the error e so that (3.65) gives a differential equation describing error evolution over time. Nonetheless, it is necessary to choose a relationship between $\dot{q}$ and e that ensures convergence of the error to zero.

Having formulated the inverse kinematics problem in an algorithmic fashion implies that the joint variables q corresponding to the assigned end-effector posture x_d are accurately obtained only when the error e is below a given tolerated threshold; the settling time depends on the dynamic features of the error differential equation. The choice of the relationship between $\dot{q}$ and e allows finding inverse kinematics algorithms with different performance.

3.7.1 Jacobian (Pseudo-)Inverse

On the assumption that matrix J_A is square and nonsingular, the choice

$$\dot{q} = J_A^{-1}(q)(\dot{x}_d + Ke) \tag{3.66}$$

leads to the equivalent linear system

$$\dot{e} + Ke = 0. \tag{3.67}$$

If K is a positive definite (usually diagonal) matrix, the system (3.67) is *asymptotically stable*. The error tends to zero along the trajectory with a convergence rate that depends on the eigenvalues of matrix K; the larger the eigenvalues, the faster the convergence. Since the scheme is practically implemented as a discrete-time system, it is reasonable to predict that an upper bound exists on the eigenvalues; depending on the sampling time, there will be a limit for the maximum eigenvalue of K under which asymptotic stability of the error system is guaranteed.

The block scheme corresponding to the inverse kinematics algorithm in (3.66) is illustrated in Figure 3.11, where $k(\cdot)$ indicates the direct kinematics function in (2.70). This scheme can be revisited in terms of the usual feedback control schemes. Specifically, it can observed that the nonlinear block $k(\cdot)$ is needed to compute x and thus the tracking error e, while the block $J_A^{-1}(q)$ has been introduced to compensate for $J_A(q)$

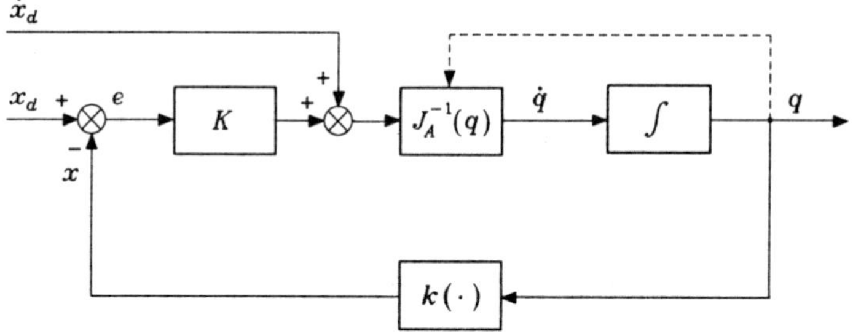

Figure 3.11 Block scheme of the inverse kinematics algorithm with Jacobian inverse.

and making the system linear. The block scheme shows the presence of a string of integrators on the forward loop and then, for a constant reference ($\dot{\boldsymbol{x}}_d = \mathbf{0}$), guarantees a null steady-state error. Further, the feedforward action provided by $\dot{\boldsymbol{x}}_d$ for a time-varying reference ensures that the error is kept to zero (in the case $\boldsymbol{e}(0) = \mathbf{0}$) along the whole trajectory, independently of the type of desired reference $\boldsymbol{x}_d(t)$. Notice, too, that (3.66), for $\dot{\boldsymbol{x}}_d = \mathbf{0}$, corresponds to the Newton method for solving a system of nonlinear equations.

In the case of a *redundant manipulator*, solution (3.66) can be generalized into

$$\dot{\boldsymbol{q}} = \boldsymbol{J}_A^{\dagger}(\dot{\boldsymbol{x}}_d + \boldsymbol{K}\boldsymbol{e}) + (\boldsymbol{I} - \boldsymbol{J}_A^{\dagger}\boldsymbol{J}_A)\dot{\boldsymbol{q}}_0 \tag{3.68}$$

which represents the algorithmic version of solution (3.50).

3.7.2 Jacobian Transpose

A computationally simpler algorithm can be derived by finding a relationship between $\dot{\boldsymbol{q}}$ and $\boldsymbol{e}$ that ensures error convergence to zero, without requiring linearization of (3.65). As a consequence, the error dynamics is governed by a nonlinear differential equation. The Lyapunov direct method can be utilized to determine a dependence $\dot{\boldsymbol{q}}(\boldsymbol{e})$ that ensures asymptotic stability of the error system. Choose as Lyapunov function candidate the positive definite quadratic form

$$V(\boldsymbol{e}) = \frac{1}{2}\boldsymbol{e}^T\boldsymbol{K}\boldsymbol{e}, \tag{3.69}$$

where $\boldsymbol{K}$ is a symmetric positive definite matrix. This function is so that

$$V(\boldsymbol{e}) > 0 \quad \forall \boldsymbol{e} \neq \mathbf{0}, \qquad V(\mathbf{0}) = 0.$$

Differentiating (3.69) with respect to time and accounting for (3.64) gives

$$\dot{V} = \boldsymbol{e}^T\boldsymbol{K}\dot{\boldsymbol{x}}_d - \boldsymbol{e}^T\boldsymbol{K}\dot{\boldsymbol{x}}. \tag{3.70}$$

In view of (3.58), it is

$$\dot{V} = \boldsymbol{e}^T\boldsymbol{K}\dot{\boldsymbol{x}}_d - \boldsymbol{e}^T\boldsymbol{K}\boldsymbol{J}_A(\boldsymbol{q})\dot{\boldsymbol{q}}. \tag{3.71}$$

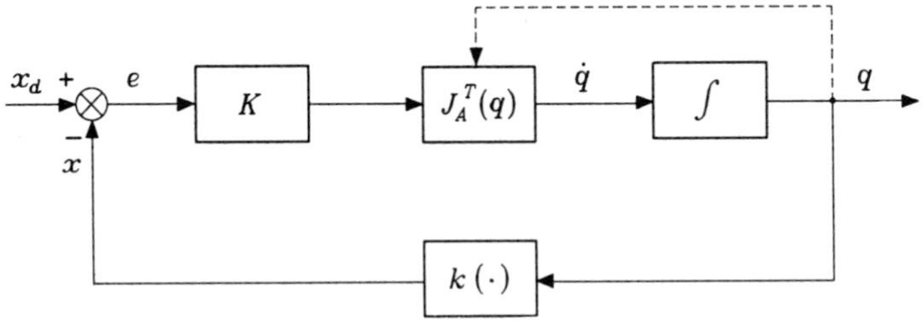

Figure 3.12 Block scheme of the inverse kinematics algorithm with Jacobian transpose.

At this point, the choice of joint velocities as

$$\dot{\boldsymbol{q}} = \boldsymbol{J}_A^T(\boldsymbol{q})\boldsymbol{K}\boldsymbol{e} \tag{3.72}$$

leads to

$$\dot{V} = \boldsymbol{e}^T\boldsymbol{K}\dot{\boldsymbol{x}}_d - \boldsymbol{e}^T\boldsymbol{K}\boldsymbol{J}_A(\boldsymbol{q})\boldsymbol{J}_A^T(\boldsymbol{q})\boldsymbol{K}\boldsymbol{e}. \tag{3.73}$$

Consider the case of a constant reference ($\dot{\boldsymbol{x}}_d = \boldsymbol{0}$). The function in (3.73) is negative definite, on the assumption of full rank for $\boldsymbol{J}_A(\boldsymbol{q})$. The condition $\dot{V} < 0$ with $V > 0$ implies that the system trajectories uniformly converge to $\boldsymbol{e} = \boldsymbol{0}$, *i.e.*, the system is *asymptotically stable*. When $\mathcal{N}(\boldsymbol{J}_A^T) \neq \emptyset$, the function in (3.73) is only negative semi-definite, since $\dot{V} = 0$ for $\boldsymbol{e} \neq \boldsymbol{0}$ with $\boldsymbol{K}\boldsymbol{e} \in \mathcal{N}(\boldsymbol{J}_A^T)$. In this case, the algorithm can get stuck at $\dot{\boldsymbol{q}} = \boldsymbol{0}$ with $\boldsymbol{e} \neq \boldsymbol{0}$. However, the example that follows will show that this situation occurs only if the assigned end-effector position is not actually reachable from the current configuration.

The resulting block scheme is illustrated in Figure 3.12, which shows the notable feature of the algorithm to require *computation only of direct kinematics functions* $\boldsymbol{k}(\boldsymbol{q})$, $\boldsymbol{J}_A^T(\boldsymbol{q})$. It can be recognized that (3.72) corresponds to the gradient method for the solution of a system on nonlinear equations.

The case when $\boldsymbol{x}_d$ is a time-varying function ($\dot{\boldsymbol{x}}_d \neq \boldsymbol{0}$) deserves a separate analysis. In order to obtain $\dot{V} < 0$ also in this case, it would be sufficient to choose a $\dot{\boldsymbol{q}}$ that depends on the (pseudo-)inverse of the Jacobian as in (3.66), recovering the asymptotic stability result derived above[5]. For the inversion scheme based on the transpose, the first term on the right-hand side of (3.73) is not canceled any more and nothing can be said about its sign. This implies that asymptotic stability along the trajectory cannot be achieved. The tracking error $\boldsymbol{e}(t)$ is, anyhow, norm-bounded; the larger the norm of $\boldsymbol{K}$, the smaller the norm of $\boldsymbol{e}$[6]. In practice, since the inversion scheme is to be implemented

[5] Notice that, anyhow, in case of kinematic singularities, it is necessary to resort to an inverse kinematics scheme that does not require inversion of the Jacobian.

[6] Notice that the negative definite term is a quadratic function of the error, while the other term is a linear function of the error. Therefore, for an error of very small norm, the linear term prevails over the quadratic term, and the norm of $\boldsymbol{K}$ shall be increased to reduce the norm of $\boldsymbol{e}$ as much as possible.

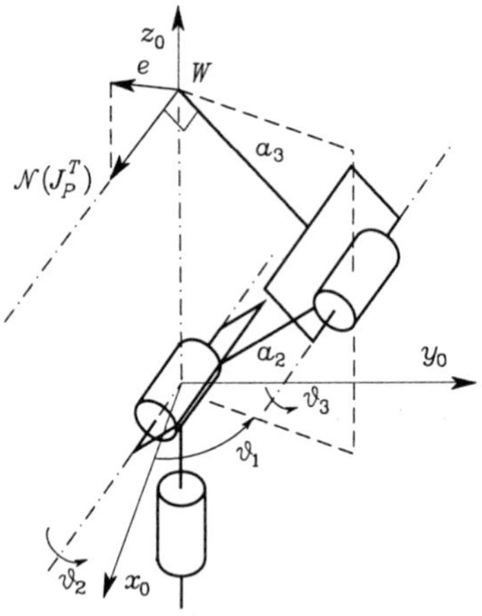

Figure 3.13 Characterization of the anthropomorphic arm at a shoulder singularity for the admissible solutions of the Jacobian transpose algorithm.

in discrete-time, there is an upper bound on the norm of $\boldsymbol{K}$ with reference to the adopted sampling time.

Example 3.4

Consider the anthropomorphic arm; a shoulder singularity occurs whenever $a_2c_2 + a_3c_{23} = 0$ (Figure 3.6). In this configuration, the transpose of the Jacobian in (3.34) is

$$\boldsymbol{J}_P^T = \begin{bmatrix} 0 & 0 & 0 \\ -c_1(a_2s_2 + a_3s_{23}) & -s_1(a_2s_2 + a_3s_{23}) & 0 \\ -a_3c_1s_{23} & -a_3s_1s_{23} & a_3c_{23} \end{bmatrix}.$$

By computing the null space of $\boldsymbol{J}_P^T$, if ν_x, ν_y and ν_z denote the components of vector $\boldsymbol{\nu}$ along the axes of the base frame, one has the result

$$\frac{\nu_y}{\nu_x} = -\frac{1}{\tan\vartheta_1} \qquad \nu_z = 0,$$

implying that the direction of $\mathcal{N}(\boldsymbol{J}_P^T)$ coincides with the direction orthogonal to the plane of the structure (Figure 3.13). The Jacobian transpose algorithm gets stuck if, with $\boldsymbol{K}$ diagonal and having all equal elements, the desired position is along the line normal to the plane of the structure at the intersection with the wrist point. On the other hand, the end effector cannot physically move from the singular configuration along such line. Instead, if the prescribed path has a nonnull component in the plane of the structure at the singularity, algorithm convergence is ensured, since in that case $\boldsymbol{Ke} \notin \mathcal{N}(\boldsymbol{J}_P^T)$.

In sum, the algorithm based on the computation of the Jacobian transpose provides a computationally efficient inverse kinematics method that can be utilized also for paths crossing kinematic singularities.

3.7.3 Orientation Error

The inverse kinematics algorithms presented in the above sections utilize the analytical Jacobian since they operate on error variables (position and orientation) that are defined in the operational space.

For what concerns the position error, it is obvious that its expression is given by

$$e_P = p_d - p(q) \tag{3.74}$$

where p_d and p denote respectively the desired and computed end-effector positions. Further, its time derivative is

$$\dot{e}_P = \dot{p}_d - \dot{p}. \tag{3.75}$$

On the other hand, for what concerns the *orientation error*, its expression depends on the particular representation of end-effector orientation; namely, Euler angles, angle and axis, and unit quaternion.

Euler Angles

The orientation error is chosen according to an expression formally analogous to (3.74), *i.e.*,

$$e_O = \phi_d - \phi(q) \tag{3.76}$$

where ϕ_d and ϕ denote respectively the desired and computed set of Euler angles. Further, its time derivative is

$$\dot{e}_O = \dot{\phi}_d - \dot{\phi}. \tag{3.77}$$

Therefore, assuming the neither kinematic nor representation singularities occur, the Jacobian inverse solution for a nonredundant manipulator is derived from (3.66), *i.e.*,

$$\dot{q} = J_A^{-1}(q) \begin{bmatrix} \dot{p}_d + K_P e_P \\ \dot{\phi}_d + K_O e_O \end{bmatrix} \tag{3.78}$$

where K_P and K_O are positive definite matrices.

As already pointed out in Section 2.10 for computation of the direct kinematics function in the form (2.70), the determination of the orientation variables from the joint variables is not easy except for simple cases (see Example 2.5). To this purpose, it is worth recalling that computation of the angles ϕ, in a minimal representation of orientation, requires computation of the rotation matrix $R = [\,n \quad s \quad a\,]$; in fact, only the dependence of R on q is known in closed form, but not that of ϕ on q. Further, the use of inverse functions (Atan2) in (2.19) and (2.21) involves a nonnegligible complexity in the computation of the analytical Jacobian, and the occurrence of representation singularities constitutes another drawback for the orientation error based on Euler angles.

Different kinds of remarks are to be made about the way to assign a time profile for the reference variables ϕ_d chosen to represent end-effector orientation. The most intuitive way to specify end-effector orientation is to refer to the orientation of the end-effector frame $(n_d,\, s_d,\, a_d)$ with respect to the base frame. Given the limitations pointed out in Section 2.10 about guaranteeing orthonormality of the unit vectors along time, it is necessary first to compute the Euler angles corresponding to the initial and final orientation of the end-effector frame via (2.19) or (2.21); only then a time evolution can be generated. Such solutions will be presented in Chapter 5.

A radical simplification of the problem at issue can be obtained for manipulators having a spherical wrist. Section 2.12.2 pointed out the possibility to solve the inverse kinematics problem for the position part separately from that for the orientation part. This result has an impact also at algorithmic level. In fact, the implementation of an inverse kinematics algorithm for determining the joint variables influencing the wrist position allows computing the time evolution of the wrist frame $\boldsymbol{R}_W(t)$. Hence, once the desired time evolution of the end-effector frame $\boldsymbol{R}_d(t)$ is given, it is sufficient to compute the Euler angles ZYZ from the matrix $\boldsymbol{R}_W^T\boldsymbol{R}_d$ by applying (2.19). As shown in Section 2.12.5, these angles are directly the joint variables of the spherical wrist.

The above considerations show that the inverse kinematics algorithms based on the analytical Jacobian are effective for kinematic structures having a spherical wrist which are of significant interest. For manipulator structures which cannot be reduced to that class, it may be appropriate to reformulate the inverse kinematics problem on the basis of a different definition of the orientation error.

Angle and Axis

If $\boldsymbol{R}_d = [\,\boldsymbol{n}_d \quad \boldsymbol{s}_d \quad \boldsymbol{a}_d\,]$ denotes the desired rotation matrix of the end-effector frame and $\boldsymbol{R} = [\,\boldsymbol{n} \quad \boldsymbol{s} \quad \boldsymbol{a}\,]$ the rotation matrix that can be computed from the joint variables, the orientation error between the two frames can be expressed as

$$\boldsymbol{e}_O = \sin\vartheta\,\boldsymbol{r} \tag{3.79}$$

where ϑ and $\boldsymbol{r}$ identify the *angle and axis* of the equivalent rotation that can be deduced from the matrix

$$\boldsymbol{R}(\vartheta,\boldsymbol{r}) = \boldsymbol{R}_d\boldsymbol{R}^T(\boldsymbol{q}), \tag{3.80}$$

describing the rotation needed to align $\boldsymbol{R}$ with $\boldsymbol{R}_d$. Notice that (3.79) gives a unique relationship for $-\pi/2 < \vartheta < \pi/2$. The angle ϑ represents the magnitude of an orientation error, and thus the above limitation is not restrictive since the tracking error is typically small for an inverse kinematics algorithm.

By comparing the off-diagonal terms of the expression of $\boldsymbol{R}(\vartheta,\boldsymbol{r})$ in (2.23) with the corresponding terms resulting on the right-hand side of (3.80), it can be found that a functional expression of the orientation error in (3.79) is

$$\boldsymbol{e}_O = \frac{1}{2}(\boldsymbol{n}(\boldsymbol{q}) \times \boldsymbol{n}_d + \boldsymbol{s}(\boldsymbol{q}) \times \boldsymbol{s}_d + \boldsymbol{a}(\boldsymbol{q}) \times \boldsymbol{a}_d); \tag{3.81}$$

the limitation on ϑ is transformed in the condition $\boldsymbol{n}^T\boldsymbol{n}_d \geq 0$, $\boldsymbol{s}^T\boldsymbol{s}_d \geq 0$, $\boldsymbol{a}^T\boldsymbol{a}_d \geq 0$.

Differentiating (3.81) with respect to time and accounting for the expression of the columns of the derivative of a rotation matrix in (3.7) gives

$$\dot{\boldsymbol{e}}_O = \boldsymbol{L}^T\boldsymbol{\omega}_d - \boldsymbol{L}\boldsymbol{\omega} \tag{3.82}$$

where

$$\boldsymbol{L} = -\frac{1}{2}\big(\boldsymbol{S}(\boldsymbol{n}_d)\boldsymbol{S}(\boldsymbol{n}) + \boldsymbol{S}(\boldsymbol{s}_d)\boldsymbol{S}(\boldsymbol{s}) + \boldsymbol{S}(\boldsymbol{a}_d)\boldsymbol{S}(\boldsymbol{a})\big). \tag{3.83}$$

At this point, by exploiting the relations (3.1) and (3.2) of the geometric Jacobian expressing $\dot{\boldsymbol{p}}$ and $\boldsymbol{\omega}$ as a function of $\dot{\boldsymbol{q}}$, (3.75) and (3.82) become

$$\dot{\boldsymbol{e}} = \begin{bmatrix} \dot{\boldsymbol{e}}_P \\ \dot{\boldsymbol{e}}_O \end{bmatrix} = \begin{bmatrix} \dot{\boldsymbol{p}}_d - \boldsymbol{J}_P(\boldsymbol{q})\dot{\boldsymbol{q}} \\ \boldsymbol{L}^T\boldsymbol{\omega}_d - \boldsymbol{L}\boldsymbol{J}_O(\boldsymbol{q})\dot{\boldsymbol{q}} \end{bmatrix} = \begin{bmatrix} \dot{\boldsymbol{p}}_d \\ \boldsymbol{L}^T\boldsymbol{\omega}_d \end{bmatrix} - \begin{bmatrix} \boldsymbol{I} & \boldsymbol{O} \\ \boldsymbol{O} & \boldsymbol{L} \end{bmatrix} \boldsymbol{J}\dot{\boldsymbol{q}}. \tag{3.84}$$

The expression in (3.84) suggests the possibility of devising inverse kinematics algorithms analogous to the ones derived above, but using the geometric Jacobian in place of the analytical Jacobian. For instance, the Jacobian inverse solution for a nonredundant nonsingular manipulator is

$$\dot{\boldsymbol{q}} = \boldsymbol{J}^{-1}(\boldsymbol{q}) \begin{bmatrix} \dot{\boldsymbol{p}}_d + \boldsymbol{K}_P\boldsymbol{e}_P \\ \boldsymbol{L}^{-1}\left(\boldsymbol{L}^T\boldsymbol{\omega}_d + \boldsymbol{K}_O\boldsymbol{e}_O\right) \end{bmatrix}. \tag{3.85}$$

It is worth remarking that the inverse kinematics solution based on (3.85) is expected to perform better than the solution based on (3.78) since it uses the geometric Jacobian in lieu of the analytical Jacobian, thus avoiding the occurrence of representation singularities.

Unit Quaternion

In order to devise an inverse kinematics algorithm based on the *unit quaternion*, a suitable orientation error shall be defined. Let $\mathcal{Q}_d = \{\eta_d, \boldsymbol{\epsilon}_d\}$ and $\mathcal{Q} = \{\eta, \boldsymbol{\epsilon}\}$ represent the quaternions associated with $\boldsymbol{R}_d$ and $\boldsymbol{R}$, respectively. The orientation error can be described by the rotation matrix $\boldsymbol{R}_d\boldsymbol{R}^T$ and, in view of (2.32), can be expressed in terms of the quaternion $\Delta\mathcal{Q} = \{\Delta\eta, \Delta\boldsymbol{\epsilon}\}$ where

$$\Delta\mathcal{Q} = \mathcal{Q}_d * \mathcal{Q}^{-1}. \tag{3.86}$$

It can be recognized that $\Delta\mathcal{Q} = \{1, \boldsymbol{0}\}$ if and only if $\boldsymbol{R}$ and $\boldsymbol{R}_d$ are aligned. Hence, it is sufficient to define the orientation error as

$$\boldsymbol{e}_O = \Delta\boldsymbol{\epsilon} = \eta(\boldsymbol{q})\boldsymbol{\epsilon}_d - \eta_d\boldsymbol{\epsilon}(\boldsymbol{q}) - \boldsymbol{S}(\boldsymbol{\epsilon}_d)\boldsymbol{\epsilon}(\boldsymbol{q}), \tag{3.87}$$

where the skew-symmetric operator $\boldsymbol{S}(\cdot)$ has been used. Notice, however, that the explicit computation of η and $\boldsymbol{\epsilon}$ from the joint variables is not possible but it requires the intermediate computation of the rotation matrix $\boldsymbol{R}$ that is available from the manipulator direct kinematics; then, the quaternion can be extracted using (2.30).

At this point, a Jacobian inverse solution can be computed as

$$\dot{\boldsymbol{q}} = \boldsymbol{J}^{-1}(\boldsymbol{q}) \begin{bmatrix} \dot{\boldsymbol{p}}_d + \boldsymbol{K}_P\boldsymbol{e}_P \\ \boldsymbol{\omega}_d + \boldsymbol{K}_O\boldsymbol{e}_O \end{bmatrix} \tag{3.88}$$

where remarkably the geometric Jacobian has been used. Substituting (3.88) into (3.3) gives (3.75) and

$$\boldsymbol{\omega}_d - \boldsymbol{\omega} + \boldsymbol{K}_O\boldsymbol{e}_O = \boldsymbol{0}. \tag{3.89}$$

It should be observed that now the orientation error equation is nonlinear in $\boldsymbol{e}_O$ since it contains the end-effector angular velocity error instead of the time derivative of the orientation error. To this purpose, it is worth considering the relationship between the time derivative of the quaternion $\mathcal{Q}$ and the angular velocity $\boldsymbol{\omega}$. This can be found to be

$$\begin{aligned} \dot{\eta} &= -\frac{1}{2}\boldsymbol{\epsilon}^T\boldsymbol{\omega} \\ \dot{\boldsymbol{\epsilon}} &= \frac{1}{2}\left(\eta\boldsymbol{I} - \boldsymbol{S}(\boldsymbol{\epsilon})\right)\boldsymbol{\omega} \end{aligned} \tag{3.90}$$

which is the so-called *quaternion propagation*. A similar relationship holds between the time derivative of $\mathcal{Q}_d$ and $\boldsymbol{\omega}_d$.

To study stability of system (3.89), consider the positive definite Lyapunov function candidate

$$V = (\eta_d - \eta)^2 + (\boldsymbol{\epsilon}_d - \boldsymbol{\epsilon})^T(\boldsymbol{\epsilon}_d - \boldsymbol{\epsilon}). \tag{3.91}$$

In view of (3.90), differentiating (3.91) with respect to time and accounting for (3.89) yields

$$\dot{V} = -\boldsymbol{e}_O^T\boldsymbol{K}_O\boldsymbol{e}_O \tag{3.92}$$

which is negative definite, implying that $\boldsymbol{e}_O$ converges to zero.

In sum, the inverse kinematics solution based on (3.88) uses the geometric Jacobian as the solution based on (3.85) but is computationally lighter.

3.7.4 A Comparison Between Inverse Kinematics Algorithms

In order to make a comparison of performance between the inverse kinematics algorithms presented above, consider the three-link planar arm in Figure 2.20 whose link lengths are $a_1 = a_2 = a_3 = 0.5\,\text{m}$. The direct kinematics for this arm is given by (2.71), while its Jacobian can be found from (3.31) by considering the three nonnull rows of interest for the operational space.

Let the arm be at the initial posture $\boldsymbol{q} = [\pi \quad -\pi/2 \quad -\pi/2]^T$ rad, corresponding to the end-effector location: $\boldsymbol{p} = [0 \quad 0.5]^T$ m, $\phi = 0\,\text{rad}$. A circular path of radius $0.25\,\text{m}$ and centre at $(0.25, 0.5)\,\text{m}$ is assigned to the end effector. Let the motion trajectory be

$$\boldsymbol{p}_d(t) = \begin{bmatrix} 0.25(1 - \cos\pi t) \\ 0.25(2 + \sin\pi t) \end{bmatrix} \qquad 0 \le t \le 4,$$

i.e., the end effector shall make two complete circles in a time of 2 s per circle. As regards end-effector orientation, initially it is required to follow the trajectory

$$\phi_d(t) = \sin\frac{\pi}{24}t \qquad 0 \le t \le 4,$$

i.e., the end effector shall attain a different orientation ($\phi_d = 0.5\,\text{rad}$) at the end of the two circles.

The inverse kinematics algorithms were implemented on a computer by adopting the Euler numerical integration scheme (3.44) with an integration time $\Delta t = 1\,\text{ms}$.

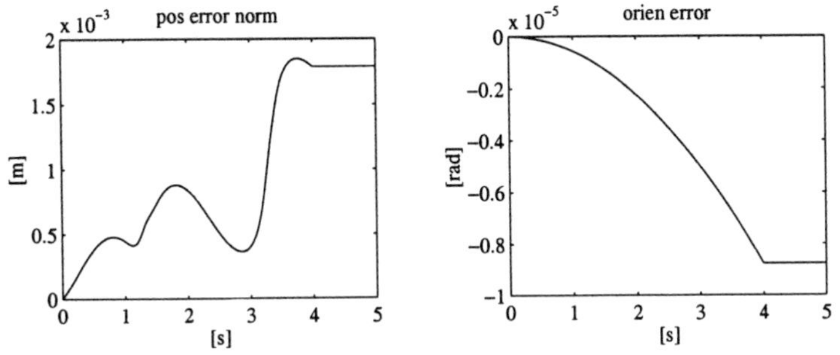

Figure 3.14 Time history of the norm of end-effector position error and orientation error with the open-loop inverse Jacobian algorithm.

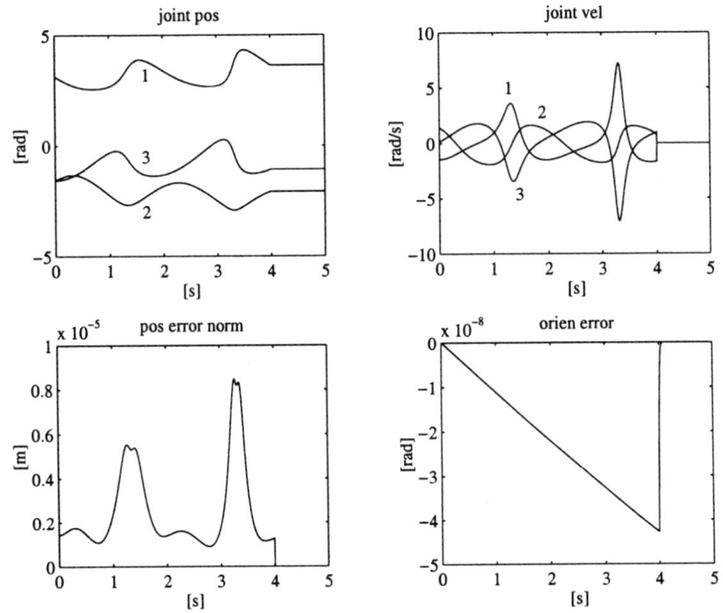

Figure 3.15 Time history of the joint positions and velocities, and of the norm of end-effector position error and orientation error with the closed-loop inverse Jacobian algorithm.

At first, the inverse kinematics along the given trajectory has been performed by using (3.43). The obtained results in Figure 3.14 show that the norm of the position error along the whole trajectory is bounded; at steady state, after $t = 4$, the error sets to a constant value in view of the typical *drift* of *open-loop* schemes. A similar drift can be observed for the orientation error.

Next, the inverse kinematics algorithm based on (3.66) using the Jacobian *inverse* has been used, with the matrix gain $\boldsymbol{K} = \text{diag}\{500, 500, 100\}$. The resulting joint positions and velocities as well as the tracking errors are shown in Figure 3.15. The

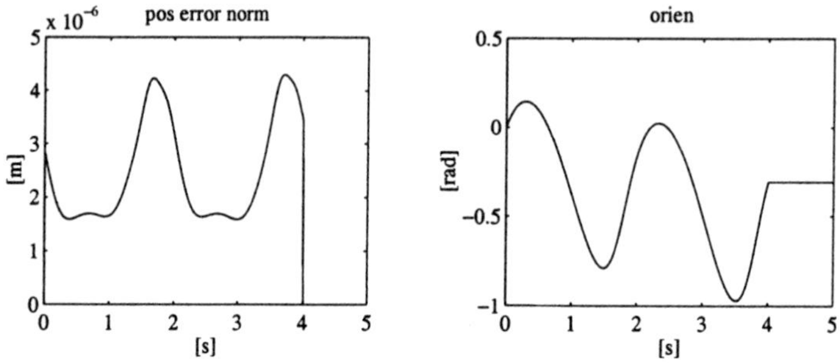

Figure 3.16 Time history of the norm of end-effector position error and orientation with the Jacobian pseudo-inverse algorithm.

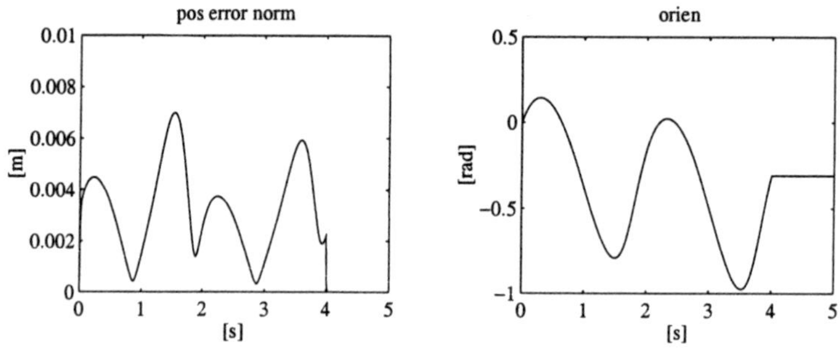

Figure 3.17 Time history of the norm of end-effector position error and orientation with the Jacobian transpose algorithm.

norm of the position error is radically decreased and converges to zero at steady state, thanks to the *closed-loop* feature of the scheme; the orientation error, too, is decreased and tends to zero at steady state.

On the other hand, if the end-effector orientation is not constrained, the operational space becomes two-dimensional and is characterized by the first two rows of the direct kinematics in (2.71) as well as by the Jacobian in (3.32); a *redundant* degree of mobility is then available. Hence, the inverse kinematics algorithm based on (3.68) using the Jacobian *pseudo-inverse* has been used with $\boldsymbol{K} = \text{diag}\{500, 500\}$. If redundancy is not exploited ($\dot{\boldsymbol{q}}_0 = \boldsymbol{0}$), the results in Figure 3.16 reveal that position tracking remains satisfactory and, of course, the end-effector orientation freely varies along the given trajectory.

With reference to the previous situation, the use of the Jacobian *transpose* algorithm based on (3.72) with $\boldsymbol{K} = \text{diag}\{500, 500\}$ gives rise to a tracking error (Figure 3.17) which is anyhow bounded and rapidly tends to zero at steady state.

In order to show the capability of handling the degree of redundancy, the algorithm based on (3.68) with $\dot{\boldsymbol{q}}_0 \neq \boldsymbol{0}$ has been used; two types of constraints have been

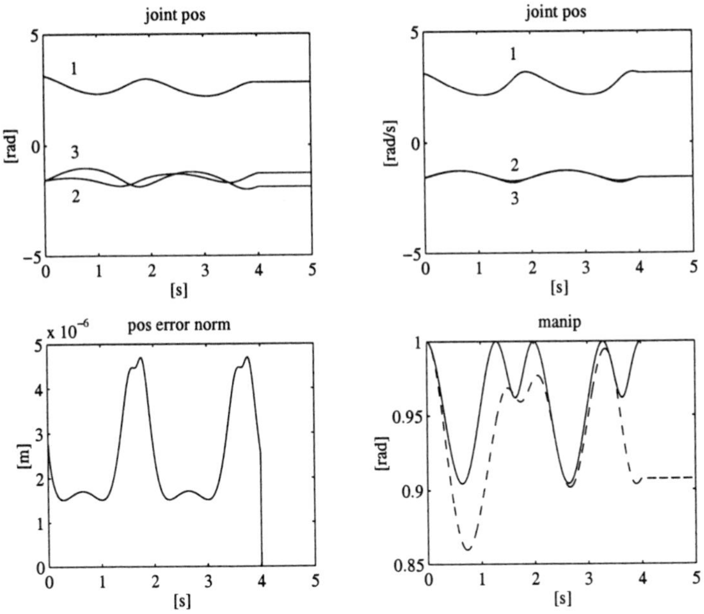

Figure 3.18 Time history of the joint positions, the norm of end-effector position error, and the manipulability measure with the Jacobian pseudo-inverse algorithm and manipulability constraint; *upper left*—with the unconstrained solution, *upper right*—with the constrained solution.

considered concerning an objective function to locally maximize. The first function is

$$w(\vartheta_2, \vartheta_3) = \frac{1}{2}(s_2^2 + s_3^2)$$

that provides a *manipulability measure*. Notice that such function is computationally simpler than the function in (3.52), but it still describes a distance from kinematic singularities in an effective way. The gain in (3.51) has been set to $k_0 = 50$. In Figure 3.18, the joint trajectories are reported for the two cases with and without ($k_0 = 0$) constraint. The addition of the constraint leads to having coincident trajectories for Joints 2 and 3. The manipulability measure in the constrained case (*continuous* line) attains larger values along the trajectory compared to the unconstrained case (*dashed* line). It is worth underlining that the tracking position error is practically the same in the two cases (Figure 3.16), since the additional joint velocity contribution is projected in the null space of the Jacobian so as not to alter the performance of the end-effector position task.

Finally, it is worth noticing that in the constrained case the resulting joint trajectories are *cyclic*, *i.e.*, they take on the same values after a period of the circular path. This does not happen for the unconstrained case, since the internal motion of the structure causes the arm to be in a different posture after one circle.

The second objective function considered is the *distance from mechanical joint limits* in (3.53). Specifically, it is assumed what follows: the first joint does not have

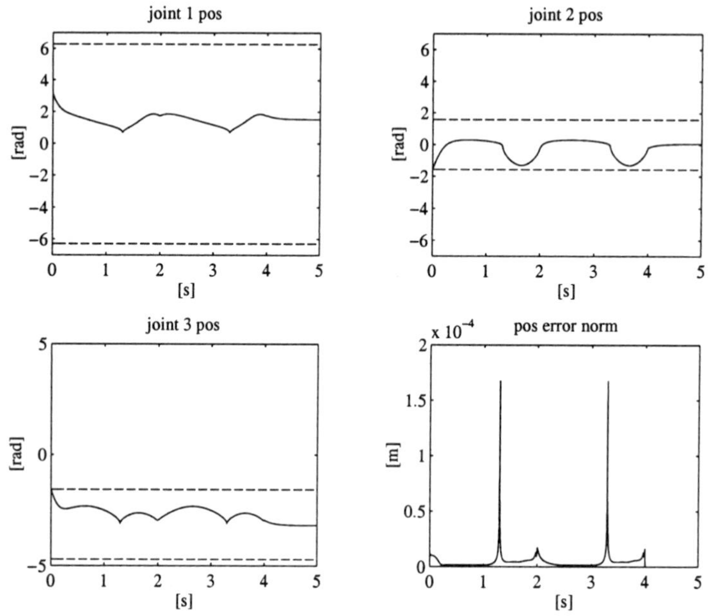

Figure 3.19 Time history of the joint positions and the norm of end-effector position error with the Jacobian pseudo-inverse algorithm and joint limit constraint (joint limits are denoted by dashed lines).

limits ($q_{1m} = -2\pi, q_{1M} = 2\pi$), the second joint has limits $q_{2m} = -\pi/2, q_{2M} = \pi/2$, and the third joint has limits $q_{3m} = -3\pi/2, q_{3M} = -\pi/2$. It is not difficult to verify that, in the unconstrained case, the trajectories of Joints 2 and 3 violate the respective limits. The gain in (3.51) has been set to $k_0 = 250$. The results in Figure 3.19 show the effectiveness of the technique with utilization of redundancy, since both Joints 2 and 3 tend to invert their motion—with respect to the unconstrained trajectories in Figure 3.18—and keep far from the minimum limit for Joint 2 and the maximum limit for Joint 3, respectively. Such an effort does not appreciably affect the position tracking error, whose norm is bounded anyhow within acceptable values.

3.8 Statics

The goal of *statics* is to determine the relationship between the generalized forces applied to the end effector and the generalized forces applied to the joints—forces for prismatic joints, torques for revolute joints—with the manipulator at an equilibrium configuration.

Let $\boldsymbol{\tau}$ denote the $(n \times 1)$ vector of joint torques and $\boldsymbol{\gamma}$ the $(r \times 1)$ vector of end-effector forces[7] where r is the dimension of the operational space of interest.

The application of the *principle of virtual work* allows determination of the required relationship. The mechanical manipulators considered are systems with time-invariant, holonomic constraints, and thus their configurations depend only on the joint variables $\boldsymbol{q}$ and not explicitly on time. This implies that virtual displacements coincide with elementary displacements.

Consider the elementary works performed by the two force systems. As for the joint torques, the elementary work associated with them is

$$dW_\tau = \boldsymbol{\tau}^T d\boldsymbol{q}. \tag{3.92}$$

As for the end-effector forces γ, if the force contributions $\boldsymbol{f}$ are separated by the moment contributions $\boldsymbol{\mu}$, the elementary work associated with them is

$$dW_\gamma = \boldsymbol{f}^T d\boldsymbol{p} + \boldsymbol{\mu}^T \boldsymbol{\omega} dt, \tag{3.94}$$

where $d\boldsymbol{p}$ is the linear displacement and $\boldsymbol{\omega} dt$ is the angular displacement[8].

By accounting for the differential kinematics relationship in (3.3) and (3.4), (3.94) can be rewritten as

$$\begin{aligned} dW_\gamma &= \boldsymbol{f}^T \boldsymbol{J}_P(\boldsymbol{q}) d\boldsymbol{q} + \boldsymbol{\mu}^T \boldsymbol{J}_O(\boldsymbol{q}) d\boldsymbol{q} \\ &= \gamma^T \boldsymbol{J}(\boldsymbol{q}) d\boldsymbol{q} \end{aligned} \tag{3.95}$$

where $\gamma = [\,\boldsymbol{f}^T \quad \boldsymbol{\mu}^T\,]^T$. Since virtual and elementary displacements coincide, the virtual works associated with the two force systems are

$$\delta W_\tau = \boldsymbol{\tau}^T \delta \boldsymbol{q} \tag{3.96}$$

$$\delta W_\gamma = \gamma^T \boldsymbol{J}(\boldsymbol{q}) \delta \boldsymbol{q}, \tag{3.97}$$

where δ is the usual symbol to indicate virtual quantities.

According to the principle of virtual work, the manipulator is at *static equilibrium* if and only if

$$\delta W_\tau - \delta W_\gamma = 0 \qquad \forall \delta \boldsymbol{q}, \tag{3.98}$$

i.e., the difference between the virtual work of the joint torques and the virtual work of the end-effector forces shall be null for all joint displacements.

From (3.97), notice that the virtual work of the end-effector forces is null for any displacement in the null space of $\boldsymbol{J}$. This implies that the joint torques associated with such displacements must be null at static equilibrium. In that case, substituting (3.96) and (3.97) in (3.98) leads to the notable result

$$\boldsymbol{\tau} = \boldsymbol{J}^T(\boldsymbol{q}) \gamma, \tag{3.99}$$

[7] Hereafter, generalized forces at the joints are often called *torques*, while generalized forces at the end effector are often called *forces*.

[8] The angular displacement has been indicated by $\boldsymbol{\omega} dt$ in view of the problems of integrability of $\boldsymbol{\omega}$ discussed in Section 3.6.

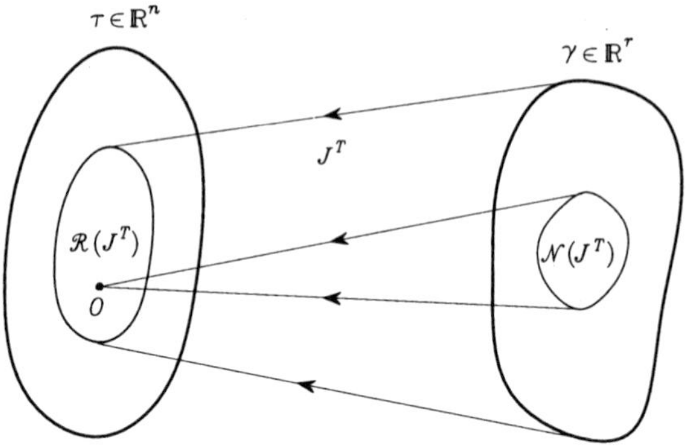

Figure 3.20 Mapping between the end-effector force space and the joint torque space.

stating that the relationship between the end-effector forces and the joint torques is established by the transpose of the manipulator geometric Jacobian.

3.8.1 Kineto-statics Duality

The statics relationship in (3.99), combined with the differential kinematics equation in (3.41), points out a property of *kineto-statics duality*. In fact, by adopting a representation similar to that of Figure 3.7 for differential kinematics, one has that (Figure 3.20):

- The range of $\boldsymbol{J}^T$ is the subspace $\mathcal{R}(\boldsymbol{J}^T)$ in $\mathbb{R}^n$ of the joint torques that can balance the end-effector forces, in the given manipulator posture.
- The null of $\boldsymbol{J}^T$ is the subspace $\mathcal{N}(\boldsymbol{J}^T)$ in $\mathbb{R}^r$ of the end-effector forces that do not require any balancing joint torques, in the given manipulator posture.

It is worth remarking that the end-effector forces $\boldsymbol{\gamma} \in \mathcal{N}(\boldsymbol{J}^T)$ are entirely absorbed by the structure in that the mechanical constraint reaction forces can balance them exactly. Hence, a manipulator at a singular configuration remains in the given posture whatever end-effector force $\boldsymbol{\gamma}$ is applied so that $\boldsymbol{\gamma} \in \mathcal{N}(\boldsymbol{J}^T)$.

The relations between the two subspaces are established by:

$$\mathcal{N}(\boldsymbol{J}) \equiv \mathcal{R}^{\perp}(\boldsymbol{J}^T) \qquad\qquad \mathcal{R}(\boldsymbol{J}) \equiv \mathcal{N}^{\perp}(\boldsymbol{J}^T)$$

and then, once the manipulator Jacobian is known, it is possible to completely characterize differential kinematics and statics in terms of the range and null spaces of the Jacobian and its transpose.

On the basis of the above duality, the inverse kinematics scheme with the Jacobian transpose in Figure 3.12 admits an interesting physical interpretation. Consider a manipulator with ideal dynamics $\boldsymbol{\tau} = \dot{\boldsymbol{q}}$ (null masses and unit viscous friction coefficients); the algorithm update law $\dot{\boldsymbol{q}} = \boldsymbol{J}^T\boldsymbol{K}\boldsymbol{e}$ plays the role of a generalized spring of stiffness constant $\boldsymbol{K}$ generating a force $\boldsymbol{K}\boldsymbol{e}$ that pulls the end effector towards the desired posture in the operational space. If this manipulator is allowed to move,

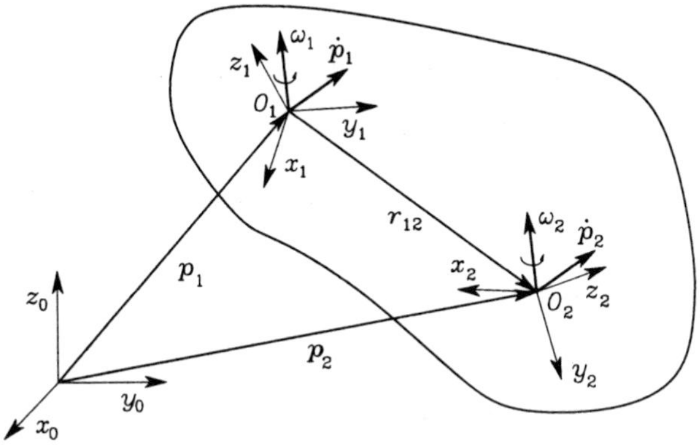

Figure 3.21 Representation of linear and angular velocities in different coordinate frames on the same rigid body.

e.g., in the case $\boldsymbol{K}\boldsymbol{e} \notin \mathcal{N}(\boldsymbol{J}^T)$, the end effector attains the desired posture and the corresponding joint variables are determined.

3.8.2 Velocity and Force Transformation

The kineto-statics duality concept presented above can be useful to characterize the transformation of velocities and forces between two coordinate frames.

Consider a reference coordinate frame O_0–$x_0y_0z_0$ and a rigid body moving with respect to such frame. Let then O_1–$x_1y_1z_1$ and O_2–$x_2y_2z_2$ be two coordinate frames attached to the body (Figure 3.21). The relationships between translational and rotational velocities of the two frames with respect to the reference frame are given by:

$$\boldsymbol{\omega}_2 = \boldsymbol{\omega}_1$$
$$\dot{\boldsymbol{p}}_2 = \dot{\boldsymbol{p}}_1 + \boldsymbol{\omega}_1 \times \boldsymbol{r}_{12}.$$

By exploiting the skew-symmetric operator $\boldsymbol{S}(\cdot)$ in (3.8), the above relations can be compactly written as

$$\begin{bmatrix} \dot{\boldsymbol{p}}_2 \\ \boldsymbol{\omega}_2 \end{bmatrix} = \begin{bmatrix} \boldsymbol{I} & -\boldsymbol{S}(\boldsymbol{r}_{12}) \\ \boldsymbol{O} & \boldsymbol{I} \end{bmatrix} \begin{bmatrix} \dot{\boldsymbol{p}}_1 \\ \boldsymbol{\omega}_1 \end{bmatrix}. \tag{3.100}$$

All vectors in (3.100) are meant to be referred to the reference frame O_0–$x_0y_0z_0$. On the other hand, if vectors are referred to their own frames, it is:

$$\boldsymbol{r}_{12} = \boldsymbol{R}_1\boldsymbol{r}_{12}^1$$

and also

$$\dot{\boldsymbol{p}}_1 = \boldsymbol{R}_1\dot{\boldsymbol{p}}_1^1 \qquad \dot{\boldsymbol{p}}_2 = \boldsymbol{R}_2\dot{\boldsymbol{p}}_2^2 = \boldsymbol{R}_1\boldsymbol{R}_2^1\dot{\boldsymbol{p}}_2^2$$
$$\boldsymbol{\omega}_1 = \boldsymbol{R}_1\boldsymbol{\omega}_1^1 \qquad \boldsymbol{\omega}_2 = \boldsymbol{R}_2\boldsymbol{\omega}_2^2 = \boldsymbol{R}_1\boldsymbol{R}_2^1\boldsymbol{\omega}_2^2.$$

Accounting for (3.100) and (3.9) gives

$$\begin{aligned}\boldsymbol{R}_1\boldsymbol{R}_2^1\dot{\boldsymbol{p}}_2^2 &= \boldsymbol{R}_1\dot{\boldsymbol{p}}_1^1 - \boldsymbol{R}_1\boldsymbol{S}(\boldsymbol{r}_{12}^1)\boldsymbol{R}_1^T\boldsymbol{R}_1\boldsymbol{\omega}_1^1\\ \boldsymbol{R}_1\boldsymbol{R}_2^1\boldsymbol{\omega}_2^2 &= \boldsymbol{R}_1\boldsymbol{\omega}_1^1.\end{aligned}$$

Eliminating the dependence on $\boldsymbol{R}_1$, which is premultiplied to each term on both sides of the previous relations, yields[9]

$$\begin{bmatrix}\dot{\boldsymbol{p}}_2^2\\ \boldsymbol{\omega}_2^2\end{bmatrix} = \begin{bmatrix}\boldsymbol{R}_1^2 & -\boldsymbol{R}_1^2\boldsymbol{S}(\boldsymbol{r}_{12}^1)\\ \boldsymbol{O} & \boldsymbol{R}_1^2\end{bmatrix}\begin{bmatrix}\dot{\boldsymbol{p}}_1^1\\ \boldsymbol{\omega}_1^1\end{bmatrix} \tag{3.101}$$

giving the sought general relationship of *velocity transformation* from one frame to another.

It may be observed that the transformation matrix in (3.101) plays the role of a true Jacobian, since it characterizes a velocity transformation, and thus (3.101) may be shortly written as

$$\boldsymbol{v}_2^2 = \boldsymbol{J}_1^2\boldsymbol{v}_1^1. \tag{3.102}$$

At this point, by virtue of the kineto-statics duality, the *force transformation* from one frame to another can be directly derived in the form

$$\boldsymbol{\gamma}_1^1 = \boldsymbol{J}_1^{2T}\boldsymbol{\gamma}_2^2 \tag{3.103}$$

which can be detailed into[10]

$$\begin{bmatrix}\boldsymbol{f}_1^1\\ \boldsymbol{\mu}_1^1\end{bmatrix} = \begin{bmatrix}\boldsymbol{R}_2^1 & \boldsymbol{O}\\ \boldsymbol{S}(\boldsymbol{r}_{12}^1)\boldsymbol{R}_2^1 & \boldsymbol{R}_2^1\end{bmatrix}\begin{bmatrix}\boldsymbol{f}_2^2\\ \boldsymbol{\mu}_2^2\end{bmatrix}. \tag{3.104}$$

Finally, notice that the above analysis is instantaneous in that, if a coordinate frame varies with respect to the other, it is necessary to recompute the Jacobian of the transformation through the computation of the related rotation matrix of one frame with respect to the other.

3.8.3 Closed Chain

As discussed in Section 2.8.3, whenever the manipulator contains a closed chain, there is a functional relationship between the joint variables. In particular, the closed chain structure is transformed into a tree-structured open chain by virtually cutting the loop at a joint. It is worth choosing such cut joint as one of the unactuated joints. Then, the constraints (2.54) or (2.55) shall be solved for a reduced number of joint variables, corresponding to the degrees of mobility of the chain. Therefore, it is reasonable to assume that at least such independent joints are actuated, while the others may or

---

[9] Recall that $\boldsymbol{R}^T\boldsymbol{R} = \boldsymbol{I}$, as in (2.4).

[10] The skew-symmetry property $\boldsymbol{S} + \boldsymbol{S}^T = \boldsymbol{O}$ is utilized.

may not be actuated. Let $\boldsymbol{q}_o = [\,\boldsymbol{q}_a^T \quad \boldsymbol{q}_u^T\,]^T$ denote the vector of joint variables of the tree-structured open chain, where $\boldsymbol{q}_a$ and $\boldsymbol{q}_u$ are the vectors of *actuated* and *unactuated* joint variables, respectively. Assume that from the above constraints it is possible to determine a functional expression

$$\boldsymbol{q}_u = \boldsymbol{q}_u(\boldsymbol{q}_a). \tag{3.105}$$

Time differentiation of (3.105) gives the relationship between joint velocities in the form

$$\dot{\boldsymbol{q}}_o = \boldsymbol{\Upsilon}\dot{\boldsymbol{q}}_a \tag{3.106}$$

where

$$\boldsymbol{\Upsilon} = \begin{bmatrix} \boldsymbol{I} \\ \dfrac{\partial \boldsymbol{q}_u}{\partial \boldsymbol{q}_a} \end{bmatrix} \tag{3.107}$$

is the transformation matrix between the two vectors of joint velocities, which in turn plays the role of a Jacobian.

At this point, according to an intuitive kineto-statics duality concept, it is possible to describe the transformation between the corresponding vectors of joint torques in the form

$$\boldsymbol{\tau}_a = \boldsymbol{\Upsilon}^T \boldsymbol{\tau}_o \tag{3.108}$$

where $\boldsymbol{\tau}_o = [\,\boldsymbol{\tau}_a^T \quad \boldsymbol{\tau}_u^T\,]^T$, with obvious meaning of the quantities.

Example 3.5

Consider the parallelogram arm of Section 2.9.2. On the assumption to actuate the two Joints $1'$ and $1''$ at the base, it is $\boldsymbol{q}_a = [\,\vartheta_{1'} \quad \vartheta_{1''}\,]^T$ and $\boldsymbol{q}_u = [\,\vartheta_{2'} \quad \vartheta_{3'}\,]^T$. Then, using (2.59), the transformation matrix in (3.107) is

$$\boldsymbol{\Upsilon} = \begin{bmatrix} 1 & 0 \\ 0 & 1 \\ -1 & 1 \\ 1 & -1 \end{bmatrix}.$$

Hence, in view of (3.108), the torque vector of the actuated joints is

$$\boldsymbol{\tau}_a = \begin{bmatrix} \tau_{1'} - \tau_{2'} + \tau_{3'} \\ \tau_{1''} + \tau_{2'} - \tau_{3'} \end{bmatrix} \tag{3.109}$$

while obviously $\boldsymbol{\tau}_u = [\,0 \quad 0\,]^T$ in agreement with the fact that both Joints $2'$ and $3'$ are unactuated.

3.9 Manipulability Ellipsoids

The differential kinematics equation in (3.41) and the statics equation in (3.99), together with the duality property, allow the definition of indices for the evaluation of manipulator performance. Such indices can be helpful both for mechanical manipulator design and for determining suitable manipulator postures to execute a given task in the current configuration.

First, it is desired to represent the attitude of a manipulator to arbitrarily change end-effector position and orientation. This capability is described in an effective manner by the *velocity manipulability ellipsoid.*

Consider the set of joint velocities of constant (unit) norm

$$\dot{\boldsymbol{q}}^T \dot{\boldsymbol{q}} = 1; \tag{3.110}$$

this equation describes the points on the surface of a sphere in the joint velocity space. It is desired to describe the operational space velocities that can be generated by the given set of joint velocities, with the manipulator in a given posture. To this purpose, one can utilize the differential kinematics equation in (3.41) solved for the joint velocities; in the general case of a redundant manipulator ($r < n$) at a nonsingular configuration, the minimum-norm solution $\dot{\boldsymbol{q}} = \boldsymbol{J}^{\dagger}(\boldsymbol{q})\boldsymbol{v}$ can be considered which, substituted into (3.110), yields

$$\boldsymbol{v}^T \big(\boldsymbol{J}^{\dagger T}(\boldsymbol{q}) \boldsymbol{J}^{\dagger}(\boldsymbol{q})\big) \boldsymbol{v} = 1.$$

Accounting for the expression of the pseudo-inverse of $\boldsymbol{J}$ in (3.48) gives

$$\boldsymbol{v}^T \big(\boldsymbol{J}(\boldsymbol{q}) \boldsymbol{J}^T(\boldsymbol{q})\big)^{-1} \boldsymbol{v} = 1, \tag{3.111}$$

which is the equation of the points on the surface of an ellipsoid in the end-effector velocity space.

The choice of the minimum-norm solution rules out the presence of internal motions for the redundant structure. If the general solution (3.50) is used for $\dot{\boldsymbol{q}}$, the points satisfying (3.110) are mapped into points inside the ellipsoid whose surface is described by (3.111).

For a nonredundant manipulator, the differential kinematics solution (3.43) is used to derive (3.111); in this case the points on the surface of the sphere in the joint velocity space are mapped into points on the surface of the ellipsoid in the end-effector velocity space.

Along the direction of the major axis of the ellipsoid, the end effector can move at large velocity, while along the direction of the minor axis small end-effector velocities are obtained. Further, the closer the ellipsoid is to a sphere—unit eccentricity—the better the end effector can move isotropically along all directions of the operational space. Hence, it can be understood why this ellipsoid is an index characterizing manipulation ability of the structure in terms of velocities.

As can be recognized from (3.111), the shape and orientation of the ellipsoid are determined by the core of its quadratic form and then by the matrix $\boldsymbol{J}\boldsymbol{J}^T$ which is in general a function of the manipulator configuration. The directions of the principal axes of the ellipsoid are determined by the eigenvectors $\boldsymbol{u}_i$, for $i = 1, \ldots, r$, of the matrix $\boldsymbol{J}\boldsymbol{J}^T$, while the dimensions of the axes are given by the singular values of $\boldsymbol{J}$,

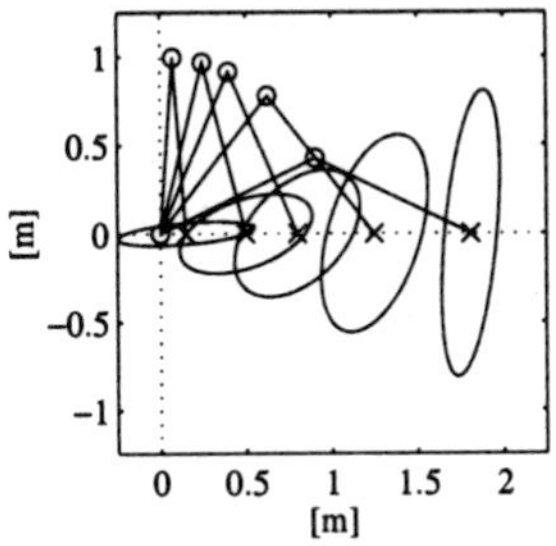

Figure 3.22 Velocity manipulability ellipses for a two-link planar arm in different postures.

$\sigma_i = \sqrt{\lambda_i(\boldsymbol{J}\boldsymbol{J}^T)}$, for $i = 1, \ldots, r$, where $\lambda_i(\boldsymbol{J}\boldsymbol{J}^T)$ denotes the generic eigenvalue of $\boldsymbol{J}\boldsymbol{J}^T$.

A global representative measure of manipulation ability can be obtained by considering the volume of the ellipsoid. This volume is proportional to the quantity

$$w(\boldsymbol{q}) = \sqrt{\det\big(\boldsymbol{J}(\boldsymbol{q})\boldsymbol{J}^T(\boldsymbol{q})\big)},$$

which is the *manipulability measure* already introduced in (3.52). In the case of a nonredundant manipulator ($r = n$), w reduces to

$$w(\boldsymbol{q}) = \big|\det\big(\boldsymbol{J}(\boldsymbol{q})\big)\big|. \tag{3.112}$$

It is easy to recognize that it is always $w > 0$, except for a manipulator at a singular configuration when $w = 0$. For this reason, this measure is usually adopted as a distance of the manipulator from singular configurations.

Example 3.6

Consider the two-link planar arm. From the expression in (3.37), the manipulability measure is in this case

$$w = \big|\det(\boldsymbol{J})\big| = a_1 a_2 |s_2|.$$

Therefore, as a function of the arm postures, the manipulability is maximum for $\vartheta_2 = \pm\pi/2$. On the other hand, for a given constant reach $a_1 + a_2$, the structure offering the maximum manipulability, independently of ϑ_1 and ϑ_2, is the one with $a_1 = a_2$.

These results have an intuitive interpretation in the human arm, if that is regarded as a two-link arm (arm + forearm). The condition $a_1 = a_2$ is satisfied with good approximation. Further, the elbow angle ϑ_2 is usually in the neighbourhood of $\pi/2$ in the execution of several tasks, such as that of writing. Hence, the human being tends to dispose the arm in the most dexterous configuration from a manipulability viewpoint.

Figure 3.22 illustrates the velocity manipulability ellipses for a certain number of postures with the tip along the horizontal axis and $a_1 = a_2 = 1$. It can be seen that

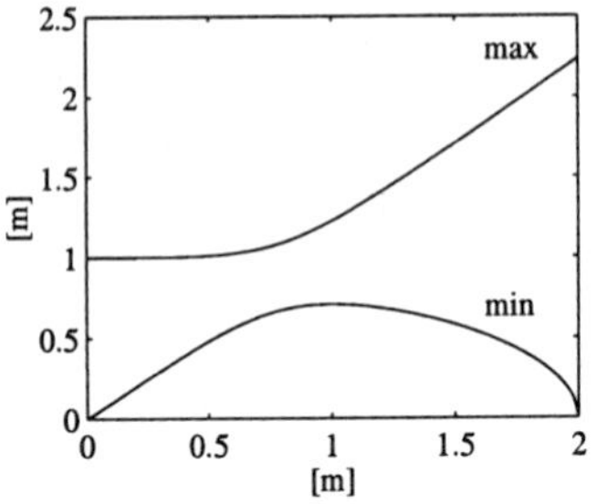

Figure 3.23 Minimum and maximum singular values of $\boldsymbol{J}$ for a two-link planar arm as a function of the arm posture.

when the arm is outstretched the ellipsoid is very thin along the vertical direction. Hence, one recovers the result anticipated in the study of singularities that the arm in this posture can generate tip velocities preferably along the vertical direction. In Figure 3.23, moreover, the behaviour of the minimum and maximum singular values of the matrix $\boldsymbol{J}$ is illustrated as a function of tip position along axis x; it can be verified that the minimum singular value is null when the manipulator is at a singularity (retracted or outstretched).

The manipulability measure w has the advantage to be easy to compute, through the determinant of matrix $\boldsymbol{J}\boldsymbol{J}^T$. However, its numerical value does not constitute an absolute measure of the actual closeness of the manipulator to a singularity. It is enough to consider the above example and take two arms of identical structure, one with links of 1 m and the other with links of 1 cm. Two different values of manipulability are obtained which differ by four orders of magnitude. Hence, in that case it is convenient to consider only $|s_2|$—eventually $|\vartheta_2|$—as the manipulability measure. In more general cases when it is not easy to find a simple, meaningful index, one can consider the ratio between the minimum and maximum singular values of the Jacobian σ_r/σ_1 which is equivalent to the inverse of the condition number of matrix $\boldsymbol{J}$. This ratio gives not only a measure of the distance from a singularity ($\sigma_r = 0$), but also a direct measure of eccentricity of the ellipsoid. The disadvantage in utilizing this index is its computational complexity; it is practically impossible to compute it in symbolic form, *i.e.*, as a function of the joint configuration, except for matrices of reduced dimension.

On the basis of the existing duality between differential kinematics and statics, it is possible to describe the manipulability of a structure not only with reference to velocities, but also with reference to forces. To be specific, one can consider the sphere in the space of joint torques

$$\boldsymbol{\tau}^T\boldsymbol{\tau} = 1 \tag{3.113}$$

which, accounting for (3.99), is mapped into the ellipsoid in the space of end-effector forces

$$\boldsymbol{\gamma}^T\left(\boldsymbol{J}(\boldsymbol{q})\boldsymbol{J}^T(\boldsymbol{q})\right)\boldsymbol{\gamma} = 1 \tag{3.114}$$

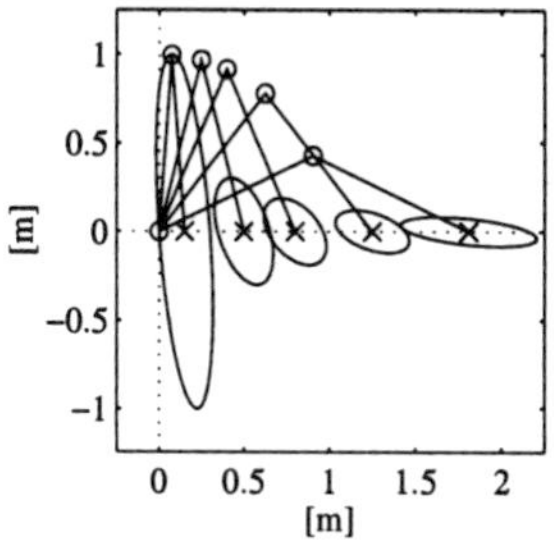

Figure 3.24 Force manipulability ellipses for a two-link planar arm in different postures.

which is defined as the *force manipulability ellipsoid*. This ellipsoid characterizes the end-effector forces that can be generated with the given set of joint torques, with the manipulator in a given posture.

As can be easily recognized from (3.114), the core of the quadratic form is constituted by the inverse of the matrix core of the velocity ellipsoid in (3.111). This feature leads to the notable result that the principal axes of the force manipulability ellipsoid coincide with the principal axes of the velocity manipulability ellipsoid, while the dimensions of the respective axes are in inverse proportion. Therefore, according to the concept of force/velocity duality, a direction along which good velocity manipulability is obtained is a direction along which poor force manipulability is obtained, and vice versa.

In Figure 3.24, the manipulability ellipses for the same postures as those of the example in Figure 3.22 are illustrated. A comparison of the shape and orientation of the ellipses confirms the force/velocity duality effect on the manipulability along different directions.

It is worth pointing out that these manipulability ellipsoids can be represented geometrically in all cases of an operational space of dimension at most three. Therefore, if it is desired to analyze manipulability in a space of greater dimension, it is worth separating the components of linear velocity (force) from those of angular velocity (moment), avoiding also problems due to nonhomogeneous dimensions of the relevant quantities (e.g., m/s vs. rad/s). For instance, for a manipulator with a spherical wrist, the manipulability analysis is naturally prone to a decoupling between arm and wrist.

An effective interpretation of the above results can be achieved by regarding the manipulator as a *mechanical transformer* of velocities and forces from the joint space to the operational space. Conservation of energy dictates that an amplification in the velocity transformation is necessarily accompanied by a reduction in the force transformation, and vice versa. The transformation ratio along a given direction is determined by the intersection of the vector along that direction with the surface of the ellipsoid. Once a unit vector $\boldsymbol{u}$ along a direction has been assigned, it is possible to compute the transformation ratio for the force manipulability ellipsoid as

$$\alpha(\boldsymbol{q}) = \left(\boldsymbol{u}^T \boldsymbol{J}(\boldsymbol{q}) \boldsymbol{J}^T(\boldsymbol{q}) \boldsymbol{u}\right)^{-1/2}, \tag{3.115}$$

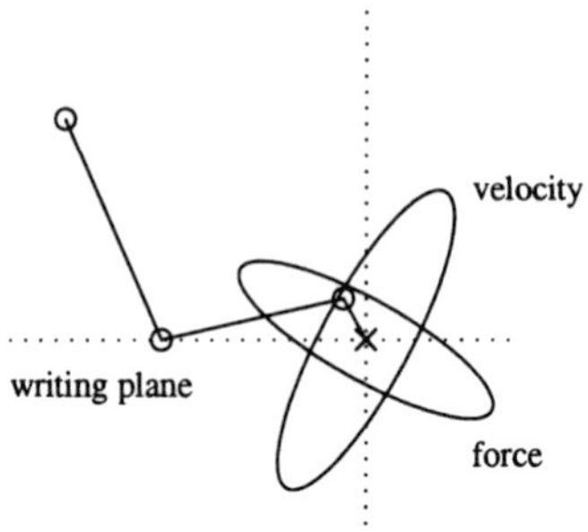

Figure 3.25 Velocity and force manipulability ellipses for a three-link planar arm in a typical configuration for a task of controlling force and velocity.

and for the velocity manipulability ellipsoid as

$$\beta(\boldsymbol{q}) = \left(\boldsymbol{u}^T\left(\boldsymbol{J}(\boldsymbol{q})\boldsymbol{J}^T(\boldsymbol{q})\right)^{-1}\boldsymbol{u}\right)^{-1/2}. \tag{3.116}$$

The manipulability ellipsoids can be conveniently utilized not only for analyzing manipulability of the structure along different directions of the operational space, but also for determining compatibility of the structure to execute a task assigned along a direction. To this purpose, it is useful to distinguish between actuation tasks and control tasks of velocity and force. In terms of the relative ellipsoid, the task of actuating a velocity (force) requires preferably a large transformation ratio along the task direction, since for a given set of joint velocities (forces) at the joints it is possible to generate a large velocity (force) at the end effector. On the other hand, for a control task it is important to have a small transformation ratio so as to gain good sensitivity to errors that may occur along the given direction.

Revisiting once again the duality between velocity manipulability ellipsoid and force manipulability ellipsoid, it can be found that an optimal direction to actuate a velocity is also an optimal direction to control a force. Analogously, a good direction to actuate a force is also a good direction to control a velocity.

To have a tangible example of the above concept, consider the typical task of writing on a horizontal surface for the human arm; this time, the arm is regarded as a three-link planar arm: arm + forearm + hand. Restricting the analysis to a two-dimensional task space (the direction vertical to the surface and the direction of the line of writing), one has to achieve fine control of the vertical force (pressing of the pen on the paper) and of the horizontal velocity (to write in good calligraphy). As a consequence, the force manipulability ellipse tends to be oriented horizontally for correct task execution. Correspondingly, the velocity manipulability ellipse tends to be oriented vertically in perfect agreement with the task requirement. In this case, from Figure 3.25 the typical configuration of the human arm when writing can be recognized.

An opposite example to the previous one is that of the human arm when throwing a load in the horizontal direction. In fact, now it is necessary to actuate a large vertical

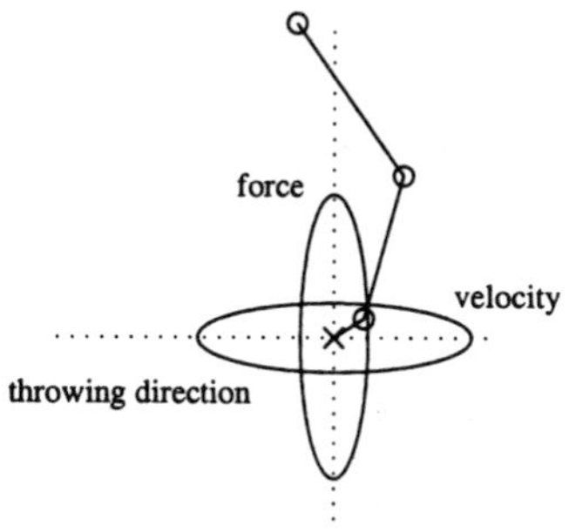

Figure 3.26 Velocity and force manipulability ellipses for a three-link planar arm in a typical configuration for a task of actuating force and velocity.

force (to sustain the weight) and a large horizontal velocity (to throw the load at a considerable distance). Differently from the above, the force (velocity) manipulability ellipse tends to be oriented vertically (horizontally) to successfully execute the task. The relative configuration in Figure 3.26 is representative of the typical attitude of the human arm when, for instance, throwing a bowl in the bowling game.

In the above two examples, it is worth pointing out that the presence of a two-dimensional operational space is certainly advantageous to try reconfiguring the structure in the best configuration compatible with the given task. In fact, the transformation ratios defined in (3.115) and (3.116) are scalar functions of the manipulator configurations that can be optimized locally according to the technique for exploiting redundant degrees of mobility previously illustrated.

Problems

3.1 Prove (3.9).

3.2 Compute the Jacobian of the cylindrical arm in Figure 2.33.

3.3 Compute the Jacobian of the SCARA manipulator in Figure 2.34.

3.4 Find the singularities of the three-link planar arm in Figure 2.20.

3.5 Find the singularities of the spherical arm in Figure 2.22.

3.6 Find the singularities of the cylindrical arm in Figure 2.33.

3.7 Find the singularities of the SCARA manipulator in Figure 2.34.

3.8 For the three-link planar arm in Figure 2.20, find an expression of the distance of the arm from a circular obstacle of given radius and coordinates.

3.9 Find the solution to the differential kinematics equation with the damped least-square inverse in (3.55).

3.10 Prove (3.60) in an alternative way, *i.e.*, by computing $\boldsymbol{S}(\boldsymbol{\omega})$ as in (3.5) starting from $\boldsymbol{R}(\boldsymbol{\phi})$ in (2.18).

3.11 With reference to (3.60), find the transformation matrix $\boldsymbol{T}(\boldsymbol{\phi})$ in the case of Roll–Pitch–Yaw angles.

3.12 Show how the inverse kinematics scheme of Figure 3.11 can be simplified in the case of a manipulator having a spherical wrist.

3.13 Find an expression of the upper bound on the norm of $\boldsymbol{e}$ for the solution (3.72) in the case $\dot{\boldsymbol{x}}_d \neq \mathbf{0}$.

3.14 Prove (3.81).

3.15 Prove (3.82) and (3.83).

3.16 Prove that the equation relating the angular velocity to the time derivative of the quaternion is given by

$$\boldsymbol{\omega} = 2\boldsymbol{S}(\boldsymbol{\epsilon})\dot{\boldsymbol{\epsilon}} + 2\eta\dot{\boldsymbol{\epsilon}} - 2\dot{\eta}\boldsymbol{\epsilon}.$$

[Hint: start showing that (2.29) can be rewritten as $\boldsymbol{R}(\eta, \boldsymbol{\epsilon}) = (2\eta^2 - 1)\boldsymbol{I} + 2\boldsymbol{\epsilon}\boldsymbol{\epsilon}^T + 2\eta\boldsymbol{S}(\boldsymbol{\epsilon})$].

3.17 Prove (3.90).

3.18 Prove that the time derivative of the Lyapunov function in (3.91) is given by (3.92).

3.19 Show that the manipulability measure defined in (3.52) is given by the product of the singular values of the Jacobian matrix.

3.20 Consider the three-link planar arm in Figure 2.20, whose link lengths are respectively 0.5 m, 0.3 m, 0.3 m. Perform a computer implementation of the inverse kinematics algorithm using the Jacobian pseudo-inverse along the operational space path given by a straight line connecting the points of coordinates $(0.8, 0.2)$ m and $(0.8, -0.2)$ m. Add a constraint aimed at avoiding link collision with a circular object located at $\boldsymbol{o} = [\,0.3 \quad 0\,]^T$ m of radius 0.1 m. The initial arm configuration is chosen so that $\boldsymbol{p}(0) = \boldsymbol{p}_d(0)$. The final time is 2 s. Use sinusoidal motion time laws. Adopt the Euler numerical integration scheme (3.44) with an integration time $\Delta t = 1$ ms.

3.21 Consider the SCARA manipulator in Figure 2.34, whose links both have a length of 0.5 m and are located at a height of 1 m from the supporting plane. Perform a computer implementation of the inverse kinematics algorithms with both Jacobian inverse and Jacobian transpose along the operational space path whose position is given by a straight line connecting the points of coordinates $(0.7, 0, 0)$ m and $(0, 0.8, 0.5)$ m, and whose orientation is given by a rotation from 0 rad to $\pi/2$ rad. The initial arm configuration is chosen so that $\boldsymbol{x}(0) = \boldsymbol{x}_d(0)$. The final time is 2 s. Use sinusoidal motion time laws. Adopt the Euler numerical integration scheme (3.44) with an integration time $\Delta t = 1$ ms.

3.22 Prove that the directions of the principal axes of the force and velocity manipulability ellipsoids coincide while their dimensions are in inverse proportion.

Bibliography

Asada H., Slotine J.-J.E. (1986) *Robot Analysis and Control.* Wiley, New York.

Balestrino A., De Maria G., Sciavicco L., Siciliano B. (1988) An algorithmic approach to coordinate transformation for robotic manipulators. *Advanced Robotics* 2:327–344.

Chiaverini S., Siciliano B. (1999) The unit quaternion: A useful tool for inverse kinematics of robot manipulators. *Systems Analysis Modelling Simulation* 35:45–60.

Chiu S.L. (1988) Task compatibility of manipulator postures. *Int. J. Robotics Research* 7(5):13–21.

Craig J.J. (1989) *Introduction to Robotics: Mechanics and Control*. 2nd ed., Addison-Wesley, Reading, Mass.

Khalil W., Dombre E. (1999) *Modélisation Identification et Commande des Robots*. 2ème éd., Hermès, Paris.

Klein C.A., Huang C.H. (1983) Review of pseudoinverse control for use with kinematically redundant manipulators. *IEEE Trans. Systems, Man, and Cybernetics* 13:245–250.

Liégeois A. (1977) Automatic supervisory control of the configuration and behavior of multibody mechanisms. *IEEE Trans. Systems, Man, and Cybernetics* 7:868–871.

Lin S.K. (1989) Singularity of a nonlinear feedback control scheme for robots. *IEEE Trans. Systems, Man, and Cybernetics* 19:134–139.

Luh J.Y.S., Walker M.W., Paul R.P.C. (1980) Resolved-acceleration control of mechanical manipulators. *IEEE Trans. Automatic Control* 25:468–474.

Maciejewski A.A., Klein C.A. (1985) Obstacle avoidance for kinematically redundant manipulators in dynamically varying environments. *Int. J. Robotics Research* 4(3):109–117.

Nakamura Y. (1991) *Advanced Robotics: Redundancy and Optimization*. Addison-Wesley, Reading, Mass.

Nakamura Y., Hanafusa H. (1986) Inverse kinematic solutions with singularity robustness for robot manipulator control. *ASME J. Dynamic Systems, Measurement, and Control* 108:163–171.

Nakamura Y., Hanafusa H. (1987) Optimal redundancy control of robot manipulators. *Int. J. Robotics Research* 6(1):32–42.

Nakamura Y., Hanafusa H., Yoshikawa T. (1987) Task-priority based redundancy control of robot manipulators. *Int. J. Robotics Research* 6(2):3–15.

Orin D.E., Schrader W.W. (1984) Efficient computation of the Jacobian for robot manipulators. *Int. J. Robotics Research* 3(4):66–75.

Salisbury J.K., Craig J.J. (1982) Articulated hands: Force control and kinematic issues. *Int. J. Robotics Research* 1(1):4–17.

Sciavicco L., Siciliano B. (1988) A solution algorithm to the inverse kinematic problem for redundant manipulators. *IEEE J. Robotics and Automation* 4:403–410.

Siciliano B. (1986) *Algoritmi di Soluzione al Problema Cinematico Inverso per Robot di Manipolazione* (in Italian). Tesi di Dottorato di Ricerca, Università degli Studi di Napoli.

Siciliano B. (1990) Kinematic control of redundant robot manipulators: A tutorial. *J. Intelligent and Robotic Systems* 3:201–212.

Spong M.W., Vidyasagar M. (1989) *Robot Dynamics and Control*. Wiley, New York.

Wampler C.W. (1986) Manipulator inverse kinematic solutions based on damped least-squares solutions. *IEEE Trans. Systems, Man, and Cybernetics* 16:93–101.

Whitney D.E. (1969) Resolved motion rate control of manipulators and human prostheses. *IEEE Trans. Man-Machine Systems* 10:47–53.

Yoshikawa T. (1985) Manipulability of robotic mechanisms. *Int. J. Robotics Research* 4(2):3–9.

Yoshikawa T. (1990) *Foundations of Robotics*. MIT Press, Cambridge, Mass.

Yuan J.S.-C. (1988) Closed-loop manipulator control using quaternion feedback. *IEEE J. Robotics and Automation* 4:434–440.

4. Dynamics

Derivation of the *dynamic model* of a manipulator plays an important role for simulation of motion, analysis of manipulator structures, and design of control algorithms. Simulating manipulator motion allows testing control strategies and motion planning techniques without the need to use a physically available system. The analysis of the dynamic model can be helpful for mechanical design of prototype arms. Computation of the forces and torques required for the execution of typical motions provides useful information for designing joints, transmissions and actuators. The goal of this chapter is to present two methods for derivation of the equations of motion of a manipulator in the *joint space*. The first method is based on the *Lagrange* formulation and is conceptually simple and systematic. The second method is based on the *Newton-Euler* formulation and allows obtaining the model in a recursive form; it is computationally more efficient since it exploits the typically open structure of the manipulator kinematic chain. The problem of *dynamic parameter identification* is also studied. The chapter ends with the derivation of the *dynamic model* of a manipulator in the *operational space* and the definition of the *dynamic manipulability ellipsoid*.

4.1 Lagrange Formulation

The dynamic model of a manipulator provides a description of the relationship between the joint actuator torques and the motion of the structure.

With *Lagrange* formulation, the equations of motion can be derived in a systematic way independently of the reference coordinate frame. Once a set of variables λ_i, $i = 1, \ldots, n$, termed *generalized coordinates*, are chosen which effectively describe the link positions of an n-degree-of-mobility manipulator, the *Lagrangian* of the mechanical system can be defined as a function of the generalized coordinates:

$$\mathcal{L} = \mathcal{T} - \mathcal{U} \tag{4.1}$$

where $\mathcal{T}$ and $\mathcal{U}$ are respectively the total *kinetic energy* and *potential energy* of the system.

The Lagrange's equations are expressed by

$$\frac{d}{dt}\frac{\partial \mathcal{L}}{\partial \dot{\lambda}_i} - \frac{\partial \mathcal{L}}{\partial \lambda_i} = \xi_i \qquad i = 1, \ldots, n \tag{4.2}$$

where ξ_i is the *generalized force* associated with the generalized coordinate λ_i.

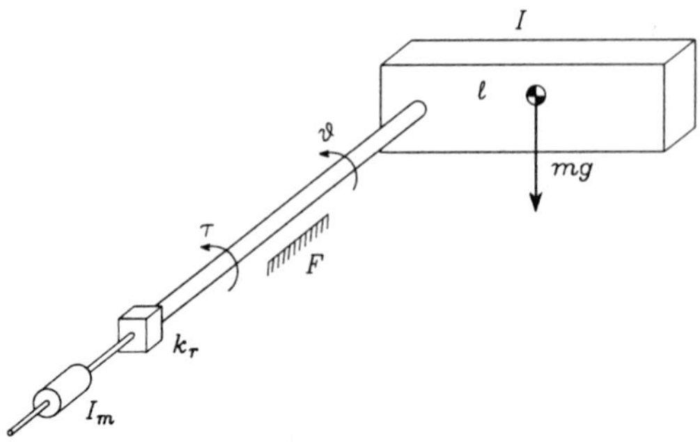

Figure 4.1 Actuated pendulum.

For a manipulator with an open kinematic chain a natural choice for the generalized coordinates is the vector of *joint variables*

$$\begin{bmatrix} \lambda_1 \\ \vdots \\ \lambda_n \end{bmatrix} = \boldsymbol{q}. \tag{4.3}$$

The contributions to the generalized forces are given by the nonconservative forces, *i.e.*, the joint actuator torques, the joint friction torques, as well as the joint torques induced by end-effector forces at the contact with the environment[1].

The equations in (4.2) establish the relations existing between the generalized forces applied to the manipulator and the joint positions, velocities and accelerations. Hence, they allow deriving the dynamic model of the manipulator starting from the determination of kinetic energy and potential energy of the mechanical system.

Example 4.1

In order to understand the Lagrange formulation technique for deriving the dynamic model, consider the simple case of a pendulum. Let τ be the actuation torque about the rotation axis; it is assumed that viscous friction occurs about the same axis (Figure 4.1). The torque is supplied by a motor with reduction gear ratio $k_r > 1$ and moment of inertia I_m about the fast shaft. Let ϑ denote the angle with respect to the reference position of the body hanging down ($\vartheta = 0$). By choosing ϑ as the generalized coordinate, the kinetic energy of the system is given by

$$\mathcal{T} = \frac{1}{2} I \dot{\vartheta}^2 + \frac{1}{2} I_m k_r^2 \dot{\vartheta}^2$$

where I is the body moment of inertia about the rotation axis. Further, the system potential energy is expressed by

$$\mathcal{U} = mg\ell(1 - \cos\vartheta)$$

[1] The term *torque* is used as a synonym of joint *generalized force*.

where m is the body mass, g is the gravity acceleration (9.81 m/s^2), and ℓ is the distance of the body centre of mass from the rotation axis. Therefore, the Lagrangian of the system is

$$\mathcal{L} = \frac{1}{2}I\dot{\vartheta}^2 + \frac{1}{2}I_m k_r^2 \dot{\vartheta}^2 - mg\ell(1 - \cos\vartheta).$$

Substituting this expression in the Lagrange's equation in (4.2) yields

$$(I + I_m k_r^2)\ddot{\vartheta} + mg\ell \sin\vartheta = \xi.$$

The generalized force ξ is given by the contributions of the actuation torque τ and of the viscous friction torque $-F\dot{\vartheta}$, *i.e.*,

$$\xi = \tau - F\dot{\vartheta},$$

leading to the complete dynamic model of the system as the second-order differential equation

$$(I + I_m k_r^2)\ddot{\vartheta} + F\dot{\vartheta} + mg\ell \sin\vartheta = \tau.$$

4.1.1 Computation of Kinetic Energy

Consider a manipulator with n *rigid links*. The total kinetic energy is given by the sum of the contributions relative to the motion of each link and the contributions relative to the motion of each joint actuator:[2]

$$\mathcal{T} = \sum_{i=1}^{n} (\mathcal{T}_{\ell_i} + \mathcal{T}_{m_i}), \tag{4.4}$$

where $\mathcal{T}_{\ell_i}$ is the kinetic energy of Link i and $\mathcal{T}_{m_i}$ is the kinetic energy of the motor actuating Joint i.

The kinetic energy contribution of Link i is given by

$$\mathcal{T}_{\ell_i} = \frac{1}{2}\int_{V_{\ell_i}} \dot{\boldsymbol{p}}_i^{*T} \dot{\boldsymbol{p}}_i^* \rho dV, \tag{4.5}$$

where $\dot{\boldsymbol{p}}_i^*$ denotes the linear velocity vector and ρ is the density of the elementary particle of volume dV; V_{ℓ_i} is the volume of Link i.

Consider the position vector $\boldsymbol{p}_i^*$ of the elementary particle and the position vector $\boldsymbol{p}_{C_i}$ of the link centre of mass, both expressed in the *base frame*. One has

$$\boldsymbol{r}_i = [\, r_{ix} \quad r_{iy} \quad r_{iz} \,]^T = \boldsymbol{p}_i^* - \boldsymbol{p}_{\ell_i} \tag{4.6}$$

[2] Link 0 is fixed and thus gives no contribution.

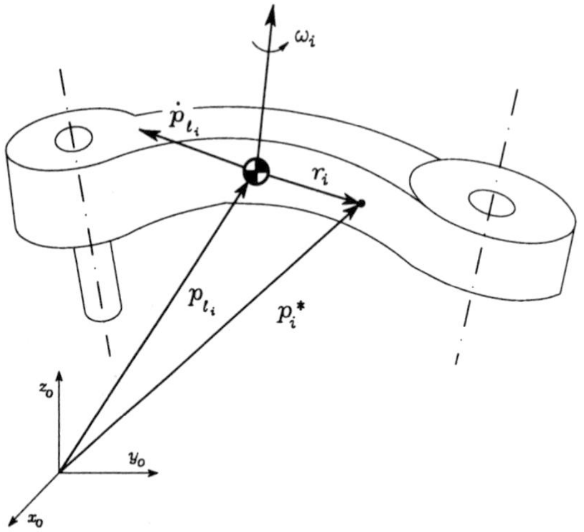

Figure 4.2 Kinematic description of Link i for Lagrange formulation.

with

$$\boldsymbol{p}_{\ell_i} = \frac{1}{m_{\ell_i}} \int_{V_{\ell_i}} \boldsymbol{p}_i^* \rho dV \tag{4.7}$$

where m_{ℓ_i} is the link mass. As a consequence, the link point velocity can be expressed as

$$\begin{aligned} \dot{\boldsymbol{p}}_i^* &= \dot{\boldsymbol{p}}_{\ell_i} + \boldsymbol{\omega}_i \times \boldsymbol{r}_i \\ &= \dot{\boldsymbol{p}}_{\ell_i} + \boldsymbol{S}(\boldsymbol{\omega}_i)\boldsymbol{r}_i, \end{aligned} \tag{4.8}$$

where $\dot{\boldsymbol{p}}_{\ell_i}$ is the linear velocity of the centre of mass and $\boldsymbol{\omega}_i$ is the angular velocity of the link (Figure 4.2).

By substituting the velocity expression (4.8) into (4.5), it can be recognized that the kinetic energy of each link is formed by the following contributions.

Translational

The contribution is

$$\frac{1}{2} \int_{V_{\ell_i}} \dot{\boldsymbol{p}}_{\ell_i}^T \dot{\boldsymbol{p}}_{\ell_i} \rho dV = \frac{1}{2} m_{\ell_i} \dot{\boldsymbol{p}}_{\ell_i}^T \dot{\boldsymbol{p}}_{\ell_i}. \tag{4.9}$$

Mutual

The contribution is

$$2\left(\frac{1}{2} \int_{V_{\ell_i}} \dot{\boldsymbol{p}}_{\ell_i}^T \boldsymbol{S}(\boldsymbol{\omega}_i)\boldsymbol{r}_i \rho dV\right) = 2\left(\frac{1}{2} \dot{\boldsymbol{p}}_{\ell_i}^T \boldsymbol{S}(\boldsymbol{\omega}_i) \int_{V_{\ell_i}} (\boldsymbol{p}_i^* - \boldsymbol{p}_{\ell_i}) \rho dV\right) = 0$$

since, by virtue of (4.7), it is

$$\int_{V_{\ell_i}} \boldsymbol{p}_i^* \rho dV = \boldsymbol{p}_{\ell_i} \int_{V_{\ell_i}} \rho dV.$$

Rotational

The contribution is

$$\frac{1}{2} \int_{V_{\ell_i}} \boldsymbol{r}_i^T \boldsymbol{S}^T(\boldsymbol{\omega}_i) \boldsymbol{S}(\boldsymbol{\omega}_i) \boldsymbol{r}_i \rho dV = \frac{1}{2} \boldsymbol{\omega}_i^T \left(\int_{V_{\ell_i}} \boldsymbol{S}^T(\boldsymbol{r}_i) \boldsymbol{S}(\boldsymbol{r}_i) \rho dV \right) \boldsymbol{\omega}_i$$

where the property $\boldsymbol{S}(\boldsymbol{\omega}_i)\boldsymbol{r}_i = -\boldsymbol{S}(\boldsymbol{r}_i)\boldsymbol{\omega}_i$ has been exploited. In view of the expression of the matrix operator $\boldsymbol{S}(\cdot)$

$$\boldsymbol{S}(\boldsymbol{r}_i) = \begin{bmatrix} 0 & -r_{iz} & r_{iy} \\ r_{iz} & 0 & -r_{ix} \\ -r_{iy} & r_{ix} & 0 \end{bmatrix},$$

it is

$$\frac{1}{2} \int_{V_{\ell_i}} \boldsymbol{r}_i^T \boldsymbol{S}^T(\boldsymbol{\omega}_i) \boldsymbol{S}(\boldsymbol{\omega}_i) \boldsymbol{r}_i \rho dV = \frac{1}{2} \boldsymbol{\omega}_i^T \boldsymbol{I}_{\ell_i} \boldsymbol{\omega}_i. \tag{4.10}$$

The matrix

$$\begin{aligned} \boldsymbol{I}_{\ell_i} &= \begin{bmatrix} \int (r_{iy}^2 + r_{iz}^2) \rho dV & -\int r_{ix} r_{iy} \rho dV & -\int r_{ix} r_{iz} \rho dV \\ * & \int (r_{ix}^2 + r_{iz}^2) \rho dV & -\int r_{iy} r_{iz} \rho dV \\ * & * & \int (r_{ix}^2 + r_{iy}^2) \rho dV \end{bmatrix} \\ &= \begin{bmatrix} I_{\ell_i xx} & -I_{\ell_i xy} & -I_{\ell_i xz} \\ * & I_{\ell_i yy} & -I_{\ell_i yz} \\ * & * & I_{\ell_i zz} \end{bmatrix}. \end{aligned} \tag{4.11}$$

is symmetric[3] and represents the *inertia tensor* relative to the centre of mass of Link i when expressed in the base frame. Notice that the position of Link i depends on the manipulator configuration and then the inertia tensor, when expressed in the base frame, is configuration-dependent. If the angular velocity of Link i is expressed with reference to a frame attached to the link (as in the Denavit-Hartenberg convention), it is

$$\boldsymbol{\omega}_i^i = \boldsymbol{R}_i^T \boldsymbol{\omega}_i$$

where $\boldsymbol{R}_i$ is the rotation matrix from Link i frame to the base frame. When referred to the link frame, the inertia tensor is constant. Let $\boldsymbol{I}_{\ell_i}^i$ denote such tensor; then it is easy to verify the following relation:

$$\boldsymbol{I}_{\ell_i} = \boldsymbol{R}_i \boldsymbol{I}_{\ell_i}^i \boldsymbol{R}_i^T. \tag{4.12}$$

---

[3] The symbol '$*$' has been used to avoid rewriting the symmetric elements.

If the axes of Link i frame coincide with the central axes of inertia, then the inertia products are null and the inertia tensor relative to the centre of mass is a diagonal matrix.

By summing the translational and rotational contributions (4.9) and (4.10), the kinetic energy of Link i is

$$\mathcal{T}_{\ell_i} = \frac{1}{2} m_{\ell_i} \dot{\boldsymbol{p}}_{\ell_i}^T \dot{\boldsymbol{p}}_{\ell_i} + \frac{1}{2} \boldsymbol{\omega}_i^T \boldsymbol{R}_i \boldsymbol{I}_{\ell_i}^i \boldsymbol{R}_i^T \boldsymbol{\omega}_i. \tag{4.13}$$

At this point, it is necessary to express the kinetic energy as a function of the generalized coordinates of the system, that are the joint variables. To this purpose, the geometric method for Jacobian computation can be applied to the intermediate link other than the end effector, yielding

$$\dot{\boldsymbol{p}}_{\ell_i} = \boldsymbol{j}_{P1}^{(\ell_i)} \dot{q}_1 + \ldots + \boldsymbol{j}_{Pi}^{(\ell_i)} \dot{q}_i = \boldsymbol{J}_P^{(\ell_i)} \dot{\boldsymbol{q}} \tag{4.14}$$

$$\boldsymbol{\omega}_i = \boldsymbol{j}_{O1}^{(\ell_i)} \dot{q}_1 + \ldots + \boldsymbol{j}_{Oi}^{(\ell_i)} \dot{q}_i = \boldsymbol{J}_O^{(\ell_i)} \dot{\boldsymbol{q}}, \tag{4.15}$$

where the contributions of the Jacobian columns relative to the joint velocities have been taken into account up to current Link i. The Jacobians to consider are then:

$$\boldsymbol{J}_P^{(\ell_i)} = \begin{bmatrix} \boldsymbol{j}_{P1}^{(\ell_i)} & \ldots & \boldsymbol{j}_{Pi}^{(\ell_i)} & \boldsymbol{0} & \ldots & \boldsymbol{0} \end{bmatrix} \tag{4.16}$$

$$\boldsymbol{J}_O^{(\ell_i)} = \begin{bmatrix} \boldsymbol{j}_{O1}^{(\ell_i)} & \ldots & \boldsymbol{j}_{Oi}^{(\ell_i)} & \boldsymbol{0} & \ldots & \boldsymbol{0} \end{bmatrix}; \tag{4.17}$$

the columns of the matrices in (4.16) and (4.17) can be computed according to (3.26), giving

$$\boldsymbol{j}_{Pj}^{(\ell_i)} = \begin{cases} \boldsymbol{z}_{j-1} & \text{for a } \textit{prismatic} \text{ joint} \\ \boldsymbol{z}_{j-1} \times (\boldsymbol{p}_{\ell_i} - \boldsymbol{p}_{j-1}) & \text{for a } \textit{revolute} \text{ joint} \end{cases} \tag{4.18}$$

$$\boldsymbol{j}_{Oj}^{(\ell_i)} = \begin{cases} \boldsymbol{0} & \text{for a } \textit{prismatic} \text{ joint} \\ \boldsymbol{z}_{j-1} & \text{for a } \textit{revolute} \text{ joint,} \end{cases} \tag{4.19}$$

where $\boldsymbol{p}_{j-1}$ is the position vector of the origin of Frame $j-1$ and $\boldsymbol{z}_{j-1}$ is the unit vector of axis z of Frame $j-1$.

In sum, the kinetic energy of Link i in (4.13) can be written as

$$\mathcal{T}_{\ell_i} = \frac{1}{2} m_{\ell_i} \dot{\boldsymbol{q}}^T \boldsymbol{J}_P^{(\ell_i)T} \boldsymbol{J}_P^{(\ell_i)} \dot{\boldsymbol{q}} + \frac{1}{2} \dot{\boldsymbol{q}}^T \boldsymbol{J}_O^{(\ell_i)T} \boldsymbol{R}_i \boldsymbol{I}_{\ell_i}^i \boldsymbol{R}_i^T \boldsymbol{J}_O^{(\ell_i)} \dot{\boldsymbol{q}}. \tag{4.20}$$

The kinetic energy contribution of the motor of Joint i can be computed in a formally analogous way to that of the link. Consider the typical case of rotary electric motors (that can actuate both revolute and prismatic joints by means of suitable transmissions). It can be assumed that the contribution of the fixed part (stator) is included in that of the link on which such motor is located, and thus the sole contribution of the rotor is to be computed.

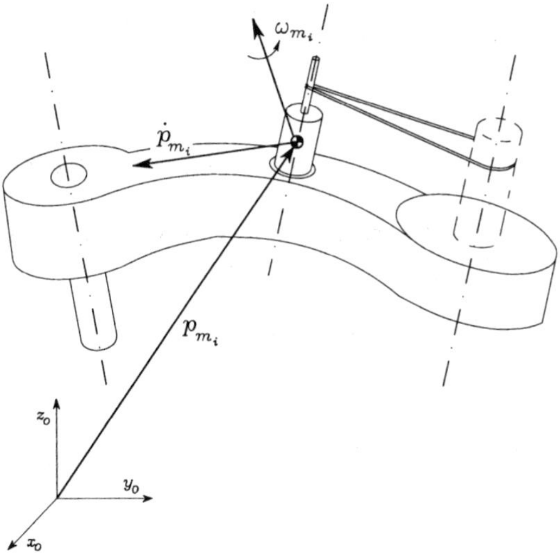

Figure 4.3 Kinematic description of Motor i.

With reference to Figure 4.3, the motor of Joint i is assumed to be located on Link $i-1$. In practice, in the design of the mechanical structure of an open kinematic chain manipulator one attempts to locate the motors as close as possible to the base of the manipulator so as to lighten the dynamic load of the first joints of the chain. The joint actuator torques are delivered by the motors by means of mechanical transmissions (gears)[4]. The contribution of the gears to the kinetic energy can be suitably included in that of the motor. It is assumed that no induced motion occurs, *i.e.*, the motion of Joint i does not actuate the motion of other joints.

The kinetic energy of Rotor i can be written as

$$\mathcal{T}_{m_i} = \frac{1}{2} m_{m_i} \dot{\boldsymbol{p}}_{m_i}^T \dot{\boldsymbol{p}}_{m_i} + \frac{1}{2} \boldsymbol{\omega}_{m_i}^T \boldsymbol{I}_{m_i} \boldsymbol{\omega}_{m_i}, \tag{4.21}$$

where m_{m_i} is the mass of the rotor, $\dot{\boldsymbol{p}}_{m_i}$ denotes the linear velocity of the centre of mass of the rotor, $\boldsymbol{I}_{m_i}$ is the inertia tensor of the rotor relative to its centre of mass, and $\boldsymbol{\omega}_{m_i}$ denotes the angular velocity of the rotor.

Let ϑ_{m_i} denote the angular position of the rotor. On the assumption of a *rigid transmission*, one has

$$k_{ri} \dot{q}_i = \dot{\vartheta}_{m_i} \tag{4.22}$$

where k_{ri} is the gear reduction ratio. Notice that, in the case of actuation of a prismatic joint, the gear reduction ratio is a dimensional quantity.

---

[4] Alternatively, the joints may be actuated by torque motors directly coupled to the rotation axis without gears.

According to the angular velocity composition rule (3.16) and the relation (4.22), the total angular velocity of the rotor is

$$\boldsymbol{\omega}_{m_i} = \boldsymbol{\omega}_{i-1} + k_{ri}\dot{q}_i \boldsymbol{z}_{m_i} \tag{4.23}$$

where $\boldsymbol{\omega}_{i-1}$ is the angular velocity of Link $i-1$ on which the motor is located, and $\boldsymbol{z}_{m_i}$ denotes the unit vector along the rotor axis.

To express the rotor kinetic energy as a function of the joint variables, it is worth expressing the linear velocity of the rotor centre of mass—similarly to (4.14)—as

$$\dot{\boldsymbol{p}}_{m_i} = \boldsymbol{J}_P^{(m_i)}\dot{\boldsymbol{q}}. \tag{4.24}$$

The Jacobian to compute is then

$$\boldsymbol{J}_P^{(m_i)} = \begin{bmatrix} \boldsymbol{j}_{P1}^{(m_i)} & \dots & \boldsymbol{j}_{P,i-1}^{(m_i)} & \boldsymbol{0} & \dots & \boldsymbol{0} \end{bmatrix} \tag{4.25}$$

whose columns are given by

$$\boldsymbol{j}_{Pj}^{(m_i)} = \begin{cases} \boldsymbol{z}_{j-1} & \text{for a } \textit{prismatic} \text{ joint} \\ \boldsymbol{z}_{j-1} \times (\boldsymbol{p}_{m_i} - \boldsymbol{p}_{j-1}) & \text{for a } \textit{revolute} \text{ joint} \end{cases} \tag{4.26}$$

where $\boldsymbol{p}_{j-1}$ is the position vector of the origin of Frame $j-1$. Notice that $\boldsymbol{j}_{Pi}^{(m_i)} = \boldsymbol{0}$ in (4.25), since the centre of mass of the rotor has been taken along its axis of rotation.

The angular velocity in (4.23) can be expressed as a function of the joint variables, *i.e.*,

$$\boldsymbol{\omega}_{m_i} = \boldsymbol{J}_O^{(m_i)}\dot{\boldsymbol{q}}. \tag{4.27}$$

The Jacobian to compute is then

$$\boldsymbol{J}_O^{(m_i)} = \begin{bmatrix} \boldsymbol{j}_{O1}^{(m_i)} & \dots & \boldsymbol{j}_{O,i-1}^{(m_i)} & \boldsymbol{j}_{Oi}^{(m_i)} & \boldsymbol{0} & \dots & \boldsymbol{0} \end{bmatrix} \tag{4.28}$$

whose columns, in view of (4.23) and (4.15), are respectively given by

$$\boldsymbol{j}_{Oj}^{(m_i)} = \begin{cases} \boldsymbol{j}_{Oj}^{(\ell_i)} & j = 1, \dots, i-1 \\ k_{ri}\boldsymbol{z}_{m_i} & j = i. \end{cases} \tag{4.29}$$

To compute the second relation in (4.29), it is sufficient to know the components of the unit vector of the rotor rotation axis $\boldsymbol{z}_{m_i}$ with respect to the base frame. Hence, the kinetic energy of Rotor i can be written as

$$\mathcal{T}_{m_i} = \frac{1}{2}m_{m_i}\dot{\boldsymbol{q}}^T\boldsymbol{J}_P^{(m_i)T}\boldsymbol{J}_P^{(m_i)}\dot{\boldsymbol{q}} + \frac{1}{2}\dot{\boldsymbol{q}}^T\boldsymbol{J}_O^{(m_i)T}\boldsymbol{R}_{m_i}\boldsymbol{I}_{m_i}^{m_i}\boldsymbol{R}_{m_i}^T\boldsymbol{J}_O^{(m_i)}\dot{\boldsymbol{q}}. \tag{4.30}$$

Finally, by summing the various contributions relative to the single links (4.20) and single rotors (4.30) as in (4.4), the total kinetic energy of the manipulator with actuators is given by the quadratic form

$$\mathcal{T} = \frac{1}{2}\sum_{i=1}^{n}\sum_{j=1}^{n} b_{ij}(\boldsymbol{q})\dot{q}_i\dot{q}_j = \frac{1}{2}\dot{\boldsymbol{q}}^T\boldsymbol{B}(\boldsymbol{q})\dot{\boldsymbol{q}} \tag{4.31}$$

where

$$\boldsymbol{B}(\boldsymbol{q}) = \sum_{i=1}^{n} \Big(m_{\ell_i} \boldsymbol{J}_P^{(\ell_i)T} \boldsymbol{J}_P^{(\ell_i)} + \boldsymbol{J}_O^{(\ell_i)T} \boldsymbol{R}_i \boldsymbol{I}_{\ell_i}^i \boldsymbol{R}_i^T \boldsymbol{J}_O^{(\ell_i)} + m_{m_i} \boldsymbol{J}_P^{(m_i)T} \boldsymbol{J}_P^{(m_i)} + \boldsymbol{J}_O^{(m_i)T} \boldsymbol{R}_{m_i} \boldsymbol{I}_{m_i}^{m_i} \boldsymbol{R}_{m_i}^T \boldsymbol{J}_O^{(m_i)} \Big) \tag{4.32}$$

is the $(n \times n)$ *inertia matrix* which is:

- *symmetric*,
- *positive definite*,
- (in general) *configuration-dependent*.

4.1.2 Computation of Potential Energy

As done for kinetic energy, the potential energy stored in the manipulator is given by the sum of the contributions relative to each link as well as to each rotor:

$$\mathcal{U} = \sum_{i=1}^{n} (\mathcal{U}_{\ell_i} + \mathcal{U}_{m_i}). \tag{4.33}$$

On the assumption of *rigid links*, the contribution due only to gravitational forces[5] is expressed by

$$\mathcal{U}_{\ell_i} = -\int_{V_{\ell_i}} \boldsymbol{g}_0^T \boldsymbol{p}_i^* \rho dV = -m_{\ell_i} \boldsymbol{g}_0^T \boldsymbol{p}_{\ell_i} \tag{4.34}$$

where $\boldsymbol{g}_0$ is the gravity acceleration vector in the base frame (e.g., $\boldsymbol{g}_0 = [\,0 \quad 0 \quad -g\,]^T$ if z is the vertical axis), and (4.7) has been utilized for the coordinates of the centre of mass of Link i. As regards the contribution of Rotor i, similarly to (4.34), one has

$$\mathcal{U}_{m_i} = -m_{m_i} \boldsymbol{g}_0^T \boldsymbol{p}_{m_i}. \tag{4.35}$$

By substituting (4.34) and (4.35) into (4.33), the *potential energy* is given by

$$\mathcal{U} = -\sum_{i=1}^{n} (m_{\ell_i} \boldsymbol{g}_0^T \boldsymbol{p}_{\ell_i} + m_{m_i} \boldsymbol{g}_0^T \boldsymbol{p}_{m_i}) \tag{4.36}$$

which reveals that potential energy, through the vectors $\boldsymbol{p}_{\ell_i}$ and $\boldsymbol{p}_{m_i}$ is a function only of the joint variables $\boldsymbol{q}$, and *not* of the joint velocities $\dot{\boldsymbol{q}}$.

---

[5] In the case of link flexibility, one would have an additional contribution due to elastic forces.

4.1.3 Equations of Motion

Having computed the total kinetic and potential energy of the system as in (4.31) and (4.36), the Lagrangian (4.1) for the manipulator can be written as

$$\begin{aligned}\mathcal{L}(\boldsymbol{q},\dot{\boldsymbol{q}}) &= \mathcal{T}(\boldsymbol{q},\dot{\boldsymbol{q}}) - \mathcal{U}(\boldsymbol{q}) \\ &= \frac{1}{2}\sum_{i=1}^{n}\sum_{j=1}^{n} b_{ij}(\boldsymbol{q})\dot{q}_i\dot{q}_j + \sum_{i=1}^{n}\left(m_{\ell_i}\boldsymbol{g}_0^T\boldsymbol{p}_{\ell_i}(\boldsymbol{q}) + m_{m_i}\boldsymbol{g}_0^T\boldsymbol{p}_{m_i}(\boldsymbol{q})\right).\end{aligned} \tag{4.37}$$

Taking the derivatives required by Lagrange's equations in (4.2) and recalling that U does not depend on $\dot{\boldsymbol{q}}$ yields

$$\begin{aligned}\frac{d}{dt}\left(\frac{\partial\mathcal{L}}{\partial\dot{q}_i}\right) = \frac{d}{dt}\left(\frac{\partial\mathcal{T}}{\partial\dot{q}_i}\right) &= \sum_{j=1}^{n} b_{ij}(\boldsymbol{q})\ddot{q}_j + \sum_{j=1}^{n}\frac{db_{ij}(\boldsymbol{q})}{dt}\dot{q}_j \\ &= \sum_{j=1}^{n} b_{ij}(\boldsymbol{q})\ddot{q}_j + \sum_{j=1}^{n}\sum_{k=1}^{n}\frac{\partial b_{ij}(\boldsymbol{q})}{\partial q_k}\dot{q}_k\dot{q}_j,\end{aligned}$$

and

$$\frac{\partial\mathcal{T}}{\partial q_i} = \frac{1}{2}\sum_{j=1}^{n}\sum_{k=1}^{n}\frac{\partial b_{jk}(\boldsymbol{q})}{\partial q_i}\dot{q}_k\dot{q}_j$$

where the indices of summation have been conveniently switched. Further, in view of (4.14) and (4.24), it is

$$\begin{aligned}\frac{\partial\mathcal{U}}{\partial q_i} &= -\sum_{j=1}^{n}\left(m_{\ell_j}\boldsymbol{g}_0^T\frac{\partial\boldsymbol{p}_{\ell_j}}{\partial q_i} + m_{m_j}\boldsymbol{g}_0^T\frac{\partial\boldsymbol{p}_{m_j}}{\partial q_i}\right) \\ &= -\sum_{j=1}^{n}\left(m_{\ell_j}\boldsymbol{g}_0^T\boldsymbol{j}_{Pi}^{(\ell_j)}(\boldsymbol{q}) + m_{m_j}\boldsymbol{g}_0^T\boldsymbol{j}_{Pi}^{(m_j)}(\boldsymbol{q})\right) = g_i(\boldsymbol{q})\end{aligned} \tag{4.38}$$

where, again, the index of summation has been changed.

As a result, the equations of motion are

$$\sum_{j=1}^{n} b_{ij}(\boldsymbol{q})\ddot{q}_j + \sum_{j=1}^{n}\sum_{k=1}^{n} h_{ijk}(\boldsymbol{q})\dot{q}_k\dot{q}_j + g_i(\boldsymbol{q}) = \xi_i \qquad i = 1,\ldots,n. \tag{4.39}$$

where

$$h_{ijk} = \frac{\partial b_{ij}}{\partial q_k} - \frac{1}{2}\frac{\partial b_{jk}}{\partial q_i}. \tag{4.40}$$

A physical interpretation of (4.39) reveals that:

- For the *acceleration terms*:

  The coefficient b_{ii} represents the moment of inertia at Joint i axis, in the current manipulator configuration, when the other joints are blocked.

The coefficient b_{ij} accounts for the effect of acceleration of Joint j on Joint j.

- For the *quadratic velocity terms*:

  The term $h_{ijj}\dot{q}_j^2$ represents the *centrifugal* effect induced on Joint i by velocity of Joint j; notice that $h_{iii} = 0$, since $\partial b_{ii}/\partial q_i = 0$.

  The term $h_{ijk}\dot{q}_j\dot{q}_k$ represents the *Coriolis* effect induced on Joint i by velocities of Joints j and k.

- For the *configuration-dependent terms*:

  The term g_i represents the moment generated at Joint i axis of the manipulator, in the current configuration, by the presence of gravity.

Some joint dynamic couplings, e.g., coefficients b_{ij} and h_{ijk}, may be reduced or eliminated when designing the structure, so as to simplify the control problem.

Regarding the nonconservative forces doing work at the manipulator joints, these are given by the *actuation torques* $\boldsymbol{\tau}$ minus the *viscous friction* torques $\boldsymbol{F}_v\dot{\boldsymbol{q}}$ and the static friction torques $\boldsymbol{f}_s(\boldsymbol{q}, \dot{\boldsymbol{q}})$: $\boldsymbol{F}_v$ denotes the $(n \times n)$ diagonal matrix of viscous friction coefficients. As a simplified model of static friction torques, one may consider the *Coulomb friction* torques $\boldsymbol{F}_s\,\mathbf{sgn}\,(\dot{\boldsymbol{q}})$, where $\boldsymbol{F}_s$ is an $(n \times n)$ diagonal matrix and $\mathbf{sgn}\,(\dot{\boldsymbol{q}})$ denotes the $(n \times 1)$ vector whose components are given by the sign functions of the single joint velocities.

If the manipulator's end effector is in contact with an environment, a portion of the actuation torques is used to balance the torques induced at the joints by the contact forces. According to a relation formally analogous to (3.99), such torques are given by $\boldsymbol{J}^T(\boldsymbol{q})\boldsymbol{h}$ where $\boldsymbol{h}$ denotes the vector of force and moment exerted by the end effector on the environment.

In sum, the equations of motion in (4.39) can be rewritten in the compact matrix form which represents the *joint space dynamic model*:

$$\boldsymbol{B}(\boldsymbol{q})\ddot{\boldsymbol{q}} + \boldsymbol{C}(\boldsymbol{q}, \dot{\boldsymbol{q}})\dot{\boldsymbol{q}} + \boldsymbol{F}_v\dot{\boldsymbol{q}} + \boldsymbol{F}_s\,\mathbf{sgn}\,(\dot{\boldsymbol{q}}) + \boldsymbol{g}(\boldsymbol{q}) = \boldsymbol{\tau} - \boldsymbol{J}^T(\boldsymbol{q})\boldsymbol{h}, \qquad (4.41)$$

where $\boldsymbol{C}$ is a suitable $(n \times n)$ matrix such that its elements c_{ij} satisfy the equation

$$\sum_{j=1}^{n} c_{ij}\dot{q}_j = \sum_{j=1}^{n}\sum_{k=1}^{n} h_{ijk}\dot{q}_k\dot{q}_j. \qquad (4.42)$$

Finally, if the manipulator structure contains a closed chain, it is advisable to compute first the dynamic model for the equivalent tree-structured open-chain manipulator. Then, from (3.108) the torques corresponding to the actuated joints can be computed, and thus the equations of motion can be cast in a form similar to (4.41) where $\boldsymbol{q} = \boldsymbol{q}_a$ is the resulting vector of generalized coordinates.

4.2 Notable Properties of Dynamic Model

In the following, two *notable properties* of the dynamic model are presented which will be useful for dynamic parameter identification as well as for deriving control algorithms.

4.2.1 Skew-symmetry of Matrix $\dot{B} - 2C$

The choice of the matrix C is not unique, since there exist several matrices C whose elements satisfy (4.42). A particular choice can be obtained by elaborating the term on the right-hand side of (4.42) and accounting for the expressions of the coefficients h_{ijk} in (4.40). To this purpose, one has

$$\begin{aligned}\sum_{j=1}^{n} c_{ij}\dot{q}_j &= \sum_{j=1}^{n}\sum_{k=1}^{n} h_{ijk}\dot{q}_k\dot{q}_j \\ &= \sum_{j=1}^{n}\sum_{k=1}^{n}\left(\frac{\partial b_{ij}}{\partial q_k} - \frac{1}{2}\frac{\partial b_{jk}}{\partial q_i}\right)\dot{q}_k\dot{q}_j.\end{aligned}$$

Splitting the first term on the right-hand side by an opportune switch of summation between j and k yields

$$\sum_{j=1}^{n} c_{ij}\dot{q}_j = \frac{1}{2}\sum_{j=1}^{n}\sum_{k=1}^{n}\frac{\partial b_{ij}}{\partial q_k}\dot{q}_k\dot{q}_j + \frac{1}{2}\sum_{j=1}^{n}\sum_{k=1}^{n}\left(\frac{\partial b_{ik}}{\partial q_j} - \frac{\partial b_{jk}}{\partial q_i}\right)\dot{q}_k\dot{q}_j.$$

As a consequence, the generic element of C is

$$c_{ij} = \sum_{k=1}^{n} c_{ijk}\dot{q}_k \tag{4.43}$$

where the coefficients

$$c_{ijk} = \frac{1}{2}\left(\frac{\partial b_{ij}}{\partial q_k} + \frac{\partial b_{ik}}{\partial q_j} - \frac{\partial b_{jk}}{\partial q_i}\right) \tag{4.44}$$

are termed *Christoffel symbols of the first type*. Notice that, in view of the symmetry of B, it is

$$c_{ijk} = c_{ikj}. \tag{4.45}$$

This choice for the matrix C leads to deriving the following notable property of the equations of motion (4.41). The matrix

$$N(q,\dot{q}) = \dot{B}(q) - 2C(q,\dot{q}) \tag{4.46}$$

is *skew-symmetric*; that is, given any $(n \times 1)$ vector w, the following relation holds:

$$w^T N(q,\dot{q})w = 0. \tag{4.47}$$

In fact, substituting the coefficient (4.44) into (4.43) gives

$$\begin{aligned}c_{ij} &= \frac{1}{2}\sum_{k=1}^{n}\frac{\partial b_{ij}}{\partial q_k}\dot{q}_k + \frac{1}{2}\sum_{k=1}^{n}\left(\frac{\partial b_{ik}}{\partial q_j} - \frac{\partial b_{jk}}{\partial q_i}\right)\dot{q}_k \\ &= \frac{1}{2}\dot{b}_{ij} + \frac{1}{2}\sum_{k=1}^{n}\left(\frac{\partial b_{ik}}{\partial q_j} - \frac{\partial b_{jk}}{\partial q_i}\right)\dot{q}_k,\end{aligned}$$

and then the expression of the generic element of the matrix $\boldsymbol{N}$ in (4.46) is

$$n_{ij} = \dot{b}_{ij} - 2c_{ij} = \sum_{k=1}^{n} \left(\frac{\partial b_{jk}}{\partial q_i} - \frac{\partial b_{ik}}{\partial q_j} \right) \dot{q}_k.$$

The result follows by observing that

$$n_{ij} = -n_{ji}.$$

An interesting property which is a direct implication of the skew-symmetry of $\boldsymbol{N}(\boldsymbol{q}, \dot{\boldsymbol{q}})$ is that, by setting $\boldsymbol{w} = \dot{\boldsymbol{q}}$,

$$\dot{\boldsymbol{q}}^T \boldsymbol{N}(\boldsymbol{q}, \dot{\boldsymbol{q}}) \dot{\boldsymbol{q}} = \boldsymbol{0}; \tag{4.48}$$

notice that (4.48) does not imply (4.47), since $\boldsymbol{N}$ is a function of $\dot{\boldsymbol{q}}$, too.

It can be shown that (4.48) holds for any choice of the matrix $\boldsymbol{C}$, since it is a result of the principle of conservation of energy (*Hamilton*). By virtue of this principle, the total time derivative of kinetic energy is balanced by the power generated by all the forces acting on the manipulator joints. For the mechanical system at issue, one may write

$$\frac{1}{2} \frac{d}{dt} \left(\dot{\boldsymbol{q}}^T \boldsymbol{B}(\boldsymbol{q}) \dot{\boldsymbol{q}} \right) = \dot{\boldsymbol{q}}^T \left(\boldsymbol{\tau} - \boldsymbol{F}_v \dot{\boldsymbol{q}} - \boldsymbol{F}_s \, \mathbf{sgn}\,(\dot{\boldsymbol{q}}) - \boldsymbol{g}(\boldsymbol{q}) - \boldsymbol{J}^T(\boldsymbol{q}) \boldsymbol{h} \right). \tag{4.49}$$

Taking the derivative on the left-hand side of (4.49) gives

$$\frac{1}{2} \dot{\boldsymbol{q}}^T \dot{\boldsymbol{B}}(\boldsymbol{q}) \dot{\boldsymbol{q}} + \dot{\boldsymbol{q}}^T \boldsymbol{B}(\boldsymbol{q}) \ddot{\boldsymbol{q}}$$

and substituting the expression of $\boldsymbol{B}(\boldsymbol{q})\ddot{\boldsymbol{q}}$ in (4.41) yields

$$\begin{aligned} \frac{1}{2} \frac{d}{dt} \left(\dot{\boldsymbol{q}}^T \boldsymbol{B}(\boldsymbol{q}) \dot{\boldsymbol{q}} \right) &= \frac{1}{2} \dot{\boldsymbol{q}}^T \left(\dot{\boldsymbol{B}}(\boldsymbol{q}) - 2\boldsymbol{C}(\boldsymbol{q}, \dot{\boldsymbol{q}}) \right) \dot{\boldsymbol{q}} \\ &\quad + \dot{\boldsymbol{q}}^T \left(\boldsymbol{\tau} - \boldsymbol{F}_v \dot{\boldsymbol{q}} - \boldsymbol{F}_s \, \mathbf{sgn}\,(\dot{\boldsymbol{q}}) - \boldsymbol{g}(\boldsymbol{q}) - \boldsymbol{J}^T(\boldsymbol{q}) \boldsymbol{h} \right). \end{aligned} \tag{4.50}$$

A direct comparison of the right-hand sides of (4.49) and (4.50) leads to the result established by (4.48).

To summarize, the relation (4.48) holds for any choice of the matrix $\boldsymbol{C}$, since it is a direct consequence of the physical properties of the system, whereas the relation (4.47) holds only for the particular choice of the elements of $\boldsymbol{C}$ as in (4.43) and (4.44).

4.2.2 Linearity in the Dynamic Parameters

An important property of the dynamic model is the *linearity* with respect to the *dynamic parameters* characterizing the manipulator links and rotors.

In order to determine such parameters, it is worth associating the kinetic and potential energy contributions of each rotor with those of the link on which it is located. Hence, by considering the union of Link i and Rotor $i+1$ (*augmented Link* i), the kinetic energy contribution is given by

$$\mathcal{T}_i = \mathcal{T}_{\ell_i} + \mathcal{T}_{m_{i+1}} \tag{4.51}$$

where

$$\mathcal{T}_{\ell_i} = \frac{1}{2} m_{\ell i} \dot{\boldsymbol{p}}_{\ell_i}^T \dot{\boldsymbol{p}}_{\ell_i} + \frac{1}{2} \boldsymbol{\omega}_i^T \boldsymbol{I}_{\ell_i} \boldsymbol{\omega}_i \tag{4.52}$$

and

$$\mathcal{T}_{m_{i+1}} = \frac{1}{2} m_{m_{i+1}} \dot{\boldsymbol{p}}_{m_{i+1}}^T \dot{\boldsymbol{p}}_{m_{i+1}} + \frac{1}{2} \boldsymbol{\omega}_{m_{i+1}}^T \boldsymbol{I}_{m_{i+1}} \boldsymbol{\omega}_{m_{i+1}}. \tag{4.53}$$

With reference to the centre of mass of the augmented link, the linear velocities of the link and rotor can be expressed according to (3.24) as:

$$\dot{\boldsymbol{p}}_{\ell_i} = \dot{\boldsymbol{p}}_{C_i} + \boldsymbol{\omega}_i \times \boldsymbol{r}_{C_i,\ell_i} \tag{4.54}$$

$$\dot{\boldsymbol{p}}_{m_{i+1}} = \dot{\boldsymbol{p}}_{C_i} + \boldsymbol{\omega}_i \times \boldsymbol{r}_{C_i,m_{i+1}} \tag{4.55}$$

with

$$\boldsymbol{r}_{C_i,\ell_i} = \boldsymbol{p}_{\ell_i} - \boldsymbol{p}_{C_i} \tag{4.56}$$

$$\boldsymbol{r}_{C_i,m_{i+1}} = \boldsymbol{p}_{m_{i+1}} - \boldsymbol{p}_{C_i}, \tag{4.57}$$

where $\boldsymbol{p}_{C_i}$ denotes the position vector of the centre of mass of augmented Link i.

Substituting (4.54) into (4.52) gives

$$\begin{aligned} \mathcal{T}_{\ell_i} &= \frac{1}{2} m_{\ell i} \dot{\boldsymbol{p}}_{C_i}^T \dot{\boldsymbol{p}}_{C_i} + \dot{\boldsymbol{p}}_{C_i}^T \boldsymbol{S}(\boldsymbol{\omega}_i) m_{\ell_i} \boldsymbol{r}_{C_i,\ell_i} \\ &\quad + \frac{1}{2} m_{\ell_i} \boldsymbol{\omega}_i^T \boldsymbol{S}^T(\boldsymbol{r}_{C_i,\ell_i}) \boldsymbol{S}(\boldsymbol{r}_{C_i,\ell_i}) \boldsymbol{\omega}_i + \frac{1}{2} \boldsymbol{\omega}_i^T \boldsymbol{I}_{\ell_i} \boldsymbol{\omega}_i. \end{aligned} \tag{4.58}$$

By virtue of *Steiner's theorem*, the matrix

$$\bar{\boldsymbol{I}}_{\ell_i} = \boldsymbol{I}_{\ell_i} + m_{\ell_i} \boldsymbol{S}^T(\boldsymbol{r}_{C_i,\ell_i}) \boldsymbol{S}(\boldsymbol{r}_{C_i,\ell_i}) \tag{4.59}$$

represents the inertia tensor relative to the overall centre of mass $\boldsymbol{p}_{C_i}$, which contains an additional contribution due to the translation of the pole with respect to which the tensor is evaluated, as in (4.56). Therefore, (4.58) can be written as

$$\mathcal{T}_{\ell_i} = \frac{1}{2} m_{\ell i} \dot{\boldsymbol{p}}_{C_i}^T \dot{\boldsymbol{p}}_{C_i} + \dot{\boldsymbol{p}}_{C_i}^T \boldsymbol{S}(\boldsymbol{\omega}_i) m_{\ell_i} \boldsymbol{r}_{C_i,\ell_i} + \frac{1}{2} \boldsymbol{\omega}_i^T \bar{\boldsymbol{I}}_{\ell_i} \boldsymbol{\omega}_i. \tag{4.60}$$

In a similar fashion, substituting (4.55) into (4.53) and exploiting (4.23) yields

$$\begin{aligned} \mathcal{T}_{m_{i+1}} &= \frac{1}{2} m_{m_{i+1}} \dot{\boldsymbol{p}}_{C_i}^T \dot{\boldsymbol{p}}_{C_i} + \dot{\boldsymbol{p}}_{C_i}^T \boldsymbol{S}(\boldsymbol{\omega}_i) m_{m_{i+1}} \boldsymbol{r}_{C_i,m_{i+1}} + \frac{1}{2} \boldsymbol{\omega}_i^T \bar{\boldsymbol{I}}_{m_{i+1}} \boldsymbol{\omega}_i \\ &\quad + k_{r,i+1} \dot{q}_{i+1} \boldsymbol{z}_{m_{i+1}}^T \boldsymbol{I}_{m_{i+1}} \boldsymbol{\omega}_i + \frac{1}{2} k_{r,i+1}^2 \dot{q}_{i+1}^2 \boldsymbol{z}_{m_{i+1}}^T \boldsymbol{I}_{m_{i+1}} \boldsymbol{z}_{m_{i+1}}, \end{aligned} \tag{4.61}$$

where

$$\bar{\boldsymbol{I}}_{m_{i+1}} = \boldsymbol{I}_{m_{i+1}} + m_{m_{i+1}} \boldsymbol{S}^T(\boldsymbol{r}_{C_i,m_{i+1}}) \boldsymbol{S}(\boldsymbol{r}_{C_i,m_{i+1}}). \tag{4.62}$$

Summing the contributions in (4.60) and (4.61) as in (4.51) gives the expression of the kinetic energy of augmented Link i in the form

$$\begin{aligned}\mathcal{T}_i = {} & \frac{1}{2} m_i \dot{\boldsymbol{p}}_{C_i}^T \dot{\boldsymbol{p}}_{C_i} + \frac{1}{2} \boldsymbol{\omega}_i^T \bar{\boldsymbol{I}}_i \boldsymbol{\omega}_i + k_{r,i+1} \dot{q}_{i+1} \boldsymbol{z}_{m_{i+1}}^T \boldsymbol{I}_{m_{i+1}} \boldsymbol{\omega}_i \\ & + \frac{1}{2} k_{r,i+1}^2 \dot{q}_{i+1}^2 \boldsymbol{z}_{m_{i+1}}^T \boldsymbol{I}_{m_{i+1}} \boldsymbol{z}_{m_{i+1}},\end{aligned} \tag{4.63}$$

where $m_i = m_{\ell_i} + m_{m_{i+1}}$ and $\bar{\boldsymbol{I}}_i = \bar{\boldsymbol{I}}_{\ell_i} + \bar{\boldsymbol{I}}_{m_{i+1}}$ are respectively the overall mass and inertia tensor. In deriving (4.63), the relations in (4.56) and (4.57) have been utilized as well as the following relation between the positions of the centres of mass:

$$m_{\ell_i} \boldsymbol{p}_{\ell_i} + m_{m_{i+1}} \boldsymbol{p}_{m_{i+1}} = m_i \boldsymbol{p}_{C_i}. \tag{4.64}$$

Notice that the first two terms on the right-hand side of (4.63) represent the kinetic energy contribution of the rotor when this is still, whereas the remaining two terms account for the rotor's own motion.

On the assumption that the rotor has a symmetric mass distribution about its axis of rotation, its inertia tensor expressed in a frame $\boldsymbol{R}_{m_i}$ with origin at the centre of mass and axis z_{m_i} aligned with the rotation axis can be written as

$$\boldsymbol{I}_{m_i}^{m_i} = \begin{bmatrix} I_{m_i xx} & 0 & 0 \\ 0 & I_{m_i yy} & 0 \\ 0 & 0 & I_{m_i zz} \end{bmatrix} \tag{4.65}$$

where $I_{m_i yy} = I_{m_i xx}$. As a consequence, the inertia tensor is invariant with respect to any rotation about axis z_{m_i} and is, anyhow, constant when referred to any frame attached to Link $i-1$.

Since the aim is to determine a set of dynamic parameters independent of the manipulator joint configuration, it is worth referring the inertia tensor of the link $\bar{\boldsymbol{I}}_i$ to frame $\boldsymbol{R}_i$ attached to the link and the inertia tensor $\boldsymbol{I}_{m_{i+1}}$ to frame $\boldsymbol{R}_{m_{i+1}}$ so that it is diagonal. In view of (4.65) one has

$$\boldsymbol{I}_{m_{i+1}} \boldsymbol{z}_{m_{i+1}} = \boldsymbol{R}_{m_{i+1}} \boldsymbol{I}_{m_{i+1}}^{m_{i+1}} \boldsymbol{R}_{m_{i+1}}^T \boldsymbol{z}_{m_{i+1}} = I_{m_{i+1}} \boldsymbol{z}_{m_{i+1}} \tag{4.66}$$

where $I_{m_{i+1}} = I_{m_{i+1}zz}$ denotes the constant scalar moment of inertia of the rotor about its rotation axis.

Therefore, the kinetic energy (4.63) becomes

$$\begin{aligned}\mathcal{T}_i = {} & \frac{1}{2} m_i \dot{\boldsymbol{p}}_{C_i}^{iT} \dot{\boldsymbol{p}}_{C_i}^i + \frac{1}{2} \boldsymbol{\omega}_i^{iT} \bar{\boldsymbol{I}}_i^i \boldsymbol{\omega}_i^i + k_{r,i+1} \dot{q}_{i+1} I_{m_{i+1}} \boldsymbol{z}_{m_{i+1}}^{iT} \boldsymbol{\omega}_i^i \\ & + \frac{1}{2} k_{r,i+1}^2 \dot{q}_{i+1}^2 I_{m_{i+1}}.\end{aligned} \tag{4.67}$$

According to the linear velocity composition rule for Link i in (3.13), one may write

$$\dot{\boldsymbol{p}}_{C_i}^i = \dot{\boldsymbol{p}}_i^i + \boldsymbol{\omega}_i^i \times \boldsymbol{r}_{i,C_i}^i \tag{4.68}$$

where all the vectors have been referred to Frame i; note that $\boldsymbol{r}^i_{i,C_i}$ is fixed in such frame. Substituting (4.68) into (4.67) gives

$$\mathcal{T}_i = \frac{1}{2}m_i\dot{\boldsymbol{p}}_i^{iT}\dot{\boldsymbol{p}}_i^i + \dot{\boldsymbol{p}}_i^{iT}\boldsymbol{S}(\boldsymbol{\omega}_i^i)m_i\boldsymbol{r}^i_{i,C_i} + \frac{1}{2}\boldsymbol{\omega}_i^{iT}\hat{\boldsymbol{I}}_i^i\boldsymbol{\omega}_i^i \\ + k_{r,i+1}\dot{q}_{i+1}I_{m_{i+1}}\boldsymbol{z}^{iT}_{m_{i+1}}\boldsymbol{\omega}_i^i + \frac{1}{2}k^2_{r,i+1}\dot{q}^2_{i+1}I_{m_{i+1}}, \tag{4.69}$$

where

$$\hat{\boldsymbol{I}}_i^i = \bar{\boldsymbol{I}}_i^i + m_i\boldsymbol{S}^T(\boldsymbol{r}^i_{i,C_i})\boldsymbol{S}(\boldsymbol{r}^i_{i,C_i}) \tag{4.70}$$

represents the inertia tensor with respect to the origin of Frame i according to Steiner's theorem.

Let $\boldsymbol{r}^i_{i,C_i} = [\,\ell_{C_ix} \quad \ell_{C_iy} \quad \ell_{C_iz}\,]^T$. The *first moment of inertia* is

$$m_i\boldsymbol{r}^i_{i,C_i} = \begin{bmatrix} m_i\ell_{C_ix} \\ m_i\ell_{C_iy} \\ m_i\ell_{C_iz} \end{bmatrix}. \tag{4.71}$$

From (4.70) the inertia tensor of augmented Link i is

$$\hat{\boldsymbol{I}}_i^i = \begin{bmatrix} \bar{I}_{ixx} + m_i(\ell^2_{C_iy} + \ell^2_{C_iz}) & -\bar{I}_{ixy} - m_i\ell_{C_ix}\ell_{C_iy} & -\bar{I}_{ixz} - m_i\ell_{C_ix}\ell_{C_iz} \\ * & \bar{I}_{iyy} + m_i(\ell^2_{C_ix} + \ell^2_{C_iz}) & -\bar{I}_{iyz} - m_i\ell_{C_iy}\ell_{C_iz} \\ * & * & \bar{I}_{izz} + m_i(\ell^2_{C_ix} + \ell^2_{C_iy}) \end{bmatrix}$$
$$= \begin{bmatrix} \hat{I}_{ixx} & -\hat{I}_{ixy} & -\hat{I}_{ixz} \\ * & \hat{I}_{iyy} & -\hat{I}_{iyz} \\ * & * & \hat{I}_{izz} \end{bmatrix}. \tag{4.72}$$

Therefore, the kinetic energy of the augmented link is linear with respect to the dynamic parameters; namely, the *mass*, the *three components of the first moment of inertia* in (4.71), the *six components of the inertia tensor* in (4.72), and the *moment of inertia of the rotor*.

As regards potential energy, it is worth referring to the centre of mass of augmented Link i defined as in (4.64), and thus the single contribution of potential energy can be written as

$$\mathcal{U}_i = -m_i\boldsymbol{g}_0^{iT}\boldsymbol{p}^i_{C_i} \tag{4.73}$$

where the vectors have been referred to Frame i. According to the relation

$$\boldsymbol{p}^i_{C_i} = \boldsymbol{p}^i_i + \boldsymbol{r}^i_{i,C_i},$$

The expression in (4.73) can be rewritten as

$$\mathcal{U}_i = -\boldsymbol{g}_0^{iT}(m_i\boldsymbol{p}^i_i + m_i\boldsymbol{r}^i_{i,C_i}), \tag{4.74}$$

that is, the potential energy of the augmented link is linear with respect to the mass and the three components of the first moment of inertia in (4.71).

By summing the contributions of kinetic energy and potential energy for all augmented links, the Lagrangian of the system (4.1) can be expressed in the form

$$\mathcal{L} = \sum_{i=1}^{n} (\boldsymbol{\beta}_{\mathcal{T}i}^T - \boldsymbol{\beta}_{\mathcal{U}i}^T)\boldsymbol{\pi}_i \tag{4.75}$$

where $\boldsymbol{\pi}_i$ is the (11×1) vector of dynamic parameters

$$\boldsymbol{\pi}_i = [\, m_i \;\; m_i\ell_{C_ix} \;\; m_i\ell_{C_iy} \;\; m_i\ell_{C_iz} \;\; \hat{I}_{ixx} \;\; \hat{I}_{ixy} \;\; \hat{I}_{ixz} \;\; \hat{I}_{iyy} \;\; \hat{I}_{iyz} \;\; \hat{I}_{izz} \;\; I_{m_i} \,]^T, \tag{4.76}$$

in which the moment of inertia of Rotor i has been associated with the parameters of Link i so as to simplify the notation.

In (4.75), $\boldsymbol{\beta}_{\mathcal{T}i}$ and $\boldsymbol{\beta}_{\mathcal{U}i}$ are two (11×1) vectors that allow writing the Lagrangian as a function of $\boldsymbol{\pi}_i$. Such vectors are a function of the generalized coordinates of the mechanical system (and also of their derivatives as regards $\boldsymbol{\beta}_{\mathcal{T}i}$). In particular, it can be shown that $\boldsymbol{\beta}_{\mathcal{T}i} = \boldsymbol{\beta}_{\mathcal{T}i}(q_1, q_2, \ldots, q_i, \dot{q}_1, \dot{q}_2, \ldots, \dot{q}_i)$ and $\boldsymbol{\beta}_{\mathcal{U}i} = \boldsymbol{\beta}_{\mathcal{U}i}(q_1, q_2, \ldots, q_i)$, *i.e.*, they do not depend on the variables of the joints subsequent to Link i.

At this point, it should be observed how the derivations required by the Lagrange's equations in (4.2) do not alter the property of linearity in the parameters, and then the generalized force at Joint i can be written as

$$\xi_i = \sum_{j=1}^{n} \boldsymbol{y}_{ij}^T \boldsymbol{\pi}_j \tag{4.77}$$

where

$$\boldsymbol{y}_{ij} = \frac{d}{dt}\frac{\partial \boldsymbol{\beta}_{\mathcal{T}j}}{\partial \dot{q}_i} - \frac{\partial \boldsymbol{\beta}_{\mathcal{T}j}}{\partial q_i} + \frac{\partial \boldsymbol{\beta}_{\mathcal{U}j}}{\partial q_i}. \tag{4.78}$$

Since the partial derivatives of $\boldsymbol{\beta}_{\mathcal{T}j}$ and $\boldsymbol{\beta}_{\mathcal{U}j}$ appearing in (4.78) vanish for $j < i$, the following notable result is obtained:

$$\begin{bmatrix} \xi_1 \\ \xi_2 \\ \vdots \\ \xi_n \end{bmatrix} = \begin{bmatrix} \boldsymbol{y}_{11}^T & \boldsymbol{y}_{12}^T & \ldots & \boldsymbol{y}_{1n}^T \\ \boldsymbol{0}^T & \boldsymbol{y}_{22}^T & \ldots & \boldsymbol{y}_{2n}^T \\ \vdots & \vdots & \ddots & \vdots \\ \boldsymbol{0}^T & \boldsymbol{0}^T & \ldots & \boldsymbol{y}_{nn}^T \end{bmatrix} \begin{bmatrix} \boldsymbol{\pi}_1 \\ \boldsymbol{\pi}_2 \\ \vdots \\ \boldsymbol{\pi}_n \end{bmatrix} \tag{4.79}$$

which constitutes the property of *linearity of the model* of a manipulator with respect to a suitable set of *dynamic parameters*.

In the simple case of no contact forces ($\boldsymbol{h} = \boldsymbol{0}$), it may be worth including the viscous friction coefficient F_{vi} and Coulomb friction coefficient F_{si} in the parameters of the vector $\boldsymbol{\pi}_i$, thus leading to a total number of 13 parameters per joint. In sum, (4.79) can be compactly written as

$$\boldsymbol{\tau} = \boldsymbol{Y}(\boldsymbol{q}, \dot{\boldsymbol{q}}, \ddot{\boldsymbol{q}})\boldsymbol{\pi}, \tag{4.80}$$

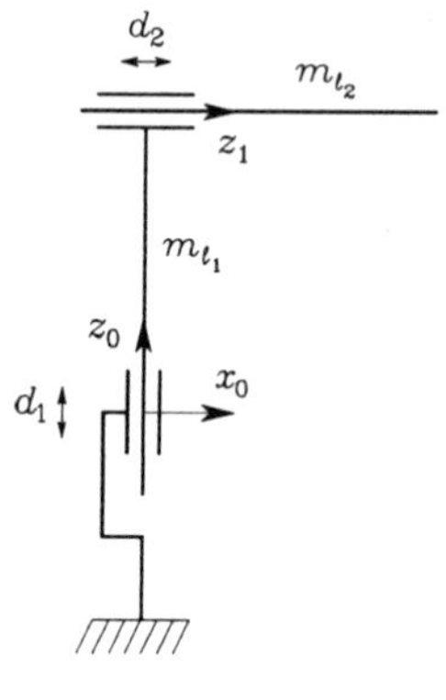

Figure 4.4 Two-link Cartesian arm.

where $\boldsymbol{\pi}$ is a $(p \times 1)$ vector of *constant* parameters and $\boldsymbol{Y}$ is an $(n \times p)$ matrix which is a *function of joint positions, velocities and accelerations*; such a matrix is usually called *regressor*. Regarding the dimension of the parameter vector, notice that $p \leq 13n$ since not all the thirteen parameters for each joint may explicitly appear in (4.80).

4.3 Dynamic Model of Simple Manipulator Structures

In the following, three examples of dynamic model computation are illustrated for simple two-degree-of-mobility manipulator structures. Two degrees of mobility, in fact, are enough to understand the physical meaning of all dynamic terms, especially the joint coupling terms. On the other hand, dynamic model computation for manipulators with more degrees of mobility would be quite tedious and prone to errors, when carried out by paper and pencil. In those cases, it is advisable to perform it with the aid of a symbolic programming software package.

4.3.1 Two-link Cartesian Arm

Consider the two-link Cartesian arm in Figure 4.4, for which the vector of generalized coordinates is $\boldsymbol{q} = [\,d_1 \quad d_2\,]^T$. Let m_{ℓ_1}, m_{ℓ_2} be the masses of the two links, and m_{m_1}, m_{m_2} the masses of the rotors of the two joint motors. Let also I_{m_1}, I_{m_2} be the moments of inertia with respect to the axes of the two rotors. It is assumed that $\boldsymbol{p}_{m_i} = \boldsymbol{p}_{i-1}$ and $\boldsymbol{z}_{m_i} = \boldsymbol{z}_{i-1}$, for $i = 1, 2$, *i.e.*, the motors are located on the joint axes with centres of mass located at the origins of the respective frames.

With the chosen coordinate frames, computation of the Jacobians in (4.16) and (4.18) yields

$$\boldsymbol{J}_P^{(\ell_1)} = \begin{bmatrix} 0 & 0 \\ 0 & 0 \\ 1 & 0 \end{bmatrix} \qquad \boldsymbol{J}_P^{(\ell_2)} = \begin{bmatrix} 0 & 1 \\ 0 & 0 \\ 1 & 0 \end{bmatrix}.$$

Obviously, in this case there are no angular velocity contributions for both links.

Computation of the Jacobians in (4.25), (4.26), (4.28), (4.29) yields

$$\boldsymbol{J}_P^{(m_1)} = \begin{bmatrix} 0 & 0 \\ 0 & 0 \\ 0 & 0 \end{bmatrix} \qquad \boldsymbol{J}_P^{(m_2)} = \begin{bmatrix} 0 & 0 \\ 0 & 0 \\ 1 & 0 \end{bmatrix}$$

$$\boldsymbol{J}_O^{(m_1)} = \begin{bmatrix} 0 & 0 \\ 0 & 0 \\ k_{r1} & 0 \end{bmatrix} \qquad \boldsymbol{J}_O^{(m_2)} = \begin{bmatrix} 0 & k_{r2} \\ 0 & 0 \\ 0 & 0 \end{bmatrix}$$

where k_{ri} is the gear reduction ratio of Motor i. It is obvious to see that $z_1 = [1 \quad 0 \quad 0]^T$, which greatly simplifies computation of the second term in (4.30).

From (4.32), the inertia matrix is

$$\boldsymbol{B} = \begin{bmatrix} m_{\ell_1} + m_{m_2} + k_{r1}^2 I_{m_1} + m_{\ell_2} & 0 \\ 0 & m_{\ell_2} + k_{r2}^2 I_{m_2} \end{bmatrix}.$$

It has to be remarked that $\boldsymbol{B}$ is *constant*, *i.e.*, it does not depend on the arm configuration. This implies also that $\boldsymbol{C} = \boldsymbol{O}$, *i.e.*, there are no contributions of centrifugal and Coriolis forces. As for the gravitational terms, since $\boldsymbol{g}_0 = [0 \quad 0 \quad -g]^T$ (g is gravity acceleration), (4.38) with the above Jacobians gives:

$$g_1 = (m_{\ell_1} + m_{m_2} + m_{\ell_2})g \qquad g_2 = 0.$$

In the absence of friction and tip contact forces, the resulting equations of motion are

$$(m_{\ell_1} + m_{m_2} + k_{r1}^2 I_{m_1} + m_{\ell_2})\ddot{d}_1 + (m_{\ell_1} + m_{m_2} + m_{\ell_2})g = \tau_1$$
$$(m_{\ell_2} + k_{r2}^2 I_{m_2})\ddot{d}_2 = \tau_2$$

where τ_1 and τ_2 denote the forces applied to the two joints. Notice that a completely decoupled dynamics has been obtained. This is a consequence not only of the Cartesian structures but also of the particular geometry; in other words, if the second joint axis were not at a right angle with the first joint axis, the resulting inertia matrix would not be diagonal.

4.3.2 Two-link Planar Arm

Consider the two-link planar arm in Figure 4.5, for which the vector of generalized coordinates is $\boldsymbol{q} = [\vartheta_1 \quad \vartheta_2]^T$. Let ℓ_1, ℓ_2 be the distances of the centres of mass of the two links from the respective joint axes. Let also m_{ℓ_1}, m_{ℓ_2} be the masses of the two links, and m_{m_1}, m_{m_2} the masses of the rotors of the two joint motors. Finally, let I_{m_1}, I_{m_2} be the moments of inertia with respect to the axes of the two rotors, and I_{ℓ_1}, I_{ℓ_2} the moments of inertia relative to the centres of mass of the two links, respectively. It

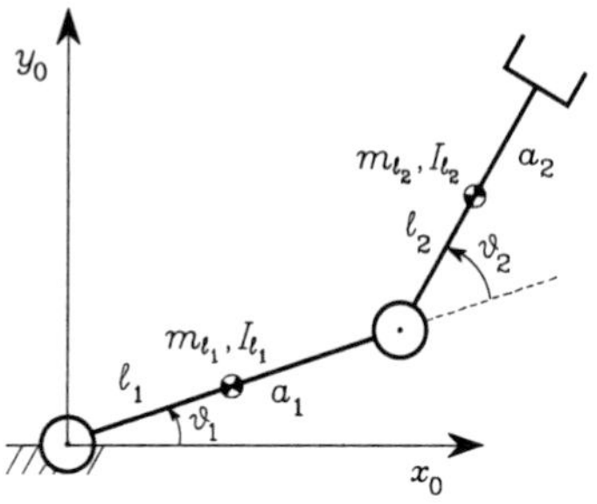

Figure 4.5 Two-link planar arm.

is assumed that $\boldsymbol{p}_{m_i} = \boldsymbol{p}_{i-1}$ and $\boldsymbol{z}_{m_i} = \boldsymbol{z}_{i-1}$, for $i = 1, 2$, *i.e.*, the motors are located on the joint axes with centres of mass located at the origins of the respective frames.

With the chosen coordinate frames, computation of the Jacobians in (4.16) and (4.18) yields

$$\boldsymbol{J}_P^{(\ell_1)} = \begin{bmatrix} -\ell_1 s_1 & 0 \\ \ell_1 c_1 & 0 \\ 0 & 0 \end{bmatrix} \qquad \boldsymbol{J}_P^{(\ell_2)} = \begin{bmatrix} -a_1 s_1 - \ell_2 s_{12} & -\ell_2 s_{12} \\ a_1 c_1 + \ell_2 c_{12} & \ell_2 c_{12} \\ 0 & 0 \end{bmatrix},$$

whereas computation of the Jacobians in (4.17) and (4.19) yields

$$\boldsymbol{J}_O^{(\ell_1)} = \begin{bmatrix} 0 & 0 \\ 0 & 0 \\ 1 & 0 \end{bmatrix} \qquad \boldsymbol{J}_O^{(\ell_2)} = \begin{bmatrix} 0 & 0 \\ 0 & 0 \\ 1 & 1 \end{bmatrix}.$$

Notice that $\boldsymbol{\omega}_i$, for $i = 1, 2$, is aligned with z_0, and thus $\boldsymbol{R}_i$ has *no* effect. It is then possible to refer to the scalar moments of inertia I_{ℓ_i}.

Computation of the Jacobians in (4.25) and (4.26) yields

$$\boldsymbol{J}_P^{(m_1)} = \begin{bmatrix} 0 & 0 \\ 0 & 0 \\ 0 & 0 \end{bmatrix} \qquad \boldsymbol{J}_P^{(m_2)} = \begin{bmatrix} -a_1 s_1 & 0 \\ a_1 c_1 & 0 \\ 0 & 0 \end{bmatrix},$$

whereas computation of the Jacobians in (4.28) and (4.29) yields

$$\boldsymbol{J}_O^{(m_1)} = \begin{bmatrix} 0 & 0 \\ 0 & 0 \\ k_{r1} & 0 \end{bmatrix} \qquad \boldsymbol{J}_O^{(m_2)} = \begin{bmatrix} 0 & 0 \\ 0 & 0 \\ 1 & k_{r2} \end{bmatrix},$$

where k_{ri} is the gear reduction ratio of Motor i.

From (4.32), the inertia matrix is

$$\boldsymbol{B}(\boldsymbol{q}) = \begin{bmatrix} b_{11}(\vartheta_2) & b_{12}(\vartheta_2) \\ b_{21}(\vartheta_2) & b_{22} \end{bmatrix}$$

$$\begin{aligned} b_{11} &= I_{\ell_1} + m_{\ell_1}\ell_1^2 + k_{r1}^2 I_{m_1} + I_{\ell_2} + m_{\ell_2}(a_1^2 + \ell_2^2 + 2a_1\ell_2 c_2) \\ &\quad + I_{m_2} + m_{m_2}a_1^2 \\ b_{12} &= b_{21} = I_{\ell_2} + m_{\ell_2}(\ell_2^2 + a_1\ell_2 c_2) + k_{r2}I_{m_2} \\ b_{22} &= I_{\ell_2} + m_{\ell_2}\ell_2^2 + k_{r2}^2 I_{m_2}. \end{aligned}$$

Compared to the previous example, the inertia matrix is now configuration-dependent and computation of Christoffel symbols as in (4.44) gives:

$$
\begin{aligned}
c_{111} &= \frac{1}{2}\frac{\partial b_{11}}{\partial q_1} = 0 \\
c_{112} = c_{121} &= \frac{1}{2}\frac{\partial b_{11}}{\partial q_2} = -m_{\ell_2}a_1\ell_2 s_2 = h \\
c_{122} &= \frac{\partial b_{12}}{\partial q_2} - \frac{1}{2}\frac{\partial b_{22}}{\partial q_1} = h \\
c_{211} &= \frac{\partial b_{21}}{\partial q_1} - \frac{1}{2}\frac{\partial b_{11}}{\partial q_2} = -h \\
c_{212} = c_{221} &= \frac{1}{2}\frac{\partial b_{22}}{\partial q_1} = 0 \\
c_{222} &= \frac{1}{2}\frac{\partial b_{22}}{\partial q_2} = 0,
\end{aligned}
$$

leading to the matrix

$$
\boldsymbol{C}(\boldsymbol{q},\dot{\boldsymbol{q}}) = \begin{bmatrix} h\dot{\vartheta}_2 & h(\dot{\vartheta}_1+\dot{\vartheta}_2) \\ -h\dot{\vartheta}_1 & 0 \end{bmatrix}.
$$

Computing the matrix $\boldsymbol{N}$ in (4.46) gives

$$
\begin{aligned}
\boldsymbol{N}(\boldsymbol{q},\dot{\boldsymbol{q}}) &= \dot{\boldsymbol{B}}(\boldsymbol{q}) - 2\boldsymbol{C}(\boldsymbol{q},\dot{\boldsymbol{q}}) \\
&= \begin{bmatrix} 2h\dot{\vartheta}_2 & h\dot{\vartheta}_2 \\ h\dot{\vartheta}_2 & 0 \end{bmatrix} - 2\begin{bmatrix} h\dot{\vartheta}_2 & h(\dot{\vartheta}_1+\dot{\vartheta}_2) \\ -h\dot{\vartheta}_1 & 0 \end{bmatrix} \\
&= \begin{bmatrix} 0 & -2h\dot{\vartheta}_1 - h\dot{\vartheta}_2 \\ 2h\dot{\vartheta}_1 + h\dot{\vartheta}_2 & 0 \end{bmatrix}
\end{aligned}
$$

that allows verifying the skew-symmetry property expressed by (4.47).

As for the gravitational terms, since $\boldsymbol{g}_0 = [\,0 \quad -g \quad 0\,]^T$, (4.38) with the above Jacobians gives:

$$
\begin{aligned}
g_1 &= (m_{\ell_1}\ell_1 + m_{m_2}a_1 + m_{\ell_2}a_1)gc_1 + m_{\ell_2}\ell_2 gc_{12} \\
g_2 &= m_{\ell_2}\ell_2 gc_{12}.
\end{aligned}
$$

In the absence of friction and tip contact forces, the resulting equations of motion are

$$
\begin{aligned}
&\big(I_{\ell_1} + m_{\ell_1}\ell_1^2 + k_{r1}^2 I_{m_1} + I_{\ell_2} + m_{\ell_2}(a_1^2 + \ell_2^2 + 2a_1\ell_2 c_2) + I_{m_2} + m_{m_2}a_1^2\big)\ddot{\vartheta}_1 \\
&\quad + \big(I_{\ell_2} + m_{\ell_2}(\ell_2^2 + a_1\ell_2 c_2) + k_{r2}I_{m_2}\big)\ddot{\vartheta}_2 \\
&\quad - 2m_{\ell_2}a_1\ell_2 s_2\dot{\vartheta}_1\dot{\vartheta}_2 - m_{\ell_2}a_1\ell_2 s_2\dot{\vartheta}_2^2 \\
&\quad + (m_{\ell_1}\ell_1 + m_{m_2}a_1 + m_{\ell_2}a_1)gc_1 + m_{\ell_2}\ell_2 gc_{12} = \tau_1 \qquad (4.81) \\
&\big(I_{\ell_2} + m_{\ell_2}(\ell_2^2 + a_1\ell_2 c_2) + k_{r2}I_{m_2}\big)\ddot{\vartheta}_1 + \big(I_{\ell_2} + m_{\ell_2}\ell_2^2 + k_{r2}^2 I_{m_2}\big)\ddot{\vartheta}_2 \\
&\quad + m_{\ell_2}a_1\ell_2 s_2\dot{\vartheta}_1^2 + m_{\ell_2}\ell_2 gc_{12} = \tau_2
\end{aligned}
$$

where τ_1 and τ_2 denote the torques applied to the joints.

Finally, it is wished to derive a parameterization of the dynamic model (4.81) according to the relation (4.80). By direct inspection of the expressions of the joint torques, it is possible to find the following parameter vector:

$$\boldsymbol{\pi} = \begin{bmatrix} \pi_1 & \pi_2 & \pi_3 & \pi_4 & \pi_5 & \pi_6 & \pi_7 & \pi_8 \end{bmatrix}^T \tag{4.82}$$

$$\begin{aligned}
\pi_1 &= m_1 = m_{\ell_1} + m_{m_2} \\
\pi_2 &= m_1 \ell_{C_1} = m_{\ell_1}(\ell_1 - a_1) \\
\pi_3 &= \hat{I}_1 = I_{\ell_1} + m_{\ell_1}(\ell_1 - a_1)^2 + I_{m_2} \\
\pi_4 &= I_{m_1} \\
\pi_5 &= m_2 = m_{\ell_2} \\
\pi_6 &= m_2 \ell_{C_2} = m_{\ell_2}(\ell_2 - a_2) \\
\pi_7 &= \hat{I}_2 = I_{\ell_2} + m_{\ell_2}(\ell_2 - a_2)^2 \\
\pi_8 &= I_{m_2},
\end{aligned}$$

where the parameters for the augmented links have been found according to (4.76). It can be recognized that the number of nonnull parameters is less than the maximum number of twenty-two parameters allowed in this case[6]. The regressor in (4.80) is

$$\boldsymbol{Y} = \begin{bmatrix} y_{11} & y_{12} & y_{13} & y_{14} & y_{15} & y_{16} & y_{17} & y_{18} \\ y_{21} & y_{22} & y_{23} & y_{24} & y_{25} & y_{26} & y_{27} & y_{28} \end{bmatrix} \tag{4.83}$$

$$\begin{aligned}
y_{11} &= a_1^2 \ddot{\vartheta}_1 + a_1 g c_1 \\
y_{12} &= 2a_1 \ddot{\vartheta}_1 + g c_1 \\
y_{13} &= \ddot{\vartheta}_1 \\
y_{14} &= k_{r1}^2 \ddot{\vartheta}_1 \\
y_{15} &= (a_1^2 + 2a_1 a_2 c_2 + a_2^2)\ddot{\vartheta}_1 + (a_1 a_2 c_2 + a_2^2)\ddot{\vartheta}_2 - 2a_1 a_2 s_2 \dot{\vartheta}_1 \dot{\vartheta}_2 \\
&\quad - a_1 a_2 s_2 \dot{\vartheta}_2^2 + a_1 g c_1 + a_2 g c_{12} \\
y_{16} &= (2a_1 c_2 + 2a_2)\ddot{\vartheta}_1 + (a_1 c_2 + 2a_2)\ddot{\vartheta}_2 - 2a_1 s_2 \dot{\vartheta}_1 \dot{\vartheta}_2 - a_1 s_2 \dot{\vartheta}_2^2 + g c_{12} \\
y_{17} &= \ddot{\vartheta}_1 + \ddot{\vartheta}_2 \\
y_{18} &= k_{r2} \ddot{\vartheta}_2
\end{aligned}$$

$$\begin{aligned}
y_{21} &= 0 \\
y_{22} &= 0 \\
y_{23} &= 0 \\
y_{24} &= 0
\end{aligned}$$

---

[6] The number of parameters can be further reduced by resorting to a more accurate inspection, which leads to finding a minimum number of five parameters; those turn out to be a linear combination of the parameters in (4.82).

$$
\begin{aligned}
y_{25} &= (a_1 a_2 c_2 + a_2^2)\ddot{\vartheta}_1 + a_2^2 \ddot{\vartheta}_2 + a_1 a_2 s_2 \dot{\vartheta}_1^2 + a_2 g c_{12} \\
y_{26} &= (a_1 c_2 + 2a_2)\ddot{\vartheta}_1 + 2a_2 \ddot{\vartheta}_2 + a_1 s_2 \dot{\vartheta}_1^2 + g c_{12} \\
y_{27} &= \ddot{\vartheta}_1 + \ddot{\vartheta}_2 \\
y_{28} &= k_{r2} \ddot{\vartheta}_1 + k_{r2}^2 \ddot{\vartheta}_2.
\end{aligned}
$$

Example 4.2

In order to understand the relative weight of the various torque contributions in the dynamic model (4.81), consider a two-link planar arm with the following data:

$$
a_1 = a_2 = 1\,\text{m} \quad \ell_1 = \ell_2 = 0.5\,\text{m} \quad m_{\ell_1} = m_{\ell_2} = 50\,\text{kg} \quad I_{\ell_1} = I_{\ell_2} = 10\,\text{kg}\cdot\text{m}^2
$$

$$
k_{r1} = k_{r2} = 100 \quad m_{m_1} = m_{m_2} = 5\,\text{kg} \quad I_{m_1} = I_{m_2} = 0.01\,\text{kg}\cdot\text{m}^2.
$$

The two links have been chosen equal to better illustrate the dynamic interaction between the two joints.

Figure 4.6 shows the time history of positions, velocities, accelerations and torques resulting from joint trajectories with typical triangular velocity profile and equal time duration. The initial arm configuration is so that the tip is located at the point $(0.2, 0)$ m with a lower elbow posture. Both joints make a rotation of $\pi/2$ rad in a time of 0.5 s.

From the time history of the single torque contributions in Figure 4.7 it can be recognized what follows:

- The inertia torque at Joint 1 due to Joint 1 acceleration follows the time history of the acceleration.
- The inertia torque at Joint 2 due to Joint 2 acceleration is piecewise constant, since the inertia moment at Joint 2 axis is constant.
- The inertia torques at each joint due to acceleration of the other joint confirm the symmetry of the inertia matrix, since the acceleration profiles are the same for both joints.
- The Coriolis effect is present only at Joint 1, since the arm tip moves with respect to the mobile frame attached to Link 1 but is fixed with respect to the frame attached to Link 2.
- The centrifugal and Coriolis torques reflect the above symmetry.

Figure 4.8 shows the time history of positions, velocities, accelerations and torques resulting from joint trajectories with typical trapezoidal velocity profile and different time duration. The initial configuration is the same as in the previous case. The two joints make a rotation so as to take the tip to the point $(1.8, 0)$ m. The acceleration time is 0.15 s and the maximum velocity is 5 rad/s for both joints.

From the time history of the single torque contributions in Figure 4.9 it can be recognized what follows:

- The inertia torque at Joint 1 due to Joint 2 acceleration is opposite to that at Joint 2 due to Joint 1 acceleration in that portion of trajectory when the two accelerations have the same magnitude but opposite sign.

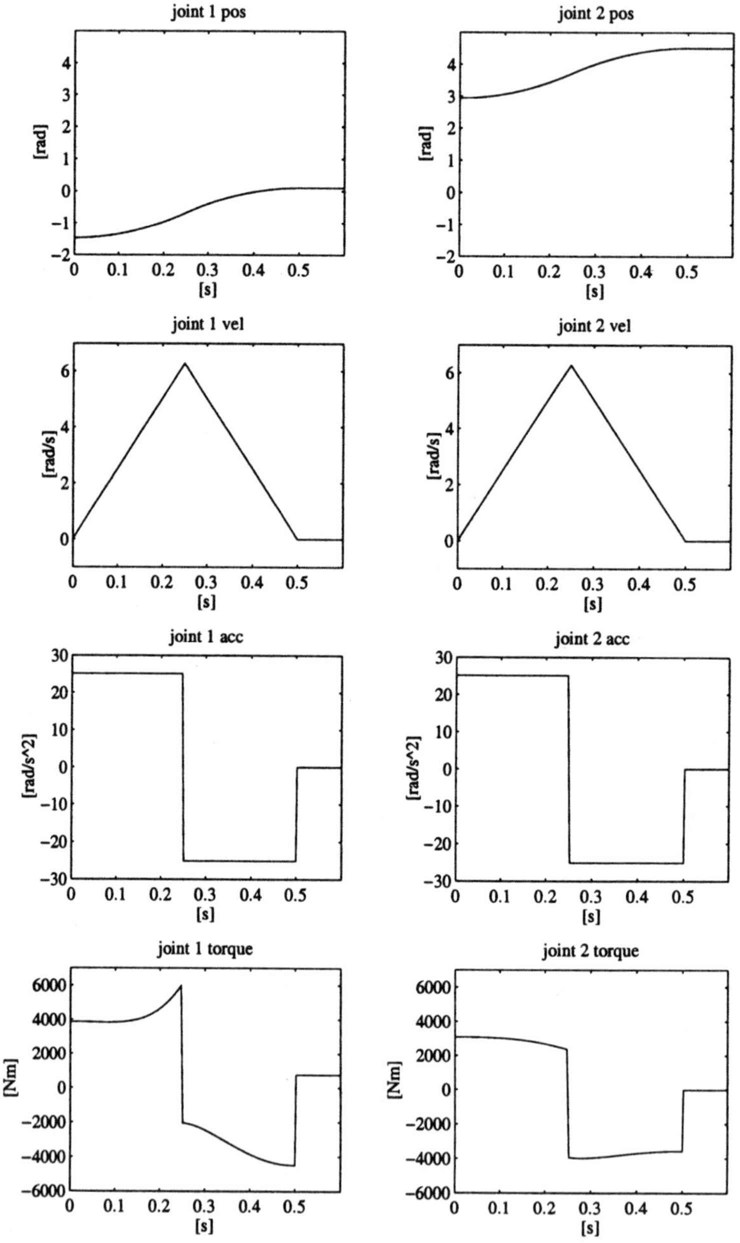

Figure 4.6 Time history of positions, velocities, accelerations and torques with joint trajectories of equal duration.

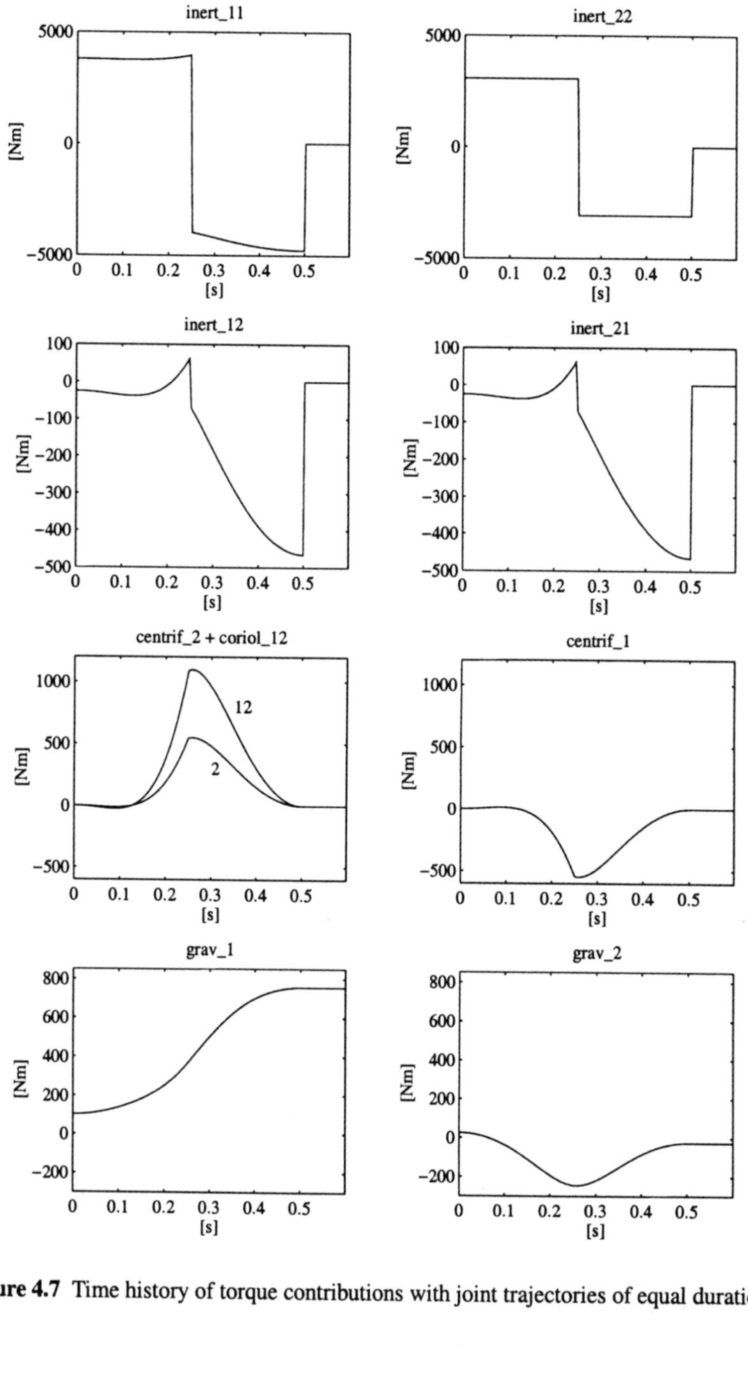

Figure 4.7 Time history of torque contributions with joint trajectories of equal duration.

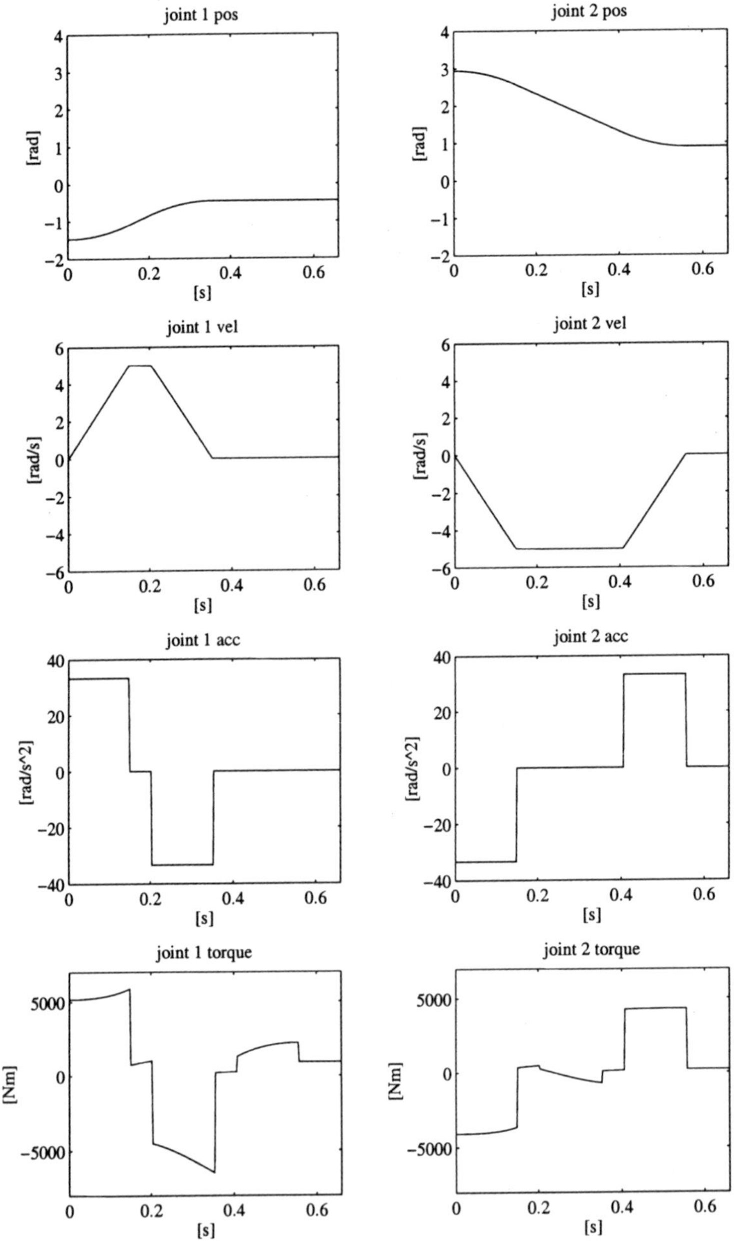

Figure 4.8 Time history of positions, velocities, accelerations and torques with joint trajectories of different duration.

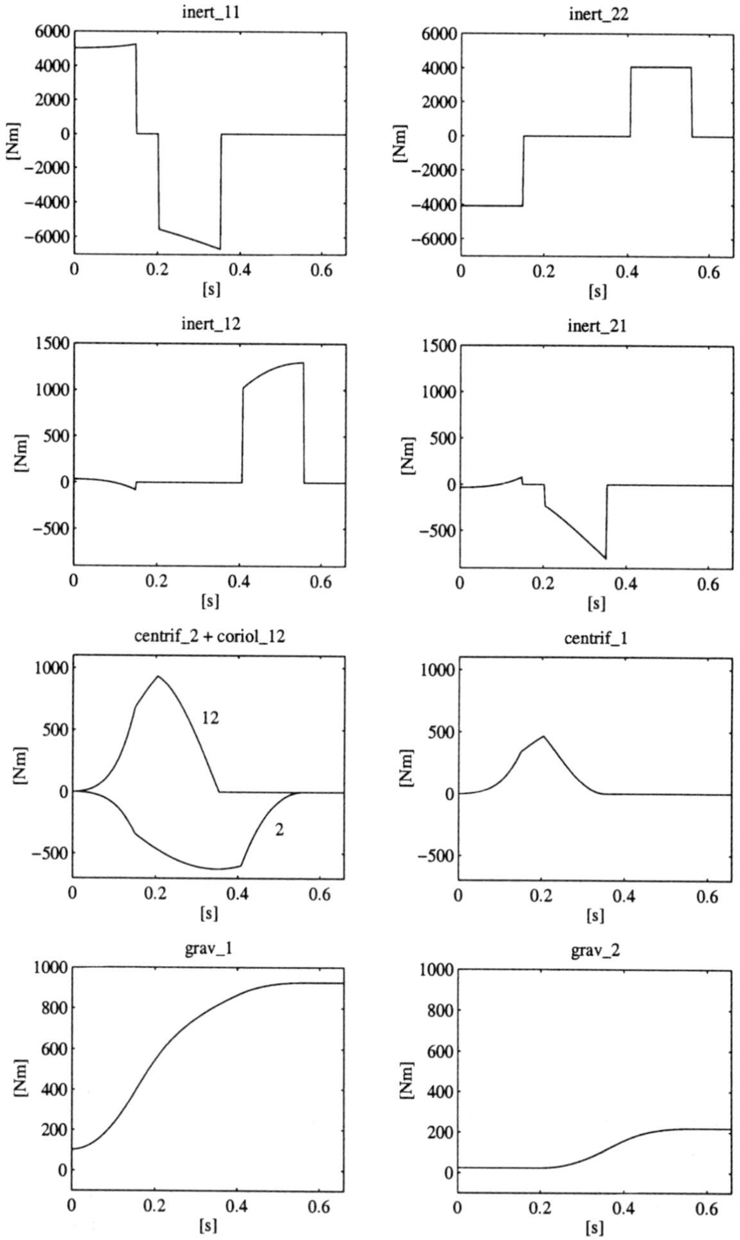

Figure 4.9 Time history of torque contributions with joint trajectories of different duration.

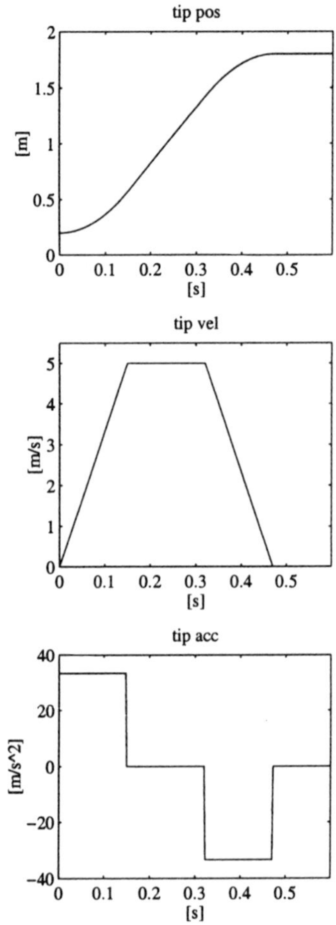

Figure 4.10 Time history of tip position, velocity and acceleration with a straight line tip trajectory along the horizontal axis.

- The different velocity profiles imply that the centrifugal effect induced at Joint 1 by Joint 2 velocity dies out later than the centrifugal effect induced at Joint 2 by Joint 1 velocity.
- The gravitational torque at Joint 2 is practically constant in the first portion of the trajectory, since Link 2 is almost kept in the same posture. As for the gravitational torque at Joint 1, instead, the centre of mass of the articulated system moves away from the origin of the axes.

Finally, Figure 4.10 shows the time history of tip position, velocity and acceleration for a trajectory with a trapezoidal velocity profile. Starting from the same initial posture as above, the arm tip makes a translation of 1.6 m along the horizontal axis; the acceleration time is 0.15 s and the maximum velocity is 5 m/s.

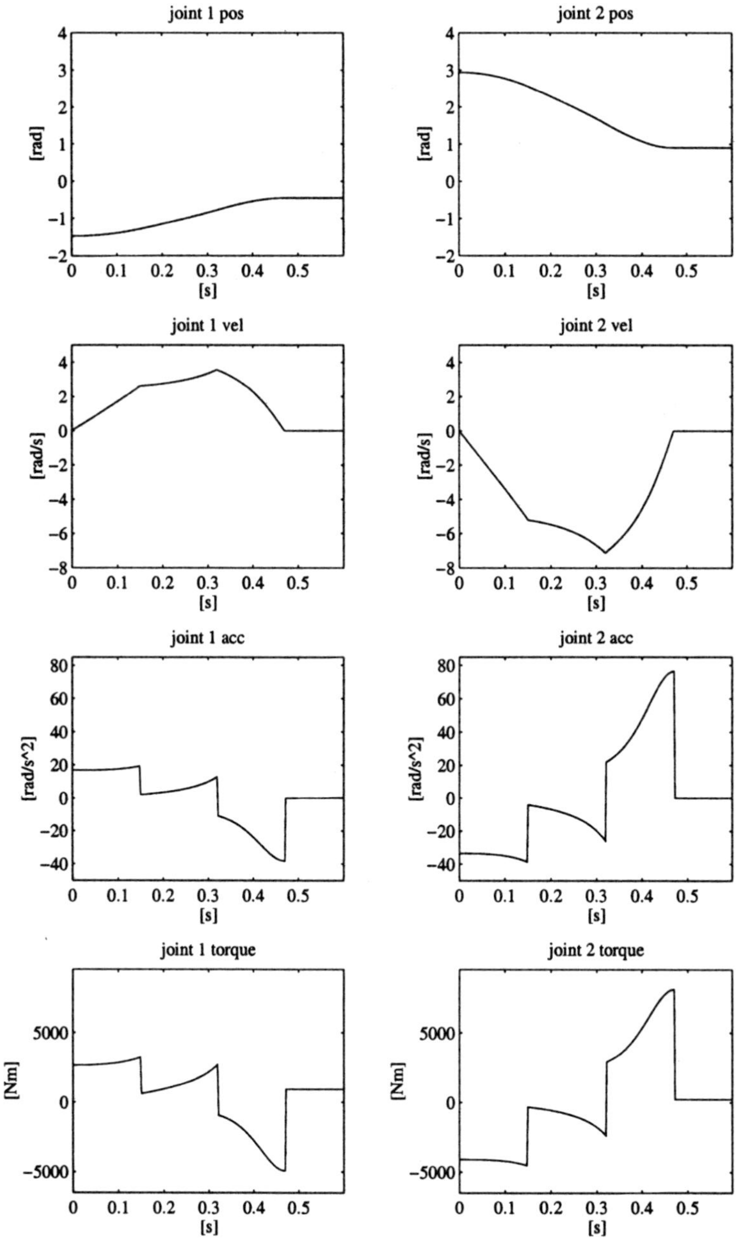

Figure 4.11 Time history of joint positions, velocities, accelerations, and torques with a straight line tip trajectory along the horizontal axis.

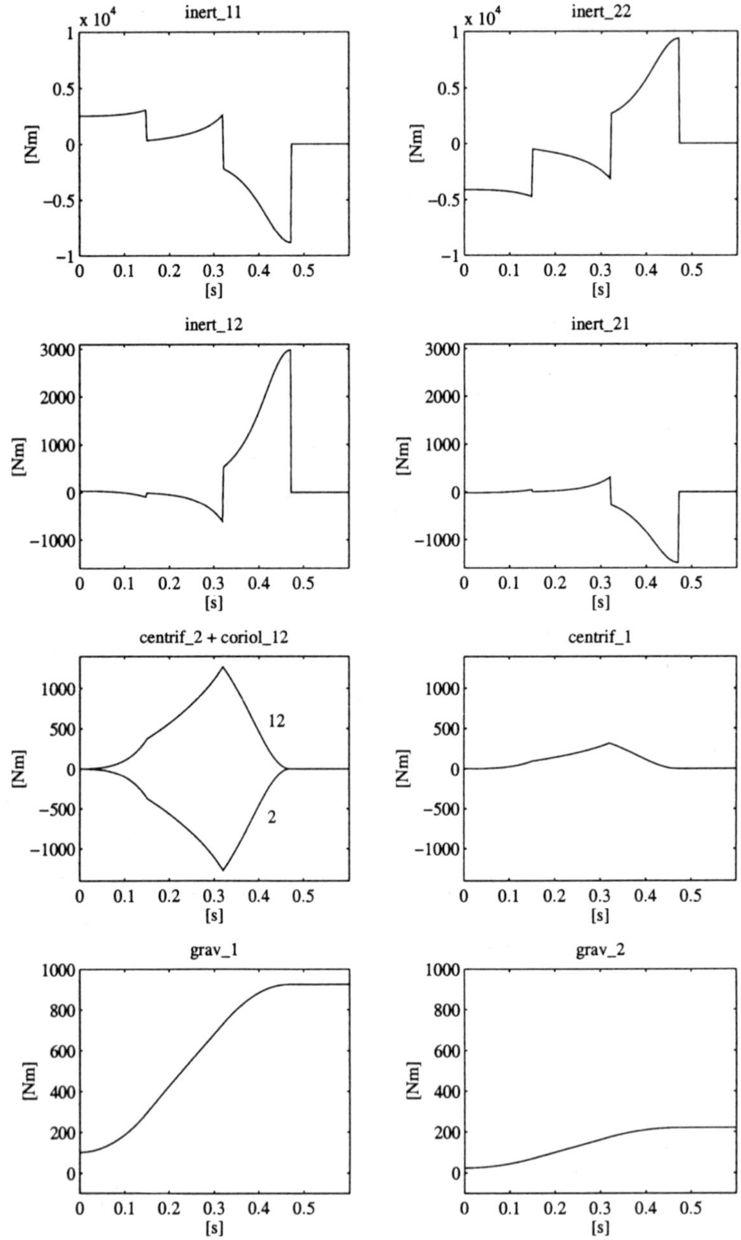

Figure 4.12 Time history of joint torque contributions with a straight line tip trajectory along the horizontal axis.

As a result of an inverse kinematics procedure, the time history of joint positions, velocities and accelerations have been computed which are illustrated in Figure 4.11, together with the joint torques that are needed to execute the assigned trajectory. It can be noticed that the time history of the represented quantities differs from the corresponding ones in the operational space, in view of the nonlinear effects introduced by kinematic relations.

For what concerns the time history of the individual torque contributions in Figure 4.12, it is possible to make a number of remarks similar to those made above for trajectories assigned directly in the joint space.

4.3.3 Parallelogram Arm

Consider the parallelogram arm in Figure 4.13. Because of the presence of the closed chain, the equivalent tree-structured open-chain arm is initially taken into account. Let $\ell_{1'}, \ell_{2'}, \ell_{3'}$ and $\ell_{1''}$ be the distances of the centres of mass of the three links along one branch of the tree, and of the single link along the other branch, from the respective joint axes. Let also $m_{\ell_{1'}}, m_{\ell_{2'}}, m_{\ell_{3'}}$ and $m_{\ell_{1''}}$ be the masses of the respective links, and $I_{\ell_{1'}}, I_{\ell_{2'}}, I_{\ell_{3'}}$ and $I_{\ell_{1''}}$ the moments of inertia relative to the centres of mass of the respective links. For the sake of simplicity, the contributions of the motors are neglected.

With the chosen coordinate frames, computation of the Jacobians in (4.16) and (4.18) yields

$$\boldsymbol{J}_P^{(\ell_{1'})} = \begin{bmatrix} -\ell_{1'}s_{1'} & 0 & 0 \\ \ell_{1'}c_{1'} & 0 & 0 \\ 0 & 0 & 0 \end{bmatrix} \qquad \boldsymbol{J}_P^{(\ell_{2'})} = \begin{bmatrix} -a_{1'}s_{1'} - \ell_{2'}s_{1'2'} & -\ell_{2'}s_{1'2'} & 0 \\ a_{1'}c_{1'} + \ell_{2'}c_{1'2'} & \ell_{2'}c_{1'2'} & 0 \\ 0 & 0 & 0 \end{bmatrix}$$

$$\boldsymbol{J}_P^{(\ell_{3'})} = \begin{bmatrix} -a_{1'}s_{1'} - a_{1''}s_{1'2'} - \ell_{3'}s_{1'2'3'} & -a_{1''}s_{1'2'} - \ell_{3'}s_{1'2'3'} & -\ell_{3'}s_{1'2'3'} \\ a_{1'}c_{1'} + a_{1''}c_{1'2'} + \ell_{3'}c_{1'2'3'} & a_{1''}c_{1'2'} + \ell_{3'}c_{1'2'3'} & \ell_{3'}c_{1'2'3'} \\ 0 & 0 & 0 \end{bmatrix}$$

and

$$\boldsymbol{J}_P^{(\ell_{1''})} = \begin{bmatrix} -\ell_{1''}s_{1''} \\ \ell_{1''}c_{1''} \\ 0 \end{bmatrix},$$

whereas computation of the Jacobians in (4.17) and (4.19) yields

$$\boldsymbol{J}_O^{(\ell_{1'})} = \begin{bmatrix} 0 & 0 & 0 \\ 0 & 0 & 0 \\ 1 & 0 & 0 \end{bmatrix} \qquad \boldsymbol{J}_O^{(\ell_{2'})} = \begin{bmatrix} 0 & 0 & 0 \\ 0 & 0 & 0 \\ 1 & 1 & 0 \end{bmatrix} \qquad \boldsymbol{J}_O^{(\ell_{3'})} = \begin{bmatrix} 0 & 0 & 0 \\ 0 & 0 & 0 \\ 1 & 1 & 1 \end{bmatrix}$$

and

$$\boldsymbol{J}_O^{(\ell_{1''})} = \begin{bmatrix} 0 \\ 0 \\ 1 \end{bmatrix}.$$

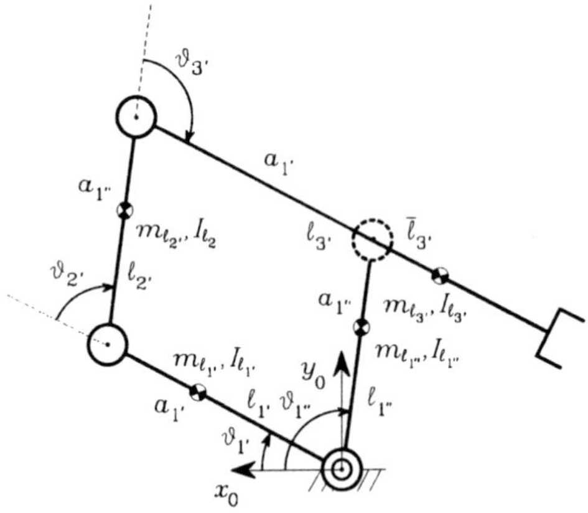

Figure 4.13 Parallelogram arm.

From (4.32), the inertia matrix of the virtual arm composed of joints $\vartheta_{1'}, \vartheta_{2'}, \vartheta_{3'}$ is

$$\boldsymbol{B}'(\boldsymbol{q}') = \begin{bmatrix} b_{1'1'}(\vartheta_{2'}, \vartheta_{3'}) & b_{1'2'}(\vartheta_{2'}, \vartheta_{3'}) & b_{1'3'}(\vartheta_{2'}, \vartheta_{3'}) \\ b_{2'1'}(\vartheta_{2'}, \vartheta_{3'}) & b_{2'2'}(\vartheta_{3'}) & b_{2'3'}(\vartheta_{3'}) \\ b_{3'1'}(\vartheta_{2'}, \vartheta_{3'}) & b_{3'2'}(\vartheta_{3'}) & b_{3'3'} \end{bmatrix}$$

$$\begin{aligned}
b_{1'1'} &= I_{\ell_{1'}} + m_{\ell_{1'}}\ell_{1'}^2 + I_{\ell_{2'}} + m_{\ell_{2'}}(a_{1'}^2 + \ell_{2'}^2 + 2a_{1'}\ell_{2'}c_{2'}) + I_{\ell_{3'}} \\
&\quad + m_{\ell_{3'}}(a_{1'}^2 + a_{1''}^2 + \ell_{3'}^2 + 2a_{1'}a_{1''}c_{2'} + 2a_{1'}\ell_{3'}c_{2'3'} + 2a_{1''}\ell_{3'}c_{3'}) \\
b_{1'2'} &= b_{2'1'} = I_{\ell_{2'}} + m_{\ell_{2'}}(\ell_{2'}^2 + a_{1'}\ell_{2'}c_{2'}) + I_{\ell_{3'}} \\
&\quad + m_{\ell_{3'}}(a_{1''}^2 + \ell_{3'}^2 + a_{1'}a_{1''}c_{2'} + a_{1'}\ell_{3'}c_{2'3'} + 2a_{1''}\ell_{3'}c_{3'}) \\
b_{1'3'} &= b_{3'1'} = I_{\ell_{3'}} + m_{\ell_{3'}}(\ell_{3'}^2 + a_{1'}\ell_{3'}c_{2'3'} + a_{1''}\ell_{3'}c_{3'}) \\
b_{2'2'} &= I_{\ell_{2'}} + m_{\ell_{2'}}\ell_{2'}^2 + I_{\ell_{3'}} + m_{\ell_{3'}}(a_{1''}^2 + \ell_{3'}^2 + 2a_{1''}\ell_{3'}c_{3'}) \\
b_{2'3'} &= b_{3'2'} = I_{\ell_{3'}} + m_{\ell_{3'}}(\ell_{3'}^2 + a_{1''}\ell_{3'}c_{3'}) \\
b_{3'3'} &= I_{\ell_{3'}} + m_{\ell_{3'}}\ell_{3'}^2
\end{aligned}$$

while the moment of inertia of the virtual arm composed of just joint $\vartheta_{1''}$ is

$$b_{1''1''} = I_{\ell_{1''}} + m_{\ell_{1''}}\ell_{1''}^2.$$

Therefore, the inertial torque contributions of the two virtual arms are respectively:

$$\tau_{i'} = \sum_{j'=1'}^{3'} b_{i'j'}\ddot{\vartheta}_{j'} \qquad \tau_{1''} = b_{1''1''}\ddot{\vartheta}_{1''}.$$

At this point, in view of (2.59) and (3.109), the inertial torque contributions at the actuated joints for the closed-chain arm turn out to be

$$\boldsymbol{\tau}_a = \boldsymbol{B}_a\ddot{\boldsymbol{q}}_a$$

where $\boldsymbol{q}_a = [\vartheta_{1'} \quad \vartheta_{1''}]^T$, $\boldsymbol{\tau}_a = [\tau_{a1} \quad \tau_{a2}]^T$ and

$$\boldsymbol{B}_a = \begin{bmatrix} b_{a11} & b_{a12} \\ b_{a21} & b_{a22} \end{bmatrix}$$

$$\begin{aligned}
b_{a11} &= I_{\ell_{1'}} + m_{\ell_{1'}}\ell_{1'}^2 + m_{\ell_{2'}}a_{1'}^2 + I_{\ell_{3'}} + m_{\ell_{3'}}\ell_{3'}^2 + m_{\ell_{3'}}a_{1'}^2 - 2a_{1'}m_{\ell_{3'}}\ell_{3'} \\
b_{a12} &= b_{a21} = \left(a_{1'}m_{\ell_{2'}}\ell_{2'} + a_{1''}m_{\ell_{3'}}(a_{1'} - \ell_{3'})\right)\cos(\vartheta_{1''} - \vartheta_{1'}) \\
b_{a22} &= I_{\ell_{1''}} + m_{\ell_{1''}}\ell_{1''}^2 + I_{\ell_{2'}} + m_{\ell_{2'}}\ell_{2'}^2 + m_{\ell_{3'}}a_{1''}^2.
\end{aligned}$$

This expression reveals the possibility of obtaining a *configuration-independent* and *decoupled* inertia matrix; to this purpose it is sufficient to design the four links of the parallelogram so that

$$\frac{m_{\ell_{3'}}\bar{\ell}_{3'}}{m_{\ell_{2'}}\ell_{2'}} = \frac{a_{1'}}{a_{1''}}$$

where $\bar{\ell}_{3'} = \ell_{3'} - a_{1'}$ is the distance of the centre of mass of Link 3′ from the axis of Joint 4. If this condition is satisfied, then the inertia matrix is diagonal ($b_{a12} = b_{a21} = 0$) with

$$\begin{aligned}
b_{a11} &= I_{\ell_{1'}} + m_{\ell_{1'}}\ell_{1'}^2 + m_{\ell_{2'}}a_{1'}^2\left(1 + \frac{\ell_{2'}\bar{\ell}_{3'}}{a_{1'}a_{1''}}\right) + I_{\ell_{3'}} \\
b_{a22} &= I_{\ell_{1''}} + m_{\ell_{1''}}\ell_{1''}^2 + I_{\ell_{2'}} + m_{\ell_{2'}}\ell_{2'}^2\left(1 + \frac{a_{1'}a_{1''}}{\ell_{2'}\bar{\ell}_{3'}}\right).
\end{aligned}$$

As a consequence, no contributions of Coriolis and centrifugal torques are obtained. Such a result could not be achieved with the previous two-link planar arm, no matter how the design parameters were chosen.

As for the gravitational terms, since $\boldsymbol{g}_0 = [0 \quad -g \quad 0]^T$, (4.38) with the above Jacobians gives:

$$\begin{aligned}
g_{1'} &= (m_{\ell_{1'}}\ell_{1'} + m_{\ell_{2'}}a_{1'} + m_{\ell_{3'}}a_{1'})gc_{1'} + (m_{\ell_{2'}}\ell_{2'} + m_{\ell_{3'}}a_{1''})gc_{1'2'} \\
&\quad + m_{\ell_{3'}}\ell_{3'}gc_{1'2'3'} \\
g_{2'} &= (m_{\ell_{2'}}\ell_{2'} + m_{\ell_{3'}}a_{1''})gc_{1'2'} + m_{\ell_{3'}}\ell_{3'}gc_{1'2'3'} \\
g_{3'} &= m_{\ell_{3'}}\ell_{3'}gc_{1'2'3'}
\end{aligned}$$

and

$$g_{1''} = m_{\ell_{1''}}\ell_{1''}gc_{1''}.$$

Composing the various contributions as done above yields

$$\boldsymbol{g}_a = \begin{bmatrix} (m_{\ell_{1'}}\ell_{1'} + m_{\ell_{2'}}a_{1'} - m_{\ell_{3'}}\bar{\ell}_{3'})gc_{1'} \\ (m_{\ell_{1''}}\ell_{1''} + m_{\ell_{2'}}\ell_{2'} + m_{\ell_{3'}}a_{1''})gc_{1''} \end{bmatrix}$$

which, together with the inertial torques, completes the derivation of the sought dynamic model.

A final comment is in order. In spite of its kinematic equivalence with the two-link planar arm, the dynamic model of the parallelogram is remarkably lighter. This property is quite advantageous for trajectory planning and control purposes. For this reason, apart from obvious considerations related to manipulation of heavy payloads, the adoption of closed kinematic chains in the design of industrial robots has received a great deal of attention.

4.4 Dynamic Parameter Identification

The use of the dynamic model for solving simulation and control problems demands the knowledge of the values of dynamic parameters of the manipulator model.

Computing such parameters from the design data of the mechanical structure is not simple. CAD modelling techniques can be adopted which allow computing the values of the inertial parameters of the various components (links, actuators and transmissions) on the basis of their geometry and type of materials employed. Nevertheless, the estimates obtained by such techniques are inaccurate because of the simplification typically introduced by geometric modelling; moreover, complex dynamic effects, such as joint friction, cannot be taken into account.

A heuristic approach could be to dismantle the various components of the manipulator and perform a series of measurements to evaluate the inertial parameters. Such technique is not easy to implement and may be troublesome to measure the relevant quantities.

In order to find accurate estimates of dynamic parameters, it is worth resorting to *identification* techniques which conveniently exploit the *property of linearity* (4.80) of the manipulator model with respect to a suitable set of *dynamic parameters*. Such techniques allow computing the parameter vector $\boldsymbol{\pi}$ from the measurements of joint torques $\boldsymbol{\tau}$ and of relevant quantities for the evaluation of the matrix $\boldsymbol{Y}$, when suitable motion trajectories are imposed to the manipulator.

On the assumption that the kinematic parameters in the matrix $\boldsymbol{Y}$ are known with good accuracy, e.g., as a result of a kinematic calibration, measurements of joint positions $\boldsymbol{q}$, velocities $\dot{\boldsymbol{q}}$ and accelerations $\ddot{\boldsymbol{q}}$ are required. Joint positions and velocities can be actually measured while numerical reconstruction of accelerations is needed; this can be performed on the basis of the position and velocity values recorded during the execution of the trajectories. The reconstructing filter does not work in real time and thus it can also be anti-causal, allowing an accurate reconstruction of the accelerations.

As regards joint torques, in the unusual case of torque sensors at the joint, these can be measured directly. Otherwise, they can be evaluated from either wrist force measurements or current measurements in the case of electric actuators.

If measurements of joint torques, positions, velocities and accelerations have been obtained at given time instants $t_1, \ldots, t_N$ along a given trajectory, one may write

$$\bar{\boldsymbol{\tau}} = \begin{bmatrix} \boldsymbol{\tau}(t_1) \\ \vdots \\ \boldsymbol{\tau}(t_N) \end{bmatrix} = \begin{bmatrix} \boldsymbol{Y}(t_1) \\ \vdots \\ \boldsymbol{Y}(t_N) \end{bmatrix} \boldsymbol{\pi} = \bar{\boldsymbol{Y}} \boldsymbol{\pi}. \tag{4.84}$$

The number of time instants sets the number of measurements to perform and shall be large enough (typically $Nn \gg p$) so as to avoid ill-conditioning of matrix $\bar{\boldsymbol{Y}}$. Solving

(4.84) by a least-squares technique leads to the solution in the form

$$\pi = (\bar{Y}^T\bar{Y})^{-1}\bar{Y}^T\bar{\tau} \tag{4.85}$$

where $(\bar{Y}^T\bar{Y})^{-1}\bar{Y}^T$ is the *left pseudo-inverse* matrix of $\bar{Y}$.

It should be noticed that, in view of the block triangular structure of matrix Y in (4.79), computation of parameter estimates could be simplified by resorting to a sequential procedure. Take the equation $\tau_n = y_{nn}^T\pi_n$ and solve it for π_n by specifying τ_n and y_{nn}^T for a given trajectory on Joint n. By iterating the procedure, the manipulator parameters can be identified on the basis of measurements performed joint by joint from the outer link to the base. Such procedure, however, may have the inconvenience to accumulate any error due to ill-conditioning of the matrices involved step by step. It may then be worth operating with a global procedure by imposing motions on all manipulator joints at the same time.

Regarding the rank of matrix $\bar{Y}$, it is possible to identify only the dynamic parameters of the manipulator that contribute to the dynamic model. Example 4.2 has indeed shown that for the two-link planar arm considered, only eight out of the twenty-two possible dynamic parameters appear in the dynamic model. Hence, there exist some dynamic parameters which, in view of the disposition of manipulator links and joints, are *non identifiable*, since for any trajectory assigned to the structure they do not contribute to the equations of motion. A direct consequence is that the columns of the matrix Y in (4.79) corresponding to such parameters are null and thus they have to be removed from the matrix itself; e.g., the resulting (2×8) matrix in (4.83).

Another issue to consider about determination of the effective number of parameters that can be identified by (4.85) is that some parameters can be *identified in linear combinations* whenever they do not appear isolated in the equations. In such case, it is necessary, for each linear combination, to remove as many columns of the matrix Y as the number of parameters in the linear combination minus one.

For the determination of the minimum number of identifiable parameters that allow direct application of the least-squares technique based on (4.85), it is possible to directly inspect the equations of the dynamic model, as long as the manipulator has few joints. Otherwise, numerical techniques based on singular value decomposition of matrix $\bar{Y}$ have to be used. If the matrix $\bar{Y}$ resulting from a series of measurements is not full-rank, one has to resort to a *damped least-squares inverse* of $\bar{Y}$ where solution accuracy depends on the weight of the damping factor.

In the above discussion, the type of trajectory imposed to the manipulator joints has not been explicitly addressed. It can be generally ascertained that the choice shall be oriented in favor of polynomial type trajectories which are sufficiently *rich* so as to allow an accurate evaluation of the identifiable parameters. This corresponds to achieving a low condition number of the matrix $\bar{Y}^T\bar{Y}$ along the trajectory. On the other hand, such trajectories shall not excite any unmodeled dynamic effects such as joint elasticity or link flexibility that would naturally lead to obtaining unreliable estimates of the dynamic parameters to identify.

Finally, it is worth observing that the technique presented above can be extended also to the identification of the dynamic parameters of an unknown payload at the manipulator's end effector. In such case, the payload can be regarded as a structural modification of the last link and one may proceed to identify the dynamic parameters of the modified link. To this purpose, if a force sensor is available at the manipulator's wrist, it is possible to directly characterize the dynamic parameters of the payload starting from force sensor measurements.

4.5 Newton-Euler Formulation

In the Lagrange formulation, the manipulator dynamic model is derived starting from the total Lagrangian of the system. On the other hand, the *Newton-Euler* formulation is based on a balance of all the forces acting on the generic link of the manipulator. This leads to a set of equations whose structure allows a recursive type of solution; a forward recursion is performed for propagating link velocities and accelerations, followed by a backward recursion for propagating forces.

Consider the generic *augmented Link* i (Link i plus motor of Joint $i+1$) of the manipulator kinematic chain (Figure 4.14). According to what was presented in Section 4.2.2, one can refer to the centre of mass C_i of the augmented link to characterize the following parameters:

m_i mass of augmented link,

$\bar{\boldsymbol{I}}_i$ inertia tensor of augmented link,

I_{m_i} moment of inertia of rotor,

$\boldsymbol{r}_{i-1,C_i}$ vector from origin of Frame $(i-1)$ to centre of mass C_i,

$\boldsymbol{r}_{i,C_i}$ vector from origin of Frame i to centre of mass C_i,

$\boldsymbol{r}_{i-1,i}$ vector from origin of Frame $(i-1)$ to origin of Frame i.

The velocities and accelerations to be considered are:

$\dot{\boldsymbol{p}}_{C_i}$ linear velocity of centre of mass C_i,

$\dot{\boldsymbol{p}}_i$ linear velocity of origin of Frame i,

$\boldsymbol{\omega}_i$ angular velocity of link,

$\boldsymbol{\omega}_{m_i}$ angular velocity of rotor,

$\ddot{\boldsymbol{p}}_{C_i}$ linear acceleration of centre of mass C_i,

$\ddot{\boldsymbol{p}}_i$ linear acceleration of origin of Frame i,

$\dot{\boldsymbol{\omega}}_i$ angular acceleration of link,

$\dot{\boldsymbol{\omega}}_{m_i}$ angular acceleration of rotor,

$\boldsymbol{g}_0$ gravity acceleration.

The forces and moments to be considered are:

$\boldsymbol{f}_i$ force exerted by Link $i-1$ on Link i,

$-\boldsymbol{f}_{i+1}$ force exerted by Link $i+1$ on Link i,

$\boldsymbol{\mu}_i$ moment exerted by Link $i-1$ on Link i with respect to origin of Frame $i-1$,

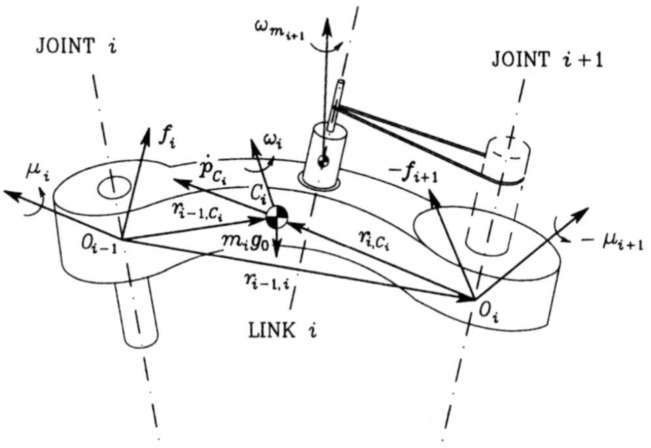

Figure 4.14 Characterization of Link i for Newton-Euler formulation.

$-\boldsymbol{\mu}_{i+1}$ moment exerted by Link $i+1$ on Link i with respect to origin of Frame i.

Initially, all the vectors and matrices are assumed to be expressed with reference to the *base frame*.

As already anticipated, the Newton-Euler formulation describes the motion of the link in terms of a balance of forces and moments acting on it.

The *Newton* equation for the *translational* motion of the centre of mass can be written as

$$\boldsymbol{f}_i - \boldsymbol{f}_{i+1} + m_i \boldsymbol{g}_0 = m_i \ddot{\boldsymbol{p}}_{C_i}. \tag{4.86}$$

The *Euler* equation for the *rotational* motion of the link (referring moments to the centre of mass) can be written as

$$\boldsymbol{\mu}_i + \boldsymbol{f}_i \times \boldsymbol{r}_{i-1,C_i} - \boldsymbol{\mu}_{i+1} - \boldsymbol{f}_{i+1} \times \boldsymbol{r}_{i,C_i} = \frac{d}{dt}(\bar{\boldsymbol{I}}_i \boldsymbol{\omega}_i + k_{r,i+1} \dot{q}_{i+1} I_{m_{i+1}} \boldsymbol{z}_{m_{i+1}}), \tag{4.87}$$

where (4.66) has been used for the angular momentum of the rotor. Notice that the gravitational force $m_i \boldsymbol{g}_0$ does not generate any moment, since it is concentrated at the centre of mass.

As pointed out in the above Lagrange formulation, it is convenient to express the inertia tensor in the current frame (constant tensor). Hence, according to (4.12), one has $\bar{\boldsymbol{I}}_i = \boldsymbol{R}_i \bar{\boldsymbol{I}}_i^i \boldsymbol{R}_i^T$, where $\boldsymbol{R}_i$ is the rotation matrix from Frame i to the base frame. Substituting this relation in the first term on the right-hand side of (4.87) yields

$$\begin{aligned} \frac{d}{dt}(\bar{\boldsymbol{I}}_i \boldsymbol{\omega}_i) &= \dot{\boldsymbol{R}}_i \bar{\boldsymbol{I}}_i^i \boldsymbol{R}_i^T \boldsymbol{\omega}_i + \boldsymbol{R}_i \bar{\boldsymbol{I}}_i^i \dot{\boldsymbol{R}}_i^T \boldsymbol{\omega}_i + \boldsymbol{R}_i \bar{\boldsymbol{I}}_i^i \boldsymbol{R}_i^T \dot{\boldsymbol{\omega}}_i \\ &= \boldsymbol{S}(\boldsymbol{\omega}_i) \boldsymbol{R}_i \bar{\boldsymbol{I}}_i^i \boldsymbol{R}_i^T \boldsymbol{\omega}_i + \boldsymbol{R}_i \bar{\boldsymbol{I}}_i^i \boldsymbol{R}_i^T \boldsymbol{S}^T(\boldsymbol{\omega}_i) \boldsymbol{\omega}_i + \boldsymbol{R}_i \bar{\boldsymbol{I}}_i^i \boldsymbol{R}_i^T \dot{\boldsymbol{\omega}}_i \\ &= \bar{\boldsymbol{I}}_i \dot{\boldsymbol{\omega}}_i + \boldsymbol{\omega}_i \times (\bar{\boldsymbol{I}}_i \boldsymbol{\omega}_i) \end{aligned} \tag{4.88}$$

where the second term represents the *gyroscopic* torque induced by the dependence of $\bar{\boldsymbol{I}}_i$ on link orientation[7]. Moreover, by observing that the unit vector $\boldsymbol{z}_{m_{i+1}}$ rotates accordingly to Link i, the derivative needed in the second term on the right-hand side of (4.87) is

$$\frac{d}{dt}(\dot{q}_{i+1} I_{m_{i+1}} \boldsymbol{z}_{m_{i+1}}) = \ddot{q}_{i+1} I_{m_{i+1}} \boldsymbol{z}_{m_{i+1}} + \dot{q}_{i+1} I_{m_{i+1}} \boldsymbol{\omega}_i \times \boldsymbol{z}_{m_{i+1}}. \tag{4.89}$$

By substituting (4.88) and (4.89) in (4.87), the resulting Euler equation is

$$\begin{aligned}\boldsymbol{\mu}_i + \boldsymbol{f}_i \times \boldsymbol{r}_{i-1,C_i} - \boldsymbol{\mu}_{i+1} - \boldsymbol{f}_{i+1} \times \boldsymbol{r}_{i,C_i} &= \bar{\boldsymbol{I}}_i \dot{\boldsymbol{\omega}}_i + \boldsymbol{\omega}_i \times (\bar{\boldsymbol{I}}_i \boldsymbol{\omega}_i) \\ &+ k_{r,i+1} \ddot{q}_{i+1} I_{m_{i+1}} \boldsymbol{z}_{m_{i+1}} + k_{r,i+1} \dot{q}_{i+1} I_{m_{i+1}} \boldsymbol{\omega}_i \times \boldsymbol{z}_{m_{i+1}}.\end{aligned} \tag{4.90}$$

The generalized force at Joint i can be computed by projecting the force $\boldsymbol{f}_i$ for a prismatic joint, or the moment $\boldsymbol{\mu}_i$ for a revolute joint, along the joint axis. In addition, there is the contribution of the rotor inertia torque $k_{ri} I_{m_i} \dot{\boldsymbol{\omega}}_{m_i}^T \boldsymbol{z}_{m_i}$. Hence, the generalized force at Joint i is expressed by:

$$\tau_i = \begin{cases} \boldsymbol{f}_i^T \boldsymbol{z}_{i-1} + k_{ri} I_{m_i} \dot{\boldsymbol{\omega}}_{m_i}^T \boldsymbol{z}_{m_i} & \text{for a } \textit{prismatic} \text{ joint} \\ \boldsymbol{\mu}_i^T \boldsymbol{z}_{i-1} + k_{ri} I_{m_i} \dot{\boldsymbol{\omega}}_{m_i}^T \boldsymbol{z}_{m_i} & \text{for a } \textit{revolute} \text{ joint.} \end{cases} \tag{4.91}$$

4.5.1 Link Acceleration

The Newton-Euler equations in (4.86) and (4.90) and the equation in (4.91) require the computation of linear and angular acceleration of Link i and Rotor i. This computation can be carried out on the basis of the relations expressing the linear and angular velocities previously derived. The equations in (3.19), (3.20), (3.23), (3.24) can be briefly rewritten as

$$\boldsymbol{\omega}_i = \begin{cases} \boldsymbol{\omega}_{i-1} & \text{for a } \textit{prismatic} \text{ joint} \\ \boldsymbol{\omega}_{i-1} + \dot{\vartheta}_i \boldsymbol{z}_{i-1} & \text{for a } \textit{revolute} \text{ joint} \end{cases} \tag{4.92}$$

and

$$\dot{\boldsymbol{p}}_i = \begin{cases} \dot{\boldsymbol{p}}_{i-1} + \dot{d}_i \boldsymbol{z}_{i-1} + \boldsymbol{\omega}_i \times \boldsymbol{r}_{i-1,i} & \text{for a } \textit{prismatic} \text{ joint} \\ \dot{\boldsymbol{p}}_{i-1} + \boldsymbol{\omega}_i \times \boldsymbol{r}_{i-1,i} & \text{for a } \textit{revolute} \text{ joint.} \end{cases} \tag{4.93}$$

As for the angular acceleration of the link, it can be seen that, for a prismatic joint, differentiating (3.19) with respect to time gives

$$\dot{\boldsymbol{\omega}}_i = \dot{\boldsymbol{\omega}}_{i-1}, \tag{4.94}$$

---

[7] In deriving (4.88), the operator $\boldsymbol{S}$ has been introduced to compute the derivative of $\boldsymbol{R}$, as in (3.7); also, the property $\boldsymbol{S}^T(\boldsymbol{\omega}_i)\boldsymbol{\omega}_i = \boldsymbol{0}$ has been utilized.

whereas, for a revolute joint, differentiating (3.23) with respect to time gives

$$\dot{\boldsymbol{\omega}}_i = \dot{\boldsymbol{\omega}}_{i-1} + \ddot{\vartheta}_i \boldsymbol{z}_{i-1} + \dot{\vartheta}_i \boldsymbol{\omega}_{i-1} \times \boldsymbol{z}_{i-1}. \tag{4.95}$$

As for the linear acceleration of the link, for a prismatic joint, differentiating (3.20) with respect to time gives

$$\begin{aligned}\ddot{\boldsymbol{p}}_i = \ddot{\boldsymbol{p}}_{i-1} &+ \ddot{d}_i \boldsymbol{z}_{i-1} + \dot{d}_i \boldsymbol{\omega}_{i-1} \times \boldsymbol{z}_{i-1} + \dot{\boldsymbol{\omega}}_i \times \boldsymbol{r}_{i-1,i} \\ &+ \boldsymbol{\omega}_i \times \dot{d}_i \boldsymbol{z}_{i-1} + \boldsymbol{\omega}_i \times (\boldsymbol{\omega}_{i-1} \times \boldsymbol{r}_{i-1,i})\end{aligned} \tag{4.96}$$

where the relation $\dot{\boldsymbol{r}}_{i-1,i} = \dot{d}_i \boldsymbol{z}_{i-1} + \boldsymbol{\omega}_{i-1} \times \boldsymbol{r}_{i-1,i}$ has been used. Hence, in view of (3.19), the equation in (4.96) can be rewritten as

$$\ddot{\boldsymbol{p}}_i = \ddot{\boldsymbol{p}}_{i-1} + \ddot{d}_i \boldsymbol{z}_{i-1} + 2\dot{d}_i \boldsymbol{\omega}_i \times \boldsymbol{z}_{i-1} + \dot{\boldsymbol{\omega}}_i \times \boldsymbol{r}_{i-1,i} + \boldsymbol{\omega}_i \times (\boldsymbol{\omega}_i \times \boldsymbol{r}_{i-1,i}). \tag{4.97}$$

Also, for a revolute joint, differentiating (3.24) with respect to time gives

$$\ddot{\boldsymbol{p}}_i = \ddot{\boldsymbol{p}}_{i-1} + \dot{\boldsymbol{\omega}}_i \times \boldsymbol{r}_{i-1,i} + \boldsymbol{\omega}_i \times (\boldsymbol{\omega}_i \times \boldsymbol{r}_{i-1,i}). \tag{4.98}$$

In sum, the equations in (4.94), (4.95), (4.97), (4.98) can be briefly rewritten as

$$\dot{\boldsymbol{\omega}}_i = \begin{cases} \dot{\boldsymbol{\omega}}_{i-1} & \text{for a } \textit{prismatic} \text{ joint} \\ \dot{\boldsymbol{\omega}}_{i-1} + \ddot{\vartheta}_i \boldsymbol{z}_{i-1} + \dot{\vartheta}_i \boldsymbol{\omega}_{i-1} \times \boldsymbol{z}_{i-1} & \text{for a } \textit{revolute} \text{ joint} \end{cases} \tag{4.99}$$

and

$$\ddot{\boldsymbol{p}}_i = \begin{cases} \ddot{\boldsymbol{p}}_{i-1} + \ddot{d}_i \boldsymbol{z}_{i-1} + 2\dot{d}_i \boldsymbol{\omega}_i \times \boldsymbol{z}_{i-1} & \\ \quad + \dot{\boldsymbol{\omega}}_i \times \boldsymbol{r}_{i-1,i} + \boldsymbol{\omega}_i \times (\boldsymbol{\omega}_i \times \boldsymbol{r}_{i-1,i}) & \text{for a } \textit{prismatic} \text{ joint} \\ \ddot{\boldsymbol{p}}_{i-1} + \dot{\boldsymbol{\omega}}_i \times \boldsymbol{r}_{i-1,i} & \\ \quad + \boldsymbol{\omega}_i \times (\boldsymbol{\omega}_i \times \boldsymbol{r}_{i-1,i}) & \text{for a } \textit{revolute} \text{ joint.} \end{cases} \tag{4.100}$$

The acceleration of the centre of mass of Link i required by the Newton equation in (4.86) can be derived from (3.13), since $\dot{\boldsymbol{r}}^i_{i,C_i} = \boldsymbol{0}$; by differentiating (3.13) with respect to time, the acceleration of the centre of mass C_i can be expressed as a function of the velocity and acceleration of the origin of Frame i, *i.e.*,

$$\ddot{\boldsymbol{p}}_{C_i} = \ddot{\boldsymbol{p}}_i + \dot{\boldsymbol{\omega}}_i \times \boldsymbol{r}_{i,C_i} + \boldsymbol{\omega}_i \times (\boldsymbol{\omega}_i \times \boldsymbol{r}_{i,C_i}). \tag{4.101}$$

Finally, the angular acceleration of the rotor can be obtained by time differentiation of (4.23), *i.e.*,

$$\dot{\boldsymbol{\omega}}_{m_i} = \dot{\boldsymbol{\omega}}_{i-1} + k_{ri} \ddot{q}_i \boldsymbol{z}_{m_i} + k_{ri} \dot{q}_i \boldsymbol{\omega}_{i-1} \times \boldsymbol{z}_{m_i}. \tag{4.102}$$

4.5.2 Recursive Algorithm

It is worth remarking that the resulting Newton-Euler equations of motion are *not* in *closed form*, since the motion of a single link is coupled to the motion of the other links through the kinematic relationship for velocities and accelerations.

Once the joint positions, velocities and accelerations are known, one can compute the link velocities and accelerations, and the Newton-Euler equations can be utilized to find the forces and moments acting on each link in a recursive fashion, starting from the force and moment applied to the end effector. On the other hand, also link and rotor velocities and accelerations can be computed recursively starting from the velocity and acceleration of the base link. In sum, a computationally *recursive algorithm* can be constructed that features a *forward recursion* relative to the propagation of *velocities and accelerations* and a *backward recursion* for the propagation of *forces and moments* along the structure.

For the forward recursion, once $\boldsymbol{q}$, $\dot{\boldsymbol{q}}$, $\ddot{\boldsymbol{q}}$, and the velocity and acceleration of the base link $\boldsymbol{\omega}_0, \ddot{\boldsymbol{p}}_0 - \boldsymbol{g}_0, \dot{\boldsymbol{\omega}}_0$ are specified, $\boldsymbol{\omega}_i, \dot{\boldsymbol{\omega}}_i, \ddot{\boldsymbol{p}}_i, \ddot{\boldsymbol{p}}_{C_i}, \dot{\boldsymbol{\omega}}_{m_i}$ can be computed using (4.92), (4.99), (4.100), (4.101), (4.102), respectively. Notice that the linear acceleration has been taken as $\ddot{\boldsymbol{p}}_0 - \boldsymbol{g}_0$ so as to incorporate the term $-\boldsymbol{g}_0$ in the computation of the acceleration of the centre of mass $\ddot{\boldsymbol{p}}_{C_i}$ via (4.100) and (4.101).

Having computed the velocities and accelerations with the forward recursion from the base link to the end effector, a backward recursion can be carried out for the forces. In detail, once $\boldsymbol{h} = [\, \boldsymbol{f}_{n+1}^T \quad \boldsymbol{\mu}_{n+1}^T \,]^T$ is given (eventually $\boldsymbol{h} = \boldsymbol{0}$), the Newton equation in (4.86) to be used for the recursion can be rewritten as

$$\boldsymbol{f}_i = \boldsymbol{f}_{i+1} + m_i \ddot{\boldsymbol{p}}_{C_i} \tag{4.103}$$

since the contribution of gravity acceleration has already been included in $\ddot{\boldsymbol{p}}_{C_i}$. Further, the Euler equation gives

$$\begin{aligned}\boldsymbol{\mu}_i = &-\boldsymbol{f}_i \times (\boldsymbol{r}_{i-1,i} + \boldsymbol{r}_{i,C_i}) + \boldsymbol{\mu}_{i+1} + \boldsymbol{f}_{i+1} \times \boldsymbol{r}_{i,C_i} + \bar{\boldsymbol{I}}_i \dot{\boldsymbol{\omega}}_i + \boldsymbol{\omega}_i \times (\bar{\boldsymbol{I}}_i \boldsymbol{\omega}_i) \\ &+ k_{r,i+1} \ddot{q}_{i+1} I_{m_{i+1}} \boldsymbol{z}_{m_{i+1}} + k_{r,i+1} \dot{q}_{i+1} I_{m_{i+1}} \boldsymbol{\omega}_i \times \boldsymbol{z}_{m_{i+1}}\end{aligned} \tag{4.104}$$

which derives from (4.90), where $\boldsymbol{r}_{i-1,C_i}$ has been expressed as the sum of the two vectors appearing already in the forward recursion. Finally, the generalized forces resulting at the joints can be computed from (4.91) as

$$\tau_i = \begin{cases} \boldsymbol{f}_i^T \boldsymbol{z}_{i-1} + k_{ri} I_{m_i} \dot{\boldsymbol{\omega}}_{m_i}^T \boldsymbol{z}_{m_i} + F_{vi} \dot{d}_i + F_{si}\,\mathrm{sgn}\,(\dot{d}_i) & \text{for a } \textit{prismatic} \text{ joint} \\ \boldsymbol{\mu}_i^T \boldsymbol{z}_{i-1} + k_{ri} I_{m_i} \dot{\boldsymbol{\omega}}_{m_i}^T \boldsymbol{z}_{m_i} + F_{vi} \dot{\vartheta}_i + F_{si}\,\mathrm{sgn}\,(\dot{\vartheta}_i) & \text{for a } \textit{revolute} \text{ joint,} \end{cases} \tag{4.105}$$

where joint viscous and Coulomb friction torques have been included.

In the above derivation, it has been assumed that all vectors were referred to the base frame. To greatly simplify computation, however, the recursion is computationally more efficient if all vectors are referred to the current frame on Link i. This implies that all vectors that need to be transformed from Frame $i+1$ into Frame i have to be multiplied by the rotation matrix $\boldsymbol{R}_{i+1}^i$, whereas all vectors that need to be transformed from Frame $i-1$ into Frame i have to be multiplied by the rotation matrix $\boldsymbol{R}_i^{i-1T}$. Therefore, the equations in (4.92), (4.99), (4.100), (4.101), (4.102), (4.103), (4.104) and (4.105) can be rewritten as:

$$\boldsymbol{\omega}_i^i = \begin{cases} \boldsymbol{R}_i^{i-1T} \boldsymbol{\omega}_{i-1}^{i-1} & \text{for a } \textit{prismatic} \text{ joint} \\ \boldsymbol{R}_i^{i-1T} (\boldsymbol{\omega}_{i-1}^{i-1} + \dot{\vartheta}_i \boldsymbol{z}_0) & \text{for a } \textit{revolute} \text{ joint} \end{cases} \tag{4.106}$$

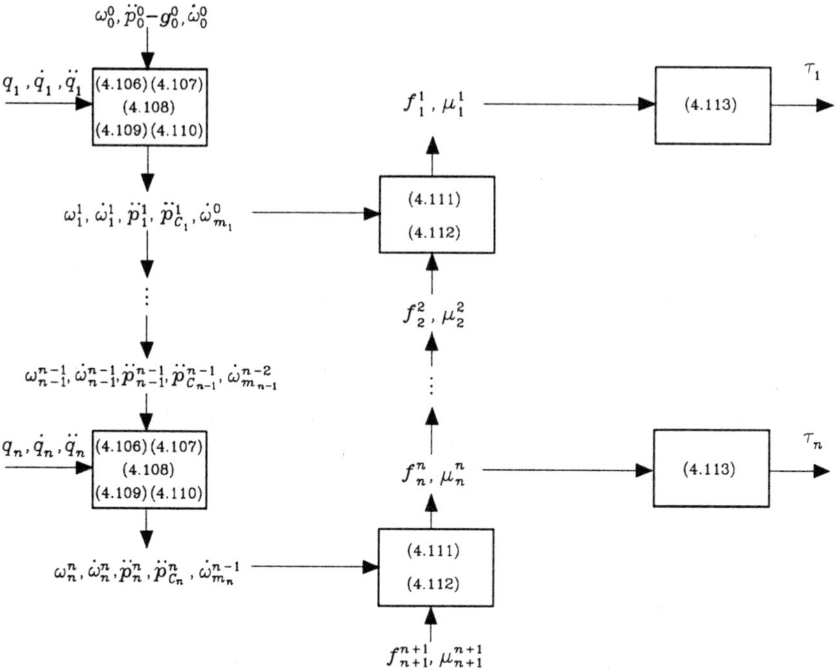

Figure 4.15 Computational structure of the Newton-Euler recursive algorithm.

$$\dot{\boldsymbol{\omega}}_i^i = \begin{cases} \boldsymbol{R}_i^{i-1T}\dot{\boldsymbol{\omega}}_{i-1}^{i-1} & \text{for a } \textit{prismatic} \text{ joint} \\ \boldsymbol{R}_i^{i-1T}(\dot{\boldsymbol{\omega}}_{i-1}^{i-1} + \ddot{\vartheta}_i \boldsymbol{z}_0 + \dot{\vartheta}_i \boldsymbol{\omega}_{i-1}^{i-1} \times \boldsymbol{z}_0) & \text{for a } \textit{revolute} \text{ joint} \end{cases} \tag{4.107}$$

$$\ddot{\boldsymbol{p}}_i^i = \begin{cases} \boldsymbol{R}_i^{i-1T}(\ddot{\boldsymbol{p}}_{i-1}^{i-1} + \ddot{d}_i \boldsymbol{z}_0) + 2\dot{d}_i \boldsymbol{\omega}_i^i \times \boldsymbol{R}_i^{i-1T}\boldsymbol{z}_0 & \\ \quad +\dot{\boldsymbol{\omega}}_i^i \times \boldsymbol{r}_{i-1,i}^i + \boldsymbol{\omega}_i^i \times (\boldsymbol{\omega}_i^i \times \boldsymbol{r}_{i-1,i}^i) & \text{for a } \textit{prismatic} \text{ joint} \\ \boldsymbol{R}_i^{i-1T}\ddot{\boldsymbol{p}}_{i-1}^{i-1} + \dot{\boldsymbol{\omega}}_i^i \times \boldsymbol{r}_{i-1,i}^i & \\ \quad +\boldsymbol{\omega}_i^i \times (\boldsymbol{\omega}_i^i \times \boldsymbol{r}_{i-1,i}^i) & \text{for a } \textit{revolute} \text{ joint} \end{cases} \tag{4.108}$$

$$\ddot{\boldsymbol{p}}_{C_i}^i = \ddot{\boldsymbol{p}}_i^i + \dot{\boldsymbol{\omega}}_i^i \times \boldsymbol{r}_{i,C_i}^i + \boldsymbol{\omega}_i^i \times (\boldsymbol{\omega}_i^i \times \boldsymbol{r}_{i,C_i}^i) \tag{4.109}$$

$$\dot{\boldsymbol{\omega}}_{m_i}^{i-1} = \dot{\boldsymbol{\omega}}_{i-1}^{i-1} + k_{ri}\ddot{q}_i \boldsymbol{z}_{m_i}^{i-1} + k_{ri}\dot{q}_i \boldsymbol{\omega}_{i-1}^{i-1} \times \boldsymbol{z}_{m_i}^{i-1} \tag{4.110}$$

$$\boldsymbol{f}_i^i = \boldsymbol{R}_{i+1}^i \boldsymbol{f}_{i+1}^{i+1} + m_i \ddot{\boldsymbol{p}}_{C_i}^i \tag{4.111}$$

$$\begin{aligned} \boldsymbol{\mu}_i^i = &-\boldsymbol{f}_i^i \times (\boldsymbol{r}_{i-1,i}^i + \boldsymbol{r}_{i,C_i}^i) + \boldsymbol{R}_{i+1}^i \boldsymbol{\mu}_{i+1}^{i+1} + \boldsymbol{R}_{i+1}^i \boldsymbol{f}_{i+1}^{i+1} \times \boldsymbol{r}_{i,C_i}^i + \bar{\boldsymbol{I}}_i^i \dot{\boldsymbol{\omega}}_i^i \\ &+ \boldsymbol{\omega}_i^i \times (\bar{\boldsymbol{I}}_i^i \boldsymbol{\omega}_i^i) + k_{r,i+1}\ddot{q}_{i+1} I_{m_{i+1}} \boldsymbol{z}_{m_{i+1}}^i + k_{r,i+1}\dot{q}_{i+1} I_{m_{i+1}} \boldsymbol{\omega}_i^i \times \boldsymbol{z}_{m_{i+1}}^i \end{aligned} \tag{4.112}$$

$$
\tau_i = \begin{cases} \boldsymbol{f}_i^{iT} \boldsymbol{R}_i^{i-1T} \boldsymbol{z}_0 + k_{ri} I_{m_i} \dot{\boldsymbol{\omega}}_{m_i}^{i-1T} \boldsymbol{z}_{m_i}^{i-1} & \\ \quad + F_{vi} \dot{d}_i + F_{si} \operatorname{sgn}(\dot{d}_i) & \text{for a } \textit{prismatic} \text{ joint} \\ \boldsymbol{\mu}_i^{iT} \boldsymbol{R}_i^{i-1T} \boldsymbol{z}_0 + k_{ri} I_{m_i} \dot{\boldsymbol{\omega}}_{m_i}^{i-1T} \boldsymbol{z}_{m_i}^{i-1} & \\ \quad + F_{vi} \dot{\vartheta}_i + F_{si} \operatorname{sgn}(\dot{\vartheta}_i) & \text{for a } \textit{revolute} \text{ joint.} \end{cases} \tag{4.113}
$$

The above equations have the advantage that the quantities $\bar{\boldsymbol{I}}_i^i, \boldsymbol{r}_{i,C_i}^i, \boldsymbol{z}_{m_i}^{i-1}$ are *constant*; further, it is $\boldsymbol{z}_0 = [0 \quad 0 \quad 1]^T$.

To summarize, for given joint positions, velocities and accelerations, the recursive algorithm is carried out in the following two phases:

- With known initial conditions $\boldsymbol{\omega}_0^0$, $\ddot{\boldsymbol{p}}_0^0 - \boldsymbol{g}_0^0$, and $\dot{\boldsymbol{\omega}}_0^0$, use (4.106), (4.107), (4.108), (4.109), (4.110), for $i = 1, \ldots, n$, to compute $\boldsymbol{\omega}_i^i, \dot{\boldsymbol{\omega}}_i^i, \ddot{\boldsymbol{p}}_i^i, \ddot{\boldsymbol{p}}_{C_i}^i, \dot{\boldsymbol{\omega}}_{m_i}^{i-1}$.
- With known terminal conditions $\boldsymbol{f}_{n+1}^{n+1}$ and $\boldsymbol{\mu}_{n+1}^{n+1}$, use (4.111) and (4.112), for $i = n, \ldots, 1$, to compute $\boldsymbol{f}_i^i$ and $\boldsymbol{\mu}_i^i$, and then (4.113) to compute τ_i.

The computational structure of the algorithm is schematically illustrated in Figure 4.15.

4.5.3 Example

In the following, an example to illustrate the single steps of the Newton-Euler algorithm is developed. Consider the two-link planar arm whose dynamic model has already been derived in Example 4.2.

Start by imposing the initial conditions for the velocities and accelerations:

$$
\ddot{\boldsymbol{p}}_0^0 - \boldsymbol{g}_0^0 = [0 \quad g \quad 0]^T \qquad \boldsymbol{\omega}_0^0 = \dot{\boldsymbol{\omega}}_0^0 = \boldsymbol{0},
$$

and the terminal conditions for the forces:

$$
\boldsymbol{f}_3^3 = \boldsymbol{0} \qquad \boldsymbol{\mu}_3^3 = \boldsymbol{0}.
$$

All quantities are referred to the current link frame. As a consequence, the following constant vectors are obtained:

$$
\boldsymbol{r}_{1,C_1}^1 = \begin{bmatrix} \ell_{C_1} \\ 0 \\ 0 \end{bmatrix} \quad \boldsymbol{r}_{0,1}^1 = \begin{bmatrix} a_1 \\ 0 \\ 0 \end{bmatrix} \quad \boldsymbol{r}_{2,C_2}^2 = \begin{bmatrix} \ell_{C_2} \\ 0 \\ 0 \end{bmatrix} \quad \boldsymbol{r}_{1,2}^2 = \begin{bmatrix} a_2 \\ 0 \\ 0 \end{bmatrix}
$$

where ℓ_{C_1} and ℓ_{C_2} are both negative quantities. The rotation matrices needed for vector transformation from one frame to another are:

$$
\boldsymbol{R}_i^{i-1} = \begin{bmatrix} c_i & -s_i & 0 \\ s_i & c_i & 0 \\ 0 & 0 & 1 \end{bmatrix}, \quad i = 1, 2 \qquad \boldsymbol{R}_3^2 = \boldsymbol{I}.
$$

Further, it is assumed that the axes of rotation of the two rotors coincide with the respective joint axes, *i.e.*, $\boldsymbol{z}_{m_i}^{i-1} = \boldsymbol{z}_0 = [0 \quad 0 \quad 1]^T$ for $i = 1, 2$.

According to (4.106)–(4.113), the Newton-Euler algorithm requires the execution of the following steps.

- Forward recursion: Link 1

$$\boldsymbol{\omega}_1^1 = \begin{bmatrix} 0 \\ 0 \\ \dot{\vartheta}_1 \end{bmatrix}$$

$$\dot{\boldsymbol{\omega}}_1^1 = \begin{bmatrix} 0 \\ 0 \\ \ddot{\vartheta}_1 \end{bmatrix}$$

$$\ddot{\boldsymbol{p}}_1^1 = \begin{bmatrix} -a_1\dot{\vartheta}_1^2 + gs_1 \\ a_1\ddot{\vartheta}_1 + gc_1 \\ 0 \end{bmatrix}$$

$$\ddot{\boldsymbol{p}}_{C_1}^1 = \begin{bmatrix} -(\ell_{C_1} + a_1)\dot{\vartheta}_1^2 + gs_1 \\ (\ell_{C_1} + a_1)\ddot{\vartheta}_1 + gc_1 \\ 0 \end{bmatrix}$$

$$\dot{\boldsymbol{\omega}}_{m_1}^0 = \begin{bmatrix} 0 \\ 0 \\ k_{r1}\ddot{\vartheta}_1 \end{bmatrix}.$$

- Forward recursion: Link 2

$$\boldsymbol{\omega}_2^2 = \begin{bmatrix} 0 \\ 0 \\ \dot{\vartheta}_1 + \dot{\vartheta}_2 \end{bmatrix}$$

$$\dot{\boldsymbol{\omega}}_2^2 = \begin{bmatrix} 0 \\ 0 \\ \ddot{\vartheta}_1 + \ddot{\vartheta}_2 \end{bmatrix}$$

$$\ddot{\boldsymbol{p}}_2^2 = \begin{bmatrix} a_1s_2\ddot{\vartheta}_1 - a_1c_2\dot{\vartheta}_1^2 - a_2(\dot{\vartheta}_1 + \dot{\vartheta}_2)^2 + gs_{12} \\ a_1c_2\ddot{\vartheta}_1 + a_2(\ddot{\vartheta}_1 + \ddot{\vartheta}_2) + a_1s_2\dot{\vartheta}_1^2 + gc_{12} \\ 0 \end{bmatrix}$$

$$\ddot{\boldsymbol{p}}_{C_2}^2 = \begin{bmatrix} a_1s_2\ddot{\vartheta}_1 - a_1c_2\dot{\vartheta}_1^2 - (\ell_{C_2} + a_2)(\dot{\vartheta}_1 + \dot{\vartheta}_2)^2 + gs_{12} \\ a_1c_2\ddot{\vartheta}_1 + (\ell_{C_2} + a_2)(\ddot{\vartheta}_1 + \ddot{\vartheta}_2) + a_1s_2\dot{\vartheta}_1^2 + gc_{12} \\ 0 \end{bmatrix}$$

$$\dot{\boldsymbol{\omega}}_{m_2}^1 = \begin{bmatrix} 0 \\ 0 \\ \ddot{\vartheta}_1 + k_{r2}\ddot{\vartheta}_2 \end{bmatrix}.$$

- Backward recursion: Link 2

$$\boldsymbol{f}_2^2 = \begin{bmatrix} m_2\big(a_1 s_2 \ddot{\vartheta}_1 - a_1 c_2 \dot{\vartheta}_1^2 - (\ell_{C_2} + a_2)(\dot{\vartheta}_1 + \dot{\vartheta}_2)^2 + g s_{12}\big) \\ m_2\big(a_1 c_2 \ddot{\vartheta}_1 + (\ell_{C_2} + a_2)(\ddot{\vartheta}_1 + \ddot{\vartheta}_2) + a_1 s_2 \dot{\vartheta}_1^2 + g c_{12}\big) \\ 0 \end{bmatrix}$$

$$\boldsymbol{\mu}_2^2 = \begin{bmatrix} * \\ * \\ \bar{I}_{2zz}(\ddot{\vartheta}_1 + \ddot{\vartheta}_2) + m_2(\ell_{C_2} + a_2)^2(\ddot{\vartheta}_1 + \ddot{\vartheta}_2) + m_2 a_1(\ell_{C_2} + a_2)c_2\ddot{\vartheta}_1 \\ +m_2 a_1(\ell_{C_2} + a_2)s_2\dot{\vartheta}_1^2 + m_2(\ell_{C_2} + a_2) g c_{12} \end{bmatrix}$$

$$\begin{aligned} \tau_2 = {} & \big(\bar{I}_{2zz} + m_2\big((\ell_{C_2} + a_2)^2 + a_1(\ell_{C_2} + a_2)c_2\big) + k_{r2} I_{m_2}\big)\ddot{\vartheta}_1 \\ & + \big(\bar{I}_{2zz} + m_2(\ell_{C_2} + a_2)^2 + k_{r2}^2 I_{m_2}\big)\ddot{\vartheta}_2 \\ & + m_2 a_1(\ell_{C_2} + a_2)s_2\dot{\vartheta}_1^2 + m_2(\ell_{C_2} + a_2) g c_{12}. \end{aligned}$$

- Backward recursion: Link 1

$$\boldsymbol{f}_1^1 = \begin{bmatrix} -m_2(\ell_{C_2} + a_2)s_2(\ddot{\vartheta}_1 + \ddot{\vartheta}_2) - m_1(\ell_{C_1} + a_1)\dot{\vartheta}_1^2 - m_2 a_1 \dot{\vartheta}_1^2 \\ -m_2(\ell_{C_2} + a_2)c_2(\dot{\vartheta}_1 + \dot{\vartheta}_2)^2 + (m_1 + m_2) g s_1 \\ m_1(\ell_{C_1} + a_1)\ddot{\vartheta}_1 + m_2 a_1 \ddot{\vartheta}_1 + m_2(\ell_{C_2} + a_2)c_2(\ddot{\vartheta}_1 + \ddot{\vartheta}_2) \\ -m_2(\ell_{C_2} + a_2)s_2(\dot{\vartheta}_1 + \dot{\vartheta}_2)^2 + (m_1 + m_2) g c_1 \\ 0 \end{bmatrix}$$

$$\boldsymbol{\mu}_1^1 = \begin{bmatrix} * \\ * \\ \bar{I}_{1zz}\ddot{\vartheta}_1 + m_2 a_1^2 \ddot{\vartheta}_1 + m_1(\ell_{C_1} + a_1)^2\ddot{\vartheta}_1 + m_2 a_1(\ell_{C_2} + a_2)c_2\ddot{\vartheta}_1 \\ +\bar{I}_{2zz}(\ddot{\vartheta}_1 + \ddot{\vartheta}_2) + m_2 a_1(\ell_{C_2} + a_2)c_2(\ddot{\vartheta}_1 + \ddot{\vartheta}_2) \\ +m_2(\ell_{C_2} + a_2)^2(\ddot{\vartheta}_1 + \ddot{\vartheta}_2) + k_{r2} I_{m_2}\ddot{\vartheta}_2 \\ +m_2 a_1(\ell_{C_2} + a_2)s_2\dot{\vartheta}_1^2 - m_2 a_1(\ell_{C_2} + a_2)s_2(\dot{\vartheta}_1 + \dot{\vartheta}_2)^2 \\ +m_1(\ell_{C_1} + a_1) g c_1 + m_2 a_1 g c_1 + m_2(\ell_{C_2} + a_2) g c_{12} \end{bmatrix}$$

$$\begin{aligned} \tau_1 = {} & \Big(\bar{I}_{1zz} + m_1(\ell_{C_1} + a_1)^2 + k_{r1}^2 I_{m_1} + \bar{I}_{2zz} \\ & \quad + m_2\big(a_1^2 + (\ell_{C_2} + a_2)^2 + 2a_1(\ell_{C_2} + a_2)c_2\big)\Big)\ddot{\vartheta}_1 \\ & + \Big(\bar{I}_{2zz} + m_2\big((\ell_{C_2} + a_2)^2 + a_1(\ell_{C_2} + a_2)c_2\big) + k_{r2} I_{m_2}\Big)\ddot{\vartheta}_2 \\ & - 2m_2 a_1(\ell_{C_2} + a_2)s_2\dot{\vartheta}_1\dot{\vartheta}_2 - m_2 a_1(\ell_{C_2} + a_2)s_2\dot{\vartheta}_2^2 \end{aligned}$$

$$+\left(m_1(\ell_{C_1}+a_1)+m_2a_1\right)gc_1+m_2(\ell_{C_2}+a_2)gc_{12}.$$

As for the moment components, those marked by the symbol '$*$' have not been computed, since they are not related to the joint torques τ_2 and τ_1.

Expressing the dynamic parameters in the above torques as a function of the link and rotor parameters as in (4.82) yields:

$$\begin{aligned}
m_1 &= m_{\ell_1}+m_{m_2}\\
m_1\ell_{C_1} &= m_{\ell_1}(\ell_1-a_1)\\
\bar{I}_{1zz}+m_1\ell_{C_1}^2 &= \hat{I}_1 = I_{\ell_1}+m_{\ell_1}(\ell_1-a_1)^2+I_{m_2}\\
m_2 &= m_{\ell_2}\\
m_2\ell_{C_2} &= m_{\ell_2}(\ell_2-a_2)\\
\bar{I}_{2zz}+m_2\ell_{C_2}^2 &= \hat{I}_2 = I_{\ell_2}+m_{\ell_2}(\ell_2-a_2)^2.
\end{aligned}$$

On the basis of these relations, it can be verified that the resulting dynamic model coincides with the model derived in (4.81) with Lagrange formulation.

4.6 Direct Dynamics and Inverse Dynamics

Both Lagrange formulation and Newton-Euler formulation allow computing the relationship between the joint torques—and, if present, the end-effector forces—and the motion of the structure. A comparison between the two approaches reveals what follows. The *Lagrange* formulation has the following advantages:

- It is *systematic* and of immediate comprehension.
- It provides the equations of motion in a compact *analytical form* containing the inertia matrix, the matrix in the centrifugal and Coriolis forces, and the vector of gravitational forces. Such a form is advantageous for *control design*.
- It is effective if it is wished to include more complex mechanical effects such as flexible link deformation.

The *Newton-Euler* formulation has the following fundamental advantage:

- It is an inherently *recursive* method that is computationally efficient.

In the study of dynamics, it is relevant to find a solution to two kinds of problems concerning computation of direct dynamics and inverse dynamics.

The *direct dynamics* problem consists of determining, for $t > t_0$, the joint accelerations $\ddot{\boldsymbol{q}}(t)$ (and thus $\dot{\boldsymbol{q}}(t)$, $\boldsymbol{q}(t)$) resulting from the given joint torques $\boldsymbol{\tau}(t)$—and the possible end-effector forces $\boldsymbol{h}(t)$—once the initial positions $\boldsymbol{q}(t_0)$ and velocities $\dot{\boldsymbol{q}}(t_0)$ are known (initial state of the system).

The *inverse dynamics* problem consists of determining the joint torques $\boldsymbol{\tau}(t)$ which are needed to generate the motion specified by the joint accelerations $\ddot{\boldsymbol{q}}(t)$, velocities $\dot{\boldsymbol{q}}(t)$, and positions $\boldsymbol{q}(t)$—once the possible end-effector forces $\boldsymbol{h}(t)$ are known.

Solving the direct dynamics problem is useful for manipulator *simulation*. Direct dynamics allows describing the motion of the real physical system in terms of the joint accelerations, when a set of assigned joint torques is applied to the manipulator; joint velocities and positions can be obtained by integrating the system of nonlinear differential equations.

Since the equations of motion obtained with Lagrange formulation give the analytical relationship between the joint torques (and the end-effector forces) and the joint positions, velocities and accelerations, these can be computed from (4.41) as

$$\ddot{\boldsymbol{q}} = \boldsymbol{B}^{-1}(\boldsymbol{q})(\boldsymbol{\tau} - \boldsymbol{\tau}') \tag{4.114}$$

where

$$\boldsymbol{\tau}'(\boldsymbol{q}, \dot{\boldsymbol{q}}) = \boldsymbol{C}(\boldsymbol{q}, \dot{\boldsymbol{q}})\dot{\boldsymbol{q}} + \boldsymbol{F}_v\dot{\boldsymbol{q}} + \boldsymbol{F}_s \,\mathbf{sgn}\,(\dot{\boldsymbol{q}}) + \boldsymbol{g}(\boldsymbol{q}) + \boldsymbol{J}^T(\boldsymbol{q})\boldsymbol{h} \tag{4.115}$$

denotes the torque contributions depending on joint positions and velocities. Therefore, for simulation of manipulator motion, once the state at the time instant t_k is known in terms of the position $\boldsymbol{q}(t_k)$ and velocity $\dot{\boldsymbol{q}}(t_k)$, the acceleration $\ddot{\boldsymbol{q}}(t_k)$ can be computed by (4.114). Then using a numerical integration method, e.g., Runge-Kutta, with integration step Δt, the velocity $\dot{\boldsymbol{q}}(t_{k+1})$ and position $\boldsymbol{q}(t_{k+1})$ at the instant $t_{k+1} = t_k + \Delta t$ can be computed.

If the equations of motion are obtained with Newton-Euler formulation, it is possible to compute direct dynamics by using a computationally more efficient method. In fact, for given $\boldsymbol{q}$ and $\dot{\boldsymbol{q}}$, the torques $\boldsymbol{\tau}'(\boldsymbol{q}, \dot{\boldsymbol{q}})$ in (4.115) can be computed as the torques given by the algorithm of Figure 4.15 with $\ddot{\boldsymbol{q}} = \boldsymbol{0}$. Further, column $\boldsymbol{b}_i$ of matrix $\boldsymbol{B}(\boldsymbol{q})$ can be computed as the torque vector given by the algorithm of Figure 4.15 with $\boldsymbol{g}_0 = \boldsymbol{0}$, $\dot{\boldsymbol{q}} = \boldsymbol{0}$, $\ddot{q}_i = 1$ and $\ddot{q}_j = 0$ for $j \neq i$; iterating this procedure for $i = 1, \ldots, n$ leads to constructing the matrix $\boldsymbol{B}(\boldsymbol{q})$. Hence, from the current values of $\boldsymbol{B}(\boldsymbol{q})$ and $\boldsymbol{\tau}'(\boldsymbol{q}, \dot{\boldsymbol{q}})$, and the given $\boldsymbol{\tau}$, the equations in (4.114) can be integrated as illustrated above.

Solving the inverse dynamics problem is useful for manipulator trajectory planning and control algorithm implementation. Once a joint trajectory is specified in terms of positions, velocities and accelerations (typically as a result of an inverse kinematics procedure), and if the end-effector forces are known, inverse dynamics allows computation of the torques to be applied to the joints to obtain the desired motion. This computation turns out to be useful both for verifying feasibility of the imposed trajectory and for compensating nonlinear terms in the dynamic model of a manipulator. To this purpose, Newton-Euler formulation provides a computationally efficient recursive method for on-line computation of inverse dynamics. Nevertheless, it can be shown that also Lagrange formulation is liable to a computationally efficient recursive implementation, though with a nonnegligible reformulation effort.

For an n-joint manipulator the *number of operations* required is:

- $O(n^2)$ for computing *direct dynamics*,
- $O(n)$ for computing *inverse dynamics*.

4.7 Operational Space Dynamic Model

As an alternative to the joint space dynamic model, the equations of motion of the system can be expressed directly in the operational space; to this purpose it is necessary to find a *dynamic model* which describes the relationship between the generalized forces acting on the manipulator and the number of minimal variables chosen to describe the end-effector position and orientation in the *operational space*.

Similarly to kinematic description of a manipulator in the operational space, the presence of redundant degrees of freedom and/or kinematic and representation singularities deserves careful attention in the derivation of an operational space dynamic model.

The determination of the dynamic model with Lagrange formulation using operational space variables allows a complete description of the system motion only in the case of a *nonredundant* manipulator, when the above variables constitute a set of *generalized coordinates* in terms of which the kinetic energy, the potential energy, and the nonconservative forces doing work on them can be expressed.

This way of proceeding does not provide a complete description of dynamics for a *redundant* manipulator; in this case, in fact, it is reasonable to expect the occurrence of *internal motions* of the structure caused by those joint generalized forces which do not affect the end-effector motion.

To develop an operational space model which can be adopted for both redundant and nonredundant manipulators, it is then convenient to start from the joint space model which is in all general. In fact, solving (4.41) for the joint accelerations, and neglecting the joint friction torques for simplicity, yields

$$\ddot{\boldsymbol{q}} = -\boldsymbol{B}^{-1}(\boldsymbol{q})\boldsymbol{C}(\boldsymbol{q},\dot{\boldsymbol{q}})\dot{\boldsymbol{q}} - \boldsymbol{B}^{-1}(\boldsymbol{q})\boldsymbol{g}(\boldsymbol{q}) + \boldsymbol{B}^{-1}(\boldsymbol{q})\boldsymbol{J}^T(\boldsymbol{q})(\boldsymbol{\gamma} - \boldsymbol{h}), \tag{4.116}$$

where the joint torques $\boldsymbol{\tau}$ have been expressed in terms of the equivalent end-effector forces $\boldsymbol{\gamma}$ according to (3.99). It is worth remarking that $\boldsymbol{h}$ represents the contribution of the end-effector forces due to contact with the environment, whereas $\boldsymbol{\gamma}$ expresses the contribution of the end-effector forces due to joint actuation.

On the other hand, the differential kinematics equation in (3.58) can be differentiated with respect to time to get the relationship between joint space and operational space accelerations, *i.e.*,

$$\ddot{\boldsymbol{x}} = \boldsymbol{J}_A(\boldsymbol{q})\ddot{\boldsymbol{q}} + \dot{\boldsymbol{J}}_A(\boldsymbol{q},\dot{\boldsymbol{q}})\dot{\boldsymbol{q}}. \tag{4.117}$$

The solution in (4.116) features the geometric Jacobian $\boldsymbol{J}$, whereas the analytical Jacobian $\boldsymbol{J}_A$ appears in (4.117). For notation uniformity, in view of (3.62), one can set

$$\boldsymbol{T}_A^T(\boldsymbol{x})\boldsymbol{\gamma} = \boldsymbol{\gamma}_A \qquad \boldsymbol{T}_A^T(\boldsymbol{x})\boldsymbol{h} = \boldsymbol{h}_A \tag{4.118}$$

where $\boldsymbol{T}_A$ is the transformation matrix between the two Jacobians. Substituting (4.116) into (4.117) and accounting for (4.118) gives

$$\ddot{\boldsymbol{x}} = -\boldsymbol{J}_A\boldsymbol{B}^{-1}\boldsymbol{C}\dot{\boldsymbol{q}} - \boldsymbol{J}_A\boldsymbol{B}^{-1}\boldsymbol{g} + \dot{\boldsymbol{J}}_A\dot{\boldsymbol{q}} + \boldsymbol{J}_A\boldsymbol{B}^{-1}\boldsymbol{J}_A^T(\boldsymbol{\gamma}_A - \boldsymbol{h}_A), \tag{4.119}$$

where the dependence on $\boldsymbol{q}$ and $\dot{\boldsymbol{q}}$ has been omitted. With the positions

$$\boldsymbol{B}_A = (\boldsymbol{J}_A\boldsymbol{B}^{-1}\boldsymbol{J}_A^T)^{-1} \tag{4.120}$$

$$\boldsymbol{C}_A\dot{\boldsymbol{x}} = \boldsymbol{B}_A\boldsymbol{J}_A\boldsymbol{B}^{-1}\boldsymbol{C}\dot{\boldsymbol{q}} - \boldsymbol{B}_A\dot{\boldsymbol{J}}_A\dot{\boldsymbol{q}} \tag{4.121}$$

$$\boldsymbol{g}_A = \boldsymbol{B}_A\boldsymbol{J}_A\boldsymbol{B}^{-1}\boldsymbol{g}, \tag{4.122}$$

The expression in (4.119) can be rewritten as

$$\boldsymbol{B}_A(\boldsymbol{x})\ddot{\boldsymbol{x}} + \boldsymbol{C}_A(\boldsymbol{x},\dot{\boldsymbol{x}})\dot{\boldsymbol{x}} + \boldsymbol{g}_A(\boldsymbol{x}) = \boldsymbol{\gamma}_A - \boldsymbol{h}_A, \tag{4.123}$$

which is formally analogous to the joint space dynamic model (4.41). Notice that the matrix $\boldsymbol{J}_A\boldsymbol{B}^{-1}\boldsymbol{J}_A^T$ is invertible if and only if $\boldsymbol{J}_A$ is full-rank, that is, in the absence of both kinematic and representation singularities.

For a nonredundant manipulator in a nonsingular configuration, the expressions in (4.120)–(4.122) become:

$$\boldsymbol{B}_A = \boldsymbol{J}_A^{-T}\boldsymbol{B}\boldsymbol{J}_A^{-1} \tag{4.120$'$}$$

$$\boldsymbol{C}_A\dot{\boldsymbol{x}} = \boldsymbol{J}_A^{-T}\boldsymbol{C}\dot{\boldsymbol{q}} - \boldsymbol{B}_A\dot{\boldsymbol{J}}_A\dot{\boldsymbol{q}} \tag{4.121$'$}$$

$$\boldsymbol{g}_A = \boldsymbol{J}_A^{-T}\boldsymbol{g}. \tag{4.122$'$}$$

As anticipated above, the main feature of the obtained model is its formal validity also for a redundant manipulator, even though the variables $\boldsymbol{x}$ do not constitute a set of generalized coordinates for the system; in this case, the matrix $\boldsymbol{B}_A$ is representative of a *kinetic pseudo-energy*.

In the remainder, the utility of the operational space dynamic model in (4.123) for solving direct and inverse dynamics problems is investigated. The following derivation is meaningful for redundant manipulators; for a nonredundant manipulator, in fact, using (4.123) does not pose specific problems as long as $\boldsymbol{J}_A$ is nonsingular (see (4.120′)–(4.122′)).

With reference to operational space, the *direct dynamics* problem consists of determining the resulting end-effector accelerations $\ddot{\boldsymbol{x}}(t)$ (and thus $\dot{\boldsymbol{x}}(t)$, $\boldsymbol{x}(t)$) from the given joint torques $\boldsymbol{\tau}(t)$ and end-effector forces $\boldsymbol{h}(t)$. For a redundant manipulator, (4.123) cannot be directly used, since (3.99) has a solution in γ only if $\boldsymbol{\tau} \in \mathcal{R}(\boldsymbol{J}^T)$. It follows that for simulation purposes, the solution to the problem is naturally obtained in the joint space; in fact, the expression in (4.41) allows computing $\boldsymbol{q}$, $\dot{\boldsymbol{q}}$, $\ddot{\boldsymbol{q}}$ which, substituted into the direct kinematics equations in (2.70), (3.58), (4.117), give $\boldsymbol{x}$, $\dot{\boldsymbol{x}}$, $\ddot{\boldsymbol{x}}$, respectively.

Formulation of an *inverse dynamics* problem in the operational space requires the determination of the joint torques $\boldsymbol{\tau}(t)$ that are needed to generate a specific motion assigned in terms of $\ddot{\boldsymbol{x}}(t)$, $\dot{\boldsymbol{x}}(t)$, $\boldsymbol{x}(t)$, for given end-effector forces $\boldsymbol{h}(t)$. A possible way of solution is to solve a complete inverse kinematics problem for (2.70), (3.58), (4.117), and then compute the required torques with the joint space inverse dynamics as in (4.41). Hence, for redundant manipulators, redundancy resolution is performed at kinematic level.

An alternative solution to the inverse dynamics problem consists of computing $\boldsymbol{\gamma}_A$ as in (4.123) and the joint torques $\boldsymbol{\tau}$ as in (3.99). In this way, however, the presence of redundant degrees of freedom is not exploited at all, since the computed torques do not generate internal motions of the structure.

If it is desired to find a formal solution that allows redundancy resolution at dynamic level, it is necessary to determine those torques corresponding to the equivalent end-effector forces computed as in (4.123). By analogy with the differential kinematics solution (3.50), the expression of the torques to be determined will feature the presence of a minimum-norm term and a homogeneous term. Since the joint torques have to be computed, it is convenient to express the model (4.123) in terms of $\boldsymbol{q}$, $\dot{\boldsymbol{q}}$, $\ddot{\boldsymbol{q}}$. By recalling the positions (4.121) and (4.122), the expression in (4.123) becomes

$$\boldsymbol{B}_A(\ddot{\boldsymbol{x}} - \dot{\boldsymbol{J}}_A\dot{\boldsymbol{q}}) + \boldsymbol{B}_A\boldsymbol{J}_A\boldsymbol{B}^{-1}\boldsymbol{C}\dot{\boldsymbol{q}} + \boldsymbol{B}_A\boldsymbol{J}_A\boldsymbol{B}^{-1}\boldsymbol{g} = \boldsymbol{\gamma}_A - \boldsymbol{h}_A$$

and, in view of (4.117),

$$\boldsymbol{B}_A\boldsymbol{J}_A\ddot{\boldsymbol{q}} + \boldsymbol{B}_A\boldsymbol{J}_A\boldsymbol{B}^{-1}\boldsymbol{C}\dot{\boldsymbol{q}} + \boldsymbol{B}_A\boldsymbol{J}_A\boldsymbol{B}^{-1}\boldsymbol{g} = \boldsymbol{\gamma}_A - \boldsymbol{h}_A. \tag{4.124}$$

By setting

$$\bar{\boldsymbol{J}}_A(\boldsymbol{q}) = \boldsymbol{B}^{-1}(\boldsymbol{q})\boldsymbol{J}_A^T(\boldsymbol{q})\boldsymbol{B}_A(\boldsymbol{q}), \tag{4.125}$$

the expression in (4.124) becomes

$$\bar{\boldsymbol{J}}_A^T(\boldsymbol{B}\ddot{\boldsymbol{q}} + \boldsymbol{C}\dot{\boldsymbol{q}} + \boldsymbol{g}) = \boldsymbol{\gamma}_A - \boldsymbol{h}_A. \tag{4.126}$$

At this point, from the joint space dynamic model in (4.41), it is easy to recognize that (4.126) can be written as

$$\bar{\boldsymbol{J}}_A^T(\boldsymbol{\tau} - \boldsymbol{J}_A^T\boldsymbol{h}_A) = \boldsymbol{\gamma}_A - \boldsymbol{h}_A,$$

from which

$$\bar{\boldsymbol{J}}_A^T\boldsymbol{\tau} = \boldsymbol{\gamma}_A. \tag{4.127}$$

The general solution to (4.127) is of the form

$$\boldsymbol{\tau} = \boldsymbol{J}_A^T(\boldsymbol{q})\boldsymbol{\gamma}_A + \big(\boldsymbol{I} - \boldsymbol{J}_A^T(\boldsymbol{q})\bar{\boldsymbol{J}}_A^T(\boldsymbol{q})\big)\boldsymbol{\tau}_0, \tag{4.128}$$

that can be derived by observing that $\boldsymbol{J}_A^T$ is a *right pseudo-inverse* of $\bar{\boldsymbol{J}}_A^T$ weighted by the inverse of the inertia matrix $\boldsymbol{B}^{-1}$. The $(n \times 1)$ vector of arbitrary torques $\boldsymbol{\tau}_0$ in (4.128) does not contribute to the end-effector forces, since it is projected in the null space of $\bar{\boldsymbol{J}}_A^T$.

To summarize, for given $\boldsymbol{x}$, $\dot{\boldsymbol{x}}$, $\ddot{\boldsymbol{x}}$ and $\boldsymbol{h}_A$, the expression in (4.123) allows computing $\boldsymbol{\gamma}_A$. Then, (4.128) gives the torques $\boldsymbol{\tau}$ which, besides executing the assigned end-effector motion, generate internal motions of the structure to be employed for handling redundancy at dynamic level through a suitable choice of $\boldsymbol{\tau}_0$.

4.8 Dynamic Manipulability Ellipsoid

The availability of the dynamic model allows formulation of the *dynamic manipulability ellipsoid* which provides a useful tool for manipulator dynamic performance analysis. This can be used for mechanical structure design as well as for seeking optimal manipulator configurations.

Consider the set of joint torques of constant (unit) norm

$$\boldsymbol{\tau}^T\boldsymbol{\tau} = 1 \tag{4.129}$$

describing the points on the surface of a sphere. It is desired to describe the operational space accelerations that can be generated by the given set of joint torques.

For studying dynamic manipulability, suppose to consider the case of a manipulator standing still ($\dot{\boldsymbol{q}} = \mathbf{0}$), not in contact with the environment ($\boldsymbol{h} = \mathbf{0}$). The simplified model is

$$\boldsymbol{B}(\boldsymbol{q})\ddot{\boldsymbol{q}} + \boldsymbol{g}(\boldsymbol{q}) = \boldsymbol{\tau}. \tag{4.130}$$

The joint accelerations $\ddot{\boldsymbol{q}}$ can be computed from the second-order differential kinematics that can be obtained by differentiating (3.35), and imposing successively $\dot{\boldsymbol{q}} = \mathbf{0}$, leading to

$$\dot{\boldsymbol{v}} = \boldsymbol{J}(\boldsymbol{q})\ddot{\boldsymbol{q}}. \tag{4.131}$$

Solving for minimum-norm accelerations only, for a *nonsingular Jacobian*, and substituting in (4.130) yields the expression of the torques

$$\boldsymbol{\tau} = \boldsymbol{B}(\boldsymbol{q})\boldsymbol{J}^{\dagger}(\boldsymbol{q})\dot{\boldsymbol{v}} + \boldsymbol{g}(\boldsymbol{q}) \tag{4.132}$$

needed to derive the ellipsoid. In fact, substituting (4.132) into (4.129) gives

$$\left(\boldsymbol{B}(\boldsymbol{q})\boldsymbol{J}^{\dagger}(\boldsymbol{q})\dot{\boldsymbol{v}} + \boldsymbol{g}(\boldsymbol{q})\right)^T \left(\boldsymbol{B}(\boldsymbol{q})\boldsymbol{J}^{\dagger}(\boldsymbol{q})\dot{\boldsymbol{v}} + \boldsymbol{g}(\boldsymbol{q})\right) = 1.$$

The vector on the right-hand side of (4.132) can be rewritten as

$$\begin{aligned}\boldsymbol{B}\boldsymbol{J}^{\dagger}\dot{\boldsymbol{v}} + \boldsymbol{g} &= \boldsymbol{B}(\boldsymbol{J}^{\dagger}\dot{\boldsymbol{v}} + \boldsymbol{B}^{-1}\boldsymbol{g}) \\ &= \boldsymbol{B}(\boldsymbol{J}^{\dagger}\dot{\boldsymbol{v}} + \boldsymbol{B}^{-1}\boldsymbol{g} + \boldsymbol{J}^{\dagger}\boldsymbol{J}\boldsymbol{B}^{-1}\boldsymbol{g} - \boldsymbol{J}^{\dagger}\boldsymbol{J}\boldsymbol{B}^{-1}\boldsymbol{g}) \\ &= \boldsymbol{B}\big(\boldsymbol{J}^{\dagger}\dot{\boldsymbol{v}} + \boldsymbol{J}^{\dagger}\boldsymbol{J}\boldsymbol{B}^{-1}\boldsymbol{g} + (\boldsymbol{I} - \boldsymbol{J}^{\dagger}\boldsymbol{J})\boldsymbol{B}^{-1}\boldsymbol{g}\big).\end{aligned} \tag{4.133}$$

where the dependence on $\boldsymbol{q}$ has been omitted. According to what was done for solving (4.131), one can neglect the contribution of the accelerations given by $\boldsymbol{B}^{-1}\boldsymbol{g}$ which are in the null space of $\boldsymbol{J}$ and then produce no end-effector acceleration. Hence, (4.133) becomes

$$\boldsymbol{B}\boldsymbol{J}^{\dagger}\dot{\boldsymbol{v}} + \boldsymbol{g} = \boldsymbol{B}\boldsymbol{J}^{\dagger}(\dot{\boldsymbol{v}} + \boldsymbol{J}\boldsymbol{B}^{-1}\boldsymbol{g}) \tag{4.134}$$

and the dynamic manipulability ellipsoid can be expressed in the form

$$(\dot{\boldsymbol{v}} + \boldsymbol{J}\boldsymbol{B}^{-1}\boldsymbol{g})^T\boldsymbol{J}^{\dagger T}\boldsymbol{B}^T\boldsymbol{B}\boldsymbol{J}^{\dagger}(\dot{\boldsymbol{v}} + \boldsymbol{J}\boldsymbol{B}^{-1}\boldsymbol{g}) = 1. \tag{4.135}$$

The core of the quadratic form $\boldsymbol{J}^{\dagger T}\boldsymbol{B}^T\boldsymbol{B}\boldsymbol{J}^{\dagger}$ depends on the geometrical and inertial characteristics of the manipulator and determines the volume and principal axes of the ellipsoid. The vector $-\boldsymbol{J}\boldsymbol{B}^{-1}\boldsymbol{g}$, describing the contribution of gravity, produces a constant translation of the centre of the ellipsoid (for each manipulator configuration) with respect to the origin of the reference frame; see the example in Figure 4.16 for a three-link planar arm.

The meaning of the dynamic manipulability ellipsoid is conceptually similar to that of the ellipsoids considered with reference to kineto-statics duality. In fact, the distance of a point on the surface of the ellipsoid from the end effector gives a measure of the

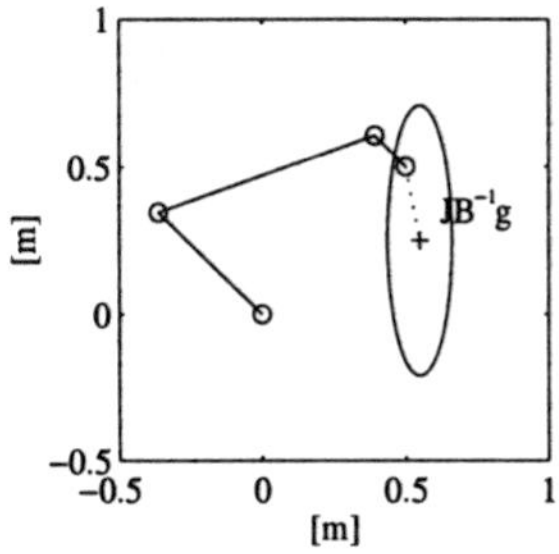

Figure 4.16 Effect of gravity on the dynamic manipulability ellipsoid for a three-link planar arm.

accelerations which can be imposed to the end effector along the given direction, with respect to the constraint (4.129). With reference to Figure 4.16, it is worth noticing how the presence of gravity acceleration allows performing larger accelerations downwards, as natural to predict.

In the case of a nonredundant manipulator, the ellipsoid reduces to

$$(\dot{\boldsymbol{v}} + \boldsymbol{J}\boldsymbol{B}^{-1}\boldsymbol{g})^T \boldsymbol{J}^{-T}\boldsymbol{B}^T\boldsymbol{B}\boldsymbol{J}^{-1}(\dot{\boldsymbol{v}} + \boldsymbol{J}\boldsymbol{B}^{-1}\boldsymbol{g}) = 1. \qquad (4.135')$$

Problems

4.1 Find the dynamic model of a two-link Cartesian arm in the case when the second joint axis forms an angle of $\pi/4$ with the first joint axis; compare the result with the model of the manipulator in Figure 4.4.

4.2 For the planar arm of Section 4.3.2, find a minimal parameterization of the dynamic model in (4.81).

4.3 Find the dynamic model of the two-link planar arm with a prismatic joint and a revolute joint in Figure 4.17 with Lagrange formulation. Then, consider the addition of a concentrated tip payload of mass m_L, and express the resulting model in a linear form with respect to a suitable set of dynamic parameters as in (4.80).

4.4 For the two-link planar arm of Figure 4.5, prove that with a different choice of the matrix $\boldsymbol{C}$, (4.48) holds true while (4.47) does not.

4.5 For the two-link planar arm of Figure 4.5, find the dynamic model with the Lagrange formulation when the absolute angles with respect to the base frame are chosen as generalized coordinates. Discuss the result in view of a comparison with the model derived in (4.81).

4.6 Compute the joint torques for the two-link planar arm of Figure 4.5 with the data and along the trajectories of Example 4.2, in the case of tip forces $\boldsymbol{f} = [\,500 \quad 500\,]^T$ N.

4.7 Find the dynamic model of the two-link planar arm with a prismatic joint and a revolute joint in Figure 4.17 by using the recursive Newton-Euler algorithm.

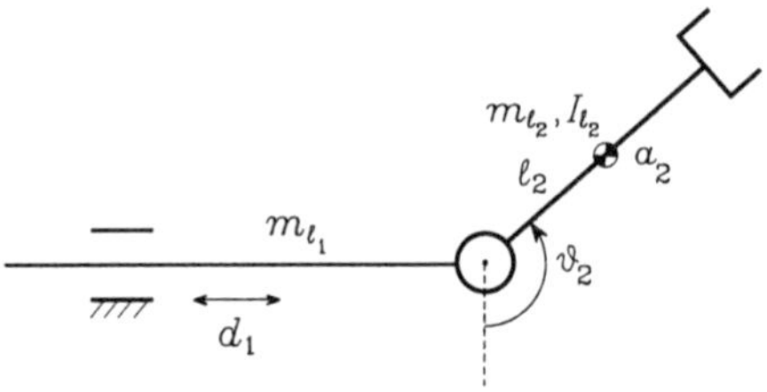

Figure 4.17 Two-link planar arm with a prismatic joint and a revolute joint.

4.8 Show that for the operational space dynamic model (4.123) a skew-symmetry property holds which is analogous to (4.47).

4.9 Show how to obtain the general solution to (4.127) in the form (4.128).

4.10 For a nonredundant manipulator, compute the relationship between the dynamic manipulability measure that can be defined for the dynamic manipulability ellipsoid and the manipulability measure defined in (3.52).

Bibliography

Armstrong M.W. (1979) Recursive solution to the equations of motion of an n-link manipulator. In *Proc. 5th World Congr. Theory of Machines and Mechanisms* Montreal, Canada, pp. 1343–1346.

Asada H., Slotine J.-J.E. (1986) *Robot Analysis and Control*. Wiley, New York, 1986.

Asada H., Youcef-Toumi K. (1984) Analysis and design of a direct-drive arm with a five-bar-link parallel drive mechanism. *ASME J. Dynamic Systems, Measurement, and Control* 106:225–230.

Bejczy A.K. (1974) *Robot Arm Dynamics and Control*. Memo. TM 33-669, Jet Propulsion Laboratory, California Institute of Technology.

Chiacchio P., Chiaverini S., Sciavicco L., Siciliano B. (1992) Influence of gravity on the manipulability ellipsoid for robot arms. *ASME J. Dynamic Systems, Measurement, and Control* 114:723–727.

Gautier M., Khalil W. (1990) Direct calculation of minimum set of inertial parameters of serial robots. *IEEE Trans. Robotics and Automation* 6:368–373.

Hollerbach J.M. (1980) A recursive Lagrangian formulation of manipulator dynamics and a comparative study of dynamics formulation complexity. *IEEE Trans. Systems, Man, and Cybernetics* 10:730–736.

Khatib O. (1987) A unified approach to motion and force control of robot manipulators: The operational space formulation. *IEEE J. Robotics and Automation* 3:43–53.

Luh J.Y.S., Walker M.W., Paul R.P.C. (1980) On-line computational scheme for mechanical manipulators. *ASME J. Dynamic Systems, Measurement, and Control* 102:69–76.

Orin D.E., McGhee R.B., Vukobratović M., Hartoch G. (1979) Kinematic and kinetic analysis of open-chain linkages utilizing Newton-Euler methods. *Mathematical Biosciences* 43:107–130.

Sciavicco L., Siciliano B., Villani L. (1996) Lagrange and Newton-Euler dynamic modeling of a gear-driven rigid robot manipulator with inclusion of motor inertia effects. *Advanced*

Robotics 10:317–334.
Silver D.B. (1982) On the equivalence of Lagrangian and Newton-Euler dynamics for manipulators. *Int. J. Robotics Research* 1(2):60–70.
Slotine J.-J.E. (1988) Putting physics in control—The example of robotics. *IEEE Control Systems Mag.* 8(6):12–18.
Spong M.W., Vidyasagar M. (1989) *Robot Dynamics and Control.* Wiley, New York.
Stepanenko Y., Vukobratović M. (1976) Dynamics of articulated open-chain active mechanisms. *Mathematical Biosciences* 28:137–170.
Uicker J.J. (1967) Dynamic force analysis of spatial linkages. *ASME J. Applied Mechanics* 34:418–424.
Vukobratović M. (1978) Dynamics of active articulated mechanisms and synthesis of artificial motion. *Mechanism and Machine Theory* 13:1–56.
Walker M.W., Orin D.E. (1982) Efficient dynamic computer simulation of robotic mechanisms. *ASME J. Dynamic Systems, Measurement, and Control* 104:205–211.
Yoshikawa T. (1985) Dynamic manipulability ellipsoid of robot manipulators. *J. Robotic Systems* 2:113–124.

5. Trajectory Planning

The previous chapters focused on mathematical modelling of mechanical manipulators in terms of kinematics, differential kinematics and statics, and dynamics. Before studying the problem of controlling a manipulation structure, it is worth presenting the main features of motion planning algorithms for the execution of specific manipulator tasks. The goal of *trajectory planning* is to generate the reference inputs to the motion control system which ensures that the manipulator executes the planned trajectories. The user typically specifies a number of parameters to describe the desired trajectory. Planning consists of generating a time sequence of the values attained by a polynomial function interpolating the desired trajectory. This chapter presents some techniques for trajectory generation both in the case when the initial and final point of the path are assigned (*point-to-point motion*), and in the case when a finite sequence of points are assigned along the path (*path motion*). First, the problem of trajectory planning in the *joint space* is considered, and then the basic concepts of trajectory planning in the *operational space* are illustrated. The chapter ends with the presentation of a technique for *dynamic scaling* a trajectory which allows adapting trajectory planning to manipulator dynamic characteristics.

5.1 Path and Trajectory

The minimal requirement for a manipulator is the capability to move from an initial posture to a final assigned posture. The transition should be characterized by motion laws requiring the actuators to exert joint generalized forces which do not violate the saturation limits and do not excite the typically unmodeled resonant modes of the structure. It is then necessary to devise planning algorithms that generate suitably *smooth* trajectories.

In order to avoid confusion between terms often used as synonyms, the difference between a path and a trajectory is to be explained. A *path* denotes the locus of points in the joint space, or in the operational space, the manipulator has to follow in the execution of the assigned motion; a path is then a pure geometric description of motion. On the other hand, a *trajectory* is a path on which a time law is specified, for instance in terms of velocities and/or accelerations at each point.

In principle, it can be conceived that the inputs to a *trajectory planning* algorithm are the path description, the path constraints, and the constraints imposed by manipulator dynamics, whereas the outputs are the joint (end-effector) trajectories in terms of a time sequence of the values attained by position, velocity and acceleration. A path can be defined either in the *joint space* or in the *operational space*. Usually, the latter is preferred since it allows a natural description of the task the manipulator has to perform.

A geometric path cannot be fully specified by the user for obvious complexity reasons. Typically, a reduced number of parameters is specified such as extremal points, possible intermediate points, and geometric primitives interpolating the points. Also, the motion time law is not typically specified at each point of the geometric path, but rather it regards the total trajectory time, the constraints on the maximum velocities and accelerations, and eventually the assignment of velocity and acceleration at points of particular interest. On the basis of the above information, the trajectory planning algorithm generates a time sequence of variables that describe end-effector position and orientation over time in respect of the imposed constraints. Since the control action on the manipulator is carried out in the joint space, a suitable inverse kinematics algorithm is to be used to reconstruct the time sequence of joint variables corresponding to the above sequence in the operational space.

Trajectory planning in the operational space naturally allows accounting for the presence of path constraints; these are due to regions of workspace which are forbidden to the manipulator, e.g., due to the presence of obstacles. In fact, such constraints are typically better described in the operational space, since their corresponding points in the joint space are difficult to compute.

With regard to motion in the neighbourhood of singular configurations and presence of redundant degrees of freedom, trajectory planning in the operational space may involve problems difficult to solve. In such cases, it may be advisable to specify the path in the joint space, still in terms of a reduced number of parameters. Hence, a time sequence of joint variables has to be generated which satisfy the constraints imposed on the trajectory.

For the sake of clarity, in the following, the case of joint space trajectory planning is treated first. The results will then be extended to the case of trajectories in the operational space.

5.2 Joint Space Trajectories

A manipulator motion is typically assigned in the operational space in terms of trajectory parameters such as the initial and final end-effector location, possible intermediate locations, and traveling time along particular geometric paths. If it is desired to plan a trajectory in the *joint space*, the values of the joint variables have to be determined first from the end-effector position and orientation specified by the user. It is then necessary to resort to an inverse kinematics algorithm, if planning is done off-line, or to directly measure the above variables, if planning is done by the teaching-by-showing technique (see Chapter 9).

The planning algorithm generates a function $\boldsymbol{q}(t)$ interpolating the given vectors of joint variables at each point, in respect of the imposed constraints.

In general, a joint space trajectory planning algorithm is required to have the following features:

- the generated trajectories be not very demanding from a computational viewpoint,
- joint positions and velocities be continuous functions of time (continuity of accelerations may be imposed, too),
- undesirable effects be minimized, e.g., nonsmooth trajectories interpolating a sequence of points on a path.

At first, the case is examined when only the initial and final points on the path and the traveling time are specified (*point-to-point motion*); the results are then generalized to the case when also intermediate points along the path are specified (*path motion*). Without loss of generality, the single joint variable $q(t)$ is considered.

5.2.1 Point-to-point Motion

In the *point-to-point motion*, the manipulator has to move from an initial to a final joint configuration in a given time t_f. In this case, the actual end-effector path is of no concern. The algorithm should generate a trajectory which, in respect to the above general requirements, is also capable to optimize some performance index when the joint is moved from one position to another.

A suggestion for choosing the motion primitive may stem from the analysis of an incremental motion problem. Let I be the moment of inertia of a rigid body about its rotation axis. It is required to take the angle q from an initial value q_i to a final value q_f in a time t_f. It is obvious that infinite solutions exist to this problem. Assuming that rotation is executed through a torque τ supplied by a motor, a solution can be found which minimizes the energy dissipated in the motor. This optimization problem can be formalized as follows. Having set $\dot{q} = \omega$, determine the solution to the differential equation

$$I\dot{\omega} = \tau$$

subject to the condition

$$\int_o^{t_f} \omega(t)dt = q_f - q_i,$$

so as to minimize the performance index

$$\int_0^{t_f} \tau^2(t)dt.$$

It can be shown that the resulting solution is of the type

$$\omega(t) = at^2 + bt + c.$$

Even though the joint dynamics cannot be described in the above simple manner[1], the

[1] In fact, recall that the moment of inertia about the joint axis is a function of manipulator configuration.

choice of a third-order polynomial function to generate a joint trajectory represents a valid solution for the problem at issue.

Therefore, to determine a joint motion, the *cubic polynomial* can be chosen

$$q(t) = a_3t^3 + a_2t^2 + a_1t + a_0, \tag{5.1}$$

resulting into a parabolic velocity profile

$$\dot{q}(t) = 3a_3t^2 + 2a_2t + a_1$$

and a linear acceleration profile

$$\ddot{q}(t) = 6a_3t + 2a_2.$$

Since four coefficients are available, it is possible to impose, besides the initial and final joint position values q_i and q_f, also the initial and final joint velocity values $\dot{q}_i$ and $\dot{q}_f$ which are usually set to zero. Determination of a specific trajectory is given by the solution to the following system of equations:

$$\begin{aligned} a_0 &= q_i \\ a_1 &= \dot{q}_i \\ a_3t_f^3 + a_2t_f^2 + a_1t_f + a_0 &= q_f \\ 3a_3t_f^2 + 2a_2t_f + a_1 &= \dot{q}_f, \end{aligned}$$

that allows computing the coefficients of the polynomial in (5.1). Figure 5.1 illustrates the time law obtained with the following data: $q_i = 0, q_f = \pi, t_f = 1$, and $\dot{q}_i = \dot{q}_f = 0$. As anticipated, velocity has a parabolic profile, while acceleration has a linear profile with initial and final discontinuity.

If it is desired to assign also the initial and final values of acceleration, six constraints have to be satisfied and then a polynomial of at least *fifth* order is needed. The motion time law for the generic joint is then given by

$$q(t) = a_5t^5 + a_4t^4 + a_3t^3 + a_2t^2 + a_1t + a_0, \tag{5.2}$$

whose coefficients can be computed, as for the previous case, by imposing the conditions for $t = 0$ and $t = t_f$ on the joint variable $q(t)$ and on its first two derivatives. With the choice (5.2), one obviously gives up minimizing the above performance index.

An alternative approach with time laws of blended polynomial type is frequently adopted in industrial practice, which allows directly verifying whether the resulting velocities and accelerations can be supported by the physical mechanical manipulator.

In this case, a *trapezoidal velocity profile* is assigned, which imposes a constant acceleration in the start phase, a cruise velocity, and a constant deceleration in the arrival phase. The resulting trajectory is formed by a linear segment connected by two parabolic segments to the initial and final positions.

As can be seen from the velocity profiles in Figure 5.2, it is assumed that both initial and final velocities are null and the segments with constant accelerations have

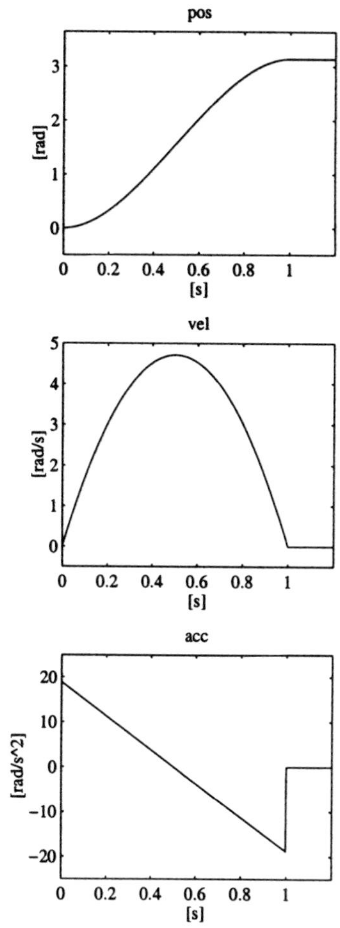

Figure 5.1 Time history of position, velocity and acceleration with a cubic polynomial time law.

the same time duration; this implies an equal magnitude $\ddot{q}_c$ in the two segments. Notice also that the above choice leads to a symmetric trajectory with respect to the average point $q_m = (q_f + q_i)/2$ at $t_m = t_f/2$.

The trajectory has to satisfy some constraints to ensure the transition from q_i to q_f in a time t_f. The velocity at the end of the parabolic segment must be equal to the (constant) velocity of the linear segment, *i.e.*,

$$\ddot{q}_c t_c = \frac{q_m - q_c}{t_m - t_c}, \tag{5.3}$$

where q_c is the value attained by the joint variable at the end of the parabolic segment at time t_c with constant acceleration $\ddot{q}_c$ (recall that $\dot{q}(0) = 0$). It is then

$$q_c = q_i + \frac{1}{2}\ddot{q}_c t_c^2. \tag{5.4}$$

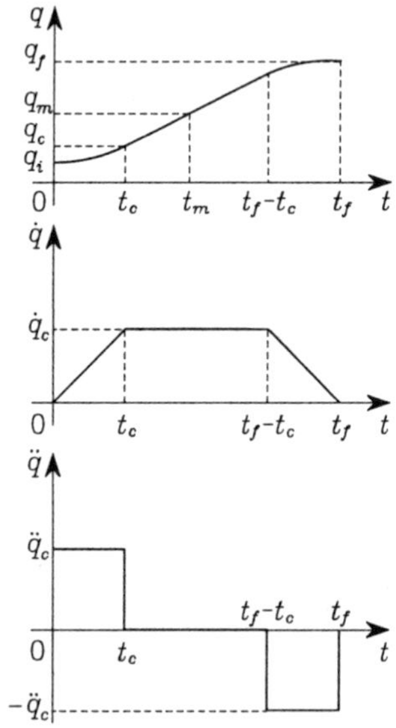

Figure 5.2 Characterization of a time law with trapezoidal velocity profile in terms of position, velocity and acceleration.

Combining (5.3) with (5.4) gives

$$\ddot{q}_c t_c^2 - \ddot{q}_c t_f t_c + q_f - q_i = 0. \tag{5.5}$$

Usually, $\ddot{q}_c$ is specified with the constraint that $\text{sgn}\,\ddot{q}_c = \text{sgn}\,(q_f - q_i)$; hence, for given t_f, q_i and q_f, the solution for t_c is computed from (5.5) as ($t_c \leq t_f/2$)

$$t_c = \frac{t_f}{2} - \frac{1}{2}\sqrt{\frac{t_f^2\ddot{q}_c - 4(q_f - q_i)}{\ddot{q}_c}}. \tag{5.6}$$

Acceleration is then subject to the constraint

$$|\ddot{q}_c| \geq \frac{4|q_f - q_i|}{t_f^2}. \tag{5.7}$$

When the acceleration $\ddot{q}_c$ is chosen so as to satisfy (5.7) with the equality sign, the resulting trajectory does not feature the constant velocity segment any more and has only the acceleration and deceleration segments (*triangular* profile).

Given q_i, q_f and t_f, and thus also an average transition velocity, the constraint in (5.7) allows imposing a value of acceleration consistent with the trajectory. Then, t_c is computed from (5.6), and the following sequence of polynomials is generated:

$$q(t) = \begin{cases} q_i + \frac{1}{2}\ddot{q}_c t^2 & 0 \leq t \leq t_c \\ q_i + \ddot{q}_c t_c (t - t_c/2) & t_c < t \leq t_f - t_c \\ q_f - \frac{1}{2}\ddot{q}_c (t_f - t)^2 & t_f - t_c < t \leq t_f. \end{cases} \tag{5.8}$$

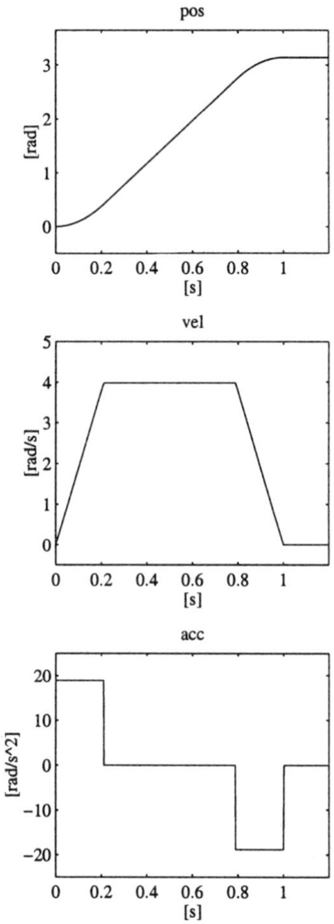

Figure 5.3 Time history of position, velocity and acceleration with a trapezoidal velocity profile time law.

Figure 5.3 illustrates a representation of the motion time law obtained by imposing the data: $q_i = 0$, $q_f = \pi$, $t_f = 1$, and $|\ddot{q}_c| = 6\pi$.

Specifying acceleration in the parabolic segment is not the only way to determine trajectories with trapezoidal velocity profile. Besides q_i, q_f and t_f, one can specify also the cruise velocity $\dot{q}_c$ which is subject to the constraint

$$\frac{|q_f - q_i|}{t_f} < |\dot{q}_c| \le \frac{2|q_f - q_i|}{t_f}. \tag{5.9}$$

By recognizing that $\dot{q}_c = \ddot{q}_c t_c$, (5.5) allows computing t_c as

$$t_c = \frac{q_i - q_f + \dot{q}_c t_f}{\dot{q}_c}, \tag{5.10}$$

and thus the resulting acceleration is

$$\ddot{q}_c = \frac{\dot{q}_c^2}{q_i - q_f + \dot{q}_c t_f}. \tag{5.11}$$

The computed values of t_c and $\ddot{q}_c$ as in (5.10) and (5.11) allow generating the sequence of polynomials expressed by (5.8).

The adoption of a trapezoidal velocity profile results in a worse performance index compared to the cubic polynomial. The decrease is, however, limited; the term $\int_0^{t_f} \tau^2 dt$ increases by 12.5% with respect to the optimal case.

5.2.2 Path Motion

In several applications, the path is described in terms of a number of points greater than two. For instance, even for the simple point-to-point motion of a pick-and-place task, it may be worth assigning two intermediate points between the initial point and the final point; suitable positions can be set for lifting off and setting down the object, so that reduced velocities are obtained with respect to direct transfer of the object. For more complex applications, it may be convenient to assign a *sequence of points* so as to guarantee better monitoring on the executed trajectories; the points are to be specified more densely in those segments of the path where obstacles have to be avoided or a high path curvature is expected. It should not be forgotten that the corresponding joint variables have to be computed from the operational space locations.

Therefore, the problem is to generate a trajectory when N points, termed *path points*, are specified and have to be reached by the manipulator at certain instants of time. For each joint variable there are N constraints, and then one might want to use an $(N-1)$-order polynomial. This choice, however, has the following disadvantages:

- It is not possible to assign the initial and final velocities.
- As the order of a polynomial increases, its oscillatory behaviour increases, and this may lead to trajectories which are not natural for the manipulator.
- Numerical accuracy for computation of polynomial coefficients decreases as order increases.
- The resulting system of constraint equations is heavy to solve.
- Polynomial coefficients depend on all the assigned points; thus, if it is desired to change a point, all of them have to be recomputed.

These drawbacks can be overcome if a suitable number of low-order *interpolating polynomials*, continuous at the path points, are considered in place of a single high-order polynomial.

According to the previous section, the interpolating polynomial of lowest order is the cubic polynomial, since it allows imposing continuity of velocities at the path points. With reference to the single joint variable, a function $q(t)$ is sought, formed by a sequence of $N-1$ cubic polynomials $\Pi_k(t)$, for $k = 1, \dots, N-1$, continuous with continuous first derivatives. The function $q(t)$ attains the values q_k for $t = t_k$ $(k = 1, \dots, N)$, and $q_1 = q_i$, $t_1 = 0$, $q_N = q_f$, $t_N = t_f$; the q_k's represent the path

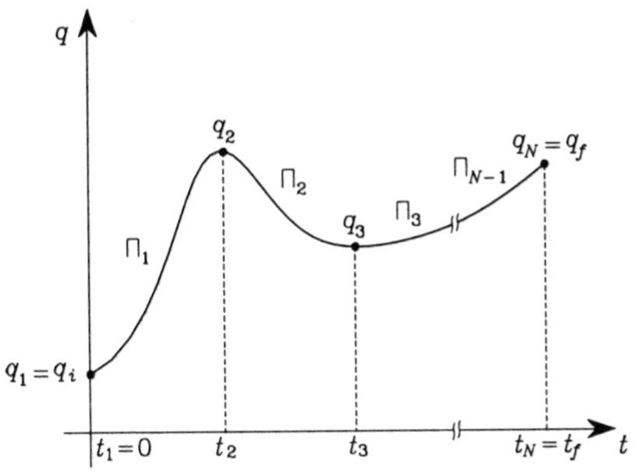

Figure 5.4 Characterization of a trajectory on a given path obtained through interpolating polynomials.

points describing the desired trajectory at $t = t_k$ (Figure 5.4). The following situations can be considered:

- Arbitrary values of $\dot{q}(t)$ are imposed at the path points.
- The values of $\dot{q}(t)$ at the path points are assigned according to a certain criterion.
- The acceleration $\ddot{q}(t)$ shall be continuous at the path points.

To simplify the problem, it is also possible to find interpolating polynomials of order less than three which determine trajectories passing nearby the path points at the given instants of time.

Interpolating Polynomials with Velocity Constraints at Path Points

This solution requires the user to be able to specify the desired velocity at each path point; the solution does not possess any novelty with respect to the above concepts.

The system of equations allowing computation of the coefficients of the $N - 1$ cubic polynomials interpolating the N path points is obtained by imposing the following conditions on the generic polynomial $\Pi_k(t)$ interpolating q_k and q_{k+1}, for $k = 1, \ldots, N - 1$:

$$\begin{aligned}
\Pi_k(t_k) &= q_k \\
\Pi_k(t_{k+1}) &= q_{k+1} \\
\dot{\Pi}_k(t_k) &= \dot{q}_k \\
\dot{\Pi}_k(t_{k+1}) &= \dot{q}_{k+1}.
\end{aligned}$$

The result is $N - 1$ systems of four equations in the four unknown coefficients of the generic polynomial; these can be solved one independently of the other. The initial and final velocities of the trajectory are typically set to zero ($\dot{q}_1 = \dot{q}_N = 0$) and continuity of velocity at the path points is ensured by setting

$$\dot{\Pi}_k(t_{k+1}) = \dot{\Pi}_{k+1}(t_{k+1})$$

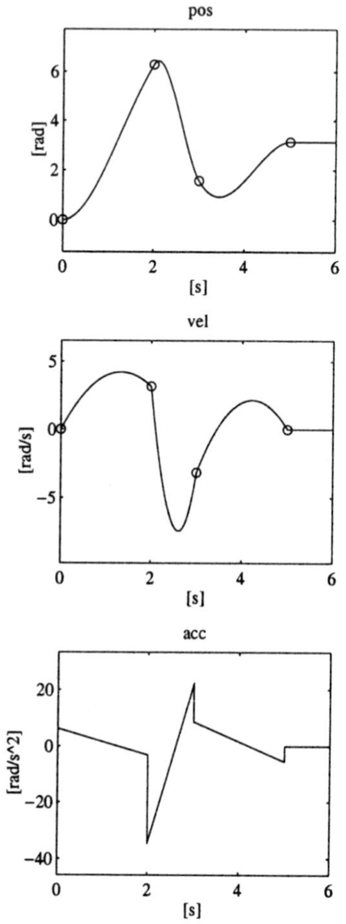

Figure 5.5 Time history of position, velocity and acceleration with a time law of interpolating polynomials with velocity constraints at path points.

for $k = 1, \ldots, N-2$.

Figure 5.5 illustrates the time history of position, velocity and acceleration obtained with the data: $q_1 = 0$, $q_2 = 2\pi$, $q_3 = \pi/2$, $q_4 = \pi$, $t_1 = 0$, $t_2 = 2$, $t_3 = 3$, $t_4 = 5$, $\dot{q}_1 = 0$, $\dot{q}_2 = \pi$, $\dot{q}_3 = -\pi$, $\dot{q}_4 = 0$. Notice the resulting discontinuity on the acceleration, since only continuity of velocity is guaranteed.

Interpolating Polynomials with Computed Velocities at Path Points

In this case, the joint velocity at a path point has to be computed according to a certain criterion. By interpolating the path points with linear segments, the relative velocities

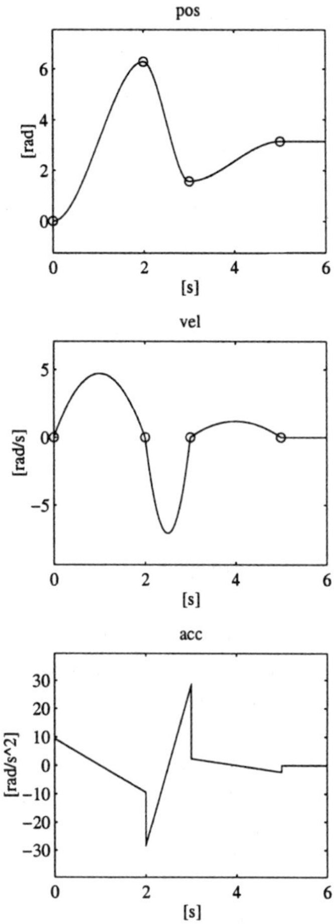

Figure 5.6 Time history of position, velocity and acceleration with a time law of interpolating polynomials with computed velocities at path points.

can be computed according to the following rules:

$$
\begin{aligned}
\dot{q}_1 &= 0 \\
\dot{q}_k &= \begin{cases} 0 & \operatorname{sgn}(v_k) \neq \operatorname{sgn}(v_{k+1}) \\ \frac{1}{2}(v_k + v_{k+1}) & \operatorname{sgn}(v_k) = \operatorname{sgn}(v_{k+1}) \end{cases} \\
\dot{q}_N &= 0,
\end{aligned}
\tag{5.12}
$$

where $v_k = (q_k - q_{k-1})/(t_k - t_{k-1})$ gives the slope of the segment in the time interval $[t_{k-1}, t_k]$. With the above settings, the determination of the interpolating polynomials is reduced to the previous case.

Figure 5.6 illustrates the time history of position, velocity and acceleration obtained with the following data: $q_1 = 0$, $q_2 = 2\pi$, $q_3 = \pi/2$, $q_4 = \pi$, $t_1 = 0$, $t_2 = 2$, $t_3 = 3$,

$t_4 = 5$, $\dot{q}_1 = 0$, $\dot{q}_4 = 0$. It is easy to recognize that the imposed sequence of path points leads to having zero velocity at the intermediate points.

Interpolating Polynomials with Continuous Accelerations at Path Points (Splines)

Both the above two solutions do not ensure continuity of accelerations at the path points. Given a sequence of N path points, also the acceleration is continuous at each t_k if four constraints are imposed; namely, two position constraints for each of the adjacent cubics and two constraints guaranteeing continuity of velocity and acceleration. The following equations have then to be satisfied:

$$\begin{aligned}\Pi_{k-1}(t_k) &= q_k\\ \Pi_{k-1}(t_k) &= \Pi_k(t_k)\\ \dot{\Pi}_{k-1}(t_k) &= \dot{\Pi}_k(t_k)\\ \ddot{\Pi}_{k-1}(t_k) &= \ddot{\Pi}_k(t_k).\end{aligned}$$

The resulting system for the N path points, including the initial and final points, cannot be solved. In fact, it is formed by $4(N-2)$ equations for the intermediate points and 6 equations for the extremal points; the position constraints for the polynomials $\Pi_0(t_1) = q_i$ and $\Pi_N(t_f) = q_f$ have to be excluded since they are not defined. Also, $\dot{\Pi}_0(t_1), \ddot{\Pi}_0(t_1), \dot{\Pi}_N(t_f), \ddot{\Pi}_N(t_f)$ do not have to be counted as polynomials since they are just the imposed values of initial and final velocities and accelerations. In sum, one has $4N-2$ equations in $4(N-1)$ unknowns.

The system can be solved only if one eliminates the two equations which allow arbitrarily assigning the initial and final acceleration values. Fourth-order polynomials should be used to include this possibility for the first and last segment.

On the other hand, if only third-order polynomials are to be used, the following deception can be operated. Two *virtual points* are introduced for which continuity constraints on position, velocity and acceleration can be imposed, without specifying the actual positions, though. It is worth remarking that the effective location of these points is irrelevant, since their position constraints regard continuity only. Hence, the introduction of two virtual points implies the determination of $N+1$ cubic polynomials.

Consider $N+2$ time instants t_k, where t_2 and t_{N+1} conventionally refer to the virtual points. The system of equations for determining the $N+1$ cubic polynomials can be found by taking the $4(N-2)$ equations:

$$\Pi_{k-1}(t_k) = q_k \tag{5.13}$$

$$\Pi_{k-1}(t_k) = \Pi_k(t_k) \tag{5.14}$$

$$\dot{\Pi}_{k-1}(t_k) = \dot{\Pi}_k(t_k) \tag{5.15}$$

$$\ddot{\Pi}_{k-1}(t_k) = \ddot{\Pi}_k(t_k) \tag{5.16}$$

for $k = 3, \ldots, N$, written for the $N-2$ intermediate path points, the 6 equations:

$$\Pi_1(t_1) = q_i \tag{5.17}$$

$$\dot{\Pi}_1(t_1) = \dot{q}_i \tag{5.18}$$

$$\ddot{\Pi}_1(t_1) = \ddot{q}_i \tag{5.19}$$

and

$$\Pi_{N+1}(t_{N+2}) = q_f \qquad (5.17')$$
$$\dot{\Pi}_{N+1}(t_{N+2}) = \dot{q}_f \qquad (5.18')$$
$$\ddot{\Pi}_{N+1}(t_{N+2}) = \ddot{q}_f \qquad (5.19')$$

written for the initial and final points, and the 6 equations:

$$\Pi_{k-1}(t_k) = \Pi_k(t_k) \qquad (5.20)$$
$$\dot{\Pi}_{k-1}(t_k) = \dot{\Pi}_k(t_k) \qquad (5.21)$$
$$\ddot{\Pi}_{k-1}(t_k) = \ddot{\Pi}_k(t_k) \qquad (5.22)$$

for $k = 2, N+1$, written for the two virtual points. The resulting system has $4(N+1)$ equations in $4(N+1)$ unknowns, that are the coefficients of the $N+1$ cubic polynomials.

The solution to the system is computationally demanding, even for low values of N. Nonetheless, the problem can be cast in a suitable form so as to solve the resulting system of equations with a computationally efficient algorithm. Since the generic polynomial $\Pi_k(t)$ is a cubic, its second derivative must be a linear function of time which then can be written as

$$\ddot{\Pi}_k(t) = \frac{\ddot{\Pi}_k(t_k)}{\Delta t_k}(t_{k+1} - t) + \frac{\ddot{\Pi}_k(t_{k+1})}{\Delta t_k}(t - t_k) \qquad k = 1, \ldots, N+1, \quad (5.23)$$

where $\Delta t_k = t_{k+1} - t_k$ indicates the time interval to reach q_{k+1} from q_k. By integrating (5.23) twice over time, the generic polynomial can be written as

$$\begin{aligned}\Pi_k(t) = {} & \frac{\ddot{\Pi}_k(t_k)}{6\Delta t_k}(t_{k+1} - t)^3 + \frac{\ddot{\Pi}_k(t_{k+1})}{6\Delta t_k}(t - t_k)^3 \\ & + \left(\frac{\Pi_k(t_{k+1})}{\Delta t_k} - \frac{\Delta t_k \ddot{\Pi}_k(t_{k+1})}{6}\right)(t - t_k) \\ & + \left(\frac{\Pi_k(t_k)}{\Delta t_k} - \frac{\Delta t_k \ddot{\Pi}_k(t_k)}{6}\right)(t_{k+1} - t) \qquad k = 1, \ldots, N+1,\end{aligned} \qquad (5.24)$$

which depends on the 4 unknowns: $\Pi_k(t_k)$, $\Pi_k(t_{k+1})$, $\ddot{\Pi}_k(t_k)$, $\ddot{\Pi}_k(t_{k+1})$.

Notice that the N variables q_k for $k \neq 2, N+1$ are given via (5.13), while continuity is imposed for q_2 and q_{N+1} via (5.20). By using (5.14), (5.17), (5.17′), the unknowns in the $N+1$ equations in (5.24) reduce to $2(N+2)$. By observing that the equations in (5.18) and (5.18′) depend on q_2 and q_{N+1}, and that $\dot{q}_i$ and $\dot{q}_f$ are given, q_2 and q_{N+1} can be computed as a function of $\ddot{\Pi}_1(t_1)$ and $\ddot{\Pi}_{N+1}(t_{N+2})$, respectively. Thus, a number of $2(N+1)$ unknowns are left.

By accounting for (5.16) and (5.22), and noticing that in (5.19) and (5.19′) $\ddot{q}_i$ and $\ddot{q}_f$ are given, the unknowns reduce to N.

At this point, (5.15) and (5.21) can be utilized to write the system of N equations in N unknowns:

$$\dot{\Pi}_1(t_2) = \dot{\Pi}_2(t_2)$$

$$\vdots$$

$$\dot{\Pi}_N(t_{N+1}) = \dot{\Pi}_{N+1}(t_{N+1}).$$

Time-differentiation of (5.24) gives both $\dot{\Pi}_k(t_{k+1})$ and $\dot{\Pi}_{k+1}(t_{k+1})$ for $k = 1, \ldots, N$, and thus it is possible to write a system of linear equations of the kind

$$\boldsymbol{A}\,[\,\ddot{\Pi}_2(t_2) \quad \ldots \quad \ddot{\Pi}_{N+1}(t_{N+1})\,]^T = \boldsymbol{b} \tag{5.25}$$

which presents a vector $\boldsymbol{b}$ of known terms and a nonsingular coefficient matrix $\boldsymbol{A}$; the solution to this system always exists and is unique. It can be shown that the matrix $\boldsymbol{A}$ has a tridiagonal band structure of the type

$$\boldsymbol{A} = \begin{bmatrix} a_{11} & a_{12} & \ldots & 0 & 0 \\ a_{21} & a_{22} & \ldots & 0 & 0 \\ \vdots & \vdots & \ddots & \vdots & \vdots \\ 0 & 0 & \ldots & a_{N-1,N-1} & a_{N-1,N} \\ 0 & 0 & \ldots & a_{N,N-1} & a_{NN} \end{bmatrix},$$

which simplifies the solution to the system. This matrix is the same for all joints, since it depends only on the time intervals Δt_k specified.

An efficient solution algorithm exists for the above system which is given by a *forward* computation followed by a *backward* computation. From the first equation, $\ddot{\Pi}_2(t_2)$ can be computed as a function of $\ddot{\Pi}_3(t_3)$ and then substituted in the second equation, which then becomes an equation in the unknowns $\ddot{\Pi}_3(t_3)$ and $\ddot{\Pi}_4(t_4)$. This is carried out forward by transforming all the equations in equations with two unknowns, except the last one which will have $\ddot{\Pi}_{N+1}(t_{N+1})$ only as unknown. At this point, all the unknowns can be determined step by step through a backward computation.

The above sequence of cubic polynomials is termed *spline* to indicate smooth functions that interpolate a sequence of given points ensuring continuity of the function and its derivatives.

Figure 5.7 illustrates the time history of position, velocity and acceleration obtained with the data: $q_1 = 0$, $q_3 = 2\pi$, $q_4 = \pi/2$, $q_6 = \pi$, $t_1 = 0$, $t_3 = 2$, $t_4 = 3$, $t_6 = 5$, $\dot{q}_1 = 0$, $\dot{q}_6 = 0$. Two different pairs of virtual points were considered at the time instants: $t_2 = 0.5$, $t_5 = 4.5$ (solid line in the figure), and $t_2 = 1.5$, $t_5 = 3.5$ (dashed line in the figure), respectively. Notice the parabolic velocity profile and the linear acceleration profile. Further, for the second pair, larger values of acceleration are obtained, since the relative time instants are closer to those of the two intermediate points.

Interpolating Linear Polynomials with Parabolic Blends

A simplification in trajectory planning can be achieved as follows. Consider the case when it is desired to interpolate N path points $q_1, \ldots, q_N$ at time instants $t_1, \ldots, t_N$

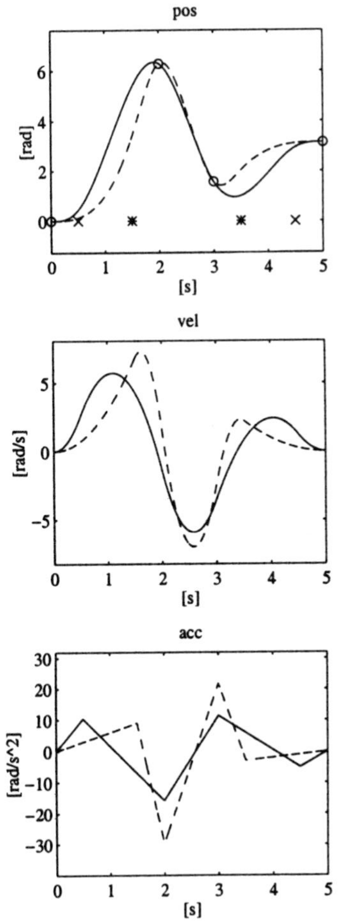

Figure 5.7 Time history of position, velocity and acceleration with a time law of cubic splines for two different pairs of virtual points.

with linear segments. To avoid discontinuity problems on the first derivative at the time instants t_k, the function $q(t)$ shall have a parabolic profile (*blend*) around t_k; as a consequence, the entire trajectory is composed by a sequence of linear and quadratic polynomials, which in turn implies that a discontinuity on $\ddot{q}(t)$ is tolerated.

Let then $\Delta t_k = t_{k+1} - t_k$ be the time distance between q_k and q_{k+1}, and $\Delta t_{k,k+1}$ be the time interval during which the trajectory interpolating q_k and q_{k+1} is a linear function of time. Let also $\dot{q}_{k,k+1}$ be the constant velocity and $\ddot{q}_k$ be the acceleration in the parabolic blend whose duration is $\Delta t'_k$. The resulting trajectory is illustrated in Figure 5.8. The values of q_k, Δt_k, and $\Delta t'_k$ are assumed to be given. Velocity and

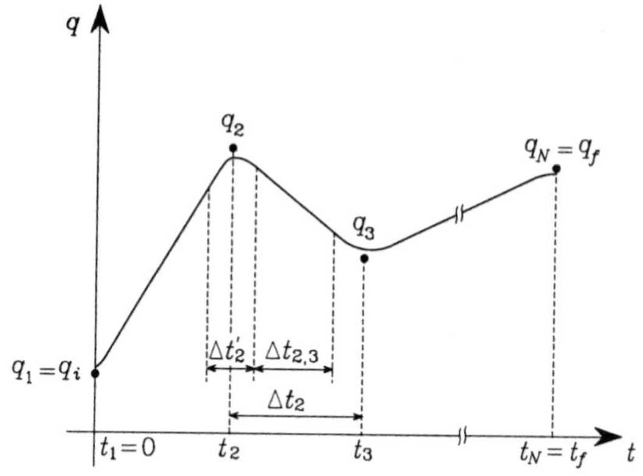

Figure 5.8 Characterization of a trajectory with interpolating linear polynomials with parabolic blends.

acceleration for the intermediate points are computed as

$$
\begin{aligned}
\dot{q}_{k-1,k} &= \frac{q_k - q_{k-1}}{\Delta t_{k-1}} \\
\ddot{q}_k &= \frac{\dot{q}_{k,k+1} - \dot{q}_{k-1,k}}{\Delta t'_k};
\end{aligned}
\tag{5.26}
$$

these equations are straightforward.

The first and last segments deserve special care. In fact, if it is desired to maintain the coincidence of the trajectory with the first and last segments, at least for a portion of time, the resulting trajectory has a longer duration given by $t_N - t_1 + (\Delta t'_1 + \Delta t'_N)/2$, where $\dot{q}_{0,1} = \dot{q}_{N,N+1} = 0$ has been imposed for computing initial and final accelerations.

Notice that $q(t)$ reaches none of the path points q_k but passes nearby (Figure 5.8). In this situation, the path points are more appropriately termed *via points*; the larger the blending acceleration, the closer the passage to a via point.

On the basis of the given q_k, Δt_k and $\Delta t'_k$, the values of $\dot{q}_{k-1,k}$ and $\ddot{q}_k$ are computed via (5.26) and a sequence of linear polynomials with parabolic blends is generated. Their expressions as a function of time are not derived here to avoid further loading of the analytic presentation.

Figure 5.9 illustrates the time history of position, velocity and acceleration obtained with the data: $q_1 = 0$, $q_2 = 2\pi$, $q_3 = \pi/2$, $q_4 = \pi$, $t_1 = 0$, $t_2 = 2$, $t_3 = 3$, $t_4 = 5$, $\dot{q}_1 = 0$, $\dot{q}_4 = 0$. Two different values for the blend times have been considered: $\Delta t'_k = 0.2$ (solid line in the figure) and $\Delta t'_k = 0.6$ (dashed line in the figure), for $k = 1, \ldots, 4$, respectively. Notice that in the first case the passage of $q(t)$ is closer to the via points, though at the expense of higher acceleration values.

The above presented technique turns out to be an application of the trapezoidal velocity profile law to the interpolation problem. If one gives up a trajectory passing near a via point at a prescribed instant of time, the use of trapezoidal velocity profiles allows developing a trajectory planning algorithm which is attractive for its simplicity.

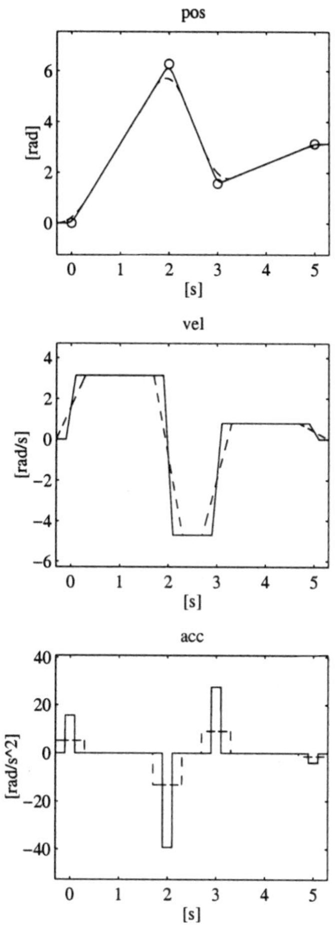

Figure 5.9 Time history of position, velocity and acceleration with a time law of interpolating linear polynomials with parabolic blends.

In particular, consider the case of one intermediate point only, and suppose that trapezoidal velocity profiles are considered as motion primitives with the possibility to specify the initial and final point and the duration of the motion only; it is assumed that $\dot{q}_i = \dot{q}_f = 0$. If two segments with trapezoidal velocity profiles were generated, the manipulator joint would certainly reach the intermediate point, but it would be forced to stop there, before continuing the motion towards the final point. A keen alternative is to start generating the second segment ahead of time with respect to the end of the first segment, using the sum of velocities (or positions) as a reference. In this way, the joint is guaranteed to reach the final position; crossing of the intermediate point at the specified instant of time is not guaranteed, though.

Figure 5.10 illustrates the time history of position, velocity and acceleration obtained with the data: $q_i = 0$, $q_f = 3\pi/2$, $t_i = 0$, $t_f = 2$. The intermediate point is

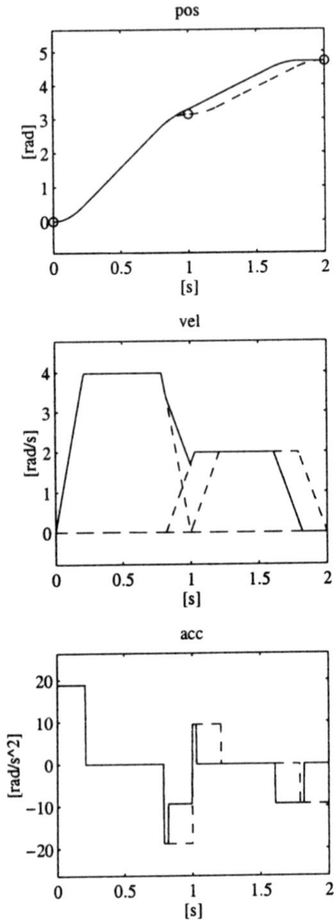

Figure 5.10 Time history of position, velocity and acceleration with a time law of interpolating linear polynomials with parabolic blends obtained by anticipating the generation of the second segment of trajectory.

located at $q = \pi$ with $t = 1$, the maximum acceleration values in the two segments are respectively $|\ddot{q}_c| = 6\pi$ and $|\ddot{q}_c| = 3\pi$, and the time anticipation is 0.18. As predicted, with time anticipation, the assigned intermediate position becomes a via point with the advantage of an overall shorter time duration. Notice, also, that velocity does not vanish at the intermediate point.

5.3 Operational Space Trajectories

A joint space trajectory planning algorithm generates a time sequence of values for the joint variables $\boldsymbol{q}(t)$ so that the manipulator is taken from the initial to the final configuration, eventually by moving through a sequence of intermediate configurations. The resulting end-effector motion is not easily predictable, in view of the nonlinear effects introduced by direct kinematics. Whenever it is desired that the end-effector motion follows a geometrically specified path in the *operational space*, it is necessary to plan trajectory execution directly in the same space. Planning can be done either by interpolating a sequence of prescribed path points or by generating the analytical motion primitive and the relative trajectory in a punctual way.

In both cases the time sequence of the values attained by the operational space variables is utilized in real time to obtain the corresponding sequence of values of the joint space variables, via an inverse kinematics algorithm. In this regard, the computational complexity induced by trajectory generation in the operational space and related kinematic inversion sets an upper limit on the maximum sampling rate to generate the above sequences. Since these sequences constitute the reference inputs to the motion control system, a linear *microinterpolation* is typically carried out. In this way, the frequency at which reference inputs are updated is increased so as to enhance dynamic performance of the system.

Whenever the path is not to be followed exactly, its characterization can be performed through the assignment of N points specifying the values of the variables $\boldsymbol{x}$ chosen to describe the end-effector location in the operational space at given time instants t_k, for $k = 1, \ldots, N$. Similarly to what was presented in the above sections, the trajectory is generated by determining a smooth interpolating vector function between the various path points. Such a function can be computed by applying to each component of $\boldsymbol{x}$ any of the interpolation techniques illustrated in Section 5.2.2 for the single joint variable.

Therefore, for given path (or via) points $\boldsymbol{x}(t_k)$, the corresponding components $x_i(t_k)$, for $i = 1, \ldots r$ (where r is the dimension of the operational space of interest) can be interpolated with a sequence of cubic polynomials, a sequence of linear polynomials with parabolic blends, and so on.

On the other hand, if the end-effector motion has to follow a prescribed trajectory of motion, this must be expressed analytically. It is then necessary to refer to motion primitives defining the geometric features of the path and time primitives defining the time law on the path itself.

5.3.1 Path Primitives

For the definition of *path primitives* it is convenient to refer to the parametric description of paths in space. Let then $\boldsymbol{p}$ be a (3×1) vector and $\boldsymbol{f}(\sigma)$ a continuous vector function defined in the interval $[\sigma_i, \sigma_f]$. Consider the equation

$$\boldsymbol{p} = \boldsymbol{f}(\sigma); \tag{5.27}$$

with reference to its geometric description, the sequence of values of $\boldsymbol{p}$ with σ varying in $[\sigma_i, \sigma_f]$ is termed *path* in space. The equation in (5.27) defines the *parametric representation* of the path Γ and the scalar σ is called *parameter*. As σ increases, the point $\boldsymbol{p}$ moves on the path in a given direction. This direction is said to be the

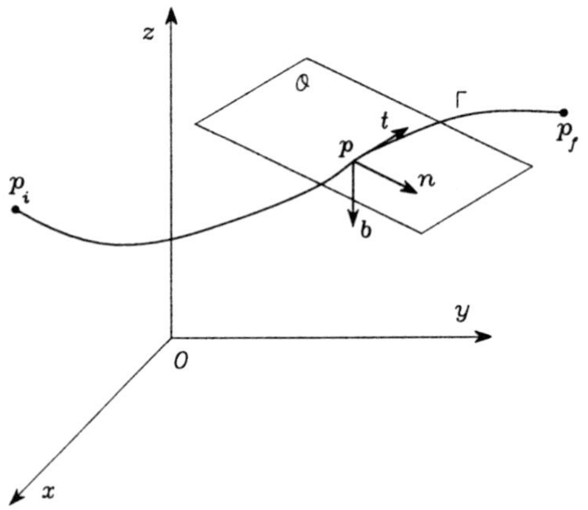

Figure 5.11 Parametric representation of a path in space.

direction induced on Γ by the parametric representation (5.27). A path is closed when $\boldsymbol{p}(\sigma_f) = \boldsymbol{p}(\sigma_i)$; otherwise it is open.

Let $\boldsymbol{p}_i$ be a point on the open path Γ on which a direction has been fixed. The *path coordinate* s of the generic point $\boldsymbol{p}$ is the length of the arc of Γ with extremes $\boldsymbol{p}$ and $\boldsymbol{p}_i$ if $\boldsymbol{p}$ follows $\boldsymbol{p}_i$, the opposite of this length if $\boldsymbol{p}$ precedes $\boldsymbol{p}_i$. The point $\boldsymbol{p}_i$ is said to be the origin of the path coordinate ($s = 0$).

From the above presentation it follows that to each value of s a well-determined path point corresponds, and then the path coordinate can be used as a parameter in a different parametric representation of the path Γ:

$$\boldsymbol{p} = \boldsymbol{f}(s); \tag{5.28}$$

the range of variation of the parameter s will be the sequence of path coordinates associated with the points of Γ.

Consider a path Γ represented by (5.28). Let $\boldsymbol{p}$ be a point corresponding to the path coordinate s. Except for special cases, $\boldsymbol{p}$ allows the definition of three unit vectors characterizing the path. The orientation of such vectors depends exclusively on the path geometry, while their direction depends also on the direction induced by (5.28) on the path.

The first of such unit vectors is the *tangent unit vector* denoted by $\boldsymbol{t}$. This vector is oriented along the direction induced on the path by s.

The second unit vector is the *normal unit vector* denoted by $\boldsymbol{n}$. This vector is oriented along the line intersecting $\boldsymbol{p}$ at a right angle with $\boldsymbol{t}$ and lies in the so-called *osculating plane* $\mathcal{O}$ (Figure 5.11); such plane is the limit position of the plane containing the unit vector $\boldsymbol{t}$ and a point $\boldsymbol{p}' \in \Gamma$ when $\boldsymbol{p}'$ tends to $\boldsymbol{p}$ along the path. The direction of $\boldsymbol{n}$ is so that the path Γ, in the neighbourhood of $\boldsymbol{p}$ with respect to the plane containing $\boldsymbol{t}$ and normal to $\boldsymbol{n}$, lies on the same side of $\boldsymbol{n}$.

The third unit vector is the *binormal unit vector* denoted by $\boldsymbol{b}$. This vector is so that the frame $(\boldsymbol{t}, \boldsymbol{n}, \boldsymbol{b})$ is right-handed (Figure 5.11). Notice that it is not always possible to uniquely define such frame.

It can be shown that the above three unit vectors are related by simple relations to the path representation Γ as a function of the path coordinate. In particular, it is:

$$\begin{aligned} \boldsymbol{t} &= \frac{d\boldsymbol{p}}{ds} \\ \boldsymbol{n} &= \frac{1}{\left\| \dfrac{d^2\boldsymbol{p}}{ds^2} \right\|} \frac{d^2\boldsymbol{p}}{ds^2} \\ \boldsymbol{b} &= \boldsymbol{t} \times \boldsymbol{n}. \end{aligned} \tag{5.29}$$

Typical path parametric representations are reported below which are useful for trajectory generation in the operational space.

Segment in Space

Consider the linear segment connecting point $\boldsymbol{p}_i$ to point $\boldsymbol{p}_f$. The parametric representation of this path is

$$\boldsymbol{p}(s) = \boldsymbol{p}_i + \frac{s}{\|\boldsymbol{p}_f - \boldsymbol{p}_i\|}(\boldsymbol{p}_f - \boldsymbol{p}_i). \tag{5.30}$$

Notice that $\boldsymbol{p}(0) = \boldsymbol{p}_i$ and $\boldsymbol{p}(\|\boldsymbol{p}_f - \boldsymbol{p}_i\|) = \boldsymbol{p}_f$. Hence, the direction induced on Γ by the parametric representation (5.30) is that going from $\boldsymbol{p}_i$ to $\boldsymbol{p}_f$. Differentiating (5.30) with respect to s gives

$$\begin{aligned} \frac{d\boldsymbol{p}}{ds} &= \frac{1}{\|\boldsymbol{p}_f - \boldsymbol{p}_i\|}(\boldsymbol{p}_f - \boldsymbol{p}_i) \\ \frac{d^2\boldsymbol{p}}{ds^2} &= \boldsymbol{0}. \end{aligned} \tag{5.31}$$

In this case it is not possible to define the frame $(\boldsymbol{t}, \boldsymbol{n}, \boldsymbol{b})$ uniquely.

Circle in Space

Consider a circle Γ in space. Before deriving its parametric representation, it is necessary to introduce its significant parameters. Suppose that the circle is specified by assigning (Figure 5.12):

- the unit vector of the circle axis $\boldsymbol{r}$,
- the position vector $\boldsymbol{d}$ of a point along the circle axis,
- the position vector $\boldsymbol{p}_i$ of a point on the circle.

With these parameters, the position vector $\boldsymbol{c}$ of the centre of the circle can be found. Let $\boldsymbol{\delta} = \boldsymbol{p}_i - \boldsymbol{d}$; for $\boldsymbol{p}_i$ not to be on the axis, *i.e.*, for the circle not to degenerate into a point, it must be

$$|\boldsymbol{\delta}^T \boldsymbol{r}| < \|\boldsymbol{\delta}\|;$$

in this case it is

$$\boldsymbol{c} = \boldsymbol{d} + (\boldsymbol{\delta}^T \boldsymbol{r})\boldsymbol{r}. \tag{5.32}$$

It is now desired to find a parametric representation of the circle as a function of the path coordinate. Notice that this representation is very simple for a suitable choice of

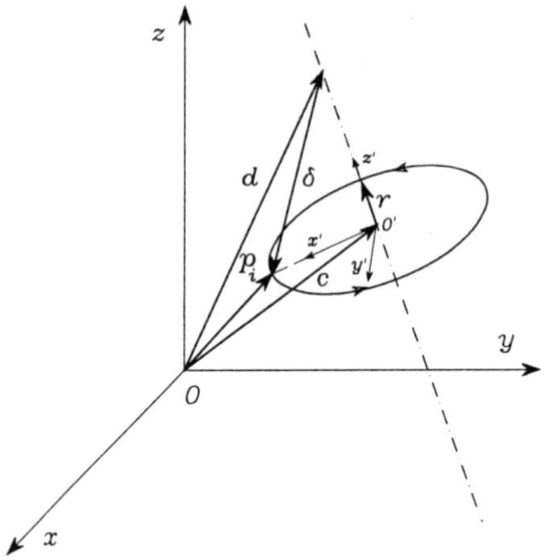

Figure 5.12 Parametric representation of a circle in space.

the reference frame. To see this, consider the frame O'–$x'y'z'$, where O' coincides with the centre of the circle, axis x' is oriented along the direction of the vector $\boldsymbol{p}_i - \boldsymbol{c}$, axis z' is oriented along $\boldsymbol{r}$ and axis y' is chosen so as to complete a right-handed frame. When expressed in this reference frame, the parametric representation of the circle is

$$\boldsymbol{p}'(s) = \begin{bmatrix} \rho\cos(s/\rho) \\ \rho\sin(s/\rho) \\ 0 \end{bmatrix}, \tag{5.33}$$

where $\rho = \|\boldsymbol{p}_i - \boldsymbol{c}\|$ is the radius of the circle and the point $\boldsymbol{p}_i$ has been assumed as the origin of the path coordinate. For a different reference frame, the path representation becomes

$$\boldsymbol{p}(s) = \boldsymbol{c} + \boldsymbol{R}\boldsymbol{p}'(s), \tag{5.34}$$

where $\boldsymbol{c}$ is expressed in the frame O–xyz and $\boldsymbol{R}$ is the rotation matrix of frame O'–$x'y'z'$ with respect to frame O–xyz which, in view of (2.3), can be written as

$$\boldsymbol{R} = [\,\boldsymbol{x}' \quad \boldsymbol{y}' \quad \boldsymbol{z}'\,];$$

$\boldsymbol{x}'$, $\boldsymbol{y}'$, $\boldsymbol{z}'$ indicate the unit vectors of the frame expressed in the frame O–xyz. Differentiating (5.34) with respect to s gives

$$\begin{aligned} \frac{d\boldsymbol{p}}{ds} &= \boldsymbol{R}\begin{bmatrix} -\sin(s/\rho) \\ \cos(s/\rho) \\ 0 \end{bmatrix} \\ \frac{d^2\boldsymbol{p}}{ds^2} &= \boldsymbol{R}\begin{bmatrix} -\cos(s/\rho)/\rho \\ -\sin(s/\rho)/\rho \\ 0 \end{bmatrix}. \end{aligned} \tag{5.35}$$

5.3.2 Position

Let $\boldsymbol{x}$ be the vector of operational space variables expressing *position* and *orientation* of the manipulator's end effector as in (2.68). Generating a trajectory in the operational space means to determine a function $\boldsymbol{x}(t)$ taking the end-effector frame from the initial to the final location in a time t_f along a given path with a specific motion time law. First, consider end-effector position. Orientation will follow.

Let $\boldsymbol{p} = \boldsymbol{f}(s)$ be the (3×1) vector of the parametric representation of the path Γ as a function of the path coordinate s; the origin of the end-effector frame moves from $\boldsymbol{p}_i$ to $\boldsymbol{p}_f$ in a time t_f. For simplicity, suppose that the origin of the path coordinate is at $\boldsymbol{p}_i$ and the direction induced on Γ is that going from $\boldsymbol{p}_i$ to $\boldsymbol{p}_f$. The path coordinate then goes from the value $s = 0$ at $t = 0$ to the value $s = s_f$ (path length) at $t = t_f$. The time law along the path is described by the function $s(t)$.

In order to find an analytic expression for $s(t)$, any of the above techniques for joint trajectory generation can be employed. In particular, either a cubic polynomial or a sequence of linear segments with parabolic blends can be chosen for $s(t)$.

It is worth making some remarks on the time evolution of $\boldsymbol{p}$ on Γ, for a given time law $s(t)$. The velocity of point $\boldsymbol{p}$ is given by the time derivative of $\boldsymbol{p}$

$$\dot{\boldsymbol{p}} = \dot{s}\frac{d\boldsymbol{p}}{ds} = \dot{s}\boldsymbol{t},$$

where $\boldsymbol{t}$ is the tangent vector to the path at point $\boldsymbol{p}$ in (5.29). Then, $\dot{s}$ represents the magnitude of the velocity vector relative to point $\boldsymbol{p}$, taken with the positive or negative sign depending on the direction of $\dot{\boldsymbol{p}}$ along $\boldsymbol{t}$. The magnitude of $\dot{\boldsymbol{p}}$ starts from zero at $t = 0$, then it varies with a parabolic or trapezoidal profile as per either of the above choices for $s(t)$, and finally it returns to zero at $t = t_f$.

As a first example, consider the segment connecting point $\boldsymbol{p}_i$ with point $\boldsymbol{p}_f$. The parametric representation of this path is given by (5.30). Velocity and acceleration of $\boldsymbol{p}$ can be easily computed by recalling the rule of differentiation of compound functions, *i.e.*,

$$\begin{aligned}\dot{\boldsymbol{p}} &= \frac{\dot{s}}{\|\boldsymbol{p}_f - \boldsymbol{p}_i\|}(\boldsymbol{p}_f - \boldsymbol{p}_i) = \dot{s}\boldsymbol{t} \\ \ddot{\boldsymbol{p}} &= \frac{\ddot{s}}{\|\boldsymbol{p}_f - \boldsymbol{p}_i\|}(\boldsymbol{p}_f - \boldsymbol{p}_i) = \ddot{s}\boldsymbol{t}.\end{aligned} \tag{5.36}$$

As a further example, consider a circle Γ in space. From the parametric representation derived above, in view of (5.35), velocity and acceleration of point $\boldsymbol{p}$ on the circle are:

$$\begin{aligned}\dot{\boldsymbol{p}} &= \boldsymbol{R}\begin{bmatrix} -\dot{s}\sin(s/\rho) \\ \dot{s}\cos(s/\rho) \\ 0 \end{bmatrix} \\ \ddot{\boldsymbol{p}} &= \boldsymbol{R}\begin{bmatrix} -\dot{s}^2\cos(s/\rho)/\rho - \ddot{s}\sin(s/\rho) \\ -\dot{s}^2\sin(s/\rho)/\rho + \ddot{s}\cos(s/\rho) \\ 0 \end{bmatrix}.\end{aligned} \tag{5.37}$$

Notice that the velocity vector is aligned with $\boldsymbol{t}$, and the acceleration vector is given by two contributions: the first one is aligned with $\boldsymbol{n}$ and represents the centripetal acceleration, while the second one is aligned with $\boldsymbol{t}$ and represents the tangential acceleration.

5.3.3 Orientation

Consider now end-effector orientation. Typically, this is specified in terms of the rotation matrix of the (time-varying) end-effector frame with respect to the base frame. As well known, the three columns of the rotation matrix represent the three unit vectors of the end-effector frame with respect to the base frame. To generate a trajectory, however, a linear interpolation on the unit vectors $\boldsymbol{n}$, $\boldsymbol{s}$, $\boldsymbol{a}$ describing the initial and final orientation does not guarantee orthonormality of the above vectors at each instant of time.

Euler Angles

In view of the above difficulty, for trajectory generation purposes, orientation is often described in terms of the Euler angles triplet $\boldsymbol{\phi} = (\varphi, \vartheta, \psi)$ for which a time law can be specified. Usually, $\boldsymbol{\phi}$ moves along the segment connecting its initial value $\boldsymbol{\phi}_i$ to its final value $\boldsymbol{\phi}_f$. Also in this case, it is convenient to choose a cubic polynomial or a linear segment with parabolic blends time law. In this way, in fact, the angular velocity $\boldsymbol{\omega}$ of the time-varying frame, which is related to $\dot{\boldsymbol{\phi}}$ by the linear relationship (3.60), will have continuous magnitude.

Therefore, for given $\boldsymbol{\phi}_i$ and $\boldsymbol{\phi}_f$, the position, velocity and acceleration profiles are:

$$\begin{aligned}
\boldsymbol{\phi} &= \boldsymbol{\phi}_i + \frac{s}{\|\boldsymbol{\phi}_f - \boldsymbol{\phi}_i\|}(\boldsymbol{\phi}_f - \boldsymbol{\phi}_i) \\
\dot{\boldsymbol{\phi}} &= \frac{\dot{s}}{\|\boldsymbol{\phi}_f - \boldsymbol{\phi}_i\|}(\boldsymbol{\phi}_f - \boldsymbol{\phi}_i) \\
\ddot{\boldsymbol{\phi}} &= \frac{\ddot{s}}{\|\boldsymbol{\phi}_f - \boldsymbol{\phi}_i\|}(\boldsymbol{\phi}_f - \boldsymbol{\phi}_i),
\end{aligned} \tag{5.38}$$

where the time law for $s(t)$ has to be specified. The three unit vectors of the end-effector frame can be computed—with reference to Euler angles ZYZ—as in (2.18), the end-effector frame angular velocity as in (3.60), and the angular acceleration by differentiating (3.60) itself.

Angle and Axis

An alternative way to generate a trajectory for orientation of clearer interpretation in the Cartesian space can be derived by resorting to the the angle and axis description presented in Section 2.5. Given two coordinate frames in the Cartesian space with the same origin and different orientation, it is always possible to determine a unit vector so that the second frame can be obtained from the first frame by a rotation of a proper angle about the axis of such unit vector.

Let $\boldsymbol{R}_i$ and $\boldsymbol{R}_f$ denote respectively the rotation matrices of the initial frame O_i–$x_i y_i z_i$ and the final frame O_f–$x_f y_f z_f$, both with respect to the base frame. The rotation matrix between the two frames can be computed by recalling that $\boldsymbol{R}_f = \boldsymbol{R}_i \boldsymbol{R}_f^i$; the expression in (2.5) allows writing

$$\boldsymbol{R}_f^i = \boldsymbol{R}_i^T \boldsymbol{R}_f = \begin{bmatrix} r_{11} & r_{12} & r_{13} \\ r_{21} & r_{22} & r_{23} \\ r_{31} & r_{32} & r_{33} \end{bmatrix}.$$

If the matrix $\boldsymbol{R}^i(t)$ is defined to describe the transition from $\boldsymbol{R}_i$ to $\boldsymbol{R}_f$, it must be $\boldsymbol{R}^i(0) = \boldsymbol{I}$ and $\boldsymbol{R}^i(t_f) = \boldsymbol{R}_f^i$. Hence, the matrix $\boldsymbol{R}_f^i$ can be expressed as the rotation matrix about a fixed axis in space; the unit vector $\boldsymbol{r}^i$ of the axis and the angle of rotation ϑ_f can be computed by using (2.25):

$$\begin{aligned} \vartheta_f &= \cos^{-1}\left(\frac{r_{11} + r_{22} + r_{33} - 1}{2}\right) \\ \boldsymbol{r}^i &= \frac{1}{2\sin\vartheta_f}\begin{bmatrix} r_{32} - r_{23} \\ r_{13} - r_{31} \\ r_{21} - r_{12} \end{bmatrix} \end{aligned} \tag{5.39}$$

for $\sin\vartheta_f \neq 0$.

The matrix $\boldsymbol{R}^i(t)$ can be interpreted as a matrix $\boldsymbol{R}^i(\vartheta(t), \boldsymbol{r}^i)$ and computed via (2.23); it is then sufficient to assign a time law to ϑ, of the type of those presented for the single joint with $\vartheta(0) = 0$ and $\vartheta(t_f) = \vartheta_f$, and compute the components of $\boldsymbol{r}^i$ from (5.39). Since $\boldsymbol{r}^i$ is constant, the resulting velocity and acceleration are respectively

$$\begin{aligned} \boldsymbol{\omega}^i &= \dot{\vartheta}\,\boldsymbol{r}^i \\ \dot{\boldsymbol{\omega}}^i &= \ddot{\vartheta}\,\boldsymbol{r}^i. \end{aligned} \tag{5.40}$$

Finally, in order to characterize the end-effector orientation trajectory with respect to the base frame, the following transformations are needed:

$$\begin{aligned} \boldsymbol{R}(t) &= \boldsymbol{R}_i \boldsymbol{R}^i(t) \\ \boldsymbol{\omega}(t) &= \boldsymbol{R}_i \boldsymbol{\omega}^i(t) \\ \dot{\boldsymbol{\omega}}(t) &= \boldsymbol{R}_i \dot{\boldsymbol{\omega}}^i(t). \end{aligned}$$

Once a path and a trajectory have been specified in the operational space in terms of $\boldsymbol{p}(t)$ and $\boldsymbol{\phi}(t)$ or $\boldsymbol{R}(t)$, inverse kinematics techniques can be used to find the corresponding trajectories in the joint space $\boldsymbol{q}(t)$.

5.4 Dynamic Scaling of Trajectories

The existence of *dynamic constraints* to be taken into account for trajectory generation has been mentioned in Section 5.1. In practice, with reference to the given trajectory time or path shape (segments with high curvature), the trajectories that can be obtained with any of the previously illustrated methods may impose too severe dynamic performance for the manipulator. A typical case is that when the required torques to generate the motion are larger than the maximum torques the actuators can supply. In this case, an infeasible trajectory has to be suitably time-scaled.

Suppose a trajectory has been generated for all the manipulator joints as $\boldsymbol{q}(t)$, for $t \in [0, t_f]$. Computing inverse dynamics allows evaluating the time history of the torques $\boldsymbol{\tau}(t)$ required for the execution of the given motion. By comparing the obtained torques with the torque limits available at the actuators, it is easy to check whether or not the trajectory is actually executable. The problem is then to seek an automatic trajectory *dynamic scaling* technique—avoiding inverse dynamics recomputation—so that the manipulator can execute the motion on the specified path with a proper time law without exceeding the torque limits.

Consider the manipulator dynamic model as given in (4.41) with $\boldsymbol{F}_v = \boldsymbol{O}, \boldsymbol{F}_s = \boldsymbol{O}$ and $\boldsymbol{h} = \boldsymbol{0}$, for simplicity. The term $\boldsymbol{C}(\boldsymbol{q}, \dot{\boldsymbol{q}})$ accounting for centrifugal and Coriolis forces has a quadratic dependence on joint velocities, and thus it can be formally rewritten as

$$\boldsymbol{C}(\boldsymbol{q}, \dot{\boldsymbol{q}})\dot{\boldsymbol{q}} = \boldsymbol{\Gamma}(\boldsymbol{q})[\dot{\boldsymbol{q}}\dot{\boldsymbol{q}}], \tag{5.41}$$

where $[\dot{\boldsymbol{q}}\dot{\boldsymbol{q}}]$ indicates the symbolic notation of the $(n(n+1)/2 \times 1)$ vector

$$[\dot{\boldsymbol{q}}\dot{\boldsymbol{q}}] = [\,\dot{q}_1^2 \quad \dot{q}_1\dot{q}_2 \quad \dots \quad \dot{q}_{n-1}\dot{q}_n \quad \dot{q}_n^2\,]^T;$$

$\boldsymbol{\Gamma}(\boldsymbol{q})$ is a proper $(n \times n(n+1)/2)$ matrix that satisfies (5.41). In view of such position, the manipulator dynamic model can be expressed as

$$\boldsymbol{B}(\boldsymbol{q}(t))\ddot{\boldsymbol{q}}(t) + \boldsymbol{\Gamma}(\boldsymbol{q}(t))[\dot{\boldsymbol{q}}(t)\dot{\boldsymbol{q}}(t)] + \boldsymbol{g}(\boldsymbol{q}(t)) = \boldsymbol{\tau}(t), \tag{5.42}$$

where the explicit dependence on time t has been shown.

Consider the new variable $\bar{\boldsymbol{q}}(r(t))$ satisfying the equation

$$\boldsymbol{q}(t) = \bar{\boldsymbol{q}}(r(t)), \tag{5.43}$$

where $r(t)$ is a strictly increasing scalar function of time with $r(0) = 0$ and $r(t_f) = \bar{t}_f$.

Differentiating (5.43) twice with respect to time provides the following relations:

$$\begin{aligned} \dot{\boldsymbol{q}} &= \dot{r}\bar{\boldsymbol{q}}'(r) \\ \ddot{\boldsymbol{q}} &= \dot{r}^2\bar{\boldsymbol{q}}''(r) + \ddot{r}\bar{\boldsymbol{q}}'(r), \end{aligned} \tag{5.44}$$

where the prime denotes the derivative with respect to r. Substituting (5.44) into (5.42) yields

$$\dot{r}^2\Big(\boldsymbol{B}(\bar{\boldsymbol{q}}(r))\bar{\boldsymbol{q}}''(r) + \boldsymbol{\Gamma}(\bar{\boldsymbol{q}}(r))[\bar{\boldsymbol{q}}'(r)\bar{\boldsymbol{q}}'(r)]\Big) + \ddot{r}\boldsymbol{B}(\bar{\boldsymbol{q}}(r))\bar{\boldsymbol{q}}'(r) + \boldsymbol{g}(\bar{\boldsymbol{q}}(r)) = \boldsymbol{\tau}. \tag{5.45}$$

In (5.42) it is possible to identify the term

$$\boldsymbol{\tau}_s(t) = \boldsymbol{B}(\boldsymbol{q}(t))\ddot{\boldsymbol{q}}(t) + \boldsymbol{\Gamma}(\boldsymbol{q}(t))[\dot{\boldsymbol{q}}(t)\dot{\boldsymbol{q}}(t)], \tag{5.46}$$

representing the torque contribution that depends on velocities and accelerations. Correspondingly, in (5.45) one can set

$$\boldsymbol{\tau}_s(t) = \dot{r}^2\Big(\boldsymbol{B}(\bar{\boldsymbol{q}}(r))\bar{\boldsymbol{q}}''(r) + \boldsymbol{\Gamma}(\bar{\boldsymbol{q}}(r))[\bar{\boldsymbol{q}}'(r)\bar{\boldsymbol{q}}'(r)]\Big) + \ddot{r}\boldsymbol{B}(\bar{\boldsymbol{q}}(r))\bar{\boldsymbol{q}}'(r). \tag{5.47}$$

By analogy with (5.46), it can be written

$$\bar{\boldsymbol{\tau}}_s(r) = \boldsymbol{B}(\bar{\boldsymbol{q}}(r))\bar{\boldsymbol{q}}''(r) + \boldsymbol{\Gamma}(\bar{\boldsymbol{q}}(r))[\bar{\boldsymbol{q}}'(r)\bar{\boldsymbol{q}}'(r)], \tag{5.48}$$

and then (5.47) becomes

$$\boldsymbol{\tau}_s(t) = \dot{r}^2\bar{\boldsymbol{\tau}}_s(r) + \ddot{r}\boldsymbol{B}(\bar{\boldsymbol{q}}(r))\bar{\boldsymbol{q}}'(r). \tag{5.49}$$

The expression in (5.49) gives the relationship between the torque contributions depending on velocities and accelerations required by the manipulator when this is subject to motions having the same path but different time laws, obtained through a time scaling of joint variables as in (5.43).

Gravitational torques have not been considered, since they are a function of the joint positions only, and thus their contribution is not influenced by time scaling.

The simplest choice for the scaling function $r(t)$ is certainly the *linear* function

$$r(t) = ct,$$

with c a positive constant. In this case, (5.49) becomes

$$\boldsymbol{\tau}_s(t) = c^2\bar{\boldsymbol{\tau}}_s(ct),$$

which reveals that a linear time scaling by c causes a scaling of the magnitude of the torques by the coefficient c^2. Let $c > 1$: (5.43) shows that the trajectory described by $\bar{\boldsymbol{q}}(r(t))$, assuming $r = ct$ as the independent variable, has a duration $\bar{t}_f > t_f$ to cover the entire path specified by $\boldsymbol{q}$. Correspondingly, the torque contributions $\bar{\boldsymbol{\tau}}_s(ct)$ computed as in (5.48) are scaled by the factor c^2 with respect to the torque contributions $\boldsymbol{\tau}_s(t)$ required to execute the original trajectory $\boldsymbol{q}(t)$.

With the use of a recursive algorithm for inverse dynamics computation, it is possible to check whether the torques exceed the allowed limits during trajectory execution; obviously, limit violation shall not be caused by the sole gravity torques. It is necessary to find the joint for which the torque has exceeded the limit more than the others, and to compute the torque contribution subject to scaling, which in turn determines the factor c^2. It is then possible to compute the time-scaled trajectory as a function of the new time variable $r = ct$ which no longer exceeds torque limits. It should be pointed out, however, that with this kind of linear scaling the entire trajectory may be penalized, even when a torque limit on a single joint is exceeded only for a short interval of time.

Problems

5.1 Compute the joint trajectory from $q(0) = 1$ to $q(2) = 4$ with null initial and final velocities and accelerations.

5.2 Compute the time law $q(t)$ for a joint trajectory with velocity profile of the type $\dot{q}(t) = k(1 - \cos(at))$ from $q(0) = 0$ to $q(2) = 3$.

5.3 Given the values for the joint variable: $q(0) = 0$, $q(2) = 2$, and $q(4) = 3$, compute the two fifth-order interpolating polynomials with continuous velocities and accelerations.

5.4 Show that the matrix $\boldsymbol{A}$ in (5.25) has a tridiagonal band structure.

5.5 Given the values for the joint variable: $q(0) = 0$, $q(2) = 2$, and $q(4) = 3$, compute the cubic interpolating spline with null initial and final velocities and accelerations.

5.6 Given the values for the joint variable: $q(0) = 0$, $q(2) = 2$, and $q(4) = 3$, find the interpolating polynomial with linear segments and parabolic blends with null initial and final velocities.

5.7 Find the motion time law $\boldsymbol{p}(t)$ for a Cartesian space straight path with trapezoidal velocity profile from $\boldsymbol{p}(0) = [\,0 \quad 0.5 \quad 0\,]^T$ to $\boldsymbol{p}(2) = [\,1 \quad -0.5 \quad 0\,]^T$.

5.8 Find the motion time law $\boldsymbol{p}(t)$ for a Cartesian space circular path with trapezoidal velocity profile from $\boldsymbol{p}(0) = [\,0 \quad 0.5 \quad 1\,]^T$ to $\boldsymbol{p}(2) = [\,0 \quad -0.5 \quad 1\,]^T$; the circle is located in the plane $x = 0$ with centre at $\boldsymbol{c} = [\,0 \quad 0 \quad 1\,]^T$ and radius $\rho = 0.5$, and is executed clockwise for an observer aligned with x.

5.9 For the two-link planar arm of Example 4.2, perform a computer implementation of dynamic linear time scaling along the trajectory of Figure 4.6, on the assumption of symmetric torque limits of 3000 N·m. Adopt a sampling time of 1 ms.

Bibliography

Brady M. (1982) Trajectory planning. In *Robot Motion: Planning and Control* M. Brady et al. (Eds.), MIT Press, Cambridge, Mass., pp. 221–243.

Craig J.J. (1989) *Introduction to Robotics: Mechanics and Control*. 2nd ed., Addison-Wesley, Reading, Mass.

De Boor C. (1978) *A Practical Guide to Splines*. Springer-Verlag, New York.

De Luca A. (1986) *A Spline Generator for Robot Arms*. Tech. rep. RAL 68, Rensselaer Polytechnic Institute, Department of Electrical, Computer, and Systems Engineering.

Fu K.S., Gonzalez R.C., Lee C.S.G. (1987) *Robotics: Control, Sensing, Vision, and Intelligence*. McGraw-Hill, New York.

Hollerbach J.M. (1984) Dynamic scaling of manipulator trajectories. *ASME J. Dynamic Systems, Measurement, and Control* 106:102–106.

Lin C.S., Chang P.R., Luh J.Y.S. (1983) Formulation and optimization of cubic polynomial joint trajectories for industrial robots. *IEEE Trans. Automatic Control* 28:1066–1073.

Paul R.P. (1972) *Modelling, Trajectory Calculation, and Servoing of a Computer Controlled Arm*. Memo. AIM 177, Stanford Artificial Intelligence Laboratory.

Paul R.P. (1979) Manipulator Cartesian path control. *IEEE Trans. Systems, Man, and Cybernetics* 9:702–711.

Paul R.P. (1981) *Robot Manipulators: Mathematics, Programming, and Control*. MIT Press, Cambridge, Mass.

Taylor R.H. (1979) Planning and execution of straight line manipulator trajectories. *IBM J. Research and Development* 23:424–436.

6. Motion Control

In the previous chapter, trajectory planning techniques have been presented which allow generating the reference inputs to the motion control system. The problem of controlling a manipulator can be formulated as that to determine the time history of the generalized forces (forces or torques) to be developed by the joint actuators so as to guarantee execution of the commanded task while satisfying given transient and steady-state requirements. The task may regard either the execution of specified motions for a manipulator operating in free space, or the execution of specified motions and contact forces for a manipulator whose end effector is constrained by the environment. In view of problem complexity, the two aspects will be treated separately; first, motion control in free space, and then interaction control in constrained space. The problem of *motion control* of a manipulator is the topic of this chapter. A number of *joint space* control techniques are presented. These can be distinguished between *decentralized control* schemes, *i.e.*, when the single manipulator joint is controlled independently of the others, and *centralized control* schemes, *i.e.*, when the dynamic interaction effects between the joints are taken into account. Finally, as a premise to the interaction control problem, the basic features of *operational space* control schemes are illustrated.

6.1 The Control Problem

Several techniques can be employed for controlling a manipulator. The technique followed, as well as the way it is implemented, may have a significant influence on the manipulator performance and then on the possible range of applications. For instance, the need for trajectory tracking control in the operational space may lead to hardware/software implementations which differ from those allowing point-to-point control where only reaching of the final position is of concern.

On the other hand, the manipulator mechanical design has an influence on the kind of control scheme utilized. For instance, the control problem of a Cartesian manipulator is substantially different from that of an anthropomorphic manipulator.

The driving system of the joints has also an effect on the type of control strategy used. If a manipulator is actuated by electric motors with reduction gears of high ratios, the presence of gears tends to linearize system dynamics and thus to decouple the joints in view of the reduction of nonlinearity effects. The price to pay, however, is the occurrence of joint friction, elasticity and backlash that may limit system performance more than it is due to configuration-dependent inertia, Coriolis and centrifugal forces,

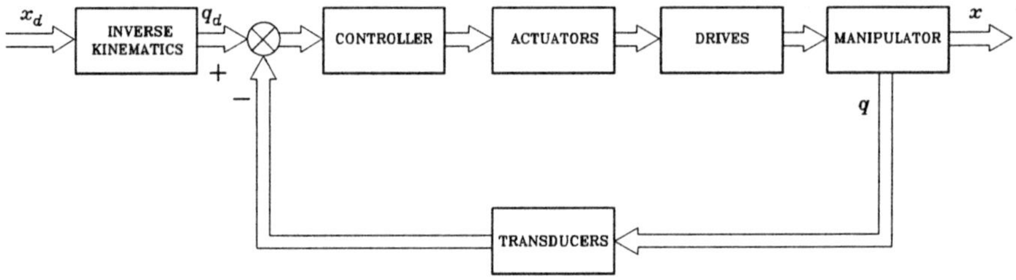

Figure 6.1 General scheme of joint space control.

and so forth. On the other hand, a robot actuated with direct drives eliminates the drawbacks due to friction, elasticity and backlash but the weight of nonlinearities and couplings between the joints becomes relevant. As a consequence, different control strategies have to be thought of to obtain high performance.

Without any concern to the specific type of mechanical manipulator, it is worth remarking that task specification (end-effector motion and forces) is usually carried out in the operational space, whereas control actions (joint actuator generalized forces) are performed in the joint space. This fact naturally leads to considering two kinds of general control schemes; namely, a *joint space control* scheme (Figure 6.1) and an *operational space control* scheme (Figure 6.2). In both schemes, the control structure has closed loops to exploit the good features provided by feedback, *i.e.*, robustness to modelling uncertainties and reduction of disturbance effects. In general terms, the following considerations shall be made.

The *joint space control* problem is actually articulated in two subproblems. First, manipulator inverse kinematics is solved to transform motion requirements from the operational space into the joint space. Then, a joint space control scheme is designed that allows tracking of the reference inputs. However, this solution has the drawback that a joint space control scheme does not influence the operational space variables which are controlled in an open-loop fashion through the manipulator mechanical structure. It is then clear that any uncertainty of the structure (construction tolerance, lack of calibration, gear backlash, elasticity) or any imprecision on the knowledge of the end-effector position relative to an object to manipulate causes a loss of accuracy on the operational space variables.

The *operational space control* problem follows a global approach that requires a greater algorithmic complexity; notice that inverse kinematics is now embedded into the feedback control loop. Its conceptual advantage regards the possibility of acting directly on operational space variables; this is somewhat only a potential advantage, since measurement of operational space variables is often performed not directly, but through the evaluation of direct kinematics functions starting from measured joint space variables.

On the above premises, in the following, joint space control schemes for manipulator motion in the free space are presented first. In the sequel, operational space control schemes will be illustrated which are logically at the basis of interaction control in constrained manipulator motion.

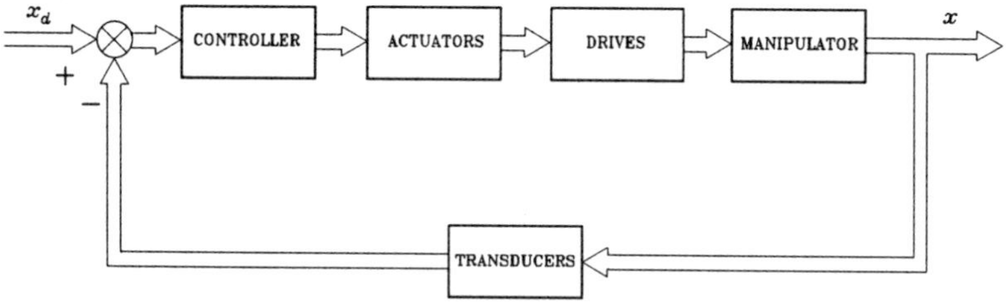

Figure 6.2 General scheme of operational space control.

6.2 Joint Space Control

In Chapter 4, it was shown that the equations of motion of a manipulator in the absence of external end-effector forces and, for simplicity, of static friction (difficult to model accurately) are described by

$$\boldsymbol{B}(\boldsymbol{q})\ddot{\boldsymbol{q}} + \boldsymbol{C}(\boldsymbol{q}, \dot{\boldsymbol{q}})\dot{\boldsymbol{q}} + \boldsymbol{F}_v\dot{\boldsymbol{q}} + \boldsymbol{g}(\boldsymbol{q}) = \boldsymbol{\tau} \tag{6.1}$$

with obvious meaning of the symbols. To control the motion of the manipulator in free space means to determine the n components of generalized forces—torques for revolute joints, forces for prismatic joints—that allow execution of a motion $\boldsymbol{q}(t)$ so that

$$\boldsymbol{q}(t) = \boldsymbol{q}_d(t)$$

as closely as possible, where $\boldsymbol{q}_d(t)$ denotes the vector of desired joint trajectory variables.

The generalized forces are supplied by the actuators through proper transmissions to transform the motion characteristics. Let $\boldsymbol{q}_m$ denote the vector of joint actuator displacements; the transmissions—assumed to be rigid and with no backlash—establish the following relationship:

$$\boldsymbol{K}_r\boldsymbol{q} = \boldsymbol{q}_m, \tag{6.2}$$

where $\boldsymbol{K}_r$ is an $(n \times n)$ diagonal matrix, whose elements are defined in (4.22) and are much greater than unity. Assuming a diagonal $\boldsymbol{K}_r$ leads to excluding the presence of kinematic couplings in the transmission, that is the motion of each actuator does not induce motion on a joint other than that actuated.

In view of (6.2), if $\boldsymbol{\tau}_m$ denotes the vector of actuator driving torques, one can write

$$\boldsymbol{\tau}_m = \boldsymbol{K}_r^{-1}\boldsymbol{\tau}. \tag{6.3}$$

By observing that the diagonal elements of $\boldsymbol{B}(\boldsymbol{q})$ are formed by constant terms and configuration-dependent terms (functions of sine and cosine for revolute joints), one can set

$$\boldsymbol{B}(\boldsymbol{q}) = \bar{\boldsymbol{B}} + \Delta\boldsymbol{B}(\boldsymbol{q}) \tag{6.4}$$

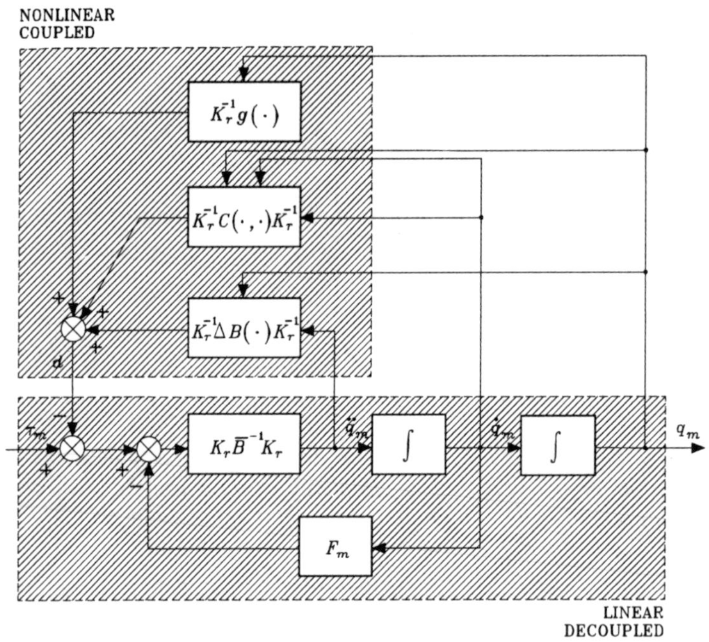

Figure 6.3 Block scheme of manipulator with drives.

where $\bar{\boldsymbol{B}}$ is the *diagonal* matrix whose constant elements represent the resulting average inertia at each joint. Substituting (6.2)–(6.4) into (6.1) yields

$$\boldsymbol{\tau}_m = \boldsymbol{K}_r^{-1}\bar{\boldsymbol{B}}\boldsymbol{K}_r^{-1}\ddot{\boldsymbol{q}}_m + \boldsymbol{F}_m\dot{\boldsymbol{q}}_m + \boldsymbol{d} \tag{6.5}$$

where

$$\boldsymbol{F}_m = \boldsymbol{K}_r^{-1}\boldsymbol{F}_v\boldsymbol{K}_r^{-1} \tag{6.6}$$

represents the matrix of viscous friction coefficients about the motor axes, and

$$\boldsymbol{d} = \boldsymbol{K}_r^{-1}\Delta\boldsymbol{B}(\boldsymbol{q})\boldsymbol{K}_r^{-1}\ddot{\boldsymbol{q}}_m + \boldsymbol{K}_r^{-1}\boldsymbol{C}(\boldsymbol{q},\dot{\boldsymbol{q}})\boldsymbol{K}_r^{-1}\dot{\boldsymbol{q}}_m + \boldsymbol{K}_r^{-1}\boldsymbol{g}(\boldsymbol{q}) \tag{6.7}$$

represents the contribution depending on the configuration.

As illustrated by the block scheme of Figure 6.3, the system of manipulator with drives is actually constituted by two subsystems; one has $\boldsymbol{\tau}_m$ as input and $\boldsymbol{q}_m$ as output, the other has $\boldsymbol{q}_m$, $\dot{\boldsymbol{q}}_m$, $\ddot{\boldsymbol{q}}_m$ as inputs, and $\boldsymbol{d}$ as output. The former is *linear* and *decoupled*, since each component of $\boldsymbol{\tau}_m$ influences only the corresponding component of $\boldsymbol{q}_m$. The latter is *nonlinear* and *coupled*, since it accounts for all those nonlinear and coupling terms of manipulator joint dynamics.

On the basis of the above scheme, several control algorithms can be derived with reference to the detail of knowledge of the dynamic model. The simplest approach that can be followed, in case of high gear reduction ratios and/or limited performance in terms of required velocities and accelerations, is to consider the component of the nonlinear interacting term $\boldsymbol{d}$ as a *disturbance* for the single joint servo.

The design of the control algorithm leads to a *decentralized* control structure, since each joint is considered independently of the others. The joint controller must guarantee good performance in terms of high disturbance rejection and enhanced trajectory tracking capabilities. The resulting control structure is substantially based on the error between the desired and actual output, while the input control torque at actuator i depends only on the error of output i.

On the other hand, when large operational speeds are required or direct-drive actuation is employed ($\boldsymbol{K}_r = \boldsymbol{I}$), the nonlinear coupling terms strongly influence system performance. Therefore, considering the effects of the components of $\boldsymbol{d}$ as a disturbance may generate large tracking errors. In this case, it is advisable to design control algorithms that take advantage of a detailed knowledge of manipulator dynamics so as to compensate for the nonlinear coupling terms of the model. In other words, it is necessary to eliminate the causes rather than to reduce the effects induced by them; that is, to generate compensating torques for the nonlinear terms in (6.7). This leads to *centralized control* algorithms that are based on the (partial or complete) knowledge of the manipulator dynamic model.

Nevertheless, it should be pointed out that these techniques still require the use of error contributions between the desired and the actual trajectory, no matter whether they are implemented in a feedback or in a feedforward fashion. This is a consequence of the fact that the considered dynamic model, even though a quite complex one, is anyhow an idealization of reality which does not include effects, such as joint Coulomb friction, gear backlash, dimension tolerance, and the simplifying assumptions in the model, e.g., link rigidity, and so on.

As pointed out above, the role of the drive system is relevant for the type of control chosen. In the case of decentralized control, the actuator will be described in terms of its own model as a velocity-controlled generator. Instead, in the case of centralized control, the actuator will have to generate torque contributions computed on the basis of a complete or reduced manipulator dynamic model; it will be then considered as a torque-controlled generator which is representative of the actuator/power amplifier system satisfying the above requirement.

6.3 Independent Joint Control

The simplest control strategy that can be thought of is one that regards the manipulator as formed by n independent systems (the n joints) and controls each joint axis as a *single-input/single-output system*. Coupling effects between joints due to varying configurations during motion are treated as disturbance inputs.

In the case of interest, the system to control is Joint i drive corresponding to the single-input/single-output system of the decoupled and linear part of the scheme in Figure 6.3. The interaction with the other joints is described by component i of the vector $\boldsymbol{d}$ in (6.7).

Assuming that the actuator is a rotary electric dc motor, the block scheme of Joint i

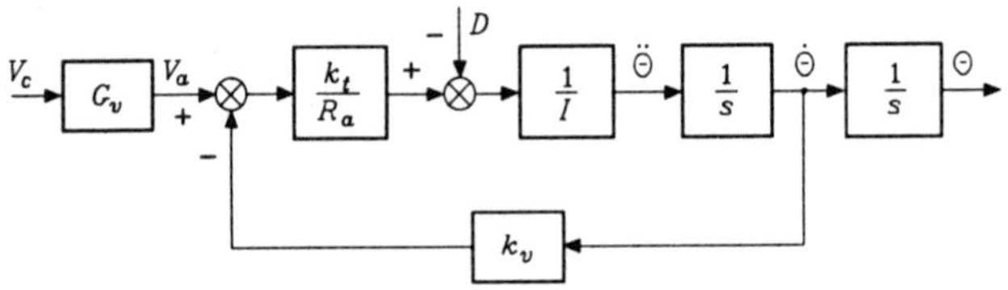

Figure 6.4 Block scheme of joint drive system.

can be represented in the domain of the complex variable s as in Figure 6.4[1]; other types of actuators can be modeled analogously. In this scheme θ is the angular variable of the motor, I is the average inertia reported to the motor axis ($I_i = \bar{b}_{ii}/k_{ri}^2$), R_a is the armature resistance (auto-inductance has been neglected), and k_t and k_v are respectively the torque and motor constants. Further, G_v denotes the voltage gain of the power amplifier, and then the reference input is not the armature voltage V_a but the input voltage V_c of the amplifier; note that the amplifier bandwidth has been assumed to be much larger than that of the controlled system. In the scheme of Figure 6.4, it has been assumed also that

$$F_m \ll \frac{k_v k_t}{R_a},$$

i.e., the mechanical viscous friction coefficient has been neglected with respect to the electrical friction coefficient[2].

The input/output transfer function of the motor can be written as

$$M(s) = \frac{k_m}{s(1 + sT_m)}, \tag{6.8}$$

where

$$k_m = \frac{1}{k_v} \qquad T_m = \frac{R_a I}{k_v k_t}$$

are respectively the velocity-to-voltage gain and time constant of the motor.

6.3.1 Feedback Control

To guide selection of the controller structure, start by noticing that an effective rejection of the disturbance d on the output θ is ensured by:

- a large value of the amplifier gain before the point of intervention of the disturbance,

---

[1] Subscript i has been dropped for notation compactness. Also, Laplace transforms of time-dependent functions are indicated by capital letters without specifying dependence on s.

[2] A complete treatment of actuators is deferred to Chapter 8.

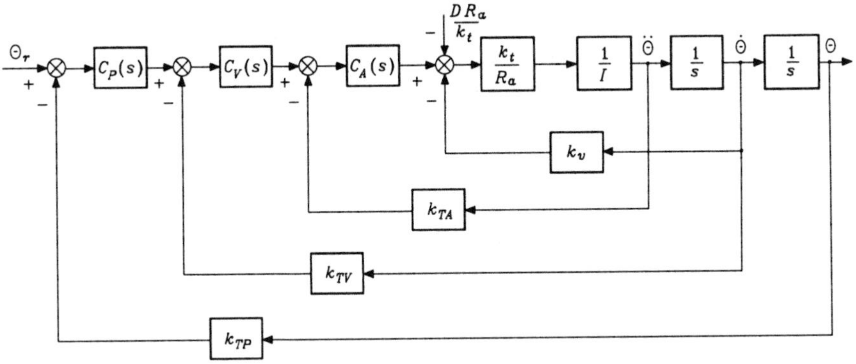

Figure 6.5 Block scheme of general independent joint control.

- the presence of an integral action in the controller so as to cancel the effect of the gravitational component on the output at steady state (constant θ).

These requisites clearly suggest the use of a *proportional-integral* (PI) control action in the forward path whose transfer function is

$$C(s) = K_c \frac{1 + sT_c}{s}; \tag{6.9}$$

this yields zero error at steady state for a constant disturbance, and the presence of the real zero at $s = -1/T_c$ offers a stabilizing action. To improve dynamic performance, it is worth choosing the controller as a cascade of elementary actions with local feedback loops closed around the disturbance.

Besides closure of a position feedback loop, the most general solution is obtained by closing inner loops on velocity and acceleration. This leads to the scheme in Figure 6.5, where $C_P(s)$, $C_V(s)$, $C_A(s)$ respectively represent *position*, *velocity*, *acceleration* controllers, and the inmost controller shall be of PI type as in (6.9) so as to obtain zero error at steady state for a constant disturbance. Further, k_{TP}, k_{TV}, k_{TA} are the respective transducer constants, and the amplifier gain G_v has been embedded in the gain of the inmost controller. In the scheme of Figure 6.5, notice that θ_r is the reference input, which is related to the desired output as

$$\theta_r = k_{TP}\theta_d;$$

further, the disturbance torque D has been suitably transformed into a voltage by the factor R_a/k_t.

In the following, a number of possible solutions that can be derived from the general scheme of Figure 6.5 are presented; at this stage, the issue arising from possible lack of measurement of physical variables is not considered yet. Three case studies are considered which differ in the number of active feedback loops.

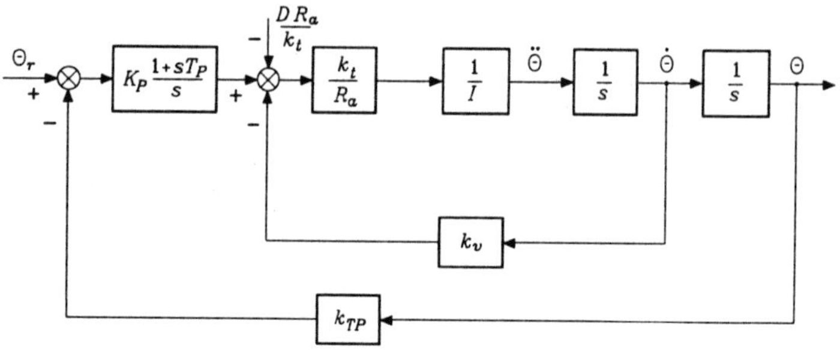

Figure 6.6 Block scheme of position feedback control.

Position Feedback

In this case, the control action is characterized by:

$$C_P(s) = K_P \frac{1 + sT_P}{s} \qquad C_V(s) = 1 \qquad C_A(s) = 1$$

$$k_{TV} = k_{TA} = 0.$$

The scheme of Figure 6.6 shows that the transfer function of the forward path is

$$P(s) = \frac{k_m K_P (1 + sT_P)}{s^2(1 + sT_m)},$$

while that of the return path is

$$H(s) = k_{TP}.$$

A root locus analysis can be performed as a function of the gain of the position loop $k_m K_P k_{TP} T_P / T_m$. Three situations are illustrated for the poles of the closed-loop system with reference to the relation between T_P and T_m (Figure 6.7). Stability of the closed-loop feedback system imposes some constraints on the choice of the parameters of the PI controller. If $T_P < T_m$, the system is inherently unstable (Figure 6.7a). Then, it must be $T_P > T_m$ (Figure 6.7b). As T_P increases, the absolute value of the real part of the two roots of the locus tending towards the asymptotes increases too, and the system has faster time response. Hence, it is convenient to render $T_P \gg T_m$ (Figure 6.7c). In any case, the real part of the dominant poles cannot be less than $-1/2T_m$.

The closed-loop input/output transfer function is

$$\frac{\Theta(s)}{\Theta_r(s)} = \frac{\dfrac{1}{k_{TP}}}{1 + \dfrac{s^2(1 + sT_m)}{k_m K_P k_{TP}(1 + sT_P)}} \tag{6.10}$$

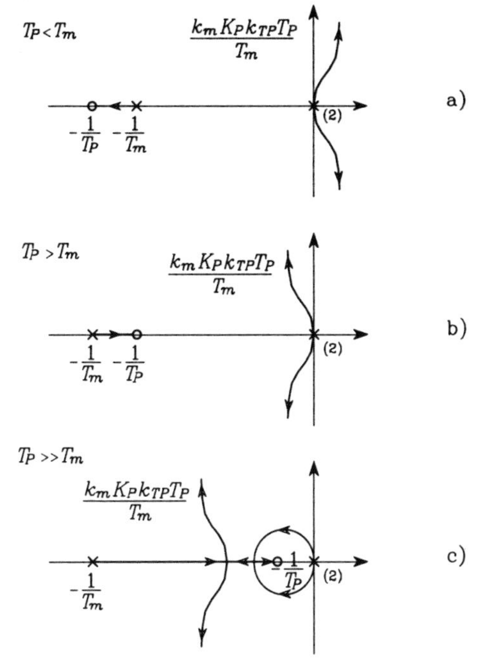

Figure 6.7 Root loci for the position feedback control scheme.

which can be expressed in the form

$$W(s) = \frac{\dfrac{1}{k_{TP}}(1 + sT_P)}{\left(1 + \dfrac{2\zeta s}{\omega_n} + \dfrac{s^2}{\omega_n^2}\right)(1 + s\tau)},$$

where ω_n and ζ are respectively the natural frequency and damping ratio of the pair of complex poles and $-1/\tau$ locates the real pole. These values are assigned to define the joint drive dynamics as a function of the constant T_P; if $T_P > T_m$, then $1/\zeta\omega_n > T_P > \tau$ (Figure 6.7b); if $T_P \gg T_m$ (Figure 6.7c), for large values of the loop gain, then $\zeta\omega_n > 1/\tau \approx 1/T_P$ and the zero at $-1/T_P$ in the transfer function $W(s)$ tends to cancel the effect of the real pole.

The closed-loop disturbance/output transfer function is

$$\frac{\Theta(s)}{D(s)} = -\frac{\dfrac{sR_a}{k_t K_P k_{TP}(1 + sT_P)}}{1 + \dfrac{s^2(1 + sT_m)}{k_m K_P k_{TP}(1 + sT_P)}}, \tag{6.11}$$

which shows that it is worth increasing K_P to reduce the effect of disturbance on the output during the transient. The function in (6.11) has two complex poles

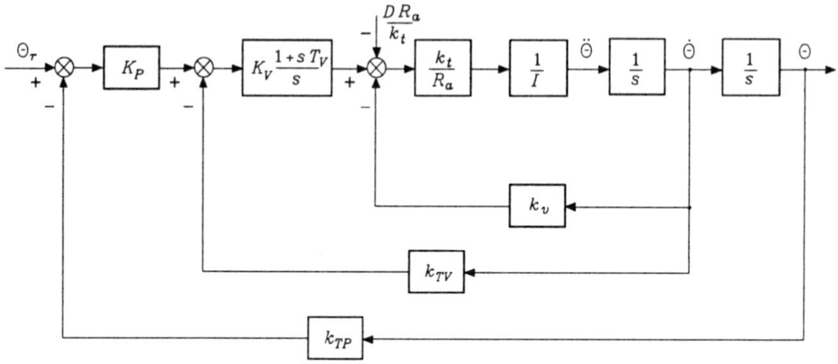

Figure 6.8 Block scheme of position and velocity feedback control.

$(-\zeta\omega_n, \pm j\sqrt{1-\zeta^2}\omega_n)$, a real pole $(-1/\tau)$, and a zero at the origin. The zero is due to the PI controller and allows canceling the effects of gravity on the angular position when θ is a constant.

In (6.11), it can be recognized that the term $K_P k_{TP}$ is the reduction factor imposed by the feedback gain on the amplitude of the output due to disturbance; hence, the quantity

$$X_R = K_P k_{TP} \tag{6.12}$$

can be interpreted as the *disturbance rejection factor*, which in turn is determined by the gain K_P. However, it is not advisable to increase K_P too much, because small damping ratios would result leading to unacceptable oscillations of the output. An estimate T_R of the *output recovery time* needed by the control system to recover the effects of the disturbance on the angular position can be evaluated by analyzing the modes of evolution of (6.11). Since $\tau \approx T_P$, such estimate is expressed by

$$T_R = \max\left\{T_P, \frac{1}{\zeta\omega_n}\right\}. \tag{6.13}$$

Position and Velocity Feedback

In this case, the control action is characterized by:

$$C_P(s) = K_P \qquad C_V(s) = K_V \frac{1+sT_V}{s} \qquad C_A(s) = 1$$

$$k_{TA} = 0.$$

To carry out a root locus analysis as a function of the velocity feedback loop gain, it is worth reducing the velocity loop in parallel to the position loop by following the usual rules for moving blocks. From the scheme in Figure 6.8 the transfer function of the forward path is

$$P(s) = \frac{k_m K_P K_V (1+sT_V)}{s^2(1+sT_m)},$$

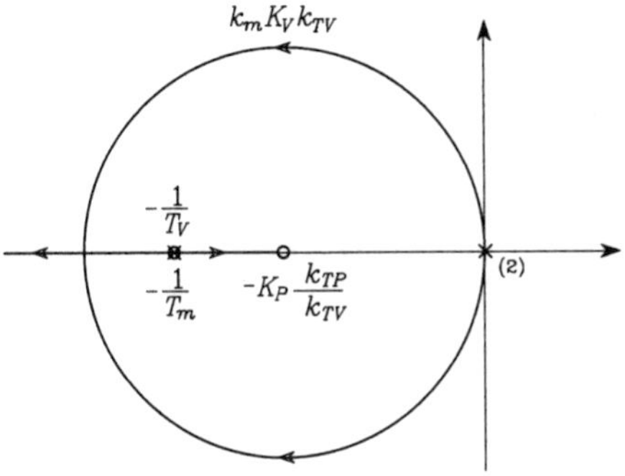

Figure 6.9 Root locus for the position and velocity feedback control scheme.

while that of the return path is

$$H(s) = k_{TP}\left(1 + s\frac{k_{TV}}{K_P k_{TP}}\right).$$

The zero of the controller at $s = -1/T_V$ can be chosen so as to cancel the effects of the real pole of the motor at $s = -1/T_m$. Then, by setting

$$T_V = T_m,$$

the poles of the closed-loop system move on the root locus as a function of the loop gain $k_m K_V k_{TV}$, as shown in Figure 6.9. By increasing the position feedback gain K_P, it is possible to confine the closed-loop poles into a region of the complex plane with large absolute values of the real part. Then, the actual location can be established by a suitable choice of K_V.

The closed-loop input/output transfer function is

$$\frac{\Theta(s)}{\Theta_r(s)} = \frac{\dfrac{1}{k_{TP}}}{1 + \dfrac{s k_{TV}}{K_P k_{TP}} + \dfrac{s^2}{k_m K_P k_{TP} K_V}}, \tag{6.14}$$

which can be compared with the typical transfer function of a second-order system

$$W(s) = \frac{\dfrac{1}{k_{TP}}}{1 + \dfrac{2\zeta s}{\omega_n} + \dfrac{s^2}{\omega_n^2}}. \tag{6.15}$$

It can be recognized that, with a suitable choice of the gains, it is possible to obtain any value of natural frequency ω_n and damping ratio ζ. Hence, if ω_n and ζ are given as design requirements, the following relations can be found:

$$K_V k_{TV} = \frac{2\zeta\omega_n}{k_m} \tag{6.16}$$

$$K_P k_{TP} K_V = \frac{\omega_n^2}{k_m}. \tag{6.17}$$

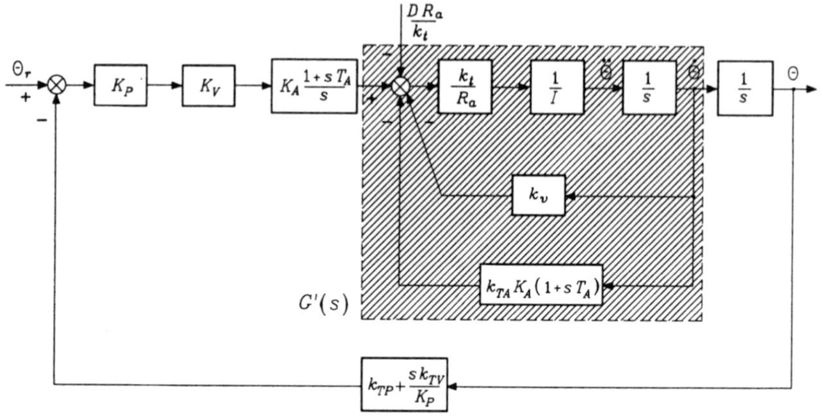

Figure 6.10 Block scheme of position, velocity and acceleration feedback control.

For given transducer constants k_{TP} and k_{TV}, once K_V has been chosen to satisfy (6.16), the value of K_P is obtained from (6.17).

The closed-loop disturbance/output transfer function is

$$\frac{\Theta(s)}{D(s)} = -\frac{\dfrac{sR_a}{k_t K_P k_{TP} K_V (1+sT_m)}}{1+\dfrac{sk_{TV}}{K_P k_{TP}}+\dfrac{s^2}{k_m K_P k_{TP} K_V}}, \tag{6.18}$$

which shows that the *disturbance rejection factor* is

$$X_R = K_P k_{TP} K_V \tag{6.19}$$

and is fixed, once K_P and K_V have been chosen via (6.16) and (6.17). Concerning disturbance dynamics, the presence of a zero at the origin introduced by the PI, of a real pole at $s = -1/T_m$, and of a pair of complex poles having real part $-\zeta\omega_n$ should be noticed. Hence, in this case, an estimate of the *output recovery time* is given by the time constant

$$T_R = \max\left\{T_m, \frac{1}{\zeta\omega_n}\right\}, \tag{6.20}$$

which reveals an improvement with respect to the previous case in (6.13), since $T_m \ll T_P$ and the real part of the dominant poles is not constrained by the inequality $\zeta\omega_n < 1/2T_m$.

Position, Velocity and Acceleration Feedback

In this case, the control action is characterized by:

$$C_P(s) = K_P \qquad C_V(s) = K_V \qquad C_A(s) = K_A\frac{1+sT_A}{s}.$$

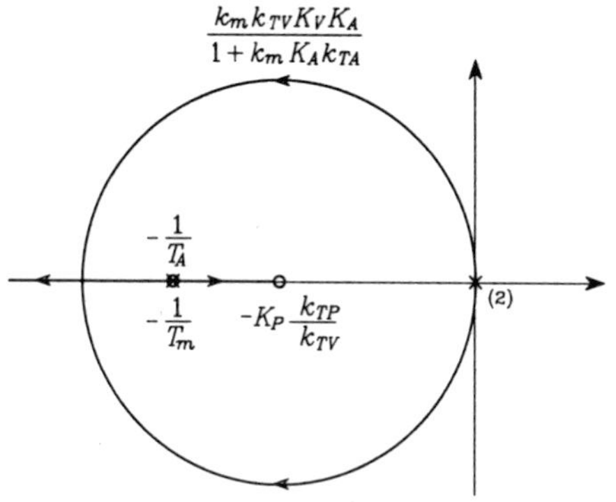

Figure 6.11 Root locus for the position, velocity and acceleration feedback control scheme.

After some manipulation, the block scheme of Figure 6.5 can be reduced to that of Figure 6.10 where $G'(s)$ indicates the following transfer function:

$$G'(s) = \frac{k_m}{(1 + k_m K_A k_{TA}) \left(1 + \frac{s T_m \left(1 + k_m K_A k_{TA} \frac{T_A}{T_m} \right)}{(1 + k_m K_A k_{TA})} \right)}.$$

The transfer function of the forward path is

$$P(s) = \frac{K_P K_V K_A (1 + s T_A)}{s^2} G'(s),$$

while that of the return path is

$$H(s) = k_{TP} \left(1 + \frac{s k_{TV}}{K_P k_{TP}} \right).$$

Also in this case, a suitable pole cancellation is worthy which can be achieved either by setting

$$T_A = T_m,$$

or by making

$$k_m K_A k_{TA} T_A \gg T_m \qquad k_m K_A k_{TA} \gg 1.$$

The two solutions are equivalent as regards dynamic performance of the control system. In both cases, the poles of the closed-loop system are constrained to move on the root locus as a function of the loop gain $k_m K_P K_V K_A / (1 + k_m K_A k_{TA})$ (Figure 6.11). A close analogy with the previous scheme can be recognized, in that the resulting closed-loop system is again of second-order type.

The closed-loop input/output transfer function is

$$\frac{\Theta(s)}{\Theta_r(s)} = \frac{\dfrac{1}{k_{TP}}}{1 + \dfrac{s k_{TV}}{K_P k_{TP}} + \dfrac{s^2(1 + k_m K_A k_{TA})}{k_m K_P k_{TP} K_V K_A}}, \tag{6.21}$$

while the closed-loop disturbance/output transfer function is

$$\frac{\Theta(s)}{D(s)} = -\frac{\dfrac{s R_a}{k_t K_P k_{TP} K_V K_A (1 + s T_A)}}{1 + \dfrac{s k_{TV}}{K_P k_{TP}} + \dfrac{s^2(1 + k_m K_A k_{TA})}{k_m K_P k_{TP} K_V K_A}}. \tag{6.22}$$

The resulting *disturbance rejection factor* is given by

$$X_R = K_P k_{TP} K_V K_A, \tag{6.23}$$

while the *output recovery time* is given by the time constant

$$T_R = \max\left\{T_A, \frac{1}{\zeta\omega_n}\right\} \tag{6.24}$$

where T_A can be made less than T_m, as pointed out above.

With reference to the transfer function in (6.15), the following relations can be established for design purposes, once ζ, ω_n, and X_R have been specified:

$$\frac{2K_P k_{TP}}{k_{TV}} = \frac{\omega_n}{\zeta} \tag{6.25}$$

$$k_m K_A k_{TA} = \frac{k_m X_R}{\omega_n^2} - 1 \tag{6.26}$$

$$K_P k_{TP} K_V K_A = X_R. \tag{6.27}$$

For given $k_{TP}, k_{TV}, k_{TA}, K_P$ is chosen to satisfy (6.25), K_A is chosen to satisfy (6.26), and then K_V is obtained from (6.27). Notice how admissible solutions for the controller typically require large values for the rejection factor X_R. Hence, in principle, not only does the acceleration feedback allow achieving any desired dynamic behaviour but, with respect to the previous case, it also allows prescribing the disturbance rejection factor as long as $k_m X_R/\omega_n^2 > 1$.

In deriving the above control schemes, the issue of measurement of feedback variables was not considered explicitly. With reference to the typical position control servos that are implemented in industrial practice, there is no problem to measure position and velocity, while a direct measurement of acceleration, in general, either is not available or is too expensive to obtain. Therefore, for the scheme of Figure 6.10, an indirect measure can be obtained by reconstructing acceleration from direct velocity

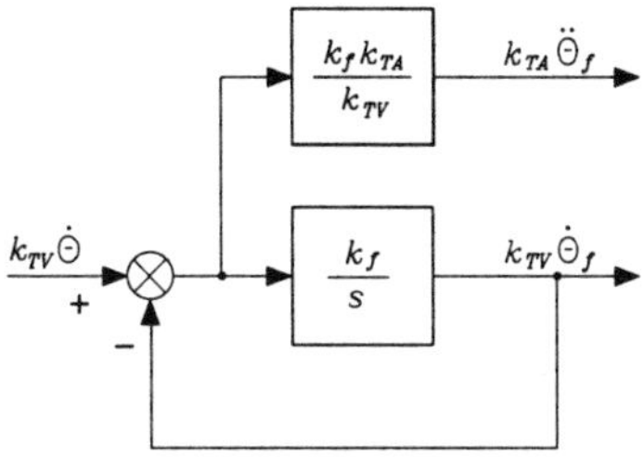

Figure 6.12 Block scheme of a first-order filter.

measurement through a first-order *filter* (Figure 6.12). The filter is characterized by a bandwidth $\omega_{3f} = k_f$. By choosing this bandwidth wide enough, the effects due to measurement lags are not appreciable, and then it is feasible to take the acceleration filter output as the quantity to feed back. Some problem may occur concerning the noise superimposed on the filtered acceleration signal, though.

Resorting to a filtering technique may be useful when only the direct position measurement is available. In this case, by means of a second-order state variable filter, it is possible to reconstruct velocity and acceleration. However, the greater lags induced by the use of a second-order filter typically degrade the performance with respect to the use of a first-order filter, because of limitations imposed on the filter bandwidth by numerical implementation of the controller and filter.

Notice that the above derivation is based on an ideal dynamic model, *i.e.*, when the effects of transmission elasticity as well as those of amplifier and motor electrical time constants are neglected. This implies that satisfaction of design requirements imposing large values of feedback gains may not be verified in practice, since the existence of unmodeled dynamics—such as electric dynamics, elastic dynamics due to non perfectly rigid transmissions, filter dynamics for the third scheme—might lead to degrading the system and eventually driving it to instability. In sum, the above solutions constitute design guidelines whose limits shall be emphasized with regard to the specific application.

6.3.2 Decentralized Feedforward Compensation

When the joint control servos are required to track reference trajectories with high values of speed and acceleration, the tracking capabilities of the scheme in Figure 6.5 are unavoidably degraded. The adoption of a *decentralized feedforward compensation* allows reducing the tracking error. Therefore, in view of the closed-loop input/output transfer functions in (6.10),(6.14),(6.21), the reference inputs to the three control structures analyzed in the previous section can be respectively modified into:

$$\Theta_r'(s) = \left(k_{TP} + \frac{s^2(1+sT_m)}{k_m K_P(1+sT_P)}\right)\Theta_d(s) \tag{6.28}$$

$$\Theta_r'(s) = \left(k_{TP} + \frac{sk_{TV}}{K_P} + \frac{s^2}{k_m K_P K_V}\right)\Theta_d(s) \tag{6.29}$$

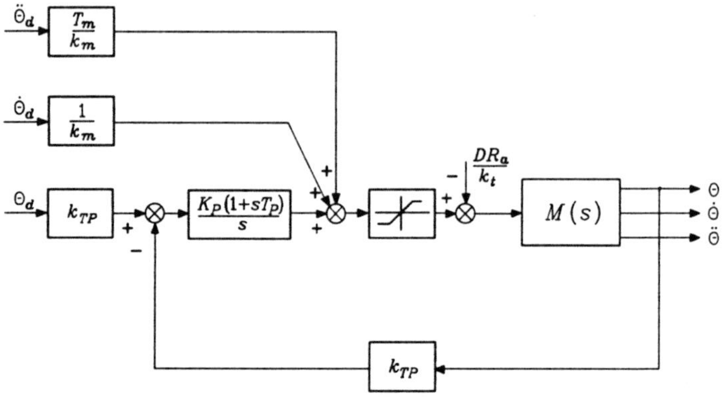

Figure 6.13 Block scheme of position feedback control with decentralized feedforward compensation.

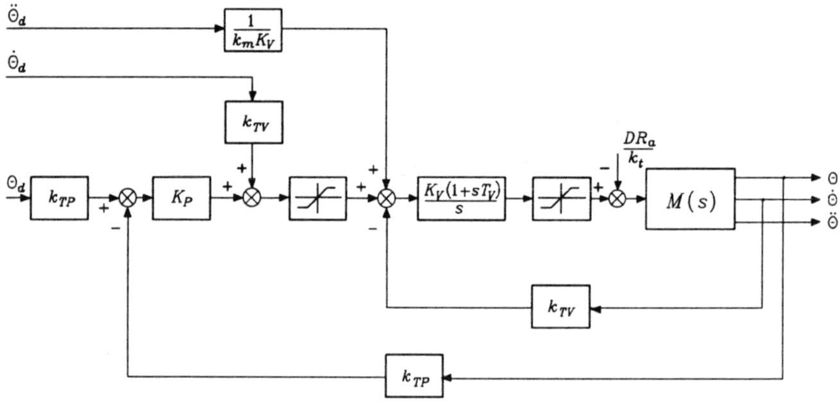

Figure 6.14 Block scheme of position and velocity feedback control with decentralized feedforward compensation.

$$\Theta_r'(s) = \left(k_{TP} + \frac{sk_{TV}}{K_P} + \frac{s^2(1 + k_m K_A k_{TA})}{k_m K_P K_V K_A}\right)\Theta_d(s); \tag{6.30}$$

in this way, tracking of the desired joint position trajectory is achieved, if not for the effect of disturbances. Notice that computing time derivatives of the desired trajectory is not a problem, once $\theta_d(t)$ is known analytically. The tracking control schemes, resulting from simple manipulation of (6.28), (6.29), (6.30) are reported respectively in Figures 6.13, 6.14, 6.15, where $M(s)$ indicates the motor transfer function in (6.8).

All the solutions allow tracking of the input trajectory within the range of validity and linearity of the employed models. It is worth noticing that, as the number of nested feedback loops increases, a less accurate knowledge of the system model is required to perform feedforward compensation. In fact, T_m and k_m are required for the scheme

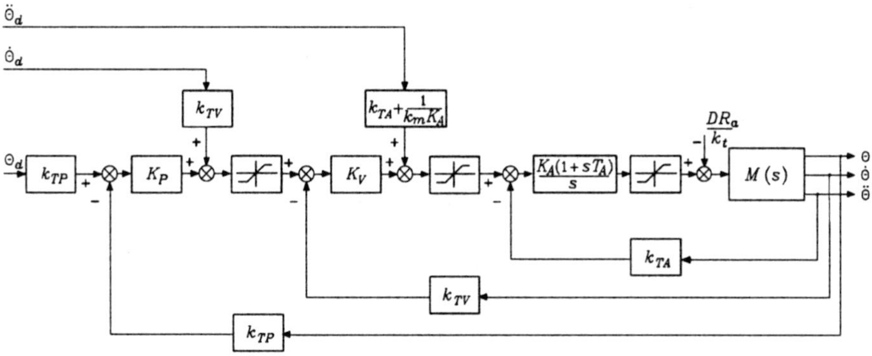

Figure 6.15 Block scheme of position, velocity and acceleration feedback control with decentralized feedforward compensation.

of Figure 6.13, only k_m is required for the scheme of Figure 6.14, and k_m again—but with reduced weight—for the scheme of Figure 6.15.

It is worth recalling that *perfect* tracking can be obtained only on the assumption of exact matching of the controller and feedforward compensation parameters with the process parameters, as well as of exact modelling and linearity of the physical system. Deviations from the ideal values cause a performance degradation that shall be analyzed case by case.

The presence of saturation blocks in the schemes of Figures 6.13, 6.14, 6.15 is to be intended as intentional nonlinearities whose function is to limit relevant physical quantities during transients; the greater the number of feedback loops, the greater the number of quantities that can be limited (velocity, acceleration, and motor voltage). To this purpose, notice that trajectory tracking is obviously lost whenever any of the above quantities saturates. This situation often occurs for industrial manipulators required to execute point-to-point motions; in this case, there is less concern about the actual trajectories followed, and the actuators are intentionally taken to operate at the current limits so as to realize the fastest possible motions.

After simple block reduction on the above schemes, it is possible to determine equivalent control structures that utilize position feedback only and *regulators with standard actions*. It should be emphasized that the two solutions are equivalent in terms of disturbance rejection and trajectory tracking. However, tuning of regulator parameters is less straightforward, and the elimination of inner feedback loops prevents the possibility of setting saturations on velocity and/or acceleration. The control structures equivalent to those of Figures 6.13, 6.14, 6.15 are illustrated in Figures 6.16, 6.17, 6.18, respectively; control actions of PI, PID, PIDD^2 type are illustrated which are respectively equivalent to the cases of: position feedback; position and velocity feedback; position, velocity and acceleration feedback.

The above schemes can incorporate the typical structure of the controllers actually implemented in the control architectures of industrial robots. In these systems it is important to choose the largest possible gains so that model inaccuracy and interaction terms do not appreciably affect positions of the single joints. As pointed out above, the

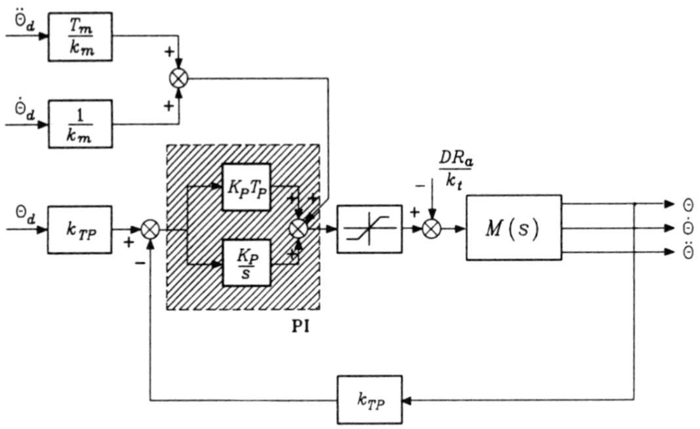

Figure 6.16 Equivalent control scheme of PI type.

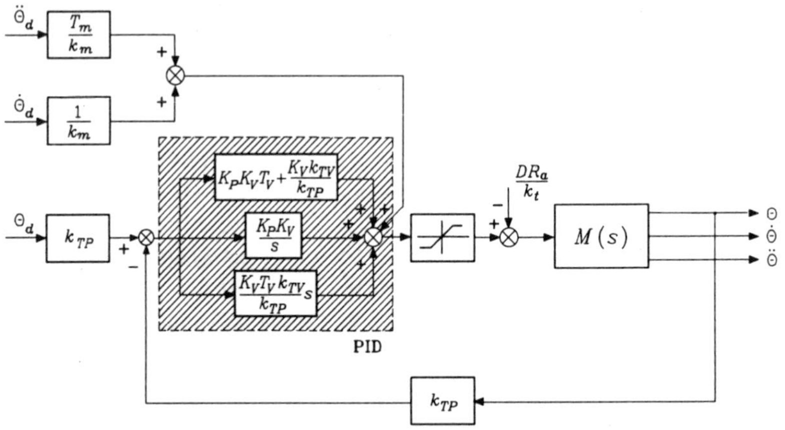

Figure 6.17 Equivalent control scheme of PID type.

upper limit on the gains is imposed by all those factors that have not been modeled, such as implementation of discrete-time controllers in lieu of the continuous-time controllers analyzed in theory, presence of finite sampling time, neglected dynamic effects (e.g., joint elasticity, structural resonance, finite transducer bandwidth), and sensor noise. In fact, the influence of such factors in the implementation of the above controllers may cause a severe system performance degradation for much too large values of feedback gains.

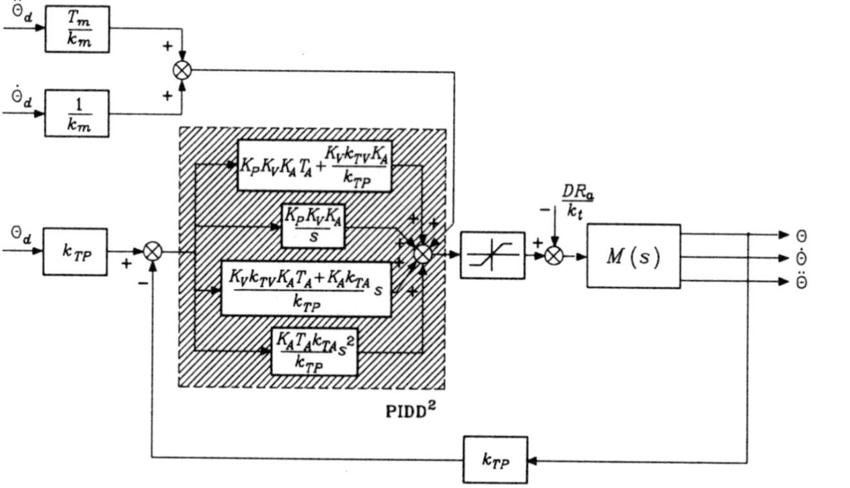

Figure 6.18 Equivalent control scheme of PIDD^2 type.

6.4 Computed Torque Feedforward Control

Define the tracking error $e(t) = \theta_d(t) - \theta(t)$. With reference to the most general scheme (Figure 6.18), the output of the PIDD^2 regulator can be written as

$$a_2\ddot{e} + a_1\dot{e} + a_0 e + a_{-1}\int^t e(\varsigma)d\varsigma$$

which describes the time evolution of the error. The constant coefficients a_2, a_1, a_0, a_{-1} are determined by the particular solution adopted. Summing the contribution of the feedforward actions and of the disturbance to this expression yields

$$\frac{1}{k_m}\dot{\theta}_d + \frac{T_m}{k_m}\ddot{\theta}_d - \frac{R_a}{k_t}d,$$

where

$$\frac{T_m}{k_m} = \frac{IR_a}{k_t} \qquad k_m = \frac{1}{k_v}.$$

The input to the motor (Figure 6.5) has then to satisfy the following equation:

$$a_2\ddot{e} + a_1\dot{e} + a_0 e + a_{-1}\int^t e(\varsigma)d\varsigma + \frac{T_m}{k_m}\ddot{\theta}_d + \frac{1}{k_m}\dot{\theta}_d - \frac{R_a}{k_t}d = \frac{T_m}{k_m}\ddot{\theta} + \frac{1}{k_m}\dot{\theta}.$$

With a suitable change of coefficients, this can be rewritten as

$$a_2'\ddot{e} + a_1'\dot{e} + a_0'e + a_{-1}'\int^t e(\varsigma)d\varsigma = \frac{R_a}{k_t}d.$$

This equation describes the *error dynamics* and shows that any physically executable trajectory is asymptotically tracked only if the disturbance term $d(t) = 0$. With the term *physically executable* it is meant that the saturation limits on the physical quantities, e.g., current and voltage in electric motors, are not violated in the execution of the desired trajectory.

The presence of the term $d(t)$ causes a tracking error whose magnitude is reduced as much as the disturbance frequency content is located off to the left of the lower limit of the bandwidth of the error system. The disturbance/error transfer function is given by

$$\frac{E(s)}{D(s)} = \frac{\frac{R_a}{k_t}s}{a'_2 s^3 + a'_1 s^2 + a'_0 s + a'_{-1}},$$

and thus the adoption of loop gains which are not realizable for the above discussed reasons is often required.

Nevertheless, even if the term $d(t)$ has been introduced as a disturbance, its expression is given by (6.7). It is then possible to add a further term to the previous feedforward actions which is able to compensate the disturbance itself rather than its effects. In other words, by taking advantage of model knowledge, the rejection effort of an independent joint control scheme can be lightened with notable simplification from the implementation viewpoint.

Let $\boldsymbol{q}_d(t)$ be the desired joint trajectory and $\boldsymbol{q}_{md}(t)$ the corresponding actuator trajectory as in (6.2). By adopting an *inverse model* strategy, the *feedforward* action $\boldsymbol{R}_a\boldsymbol{K}_t^{-1}\boldsymbol{d}_d$ can be introduced with

$$\boldsymbol{d}_d = \boldsymbol{K}_r^{-1}\Delta\boldsymbol{B}(\boldsymbol{q}_d)\boldsymbol{K}_r^{-1}\ddot{\boldsymbol{q}}_{md} + \boldsymbol{K}_r^{-1}\boldsymbol{C}(\boldsymbol{q}_d,\dot{\boldsymbol{q}}_d)\boldsymbol{K}_r^{-1}\dot{\boldsymbol{q}}_{md} + \boldsymbol{K}_r^{-1}\boldsymbol{g}(\boldsymbol{q}_d), \quad (6.31)$$

where $\boldsymbol{R}_a$ and $\boldsymbol{K}_t$ denote the diagonal matrices of armature resistances and torque constants of the actuators. This action tends to compensate the actual disturbance expressed by (6.7) and in turn allows the control system to operate in a better condition.

This solution is illustrated in the scheme of Figure 6.19, which conceptually describes the control system of a manipulator with *computed torque* control. The feedback control system is representative of the n independent joint control servos; it is *decentralized*, since controller i elaborates references and measurements that refer to single Joint i. The interactions between the various joints, expressed by $\boldsymbol{d}$, are compensated by a *centralized* action whose function is to generate a feedforward action that depends on the joint references as well as on the manipulator dynamic model. This action compensates the nonlinear coupling terms due to inertial, Coriolis, centrifugal, and gravitational forces that depend on the structure and, as such, vary during manipulator motion.

Although the residual disturbance term $\widetilde{\boldsymbol{d}} = \boldsymbol{d}_d - \boldsymbol{d}$ vanishes only in the ideal case of perfect tracking ($\boldsymbol{q} = \boldsymbol{q}_d$) and exact dynamic modelling, $\widetilde{\boldsymbol{d}}$ is representative of interaction disturbances of considerably reduced magnitude with respect to $\boldsymbol{d}$. Hence, the computed torque technique has the advantage to alleviate the disturbance rejection task for the feedback control structure and in turn allows limited gains. Notice that expression (6.31) in general imposes a computationally demanding burden on the centralized part of the controller. Therefore, in those applications where the desired

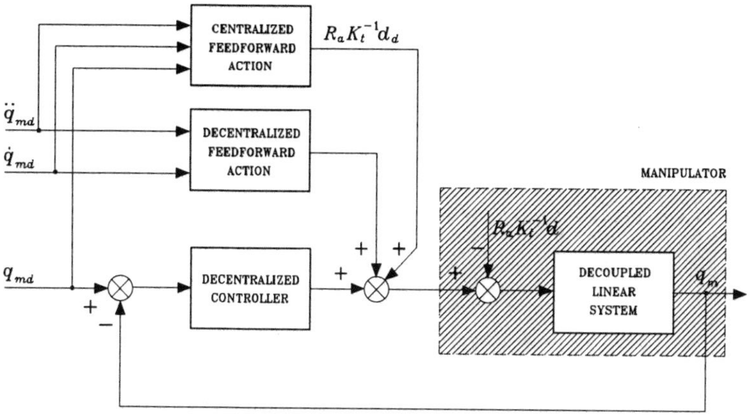

Figure 6.19 Block scheme of computed torque feedforward control.

trajectory is generated in real time with regard to heteroceptive sensory data and commands from higher hierarchical levels of the robot control architecture[3], on-line computation of the centralized feedforward action may require too much time[4].

Since the actual controller is to be implemented on a computer with a finite sampling time, torque computation has to be carried out during this interval of time; in order not to degrade dynamic system performance, typical sampling times are of the order of the millisecond.

Therefore, it may be worth performing only a *partial* feedforward action so as to compensate those terms of (6.31) that give the most relevant contributions during manipulator motion. Since inertial and gravitational terms dominate velocity-dependent terms (at operational joint speeds not greater than a few radians per second), a partial compensation can be achieved by computing only the gravitational torques and the inertial torques due to the diagonal elements of the inertia matrix. In this way, only the terms depending on the global manipulator configuration are compensated while those deriving from motion interaction with the other joints are not.

Finally, it should be pointed out that, for repetitive trajectories, the above compensating contributions can be computed off-line and properly stored on the basis of a trade-off solution between memory capacity and computational requirements of the control architecture.

---

[3] See also Chapter 9.

[4] In this regard, the problem of real-time computation of compensating torques can be solved by resorting to efficient recursive formulations of manipulator inverse dynamics, such as the Newton-Euler algorithm presented in Chapter 4.

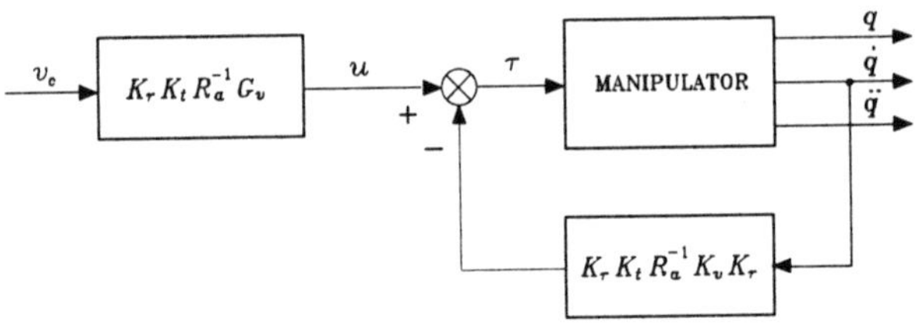

Figure 6.20 Block scheme of the manipulator and drives system as a voltage-controlled system.

6.5 Centralized Control

In the previous sections several techniques have been discussed that allow designing independent joint controllers. These are based on a single-input/single-output approach, since interaction and coupling effects between the joints have been considered as disturbances acting on each single joint drive system. However, as shown by the dynamic model (6.1), the manipulator is not a set of n decoupled system but it is a multivariable system with n inputs (joint torques) and n outputs (joint positions) interacting between them by means of nonlinear relations.

In order to follow a methodological approach which is consistent with control design, it is necessary to treat the control problem in the context of nonlinear multivariable systems. This approach will obviously account for the manipulator *dynamic model* and lead to finding *nonlinear centralized control* laws, whose implementation is needed for high manipulator dynamic performance. On the other hand, the above computed torque control can be interpreted in this framework, since it provides a model-based nonlinear control term to enhance trajectory tracking performance. Notice, however, that this action is inherently performed off line, as it is computed on the time history of the desired trajectory and not of the actual one.

For the following derivation, it is worth rewriting the mathematical model of the manipulator with drives in a more suitable form. The manipulator is described by (6.1)

$$\boldsymbol{B}(\boldsymbol{q})\ddot{\boldsymbol{q}} + \boldsymbol{C}(\boldsymbol{q}, \dot{\boldsymbol{q}})\dot{\boldsymbol{q}} + \boldsymbol{F}_v\dot{\boldsymbol{q}} + \boldsymbol{g}(\boldsymbol{q}) = \boldsymbol{\tau},$$

while the transmissions are described by (6.2)

$$\boldsymbol{K}_r\boldsymbol{q} = \boldsymbol{q}_m.$$

With reference to (6.3) and the block scheme of Figure 6.4, the n driving systems can be described in compact matrix form by the equations:

$$\boldsymbol{K}_r^{-1}\boldsymbol{\tau} = \boldsymbol{K}_t\boldsymbol{i}_a \tag{6.32}$$

$$\boldsymbol{v}_a = \boldsymbol{R}_a\boldsymbol{i}_a + \boldsymbol{K}_v\dot{\boldsymbol{q}}_m \tag{6.33}$$

$$\boldsymbol{v}_a = \boldsymbol{G}_v\boldsymbol{v}_c. \tag{6.34}$$

In (6.32), $\boldsymbol{K}_t$ is the diagonal matrix of torque constants and $\boldsymbol{i}_a$ is the vector of armature currents of the n motors; in (6.33), $\boldsymbol{v}_a$ is the vector of armature voltages, $\boldsymbol{R}_a$ is the diagonal matrix of armature resistances, and $\boldsymbol{K}_v$ is the diagonal matrix of voltage constants of the n motors; in (6.34), $\boldsymbol{G}_v$ is the diagonal matrix of gains of the n amplifiers and $\boldsymbol{v}_c$ is the vector of control voltages of the n servomotors.

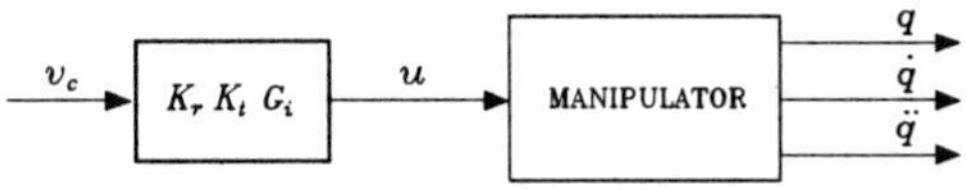

Figure 6.21 Block scheme of the manipulator and drives system as a torque-controlled system.

On reduction of (6.1), (6.2), (6.32), (6.33), (6.34), the dynamic model of the system given by the manipulator and drives is described by

$$\boldsymbol{B}(\boldsymbol{q})\ddot{\boldsymbol{q}} + \boldsymbol{C}(\boldsymbol{q}, \dot{\boldsymbol{q}})\dot{\boldsymbol{q}} + \boldsymbol{F}\dot{\boldsymbol{q}} + \boldsymbol{g}(\boldsymbol{q}) = \boldsymbol{u} \tag{6.35}$$

where the following positions have been made:

$$\boldsymbol{F} = \boldsymbol{F}_v + \boldsymbol{K}_r\boldsymbol{K}_t\boldsymbol{R}_a^{-1}\boldsymbol{K}_v\boldsymbol{K}_r \tag{6.36}$$

$$\boldsymbol{u} = \boldsymbol{K}_r\boldsymbol{K}_t\boldsymbol{R}_a^{-1}\boldsymbol{G}_v\boldsymbol{v}_c. \tag{6.37}$$

In (6.36), $\boldsymbol{F}$ is the diagonal matrix accounting for all viscous (mechanical and electrical) damping terms, and, in (6.37), $\boldsymbol{u}$ is the vector which is taken as control input to the system. Notice that the actual torques that determine the motion of the system of manipulator with drives can be obtained by subtracting to (6.37) the term $\boldsymbol{K}_r\boldsymbol{K}_t\boldsymbol{R}_a^{-1}\boldsymbol{K}_v\boldsymbol{K}_r\dot{\boldsymbol{q}}$ due to electrical friction. The overall system is then *voltage-controlled* and the corresponding block scheme is illustrated in Figure 6.20. In this case, each component of $\boldsymbol{v}_c$ corresponds to the control voltage of any of the previous independent joint control schemes.

If the actuators have to provide torque contributions computed on the basis of a complete or reduced manipulator model, the design of $\boldsymbol{u}$ in (6.35) depends on the matrices $\boldsymbol{K}_t$, $\boldsymbol{K}_v$ and $\boldsymbol{R}_a$ of the motors, which are influenced by the operating conditions. To reduce sensitivity to parameter variations, it is worth considering driving systems characterized by a torque (current) control rather than by a voltage control. In this case the actuators behave as *torque-controlled generators*; the equation in (6.33) becomes meaningless and (6.34) is replaced with

$$\boldsymbol{i}_a = \boldsymbol{G}_i\boldsymbol{v}_c, \tag{6.34$'$}$$

which gives a proportional relation between the armature currents $\boldsymbol{i}_a$ (and thus the torques) and the control voltages $\boldsymbol{v}_c$ established by the constant matrix $\boldsymbol{G}_i$. As a consequence, (6.36) and (6.37) become

$$\boldsymbol{F} = \boldsymbol{F}_v \tag{6.36$'$}$$

$$\boldsymbol{u} = \boldsymbol{K}_r\boldsymbol{K}_t\boldsymbol{G}_i\boldsymbol{v}_c = \boldsymbol{\tau}, \tag{6.37$'$}$$

which show a reduced dependence of $\boldsymbol{u}$ and $\boldsymbol{F}$ on the motor parameters. The overall system is now *torque-controlled* and the resulting block scheme is illustrated in Figure 6.21.

In the remainder, the problem of finding control laws $\boldsymbol{u}$ that ensure a given performance for the system of manipulator with drives is considered. Since (6.37$'$) can be considered as a constant proportional relation between $\boldsymbol{v}_c$ and $\boldsymbol{u}$, the centralized control algorithms that follow directly refer to the generation of control torques $\boldsymbol{u}$.

6.5.1 PD Control with Gravity Compensation

Let a *constant* equilibrium posture be assigned for the system as the vector of desired joint variables $\boldsymbol{q}_d$. It is desired to find the structure of the controller which ensures global asymptotic stability of the above posture.

The determination of the control input which stabilizes the system around the equilibrium posture is based on the Lyapunov direct method.

Take the vector $[\,\widetilde{\boldsymbol{q}}^T \quad \dot{\boldsymbol{q}}^T\,]^T$ as the system state, where

$$\widetilde{\boldsymbol{q}} = \boldsymbol{q}_d - \boldsymbol{q} \tag{6.38}$$

represents the error between the desired and the actual posture. Choose the following positive definite quadratic form as Lyapunov function candidate:

$$V(\dot{\boldsymbol{q}}, \widetilde{\boldsymbol{q}}) = \frac{1}{2}\dot{\boldsymbol{q}}^T \boldsymbol{B}(\boldsymbol{q})\dot{\boldsymbol{q}} + \frac{1}{2}\widetilde{\boldsymbol{q}}^T \boldsymbol{K}_P \widetilde{\boldsymbol{q}} > 0 \qquad \forall \dot{\boldsymbol{q}}, \widetilde{\boldsymbol{q}} \neq \boldsymbol{0} \tag{6.39}$$

where $\boldsymbol{K}_P$ is an $(n \times n)$ symmetric positive definite matrix. An energy-based interpretation of (6.39) reveals a first term expressing the system kinetic energy and a second term expressing the potential energy stored in the system of equivalent stiffness $\boldsymbol{K}_P$ provided by the n position feedback loops.

Differentiating (6.39) with respect to time, and recalling that $\boldsymbol{q}_d$ is constant, yields

$$\dot{V} = \dot{\boldsymbol{q}}^T \boldsymbol{B}(\boldsymbol{q})\ddot{\boldsymbol{q}} + \frac{1}{2}\dot{\boldsymbol{q}}^T \dot{\boldsymbol{B}}(\boldsymbol{q})\dot{\boldsymbol{q}} - \dot{\boldsymbol{q}}^T \boldsymbol{K}_P \widetilde{\boldsymbol{q}}. \tag{6.40}$$

Solving (6.35) for $\boldsymbol{B}\ddot{\boldsymbol{q}}$ and substituting it in (6.40) gives

$$\dot{V} = \frac{1}{2}\dot{\boldsymbol{q}}^T \big(\dot{\boldsymbol{B}}(\boldsymbol{q}) - 2\boldsymbol{C}(\boldsymbol{q}, \dot{\boldsymbol{q}})\big)\dot{\boldsymbol{q}} - \dot{\boldsymbol{q}}^T \boldsymbol{F}\dot{\boldsymbol{q}} + \dot{\boldsymbol{q}}^T \big(\boldsymbol{u} - \boldsymbol{g}(\boldsymbol{q}) - \boldsymbol{K}_P \widetilde{\boldsymbol{q}}\big). \tag{6.41}$$

The first term on the right-hand side is null since the matrix $\boldsymbol{N} = \dot{\boldsymbol{B}} - 2\boldsymbol{C}$ satisfies (4.48). The second term is negative definite. Then, the choice

$$\boldsymbol{u} = \boldsymbol{g}(\boldsymbol{q}) + \boldsymbol{K}_P \widetilde{\boldsymbol{q}}, \tag{6.42}$$

describing a controller with compensation of gravitational terms and a proportional action, leads to a negative semi-definite $\dot{V}$ since

$$\dot{V} = 0 \qquad \dot{\boldsymbol{q}} = \boldsymbol{0}, \forall \widetilde{\boldsymbol{q}}.$$

This result can be obtained also by taking the control law

$$\boldsymbol{u} = \boldsymbol{g}(\boldsymbol{q}) + \boldsymbol{K}_P \widetilde{\boldsymbol{q}} - \boldsymbol{K}_D \dot{\boldsymbol{q}}, \tag{6.43}$$

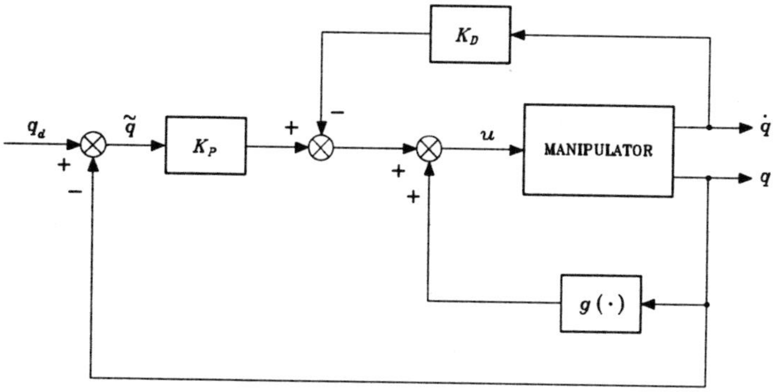

Figure 6.22 Block scheme of joint space PD control with gravity compensation.

with $\boldsymbol{K}_D$ positive definite, corresponding to a *nonlinear compensation action of gravitational terms* with a *linear proportional-derivative* (PD) *action*. In fact, substituting (6.43) into (6.41) gives

$$\dot{V} = -\dot{\boldsymbol{q}}^T(\boldsymbol{F} + \boldsymbol{K}_D)\dot{\boldsymbol{q}}, \tag{6.44}$$

which reveals that the introduction of the derivative term causes an increase of the absolute values of $\dot{V}$ along the system trajectories, and then it gives an improvement of system time response. Notice that the inclusion of a derivative action in the controller, as in (6.43), is crucial when direct-drive manipulators are considered. In that case, in fact, mechanical viscous damping is practically null, and current control does not allow exploiting the electrical viscous damping provided by voltage-controlled actuators.

According to the above, the function candidate V decreases as long as $\dot{\boldsymbol{q}} \neq \boldsymbol{0}$ for all system trajectories. It can be shown that the system reaches an *equilibrium posture*. To find such posture, notice that $\dot{V} \equiv 0$ only if $\dot{\boldsymbol{q}} \equiv \boldsymbol{0}$. System dynamics under control (6.43) is given by

$$\boldsymbol{B}(\boldsymbol{q})\ddot{\boldsymbol{q}} + \boldsymbol{C}(\boldsymbol{q}, \dot{\boldsymbol{q}})\dot{\boldsymbol{q}} + \boldsymbol{F}\dot{\boldsymbol{q}} + \boldsymbol{g}(\boldsymbol{q}) = \boldsymbol{g}(\boldsymbol{q}) + \boldsymbol{K}_P\tilde{\boldsymbol{q}} - \boldsymbol{K}_D\dot{\boldsymbol{q}}. \tag{6.45}$$

At the equilibrium ($\dot{\boldsymbol{q}} \equiv \boldsymbol{0}$, $\ddot{\boldsymbol{q}} \equiv \boldsymbol{0}$) it is

$$\boldsymbol{K}_P\tilde{\boldsymbol{q}} = \boldsymbol{0}, \tag{6.46}$$

and then

$$\tilde{\boldsymbol{q}} = \boldsymbol{q}_d - \boldsymbol{q} \equiv \boldsymbol{0}$$

is the sought equilibrium posture. The above derivation rigorously shows that any manipulator equilibrium posture is *globally asymptotically stable* under a controller with a PD linear action and a nonlinear gravity compensating action. Stability is ensured for any choice of $\boldsymbol{K}_P$ and $\boldsymbol{K}_D$, as long as these are positive definite matrices. The resulting block scheme is shown in Figure 6.22.

The control law requires the on-line computation of the term $\boldsymbol{g}(\boldsymbol{q})$. If compensation is imperfect, the above discussion does not lead to the same result; this aspect will

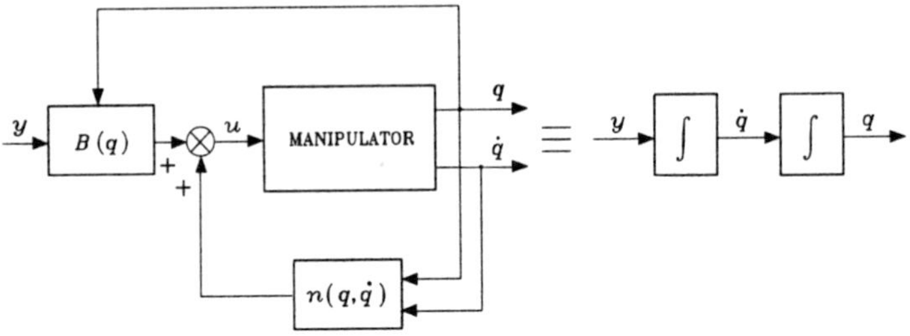

Figure 6.23 Global linearization performed by inverse dynamics control.

be revisited later with reference to robustness of controllers performing nonlinear compensation.

6.5.2 Inverse Dynamics Control

Consider now the problem of tracking a joint space trajectory. The reference framework is that of control of nonlinear multivariable systems. The dynamic model of an n-joint manipulator is expressed by (6.35) which can be rewritten as

$$\boldsymbol{B}(\boldsymbol{q})\ddot{\boldsymbol{q}} + \boldsymbol{n}(\boldsymbol{q}, \dot{\boldsymbol{q}}) = \boldsymbol{u}, \tag{6.47}$$

where for simplicity it has been set

$$\boldsymbol{n}(\boldsymbol{q}, \dot{\boldsymbol{q}}) = \boldsymbol{C}(\boldsymbol{q}, \dot{\boldsymbol{q}})\dot{\boldsymbol{q}} + \boldsymbol{F}\dot{\boldsymbol{q}} + \boldsymbol{g}(\boldsymbol{q}). \tag{6.48}$$

The approach that follows is founded on the idea to find a control vector $\boldsymbol{u}$, as a function of the system state, which is capable to realize an input/output relationship of linear type; in other words, it is desired to perform not a local linearization but a *global linearization* of system dynamics obtained by means of a *nonlinear state feedback*. The possibility of finding such a linearizing controller is guaranteed by the particular form of system dynamics. In fact, the equation in (6.47) is linear in the control $\boldsymbol{u}$ and has a full-rank matrix $\boldsymbol{B}(\boldsymbol{q})$ which can be inverted for any manipulator configuration.

Taking the control $\boldsymbol{u}$ as a function of the manipulator state in the form

$$\boldsymbol{u} = \boldsymbol{B}(\boldsymbol{q})\boldsymbol{y} + \boldsymbol{n}(\boldsymbol{q}, \dot{\boldsymbol{q}}) \tag{6.49}$$

leads to the system described by

$$\ddot{\boldsymbol{q}} = \boldsymbol{y},$$

where $\boldsymbol{y}$ represents a new input vector whose expression is to be determined yet; the resulting block scheme is shown in Figure 6.23. The nonlinear control law in (6.49) is termed *inverse dynamics control* since it is based on the computation of manipulator inverse dynamics. The system under control (6.49) is *linear* and *decoupled* with respect to the new input $\boldsymbol{y}$. In other words, the component y_i influences, with a double integrator relationship, only the joint variable q_i, independently of the motion of the other joints.

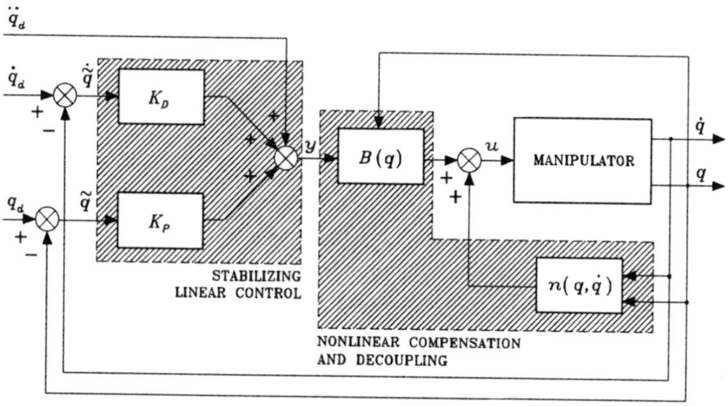

Figure 6.24 Block scheme of joint space inverse dynamics control.

In view of the choice (6.49), the manipulator control problem is reduced to that of finding a stabilizing control law $\boldsymbol{y}$. To this purpose, the choice

$$\boldsymbol{y} = -\boldsymbol{K}_P\boldsymbol{q} - \boldsymbol{K}_D\dot{\boldsymbol{q}} + \boldsymbol{r} \tag{6.50}$$

leads to the system of second-order equations

$$\ddot{\boldsymbol{q}} + \boldsymbol{K}_D\dot{\boldsymbol{q}} + \boldsymbol{K}_P\boldsymbol{q} = \boldsymbol{r} \tag{6.51}$$

which, on the assumption of positive definite matrices $\boldsymbol{K}_P$ and $\boldsymbol{K}_D$, is asymptotically stable. Choosing $\boldsymbol{K}_P$ and $\boldsymbol{K}_D$ as *diagonal* matrices of the type

$$\boldsymbol{K}_P = \text{diag}\{\omega_{n1}^2, \dots, \omega_{nn}^2\} \qquad \boldsymbol{K}_D = \text{diag}\{2\zeta_1\omega_{n1}, \dots, 2\zeta_n\omega_{nn}\},$$

gives a decoupled system. The reference component r_i influences only the joint variable q_i with a second-order input/output relationship characterized by a natural frequency ω_{ni} and a damping ratio ζ_i.

Given any desired trajectory $\boldsymbol{q}_d(t)$, tracking of this trajectory for the output $\boldsymbol{q}(t)$ is ensured by choosing

$$\boldsymbol{r} = \ddot{\boldsymbol{q}}_d + \boldsymbol{K}_D\dot{\boldsymbol{q}}_d + \boldsymbol{K}_P\boldsymbol{q}_d. \tag{6.52}$$

In fact, substituting (6.52) into (6.51) gives the homogeneous second-order differential equation

$$\ddot{\tilde{\boldsymbol{q}}} + \boldsymbol{K}_D\dot{\tilde{\boldsymbol{q}}} + \boldsymbol{K}_P\tilde{\boldsymbol{q}} = \boldsymbol{0} \tag{6.53}$$

expressing the dynamics of position error (6.38) while tracking the given trajectory. Such error occurs only if $\tilde{\boldsymbol{q}}(0)$ and/or $\dot{\tilde{\boldsymbol{q}}}(0)$ are different from zero and converges to zero with a speed depending on the matrices $\boldsymbol{K}_P$ and $\boldsymbol{K}_D$ chosen.

The resulting block scheme is illustrated in Figure 6.24, in which two feedback loops are represented; an inner loop based on the manipulator dynamic model and an outer loop operating on the tracking error. The function of the *inner loop* is to obtain a *linear and decoupled input/output relationship*, whereas the *outer loop* is required to *stabilize the overall system*. The controller design for the outer loop is simplified since it operates on a linear and time-invariant system. Notice that the implementation of this control scheme requires computation of the inertia matrix $\boldsymbol{B}(\boldsymbol{q})$ and of the vector of Coriolis, centrifugal, gravitational, and damping terms $\boldsymbol{n}(\boldsymbol{q}, \dot{\boldsymbol{q}})$ in (6.48). Differently from computed torque control, these terms must be computed *on line* since control is now based on nonlinear feedback of the current system state, and thus it is not possible to precompute the terms off line as for the previous technique.

The above technique of nonlinear compensation and decoupling is very attractive from a control viewpoint since the nonlinear and coupled manipulator dynamics is replaced with n linear and decoupled second-order subsystems. Nonetheless, this technique is based on the assumption of perfect cancellation of dynamic terms, and then it is quite natural to raise questions about sensitivity and robustness problems due to unavoidably imperfect compensation.

Implementation of inverse dynamics control laws indeed requires that parameters of the system dynamic model are accurately known and the complete equations of motion are computed in real time. These conditions are difficult to verify in practice. On one hand, the model is usually known with a certain degree of uncertainty due to imperfect knowledge of manipulator mechanical parameters, existence of unmodeled dynamics, and model dependence on end-effector payloads not exactly known and thus not perfectly compensated. On the other hand, inverse dynamics computation is to be performed at sampling times of the order of the millisecond so as to ensure that the assumption of operating in the continuous time domain is realistic. This may pose severe constraints on the hardware/software architecture of the control system. In such cases, it may be advisable to lighten the computation of inverse dynamics and compute only the dominant terms.

On the basis of the above remarks, from an implementation viewpoint, *compensation* may be *imperfect* both for model uncertainty and for the approximations made in on-line computation of inverse dynamics. In the following, two control techniques are presented which are aimed at counteracting the effects of imperfect compensation. The first one consists of the introduction of an additional term to an inverse dynamics controller which provides *robustness* to the control system by counteracting the effects of the approximations made in on-line computation of inverse dynamics. The second one *adapts* the parameters of the model used for inverse dynamics computation to those of the true manipulator dynamic model.

6.5.3 Robust Control

In the case of *imperfect compensation*, it is reasonable to assume in (6.47) a control vector expressed by

$$\boldsymbol{u} = \widehat{\boldsymbol{B}}(\boldsymbol{q})\boldsymbol{y} + \widehat{\boldsymbol{n}}(\boldsymbol{q}, \dot{\boldsymbol{q}}) \tag{6.54}$$

where $\widehat{\boldsymbol{B}}$ and $\widehat{\boldsymbol{n}}$ represent the adopted computational model in terms of estimates of the terms in the dynamic model. The error on the estimates, *i.e.*, the *uncertainty*, is represented by

$$\widetilde{\boldsymbol{B}} = \widehat{\boldsymbol{B}} - \boldsymbol{B} \qquad\qquad \widetilde{\boldsymbol{n}} = \widehat{\boldsymbol{n}} - \boldsymbol{n} \tag{6.55}$$

and is due to imperfect model compensation as well as to intentional simplification in inverse dynamics computation. Notice that by setting $\widehat{\boldsymbol{B}} = \bar{\boldsymbol{B}}$ (where $\bar{\boldsymbol{B}}$ is the diagonal matrix of average inertia at the joint axes) and $\widehat{\boldsymbol{n}} = \boldsymbol{0}$, the above decentralized control scheme is recovered where the control action $\boldsymbol{y}$ can be of the general PID type computed on the error.

Using (6.54) as a nonlinear control law gives

$$\boldsymbol{B}\ddot{\boldsymbol{q}} + \boldsymbol{n} = \widehat{\boldsymbol{B}}\boldsymbol{y} + \widehat{\boldsymbol{n}} \tag{6.56}$$

where functional dependence has been omitted. Since the inertia matrix $\boldsymbol{B}$ is invertible, it is

$$\ddot{\boldsymbol{q}} = \boldsymbol{y} + (\boldsymbol{B}^{-1}\widehat{\boldsymbol{B}} - \boldsymbol{I})\boldsymbol{y} + \boldsymbol{B}^{-1}\widetilde{\boldsymbol{n}} = \boldsymbol{y} - \boldsymbol{\eta} \tag{6.57}$$

where

$$\boldsymbol{\eta} = (\boldsymbol{I} - \boldsymbol{B}^{-1}\widehat{\boldsymbol{B}})\boldsymbol{y} - \boldsymbol{B}^{-1}\widetilde{\boldsymbol{n}}. \tag{6.58}$$

Taking as above

$$\boldsymbol{y} = \ddot{\boldsymbol{q}}_d + \boldsymbol{K}_D(\dot{\boldsymbol{q}}_d - \dot{\boldsymbol{q}}) + \boldsymbol{K}_P(\boldsymbol{q}_d - \boldsymbol{q})$$

leads to

$$\ddot{\widetilde{\boldsymbol{q}}} + \boldsymbol{K}_D\dot{\widetilde{\boldsymbol{q}}} + \boldsymbol{K}_P\widetilde{\boldsymbol{q}} = \boldsymbol{\eta}. \tag{6.59}$$

The system described by (6.59) is still nonlinear and coupled, since $\boldsymbol{\eta}$ is a nonlinear function of $\widetilde{\boldsymbol{q}}$ and $\dot{\widetilde{\boldsymbol{q}}}$; error convergence to zero is not ensured by the term on the left-hand side only.

To find control laws ensuring error convergence to zero while tracking a trajectory even in the face of uncertainties, a linear PD control is no longer sufficient. To this purpose, the Lyapunov direct method can be utilized again for the design of an outer feedback loop on the error which be *robust* to the uncertainty $\boldsymbol{\eta}$.

Let the desired trajectory $\boldsymbol{q}_d(t)$ be assigned in the joint space and let $\widetilde{\boldsymbol{q}} = \boldsymbol{q}_d - \boldsymbol{q}$ be the position error. Its first time-derivative is $\dot{\widetilde{\boldsymbol{q}}} = \dot{\boldsymbol{q}}_d - \dot{\boldsymbol{q}}$, while its second time-derivative in view of (6.57) is

$$\ddot{\widetilde{\boldsymbol{q}}} = \ddot{\boldsymbol{q}}_d - \boldsymbol{y} + \boldsymbol{\eta}. \tag{6.60}$$

By taking

$$\boldsymbol{\xi} = \begin{bmatrix} \widetilde{\boldsymbol{q}} \\ \dot{\widetilde{\boldsymbol{q}}} \end{bmatrix} \tag{6.61}$$

as the system state, the following first-order differential matrix equation is obtained:

$$\dot{\boldsymbol{\xi}} = \boldsymbol{H}\boldsymbol{\xi} + \boldsymbol{D}(\ddot{\boldsymbol{q}}_d - \boldsymbol{y} + \boldsymbol{\eta}), \tag{6.62}$$

where $\boldsymbol{H}$ and $\boldsymbol{D}$ are block matrices of dimensions $(2n \times 2n)$ and $(2n \times n)$, respectively:

$$\boldsymbol{H} = \begin{bmatrix} \boldsymbol{O} & \boldsymbol{I} \\ \boldsymbol{O} & \boldsymbol{O} \end{bmatrix} \qquad \boldsymbol{D} = \begin{bmatrix} \boldsymbol{O} \\ \boldsymbol{I} \end{bmatrix}. \tag{6.63}$$

Then, the problem of tracking a given trajectory can be regarded as the problem of finding a control law $\boldsymbol{y}$ which stabilizes the nonlinear time-varying error system (6.62).

Control design is based on the assumption that, even though the uncertainty η is unknown, an estimate on its range of variation is available. The sought control law $\boldsymbol{y}$ shall guarantee asymptotic stability for any η varying in the above range. By recalling that η in (6.58) is a function of $\boldsymbol{q}$, $\dot{\boldsymbol{q}}$, $\ddot{\boldsymbol{q}}_d$, the following assumptions are made:

$$\sup_{t\geq 0} \|\ddot{\boldsymbol{q}}_d\| < Q_M < \infty \qquad \forall \ddot{\boldsymbol{q}}_d \tag{6.64}$$

$$\|\boldsymbol{I} - \boldsymbol{B}^{-1}(\boldsymbol{q})\widehat{\boldsymbol{B}}(\boldsymbol{q})\| \leq \alpha \leq 1 \quad \forall \boldsymbol{q} \tag{6.65}$$

$$\|\widetilde{\boldsymbol{n}}\| \leq \Phi < \infty \qquad \forall \boldsymbol{q}, \dot{\boldsymbol{q}}. \tag{6.66}$$

Assumption (6.64) is practically satisfied since any planned trajectory cannot require infinite accelerations.

Regarding assumption (6.65), since $\boldsymbol{B}$ is a positive definite matrix with upper and lower limited norms, the following inequality holds:

$$0 < B_m \leq \|\boldsymbol{B}^{-1}(\boldsymbol{q})\| \leq B_M < \infty \qquad \forall \boldsymbol{q}, \tag{6.67}$$

and then a choice for $\widehat{\boldsymbol{B}}$ always exists which satisfies (6.65). In fact, by setting

$$\widehat{\boldsymbol{B}} = \frac{2}{B_M + B_m}\boldsymbol{I},$$

from (6.65) it is

$$\|\boldsymbol{B}^{-1}\widehat{\boldsymbol{B}} - \boldsymbol{I}\| \leq \frac{B_M - B_m}{B_M + B_m} = \alpha < 1. \tag{6.68}$$

If $\widehat{\boldsymbol{B}}$ is a more accurate estimate of the inertia matrix, the inequality is satisfied with values of α that can be made arbitrarily small (in the limit, it is $\widehat{\boldsymbol{B}} = \boldsymbol{B}$ and $\alpha = 0$).

Finally, concerning assumption (6.66), observe that $\widetilde{\boldsymbol{n}}$ is a function of $\boldsymbol{q}$ and $\dot{\boldsymbol{q}}$. For revolute joints a periodical dependence on $\boldsymbol{q}$ is obtained, while for prismatic joints a linear dependence is obtained, but the joint ranges are limited and then the above contribution is also limited. On the other hand, regarding the dependence on $\dot{\boldsymbol{q}}$, unbounded velocities for an unstable system may arise in the limit, but in reality saturations exist on the maximum velocities of the motors. In sum, assumption (6.66) can be realistically satisfied, too.

With reference to (6.57), choose now

$$\boldsymbol{y} = \ddot{\boldsymbol{q}}_d + \boldsymbol{K}_D\dot{\widetilde{\boldsymbol{q}}} + \boldsymbol{K}_P\widetilde{\boldsymbol{q}} + \boldsymbol{w} \tag{6.69}$$

where the PD term ensures stabilization of the error dynamic system matrix, $\ddot{\boldsymbol{q}}_d$ provides a feedforward term, and the term $\boldsymbol{w}$ is to be chosen to guarantee robustness to the effects of uncertainty described by η in (6.58).

Using (6.69) and setting $\boldsymbol{K} = [\,\boldsymbol{K}_P \quad \boldsymbol{K}_D\,]$ yields

$$\dot{\boldsymbol{\xi}} = \widetilde{\boldsymbol{H}}\boldsymbol{\xi} + \boldsymbol{D}(\eta - \boldsymbol{w}), \tag{6.70}$$

where

$$\widetilde{\boldsymbol{H}} = (\boldsymbol{H} - \boldsymbol{D}\boldsymbol{K}) = \begin{bmatrix} \boldsymbol{O} & \boldsymbol{I} \\ -\boldsymbol{K}_P & -\boldsymbol{K}_D \end{bmatrix}$$

is a matrix whose eigenvalues all have negative real parts—being $\boldsymbol{K}_P$ and $\boldsymbol{K}_D$ positive definite—which allows prescribing the desired error system dynamics. In fact, by choosing $\boldsymbol{K}_P = \text{diag}\{\omega_{n1}^2, \ldots, \omega_{nn}^2\}$ and $\boldsymbol{K}_D = \text{diag}\{2\zeta_1\omega_{n1}, \ldots, 2\zeta_n\omega_{nn}\}$, n decoupled equations are obtained as regards the linear part. If the uncertainty term vanishes, it is obviously $\boldsymbol{w} = \boldsymbol{0}$ and the above result with an exact inverse dynamics controller is recovered ($\widehat{\boldsymbol{B}} = \boldsymbol{B}$ and $\widehat{\boldsymbol{n}} = \boldsymbol{n}$).

To determine $\boldsymbol{w}$, consider the following positive definite quadratic form as Lyapunov function candidate:

$$V(\boldsymbol{\xi}) = \boldsymbol{\xi}^T \boldsymbol{Q} \boldsymbol{\xi} > 0 \qquad \forall \boldsymbol{\xi} \neq \boldsymbol{0}, \tag{6.71}$$

where $\boldsymbol{Q}$ is a $(2n \times 2n)$ positive definite matrix. The derivative of V along the trajectories of the error system (6.70) is

$$\begin{aligned} \dot{V} &= \dot{\boldsymbol{\xi}}^T \boldsymbol{Q} \boldsymbol{\xi} + \boldsymbol{\xi}^T \boldsymbol{Q} \dot{\boldsymbol{\xi}} \\ &= \boldsymbol{\xi}^T (\widetilde{\boldsymbol{H}}^T \boldsymbol{Q} + \boldsymbol{Q} \widetilde{\boldsymbol{H}}) \boldsymbol{\xi} + 2\boldsymbol{\xi}^T \boldsymbol{Q} \boldsymbol{D} (\boldsymbol{\eta} - \boldsymbol{w}). \end{aligned} \tag{6.72}$$

Since $\widetilde{\boldsymbol{H}}$ has eigenvalues with all negative real parts, it is well-known that for any symmetric positive definite matrix $\boldsymbol{P}$, the equation

$$\widetilde{\boldsymbol{H}}^T \boldsymbol{Q} + \boldsymbol{Q} \widetilde{\boldsymbol{H}} = -\boldsymbol{P} \tag{6.73}$$

gives a unique solution $\boldsymbol{Q}$ which is symmetric positive definite as well. In view of this, (6.72) becomes

$$\dot{V} = -\boldsymbol{\xi}^T \boldsymbol{P} \boldsymbol{\xi} + 2\boldsymbol{\xi}^T \boldsymbol{Q} \boldsymbol{D} (\boldsymbol{\eta} - \boldsymbol{w}). \tag{6.74}$$

The first term on the right-hand side of (6.74) is negative definite and then the solutions converge if $\boldsymbol{\xi} \in \mathcal{N}(\boldsymbol{D}^T \boldsymbol{Q})$. If instead $\boldsymbol{\xi} \notin \mathcal{N}(\boldsymbol{D}^T \boldsymbol{Q})$, the control $\boldsymbol{w}$ must be chosen so as to render the second term in (6.74) less than or equal to zero. By setting $\boldsymbol{z} = \boldsymbol{D}^T \boldsymbol{Q} \boldsymbol{\xi}$, the second term in (6.74) can be rewritten as $\boldsymbol{z}^T (\boldsymbol{\eta} - \boldsymbol{w})$. Adopting the control law

$$\boldsymbol{w} = \frac{\rho}{\|\boldsymbol{z}\|} \boldsymbol{z} \qquad \rho > 0 \tag{6.75}$$

gives[5]

$$\begin{aligned} \boldsymbol{z}^T (\boldsymbol{\eta} - \boldsymbol{w}) &= \boldsymbol{z}^T \boldsymbol{\eta} - \frac{\rho}{\|\boldsymbol{z}\|} \boldsymbol{z}^T \boldsymbol{z} \\ &\leq \|\boldsymbol{z}\| \|\boldsymbol{\eta}\| - \rho \|\boldsymbol{z}\| \\ &= \|\boldsymbol{z}\| (\|\boldsymbol{\eta}\| - \rho). \end{aligned} \tag{6.76}$$

[5] Notice that it is necessary to divide $\boldsymbol{z}$ by the norm of $\boldsymbol{z}$ so as to obtain a linear dependence on $\boldsymbol{z}$ of the term containing the control $\boldsymbol{z}^T \boldsymbol{w}$, and thus to effectively counteract, for $\boldsymbol{z} \to \boldsymbol{0}$, the term containing the uncertainty $\boldsymbol{z}^T \boldsymbol{\eta}$ which is linear in $\boldsymbol{z}$.

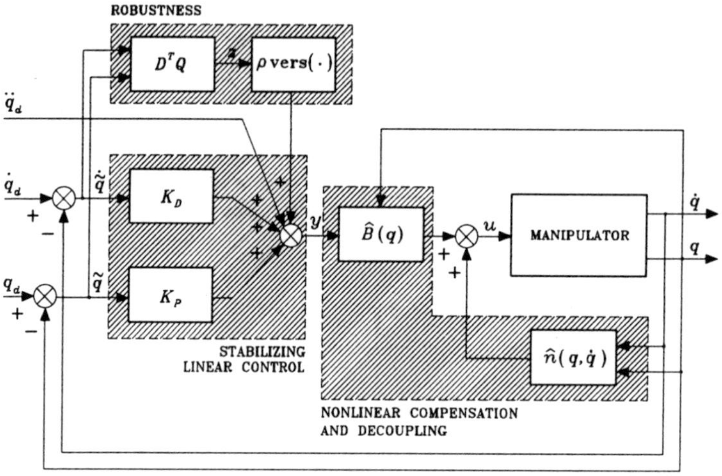

Figure 6.25 Block scheme of joint space robust control.

Then, if ρ is chosen so that

$$\rho \geq \|\boldsymbol{\eta}\| \qquad \forall \boldsymbol{q}, \dot{\boldsymbol{q}}, \ddot{\boldsymbol{q}}_d, \tag{6.77}$$

control (6.75) ensures that $\dot{V}$ is less than zero along all error system trajectories.

In order to satisfy (6.77), notice that, in view of the definition of $\boldsymbol{\eta}$ in (6.58) and of assumptions (6.64)–(6.66), and being $\|\boldsymbol{w}\| = \rho$, it is

$$\begin{aligned} \|\boldsymbol{\eta}\| &\leq \|\boldsymbol{I} - \boldsymbol{B}^{-1}\widehat{\boldsymbol{B}}\|\big(\|\ddot{\boldsymbol{q}}_d\| + \|\boldsymbol{K}\|\,\|\boldsymbol{\xi}\| + \|\boldsymbol{w}\|\big) + \|\boldsymbol{B}^{-1}\|\,\|\widetilde{\boldsymbol{n}}\| \\ &\leq \alpha Q_M + \alpha\|\boldsymbol{K}\|\,\|\boldsymbol{\xi}\| + \alpha\rho + B_M\Phi. \end{aligned} \tag{6.78}$$

Therefore, setting

$$\rho \geq \frac{1}{1-\alpha}\big(\alpha Q_M + \alpha\|\boldsymbol{K}\|\|\boldsymbol{\xi}\| + B_M\Phi\big) \tag{6.79}$$

gives

$$\dot{V} = -\boldsymbol{\xi}^T\boldsymbol{P}\boldsymbol{\xi} + 2\boldsymbol{z}^T\left(\boldsymbol{\eta} - \frac{\rho}{\|\boldsymbol{z}\|}\boldsymbol{z}\right) < 0 \qquad \forall\boldsymbol{\xi} \neq \boldsymbol{0}. \tag{6.80}$$

The resulting block scheme is illustrated in Figure 6.25.

To summarize, the presented approach has lead to finding a *control* law which is formed by three different contributions:

- The term $\widehat{\boldsymbol{B}}\boldsymbol{y} + \widehat{\boldsymbol{n}}$ ensures *an approximate compensation of nonlinear effects and joint decoupling*.
- The term $\ddot{\boldsymbol{q}}_d + \boldsymbol{K}_D\dot{\widetilde{\boldsymbol{q}}} + \boldsymbol{K}_P\widetilde{\boldsymbol{q}}$ introduces *a linear feedforward action* ($\ddot{\boldsymbol{q}}_d + \boldsymbol{K}_D\dot{\boldsymbol{q}}_d + \boldsymbol{K}_P\boldsymbol{q}_d$) and *a linear feedback action* ($-\boldsymbol{K}_D\dot{\boldsymbol{q}} - \boldsymbol{K}_P\boldsymbol{q}$) *which stabilizes the error system dynamics.*

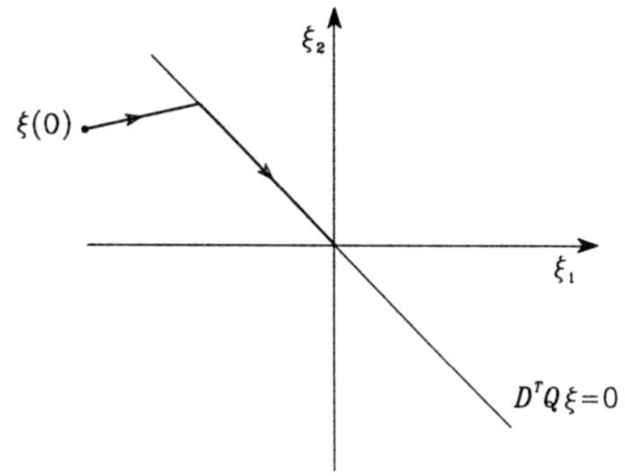

Figure 6.26 Error trajectory with robust control.

- The term $\boldsymbol{w} = (\rho/\|\boldsymbol{z}\|)\boldsymbol{z}$ represents the *robust contribution that counteracts the indeterminacy* $\widetilde{\boldsymbol{B}}$ and $\widetilde{\boldsymbol{n}}$ in computing the nonlinear terms that depend on the manipulator state; the greater the uncertainty, the greater the positive scalar ρ. The resulting control law is of the *unit vector* type, since it is described by a vector of magnitude ρ aligned with the unit vector of $\boldsymbol{z} = \boldsymbol{D}^T\boldsymbol{Q}\boldsymbol{\xi}$, $\forall\boldsymbol{\xi}$.

All the resulting trajectories under the above robust control reach the subspace $\boldsymbol{z} = \boldsymbol{D}^T\boldsymbol{Q}\boldsymbol{\xi} = \boldsymbol{0}$ that depends on the matrix $\boldsymbol{Q}$ in the Lyapunov function V. On this *attractive* subspace, termed *sliding subspace*, the control $\boldsymbol{w}$ is ideally commuted at an infinite frequency and all error components tend to zero with a transient depending on the matrices $\boldsymbol{Q}$, $\boldsymbol{K}_P$, $\boldsymbol{K}_D$. A characterization of an error trajectory in the two-dimensional case is given in Figure 6.26. Notice that in the case $\boldsymbol{\xi}(0) \neq \boldsymbol{0}$, with $\boldsymbol{\xi}(0) \notin \mathcal{N}(\boldsymbol{D}^T\boldsymbol{Q})$, the trajectory is attracted on the sliding hyperplane (a line) $\boldsymbol{z} = \boldsymbol{0}$ and tends towards the origin of the error state space with a time evolution governed by ρ.

In reality, the physical limits on the elements employed in the controller impose a control signal that commutes at a finite frequency, and the trajectories oscillate around the sliding subspace with a magnitude as low as the frequency is high.

Elimination of these high-frequency components (*chattering*) can be achieved by adopting a robust control law which, even if it does not guarantee error convergence to zero, ensures *bounded-norm* errors. A control law of this type is

$$\boldsymbol{w} = \begin{cases} \dfrac{\rho}{\|\boldsymbol{z}\|}\boldsymbol{z} & \text{for } \|\boldsymbol{z}\| \geq \epsilon \\[2ex] \dfrac{\rho}{\epsilon}\boldsymbol{z} & \text{for } \|\boldsymbol{z}\| < \epsilon. \end{cases} \tag{6.81}$$

In order to provide an intuitive interpretation of this law, notice that (6.81) gives a null control input when the error is in the null space of matrix $\boldsymbol{D}^T\boldsymbol{Q}$. On the other hand, (6.75) has an equivalent gain tending to infinity when $\boldsymbol{z}$ tends to the null vector, thus generating a control input of limited magnitude. Since these inputs commute at an infinite frequency, they force the error system dynamics to stay on the sliding subspace. With reference to the above example, control law (6.81) gives rise to an hyperplane

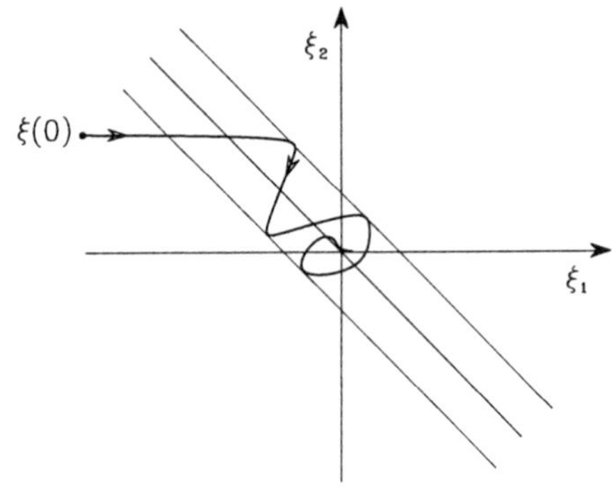

Figure 6.27 Error trajectory with robust control and chattering elimination.

$z = 0$ which is no longer attractive, and the error is allowed to vary within a boundary layer whose thickness depends on ϵ (Figure 6.27).

The introduction of a contribution based on the computation of a suitable linear combination of the generalized error confers robustness to a control scheme based on nonlinear compensation. Even if the manipulator is accurately modeled, indeed, an exact nonlinear compensation may be computationally demanding, and thus it may require either a sophisticated hardware architecture or an increase of the sampling time needed to compute the control law. The solution then becomes weak from an engineering viewpoint, due either to infeasible costs of the control architecture or to poor performance at decreased sampling rates. Therefore, considering a partial knowledge of the manipulator dynamic model with an accurate, pondered estimate of uncertainty may suggest robust control solutions of the kind presented above. It is understood that an estimate of the uncertainty shall be found so as to impose control inputs which the mechanical structure can bear.

6.5.4 Adaptive Control

The computational model employed for computing inverse dynamics typically has the same structure as that of the true manipulator dynamic model, but parameter estimate uncertainty does exist. In this case, it is possible to devise solutions that allow *adapting on line the computational model to the dynamic model*, thus performing a control scheme of the inverse dynamics type.

The possibility of finding adaptive control laws is ensured by the property of *linearity in the parameters* of the dynamic model of a manipulator. In fact, it is always possible to express the nonlinear equations of motion in a linear form with respect to a suitable set of constant dynamic parameters as in (4.80). The equation in (6.35) can then be written as

$$\boldsymbol{B}(\boldsymbol{q})\ddot{\boldsymbol{q}} + \boldsymbol{C}(\boldsymbol{q},\dot{\boldsymbol{q}})\dot{\boldsymbol{q}} + \boldsymbol{F}\dot{\boldsymbol{q}} + \boldsymbol{g}(\boldsymbol{q}) = \boldsymbol{Y}(\boldsymbol{q},\dot{\boldsymbol{q}},\ddot{\boldsymbol{q}})\boldsymbol{\pi} = \boldsymbol{u}, \tag{6.82}$$

where $\boldsymbol{\pi}$ is a $(p \times 1)$ vector of constant parameters and $\boldsymbol{Y}$ is an $(n \times p)$ matrix which is a function of joint positions, velocities and accelerations. This property of linearity in the dynamic parameters is fundamental for deriving adaptive control laws, among which the technique illustrated below is one of the simplest.

At first, a control scheme which can be derived through a combined computed torque/inverse dynamics approach is illustrated. The computational model is assumed to coincide with the dynamic model.

Consider the control law

$$\boldsymbol{u} = \boldsymbol{B}(\boldsymbol{q})\ddot{\boldsymbol{q}}_r + \boldsymbol{C}(\boldsymbol{q}, \dot{\boldsymbol{q}})\dot{\boldsymbol{q}}_r + \boldsymbol{F}\dot{\boldsymbol{q}}_r + \boldsymbol{g}(\boldsymbol{q}) + \boldsymbol{K}_D\boldsymbol{\sigma}, \tag{6.83}$$

with $\boldsymbol{K}_D$ a positive definite matrix. The choice

$$\dot{\boldsymbol{q}}_r = \dot{\boldsymbol{q}}_d + \boldsymbol{\Lambda}\widetilde{\boldsymbol{q}} \qquad \ddot{\boldsymbol{q}}_r = \ddot{\boldsymbol{q}}_d + \boldsymbol{\Lambda}\dot{\widetilde{\boldsymbol{q}}}, \tag{6.84}$$

with $\boldsymbol{\Lambda}$ a positive definite (usually diagonal) matrix, allows expressing the nonlinear compensation and decoupling terms as a function of the desired velocity and acceleration, corrected by the current state ($\boldsymbol{q}$ and $\dot{\boldsymbol{q}}$) of the manipulator. In fact, notice that the term $\dot{\boldsymbol{q}}_r = \dot{\boldsymbol{q}}_d + \boldsymbol{\Lambda}\widetilde{\boldsymbol{q}}$ weighs the contribution that depends on velocity not only on the basis of the desired velocity but also on the basis of the position tracking error. A similar argument holds also for the acceleration contribution, where a term depending on the velocity tracking error is considered besides the desired acceleration.

The term $\boldsymbol{K}_D\boldsymbol{\sigma}$ is equivalent to a PD action on the error if $\boldsymbol{\sigma}$ is taken as

$$\boldsymbol{\sigma} = \dot{\boldsymbol{q}}_r - \dot{\boldsymbol{q}} = \dot{\widetilde{\boldsymbol{q}}} + \boldsymbol{\Lambda}\widetilde{\boldsymbol{q}}. \tag{6.85}$$

Substituting (6.83) into (6.82) and accounting for (6.85) yields

$$\boldsymbol{B}(\boldsymbol{q})\dot{\boldsymbol{\sigma}} + \boldsymbol{C}(\boldsymbol{q}, \dot{\boldsymbol{q}})\boldsymbol{\sigma} + \boldsymbol{F}\boldsymbol{\sigma} + \boldsymbol{K}_D\boldsymbol{\sigma} = \boldsymbol{0}. \tag{6.86}$$

Consider the Lyapunov function candidate

$$V(\boldsymbol{\sigma}, \widetilde{\boldsymbol{q}}) = \frac{1}{2}\boldsymbol{\sigma}^T\boldsymbol{B}(\boldsymbol{q})\boldsymbol{\sigma} + \frac{1}{2}\widetilde{\boldsymbol{q}}^T\boldsymbol{M}\widetilde{\boldsymbol{q}} > 0 \qquad \forall \boldsymbol{\sigma}, \widetilde{\boldsymbol{q}} \neq \boldsymbol{0}, \tag{6.87}$$

where $\boldsymbol{M}$ is an $(n \times n)$ symmetric positive definite matrix; the introduction of the second term in (6.87) is necessary to obtain a Lyapunov function of the entire system state which vanishes for $\widetilde{\boldsymbol{q}} = \boldsymbol{0}$ and $\dot{\widetilde{\boldsymbol{q}}} = \boldsymbol{0}$. The time derivative of V along the trajectories of system (6.86) is

$$\begin{aligned} \dot{V} &= \boldsymbol{\sigma}^T\boldsymbol{B}(\boldsymbol{q})\dot{\boldsymbol{\sigma}} + \frac{1}{2}\boldsymbol{\sigma}^T\dot{\boldsymbol{B}}(\boldsymbol{q})\boldsymbol{\sigma} + \widetilde{\boldsymbol{q}}^T\boldsymbol{M}\dot{\widetilde{\boldsymbol{q}}} \\ &= -\boldsymbol{\sigma}^T(\boldsymbol{F} + \boldsymbol{K}_D)\boldsymbol{\sigma} + \widetilde{\boldsymbol{q}}^T\boldsymbol{M}\dot{\widetilde{\boldsymbol{q}}}, \end{aligned} \tag{6.88}$$

where the skew-symmetry property (4.47) of the matrix $\boldsymbol{N} = \dot{\boldsymbol{B}} - 2\boldsymbol{C}$ has been exploited. In view of the expression of $\boldsymbol{\sigma}$ in (6.85), with diagonal $\boldsymbol{\Lambda}$ and $\boldsymbol{K}_D$, it is convenient to choose $\boldsymbol{M} = 2\boldsymbol{\Lambda}\boldsymbol{K}_D$; this leads to

$$\dot{V} = -\boldsymbol{\sigma}^T\boldsymbol{F}\boldsymbol{\sigma} - \dot{\widetilde{\boldsymbol{q}}}^T\boldsymbol{K}_D\dot{\widetilde{\boldsymbol{q}}} - \widetilde{\boldsymbol{q}}^T\boldsymbol{\Lambda}\boldsymbol{K}_D\boldsymbol{\Lambda}\widetilde{\boldsymbol{q}}. \tag{6.89}$$

This expression shows that the time derivative is negative definite since it vanishes only if $\widetilde{q} \equiv \mathbf{0}$ and $\dot{\widetilde{q}} \equiv \mathbf{0}$; thus, it follows that the origin of the state space $[\, \widetilde{q}^T \quad \sigma^T \,]^T = \mathbf{0}$ is *globally asymptotically stable*. It is worth noticing that, differently from the robust control case, the error trajectory tends to the subspace $\sigma = \mathbf{0}$ without the need of a high-frequency control.

On the basis of this notable result, the *control* law can be made *adaptive* with respect to the vector of parameters π.

Suppose that the computational model has the same structure as that of the manipulator dynamic model, but its parameters are not known exactly. The control law (6.83) is then modified into

$$\begin{aligned} u &= \widehat{B}(q)\ddot{q}_r + \widehat{C}(q,\dot{q})\dot{q}_r + \widehat{F}\dot{q}_r + \widehat{g} + K_D\sigma \\ &= Y(q,\dot{q},\dot{q}_r,\ddot{q}_r)\widehat{\pi} + K_D\sigma, \end{aligned} \tag{6.90}$$

where $\widehat{\pi}$ represents the available estimate on the parameters and, accordingly, $\widehat{B}$, $\widehat{C}$, $\widehat{F}$, $\widehat{g}$ denote the estimated terms in the dynamic model. Substituting control (6.90) into (6.82) gives

$$\begin{aligned} B(q)\dot{\sigma} + C(q,\dot{q})\sigma + F\sigma + K_D\sigma &= -\widetilde{B}(q)\ddot{q}_r - \widetilde{C}(q,\dot{q})\dot{q}_r - \widetilde{F}\dot{q}_r - \widetilde{g}(q) \\ &= -Y(q,\dot{q},\dot{q}_r,\ddot{q}_r)\widetilde{\pi}, \end{aligned} \tag{6.91}$$

where the property of linearity in the error parameter vector

$$\widetilde{\pi} = \widehat{\pi} - \pi \tag{6.92}$$

has been conveniently exploited. In view of (6.55), the modelling error is characterized by:

$$\widetilde{B} = \widehat{B} - B \qquad \widetilde{C} = \widehat{C} - C \qquad \widetilde{F} = \widehat{F} - F \qquad \widetilde{g} = \widehat{g} - g. \tag{6.93}$$

It is worth remarking that, in view of position (6.84), the matrix Y does not depend on the actual joint accelerations but only on their desired values; this avoids problems due to direct measurement of acceleration.

At this point, modify the Lyapunov function candidate in (6.87) into the form

$$V(\sigma,\widetilde{q},\widetilde{\pi}) = \frac{1}{2}\sigma^T B(q)\sigma + \widetilde{q}^T \Lambda K_D \widetilde{q} + \frac{1}{2}\widetilde{\pi}^T K_\pi \widetilde{\pi} > 0 \quad \forall \sigma,\widetilde{q},\widetilde{\pi} \neq \mathbf{0}, \tag{6.94}$$

which features an additional term accounting for the parameter error (6.92), with K_π symmetric positive definite. The time derivative of V along the trajectories of system (6.91) is

$$\dot{V} = -\sigma^T F\sigma - \dot{\widetilde{q}}^T K_D \dot{\widetilde{q}} - \widetilde{q}^T \Lambda K_D \Lambda \widetilde{q} + \widetilde{\pi}^T \big(K_\pi \dot{\widehat{\pi}} - Y^T(q,\dot{q},\dot{q}_r,\ddot{q}_r)\sigma\big). \tag{6.95}$$

If the estimate of the parameter vector is updated as in the adaptive law

$$\dot{\widehat{\pi}} = K_\pi^{-1} Y^T(q,\dot{q},\dot{q}_r,\ddot{q}_r)\sigma, \tag{6.96}$$

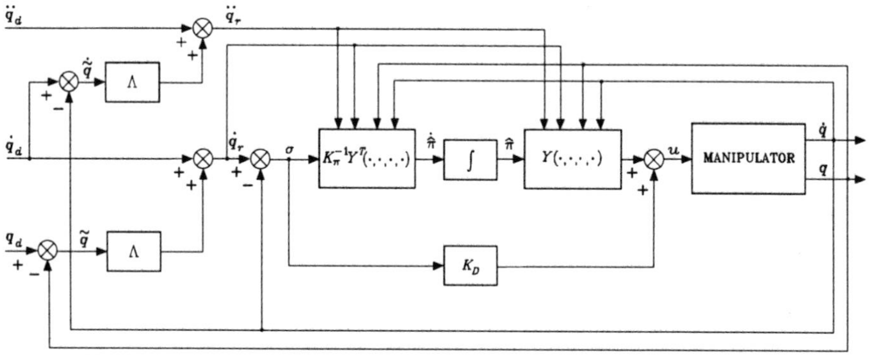

Figure 6.28 Block scheme of joint space adaptive control.

the expression in (6.95) becomes

$$\dot{V} = -\boldsymbol{\sigma}^T \boldsymbol{F}\boldsymbol{\sigma} - \dot{\tilde{\boldsymbol{q}}}^T \boldsymbol{K}_D \dot{\tilde{\boldsymbol{q}}} - \tilde{\boldsymbol{q}}^T \boldsymbol{\Lambda} \boldsymbol{K}_D \boldsymbol{\Lambda} \tilde{\boldsymbol{q}}$$

since $\dot{\hat{\boldsymbol{\pi}}} = \dot{\tilde{\boldsymbol{\pi}}}$ ($\boldsymbol{\pi}$ is constant).

By an argument similar to above, it is not difficult to show that the trajectories of the manipulator described by the model

$$\boldsymbol{B}(\boldsymbol{q})\ddot{\boldsymbol{q}} + \boldsymbol{C}(\boldsymbol{q}, \dot{\boldsymbol{q}})\dot{\boldsymbol{q}} + \boldsymbol{F}\dot{\boldsymbol{q}} + \boldsymbol{g}(\boldsymbol{q}) = \boldsymbol{u},$$

under the control law

$$\boldsymbol{u} = \boldsymbol{Y}(\boldsymbol{q}, \dot{\boldsymbol{q}}, \dot{\boldsymbol{q}}_r, \ddot{\boldsymbol{q}}_r)\hat{\boldsymbol{\pi}} + \boldsymbol{K}_D(\dot{\tilde{\boldsymbol{q}}} + \boldsymbol{\Lambda}\tilde{\boldsymbol{q}})$$

and the parameter adaptive law

$$\dot{\hat{\boldsymbol{\pi}}} = \boldsymbol{K}_\pi^{-1} \boldsymbol{Y}^T(\boldsymbol{q}, \dot{\boldsymbol{q}}, \dot{\boldsymbol{q}}_r, \ddot{\boldsymbol{q}}_r)(\dot{\tilde{\boldsymbol{q}}} + \boldsymbol{\Lambda}\tilde{\boldsymbol{q}}),$$

globally asymptotically converge to $\boldsymbol{\sigma} = \mathbf{0}$ and $\tilde{\boldsymbol{q}} = \mathbf{0}$, which implies *convergence to zero of* $\tilde{\boldsymbol{q}}$, $\dot{\tilde{\boldsymbol{q}}}$, and *boundedness of* $\hat{\boldsymbol{\pi}}$. The equation in (6.91) shows that asymptotically it is

$$\boldsymbol{Y}(\boldsymbol{q}, \dot{\boldsymbol{q}}, \dot{\boldsymbol{q}}_r, \ddot{\boldsymbol{q}}_r)(\hat{\boldsymbol{\pi}} - \boldsymbol{\pi}) = \mathbf{0}. \tag{6.97}$$

This equation does not imply that $\hat{\boldsymbol{\pi}}$ tends to $\boldsymbol{\pi}$; indeed, convergence of parameters to their true values depends on the structure of the matrix $\boldsymbol{Y}(\boldsymbol{q}, \dot{\boldsymbol{q}}, \dot{\boldsymbol{q}}_r, \ddot{\boldsymbol{q}}_r)$ and then on the desired and actual trajectories. Nonetheless, the followed approach is aimed at solving a *direct* adaptive control problem, *i.e.*, finding a control law that ensures limited tracking errors, and not at determining the actual parameters of the system (as in an indirect adaptive control problem). The resulting block scheme is illustrated in Figure 6.28. To summarize, the above control law is formed by three different contributions:

- The term $\boldsymbol{Y}\hat{\boldsymbol{\pi}}$ describes a control action of inverse dynamics type which ensures an *approximate compensation of nonlinear effects and joint decoupling*.

- The term $\boldsymbol{K}_D\boldsymbol{\sigma}$ introduces a *stabilizing linear control action of PD type on the tracking error.*
- The vector of parameter estimates $\widehat{\boldsymbol{\pi}}$ is updated by an *adaptive law* of gradient type so as to ensure asymptotic compensation of the terms in the manipulator dynamic model; the matrix $\boldsymbol{K}_\pi$ determines the convergence rate of parameters to their asymptotic values.

Notice that, with $\boldsymbol{\sigma} \approx \mathbf{0}$, the control law (6.90) is equivalent to a pure inverse dynamics compensation of the computed torque type on the basis of desired velocities and accelerations; this is made possible by the fact that $\boldsymbol{Y}\widehat{\boldsymbol{\pi}} \approx \boldsymbol{Y}\boldsymbol{\pi}$.

The control law with parameter adaptation requires the availability of a complete computational model and it does not feature any action aimed at reducing the effects of external disturbances. Therefore, a performance degradation is expected whenever unmodeled dynamic effects, e.g., when a reduced computational model is used, or external disturbances occur. In both cases, the effects induced on the output variables are attributed by the controller to parameter estimate mismatching; as a consequence, the control law attempts to counteract those effects by acting on quantities that did not provoke them originally.

On the other hand, robust control techniques provide a natural rejection to external disturbances, although they are sensitive to unmodeled dynamics; this rejection is provided by a high-frequency commuted control action that constrains the error trajectories to stay on the sliding subspace. The resulting inputs to the mechanical structure may be unacceptable. This inconvenience is in general not observed with the adoption of adaptive control techniques whose action has a naturally smooth time behaviour.

6.6 Operational Space Control

In all the above control schemes, it was always assumed that the desired trajectory is available in terms of the time sequence of the values of joint position, velocity and acceleration. Accordingly, the error for the control schemes was expressed in the joint space.

As often pointed out, motion specifications are usually assigned in the operational space, and then an inverse kinematics algorithm has to be utilized to transform operational space references into the corresponding joint space references. The process of kinematic inversion has an increasing computational load when, besides inversion of direct kinematics, also inversion of first-order and second-order differential kinematics is required to transform the desired time history of end-effector position, velocity and acceleration into the corresponding quantities at the joint level. It is for this reason that current industrial robot control systems compute the joint positions through kinematics inversion, and then perform a numerical differentiation to compute velocities and accelerations.

A different approach consists of considering control schemes developed directly in the operational space. If the motion is specified in terms of operational space variables, the measured joint space variables can be transformed into the corresponding operational space variables through direct kinematics relations. Comparing the desired

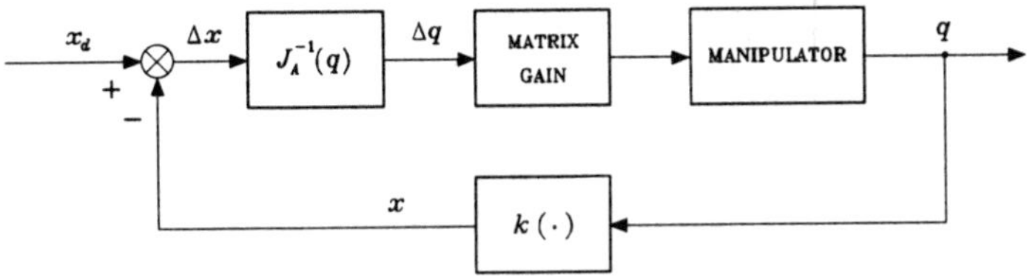

Figure 6.29 Block scheme of Jacobian inverse control.

input with the reconstructed variables allows designing feedback control loops where trajectory inversion is replaced with a suitable coordinate transformation embedded in the feedback loop.

All operational space control schemes present considerable computational requirements, in view of the necessity to perform a number of computations in the feedback loop which are somewhat representative of inverse kinematics functions. With reference to a numerical implementation, the presence of a computationally demanding load requires sampling times that may lead to degrading the performance of the overall control system.

In the face of the above limitations, it is worth presenting *operational space control* schemes, whose utilization becomes necessary when the problem of controlling interaction between the manipulator and the environment is of concern. In fact, joint space control schemes suffice only for motion control in the free space. When the manipulator's end effector is constrained by the environment, e.g., in the case of end-effector in contact with an elastic environment, it is necessary to control both positions and contact forces and it is convenient to refer to operational space control schemes. Hence, below some solutions are presented; these are worked out for motion control, but they constitute the premise for the interaction control strategies that will be illustrated in the next chapter.

6.6.1 General Schemes

As pointed out above, operational space control schemes are based on a direct comparison of the inputs, specifying operational space trajectories, with the measurements of the corresponding manipulator outputs. It follows that the control system shall incorporate some actions that allow passing from the operational space, in which the error is specified, to the joint space, in which control generalized forces are developed.

A possible control scheme that can be devised is the so-called *Jacobian inverse control* (Figure 6.29). In this scheme, the end-effector location in the operational space is compared with the corresponding desired quantity, and then an operational space deviation $\Delta \boldsymbol{x}$ can be computed. Assuming that this deviation is sufficiently small for a good control system, $\Delta \boldsymbol{x}$ can be transformed into a corresponding joint space deviation $\Delta \boldsymbol{q}$ through the inverse of the manipulator Jacobian. Then, the control input generalized forces can be computed on the basis of this deviation through a suitable feedback matrix gain. The result is a presumable reduction of $\Delta \boldsymbol{q}$ and in turn of $\Delta \boldsymbol{x}$.

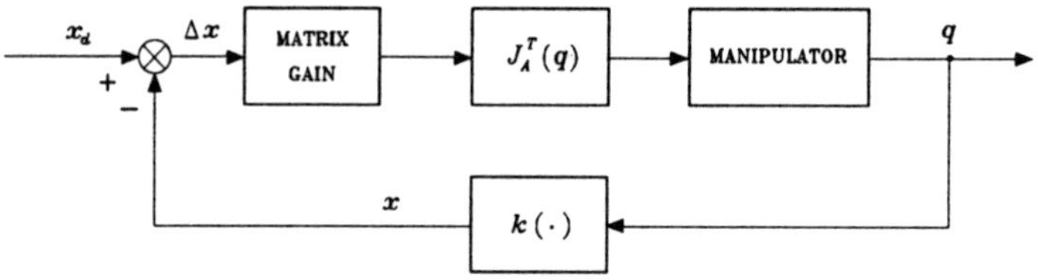

Figure 6.30 Block scheme of Jacobian transpose control.

In other words, the Jacobian inverse control leads to an overall system that intuitively behaves like a mechanical system with a generalized n-dimensional spring in the joint space, whose constant stiffness is determined by the feedback matrix gain. The role of such system is to take the deviation $\Delta \boldsymbol{q}$ to zero. If the matrix gain is diagonal, the generalized spring corresponds to n independent elastic elements, one for each joint.

A conceptually analogous scheme is the so-called *Jacobian transpose control* (Figure 6.30). In this case, the operational space error is treated first through a matrix gain. The output of this block can then be considered as the elastic force generated by a generalized spring whose function in the operational space is that to reduce or to cancel the position deviation $\Delta \boldsymbol{x}$. In other words, the resulting force drives the end effector along a direction so as to reduce $\Delta \boldsymbol{x}$. This operational space force has then to be transformed into the joint space generalized forces, through the transpose of the Jacobian, so as to realize the described behaviour.

Both Jacobian inverse and transpose control schemes have been derived in an intuitive fashion. Hence, there is no guarantee that such schemes are effective in terms of stability and trajectory tracking accuracy. These problems can be faced by presenting two mathematical solutions below, which will be shown to be substantially equivalent to the above schemes.

6.6.2 PD Control with Gravity Compensation

By analogy with joint space stability analysis, given a *constant* end-effector location $\boldsymbol{x}_d$, it is desired to find the control structure so that the operational space error

$$\widetilde{\boldsymbol{x}} = \boldsymbol{x}_d - \boldsymbol{x} = \boldsymbol{x}_d - \boldsymbol{k}(\boldsymbol{q}) \tag{6.98}$$

tends asymptotically to zero. Choose the following positive definite quadratic form as a Lyapunov function candidate:

$$V(\dot{\boldsymbol{q}}, \widetilde{\boldsymbol{x}}) = \frac{1}{2}\dot{\boldsymbol{q}}^T \boldsymbol{B}(\boldsymbol{q})\dot{\boldsymbol{q}} + \frac{1}{2}\widetilde{\boldsymbol{x}}^T \boldsymbol{K}_P \widetilde{\boldsymbol{x}} > 0 \qquad \forall \dot{\boldsymbol{q}}, \widetilde{\boldsymbol{x}} \neq \boldsymbol{0}, \tag{6.99}$$

with $\boldsymbol{K}_P$ a symmetric positive definite matrix. Differentiating (6.99) with respect to time gives

$$\dot{V} = \dot{\boldsymbol{q}}^T \boldsymbol{B}(\boldsymbol{q})\ddot{\boldsymbol{q}} + \frac{1}{2}\dot{\boldsymbol{q}}^T \dot{\boldsymbol{B}}(\boldsymbol{q})\dot{\boldsymbol{q}} + \dot{\widetilde{\boldsymbol{x}}}^T \boldsymbol{K}_P \widetilde{\boldsymbol{x}}.$$

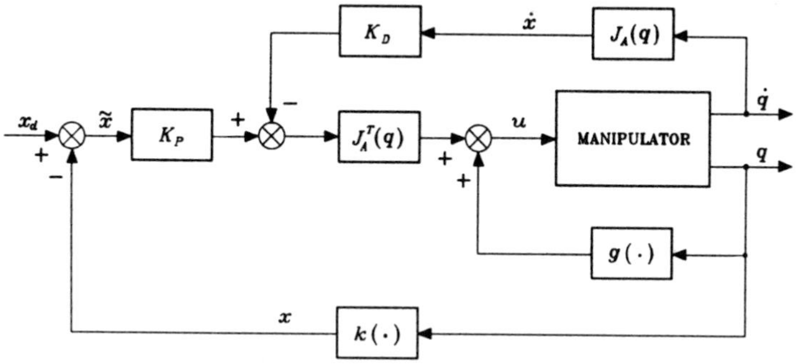

Figure 6.31 Block scheme of operational space PD control with gravity compensation.

Since $\dot{\boldsymbol{x}}_d = \boldsymbol{0}$, in view of (3.58) it is

$$\dot{\widetilde{\boldsymbol{x}}} = -\boldsymbol{J}_A(\boldsymbol{q})\dot{\boldsymbol{q}}$$

and then

$$\dot{V} = \dot{\boldsymbol{q}}^T\boldsymbol{B}(\boldsymbol{q})\ddot{\boldsymbol{q}} + \frac{1}{2}\dot{\boldsymbol{q}}^T\dot{\boldsymbol{B}}(\boldsymbol{q})\dot{\boldsymbol{q}} - \dot{\boldsymbol{q}}^T\boldsymbol{J}_A^T(\boldsymbol{q})\boldsymbol{K}_P\widetilde{\boldsymbol{x}}. \tag{6.100}$$

By recalling the expression of the joint space manipulator dynamic model in (6.35) and the property in (4.48), the expression in (6.100) becomes

$$\dot{V} = -\dot{\boldsymbol{q}}^T\boldsymbol{F}\dot{\boldsymbol{q}} + \dot{\boldsymbol{q}}^T\big(\boldsymbol{u} - \boldsymbol{g}(\boldsymbol{q}) - \boldsymbol{J}_A^T(\boldsymbol{q})\boldsymbol{K}_P\widetilde{\boldsymbol{x}}\big). \tag{6.101}$$

This equation suggests the structure of the controller; in fact, by choosing the control law

$$\boldsymbol{u} = \boldsymbol{g}(\boldsymbol{q}) + \boldsymbol{J}_A^T(\boldsymbol{q})\boldsymbol{K}_P\widetilde{\boldsymbol{x}} - \boldsymbol{J}_A^T\boldsymbol{K}_D\boldsymbol{J}_A(\boldsymbol{q})\dot{\boldsymbol{q}}, \tag{6.102}$$

with $\boldsymbol{K}_D$ positive definite, (6.101) becomes

$$\dot{V} = -\dot{\boldsymbol{q}}^T\boldsymbol{F}\dot{\boldsymbol{q}} - \dot{\boldsymbol{q}}^T\boldsymbol{J}_A^T(\boldsymbol{q})\boldsymbol{K}_D\boldsymbol{J}_A(\boldsymbol{q})\dot{\boldsymbol{q}}. \tag{6.103}$$

As can be seen from Figure 6.31, the resulting block scheme reveals an analogy with the scheme of Figure 6.30. Control law (6.102) performs a *nonlinear compensating action of joint space gravitational forces* and an *operational space linear PD control action*. The last term has been introduced to enhance system damping; in particular, if measurement of $\dot{\boldsymbol{x}}$ is deduced from that of $\dot{\boldsymbol{q}}$, one can simply choose the derivative term as $-\boldsymbol{K}_D\dot{\boldsymbol{q}}$.

The expression in (6.103) shows that, for any system trajectory, the Lyapunov function decreases as long as $\dot{\boldsymbol{q}} \neq \boldsymbol{0}$. The system then reaches an *equilibrium posture*. By a stability argument similar to that in the joint space (see Equations (6.44)–(6.46)), this posture is determined by

$$\boldsymbol{J}_A^T(\boldsymbol{q})\boldsymbol{K}_P\widetilde{\boldsymbol{x}} = \boldsymbol{0}. \tag{6.104}$$

From (6.104) it can be recognized that, on the assumption of *full-rank* Jacobian, it is

$$\widetilde{\boldsymbol{x}} = \boldsymbol{x}_d - \boldsymbol{x} = \mathbf{0},$$

i.e., the sought result.

If measurements of $\boldsymbol{x}$ and $\dot{\boldsymbol{x}}$ are made directly in the operational space, $\boldsymbol{k}(\boldsymbol{q})$ and $\boldsymbol{J}_A(\boldsymbol{q})$ in the scheme of Figure 6.31 are just indicative of direct kinematics functions; it is, however, necessary to measure $\boldsymbol{q}$ to update both $\boldsymbol{J}_A^T(\boldsymbol{q})$ and $\boldsymbol{g}(\boldsymbol{q})$ on line. If measurements of operational space quantities are indirect, the controller has to compute the direct kinematics functions, too.

6.6.3 Inverse Dynamics Control

Consider now the problem of tracking an operational space trajectory. Recall the manipulator dynamic model in the form (6.47)

$$\boldsymbol{B}(\boldsymbol{q})\ddot{\boldsymbol{q}} + \boldsymbol{n}(\boldsymbol{q}, \dot{\boldsymbol{q}}) = \boldsymbol{u}$$

where $\boldsymbol{n}$ is given by (6.48). As in (6.49), the choice of the *inverse dynamics linearizing control*

$$\boldsymbol{u} = \boldsymbol{B}(\boldsymbol{q})\boldsymbol{y} + \boldsymbol{n}(\boldsymbol{q}, \dot{\boldsymbol{q}})$$

leads to the system of double integrators

$$\ddot{\boldsymbol{q}} = \boldsymbol{y}. \tag{6.105}$$

The new control input $\boldsymbol{y}$ is to be designed so as to allow tracking of a trajectory specified by $\boldsymbol{x}_d(t)$. To this purpose, the second-order differential equation in the form (4.117)

$$\ddot{\boldsymbol{x}} = \boldsymbol{J}_A(\boldsymbol{q})\ddot{\boldsymbol{q}} + \dot{\boldsymbol{J}}_A(\boldsymbol{q}, \dot{\boldsymbol{q}})\dot{\boldsymbol{q}}$$

suggests, for a nonredundant manipulator, the choice of the control law

$$\boldsymbol{y} = \boldsymbol{J}_A^{-1}(\boldsymbol{q})\big(\ddot{\boldsymbol{x}}_d + \boldsymbol{K}_D\dot{\widetilde{\boldsymbol{x}}} + \boldsymbol{K}_P\widetilde{\boldsymbol{x}} - \dot{\boldsymbol{J}}_A(\boldsymbol{q}, \dot{\boldsymbol{q}})\dot{\boldsymbol{q}}\big) \tag{6.106}$$

with $\boldsymbol{K}_P$ and $\boldsymbol{K}_D$ positive definite (diagonal) matrices. In fact, substituting (6.106) into (6.105) gives

$$\ddot{\widetilde{\boldsymbol{x}}} + \boldsymbol{K}_D\dot{\widetilde{\boldsymbol{x}}} + \boldsymbol{K}_P\widetilde{\boldsymbol{x}} = \mathbf{0} \tag{6.107}$$

which describes the operational space error dynamics, with $\boldsymbol{K}_P$ and $\boldsymbol{K}_D$ determining the error convergence rate to zero. The resulting inverse dynamics control scheme is reported in Figure 6.32, which confirms the anticipated analogy with the scheme of Figure 6.29. Again in this case, besides $\boldsymbol{x}$ and $\dot{\boldsymbol{x}}$, also $\boldsymbol{q}$ and $\dot{\boldsymbol{q}}$ are to be measured.

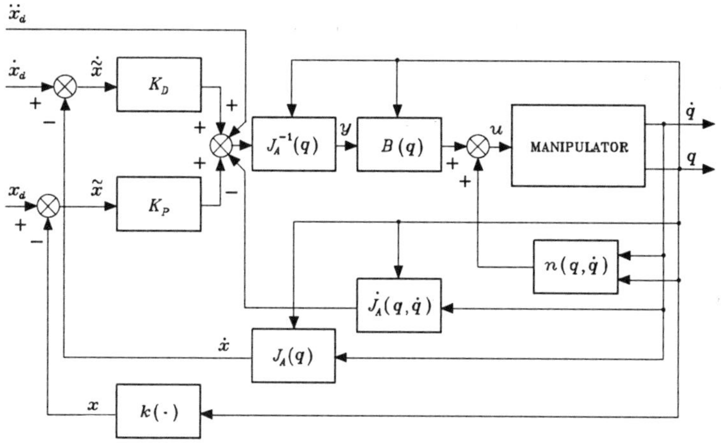

Figure 6.32 Block scheme of operational space inverse dynamics control.

If measurements of x and $\dot{x}$ are indirect, the controller must compute the direct kinematics functions $\boldsymbol{k}(\boldsymbol{q})$ and $\boldsymbol{J}_A(\boldsymbol{q})$ on line.

A critical analysis of the schemes in Figures 6.31 and 6.32 reveals that the design of an operational space controller always requires computation of manipulator Jacobian. As a consequence, controlling a manipulator in the operational space is in general more complex than controlling it in the joint space. In fact, the presence of *singularities* and/or *redundancy* influences the Jacobian, and the induced effects are somewhat difficult to handle with an operational space controller. For instance, if a singularity occurs for the scheme of Figure 6.31 and the error enters the null space of the Jacobian, the manipulator gets stuck at a different configuration from the desired one. This problem is even more critical for the scheme of Figure 6.32 which would require the computation of a DLS-inverse of the Jacobian. Yet, for a redundant manipulator, a joint space control scheme is naturally transparent to this situation, since redundancy has been already solved by inverse kinematics, whereas an operational space control scheme should incorporate a redundancy handling technique inside the feedback loop.

As a final remark, the above operational space control schemes have been derived with reference to a minimal description of orientation in terms of Euler angles. It is understood that, similarly to what presented in Section 3.7.3 for inverse kinematics algorithms, it is possible to adopt different definitions of orientation error, e.g., based on the angle and axis or the unit quaternion. The advantage is the use of the geometric Jacobian in lieu of the analytical Jacobian. The price to pay, however, is a more complex analysis of the stability and convergence characteristics of the closed-loop system. Even the inverse dynamics control scheme will not lead to a homogeneous error equation, and a Lyapunov argument shall be invoked to ascertain its stability.

6.7 A Comparison Between Various Control Schemes

In order to make a comparison between the various control schemes presented, consider the two-link planar arm with the same data of Example 4.2:

$$a_1 = a_2 = 1\,\text{m} \quad \ell_1 = \ell_2 = 0.5\,\text{m} \quad m_{\ell_1} = m_{\ell_2} = 50\,\text{kg} \quad I_{\ell_1} = I_{\ell_2} = 10\,\text{kg}\cdot\text{m}^2$$

$$k_{r1} = k_{r2} = 100 \qquad m_{m_1} = m_{m_2} = 5\,\text{kg} \qquad I_{m_1} = I_{m_2} = 0.01\,\text{kg}\cdot\text{m}^2.$$

The arm is assumed to be driven by two equal actuators with the following data:

$$F_{m_1} = F_{m_2} = 0.01\,\text{N}\cdot\text{m}\cdot\text{s/rad} \qquad R_{a1} = R_{a2} = 10\,\text{ohm}$$

$$k_{t1} = k_{t2} = 2\,\text{N}\cdot\text{m/A} \qquad k_{v1} = k_{v2} = 2\,\text{V}\cdot\text{s/rad};$$

it can be verified that $F_{m_i} \ll k_{vi}k_{ti}/R_{ai}$ for $i = 1, 2$.

The desired tip trajectories have a typical trapezoidal velocity profile, and thus it is anticipated that sharp torque variations will be induced. The tip path is a motion of 1.6 m along the horizontal axis, as in the path of Example 4.2. In the first case (*fast* trajectory), the acceleration time is 0.6 s and the maximum velocity is 1 m/s. In the second case (*slow* trajectory), the acceleration time is 0.6 s and the maximum velocity is 0.25 m/s. The motion of the controlled arm was simulated on a computer, by adopting a discrete-time implementation of the controller with a sampling time of 1 ms.

The following control schemes in the joint space and in the operational space have been utilized; an (analytic) inverse kinematics solution has been implemented to generate the reference inputs to the joint space control schemes.

A. Independent joint control with position and velocity feedback (Figure 6.8) with the following data for each joint servo:

$$K_P = 5 \qquad K_V = 10 \qquad k_{TP} = k_{TV} = 1,$$

corresponding to $\omega_n = 5$ rad/s and $\zeta = 0.5$.

B. Independent joint control with position, velocity and acceleration feedback (Figure 6.10) with the following data for each joint servo:

$$K_P = 5 \qquad K_V = 10 \qquad K_A = 2 \qquad k_{TP} = k_{TV} = k_{TA} = 1,$$

corresponding to $\omega_n = 5$ rad/s, $\zeta = 0.5$, $X_R = 100$. To reconstruct acceleration, a first-order filter has been utilized (Figure 6.12) characterized by $\omega_{3f} = 100$ rad/s.

C. As in scheme A with the addition of a decentralized feedforward action (Figure 6.14).

D. As in scheme B with the addition of a decentralized feedforward action (Figure 6.15).

E. Joint space computed torque control (Figure 6.19) with feedforward compensation of the diagonal terms of the inertia matrix and of gravitational terms, and decentralized feedback controllers as in scheme A.

F. Joint space PD control with gravity compensation (Figure 6.22), modified by the addition of a feedforward velocity term $\boldsymbol{K}_D\dot{\boldsymbol{q}}_d$, with the following data:

$$\boldsymbol{K}_P = 3750\boldsymbol{I} \qquad \boldsymbol{K}_D = 750\boldsymbol{I}.$$

G. Joint space inverse dynamics control (Figure 6.24) with the following data:

$$\boldsymbol{K}_P = 25\boldsymbol{I} \qquad \boldsymbol{K}_D = 5\boldsymbol{I}.$$

H. Joint space robust control (Figure 6.25), on the assumption of constant inertia ($\widehat{\boldsymbol{B}} = \bar{\boldsymbol{B}}$) and compensation of friction and gravity ($\widehat{\boldsymbol{n}} = \boldsymbol{F}_v\dot{\boldsymbol{q}} + \boldsymbol{g}$), with the following data:

$$\boldsymbol{K}_P = 25\boldsymbol{I} \qquad \boldsymbol{K}_D = 5\boldsymbol{I} \qquad \boldsymbol{P} = \boldsymbol{I} \qquad \rho = 70 \qquad \epsilon = 0.004.$$

I. As in case H with $\epsilon = 0.01$.

J. Joint space adaptive control (Figure 6.28) with a parameterization of the arm dynamic model (4.81) as in (4.82) and (4.83). The initial estimate of the vector $\widehat{\boldsymbol{\pi}}$ is computed on the basis of the nominal parameters. The arm is supposed to carry a load which causes the following variations on the second link parameters:

$$\Delta m_2 = 10\,\text{kg} \qquad \Delta m_2\ell_{C2} = 11\,\text{kg}\cdot\text{m} \qquad \Delta\hat{I}_2 = 12.12\,\text{kg}\cdot\text{m}^2.$$

This information is obviously utilized only to update the simulated arm model. Further, the following data are set:

$$\boldsymbol{\Lambda} = 5\boldsymbol{I} \qquad \boldsymbol{K}_D = 750\boldsymbol{I} \qquad \boldsymbol{K}_\pi = 0.01\boldsymbol{I}.$$

K. Operational space PD control with gravity compensation (Figure 6.31), modified by the addition of a feedforward velocity term $\boldsymbol{K}_D\dot{\boldsymbol{x}}_d$, with the following data:

$$\boldsymbol{K}_P = 16250\boldsymbol{I} \qquad \boldsymbol{K}_D = 3250\boldsymbol{I}.$$

L. Operational space inverse dynamics control (Figure 6.32) with the following data:

$$\boldsymbol{K}_P = 25\boldsymbol{I} \qquad \boldsymbol{K}_D = 5\boldsymbol{I}.$$

It is worth remarking that the adopted model of the dynamic system of arm with drives is that described by (6.35). In the decentralized control schemes A–E, the joints have been voltage-controlled as in the block scheme of Figure 6.20, with unit amplifier gains ($\boldsymbol{G}_v = \boldsymbol{I}$). On the other hand, in the centralized control schemes F–L, the joints

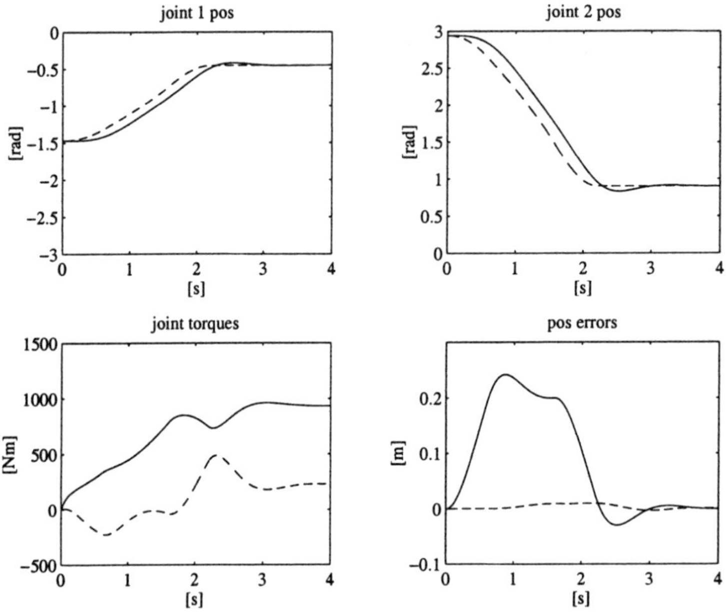

Figure 6.33 Time history of the joint positions and torques and of the tip position errors for the *fast* trajectory with control scheme A.

have been current-controlled as in the block scheme of Figure 6.21, with unit amplifier gains ($\boldsymbol{G}_i = \boldsymbol{I}$).

Regarding the parameters of the various controllers, these have been chosen in such a way as to allow a significant comparison of the performance of each scheme in response to congruent control actions. In particular, it can be observed that:

- The dynamic behaviour of the joints is the same for schemes A–E.
- The gains of the PD actions in schemes F, J and K have been tuned so as to obtain response times similar to those of schemes A–E.
- The gains of the PD actions in schemes G, H, I and L have been chosen so as to obtain the same natural frequency and damping ratios as those of schemes A–E.

The results obtained with the various control schemes are illustrated in Figures 6.33–6.41 for the *fast* trajectory and in Figures 6.42–6.50 for the *slow* trajectory, respectively. In the case of two quantities represented in the same plot notice that:

- For the joint trajectories, the dashed line indicates the reference trajectory obtained from the tip trajectory via inverse kinematics, while the solid line indicates the actual trajectory followed by the arm.
- For the joint torques, the solid line refers to Joint 1 while the dashed line refers to Joint 2.

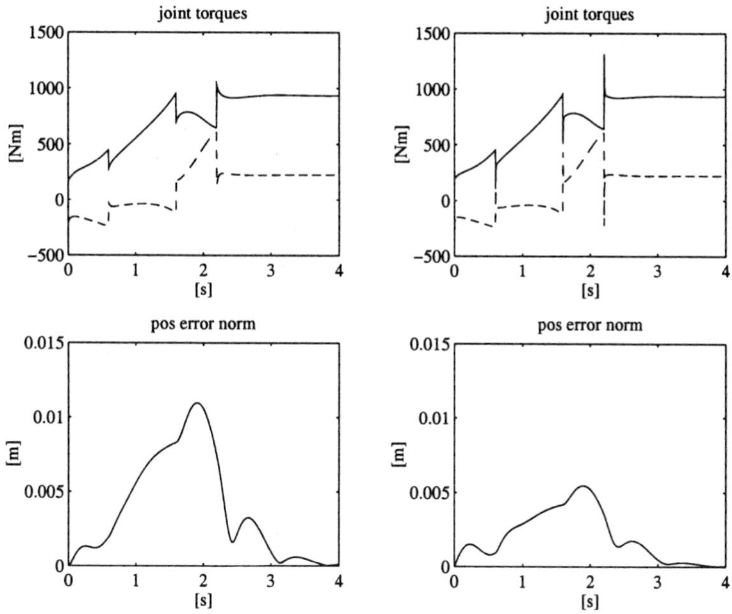

Figure 6.34 Time history of the joint torques and of the norm of tip position error for the *fast* trajectory; *left*—with control scheme C, *right*—with control scheme D.

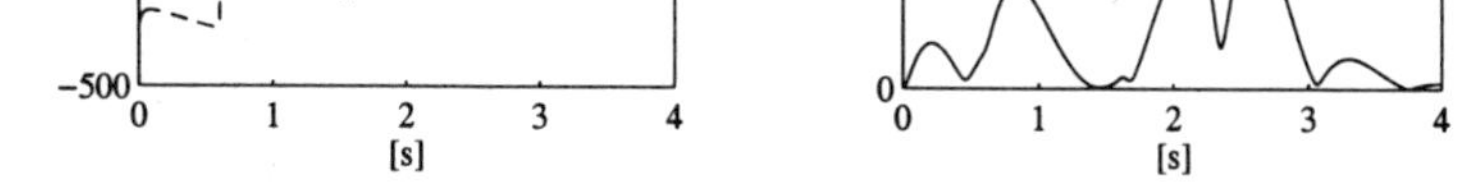

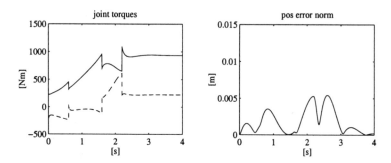

Figure 6.35 Time history of the joint torques and of the norm of tip position error for the *fast* trajectory with control scheme E.

- For the tip position error, the solid line indicates the error component along the horizontal axis while the dashed line indicates the error component along the vertical axis.

Finally, the representation scales have been made as uniform as possible in order to allow a more direct comparison of the results.

Regarding performance of the various control schemes for the *fast* trajectory, the obtained results allow drawing the following considerations.

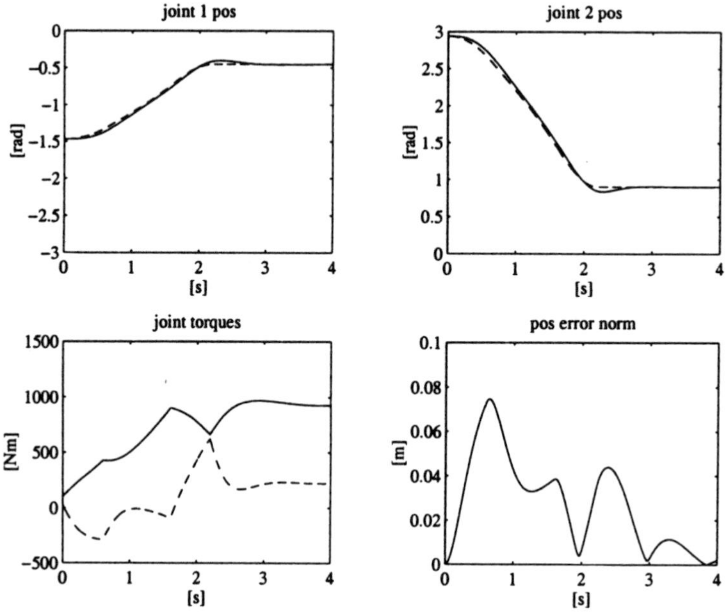

Figure 6.36 Time history of the joint positions and torques and of the norm of tip position error for the *fast* trajectory with control scheme F.

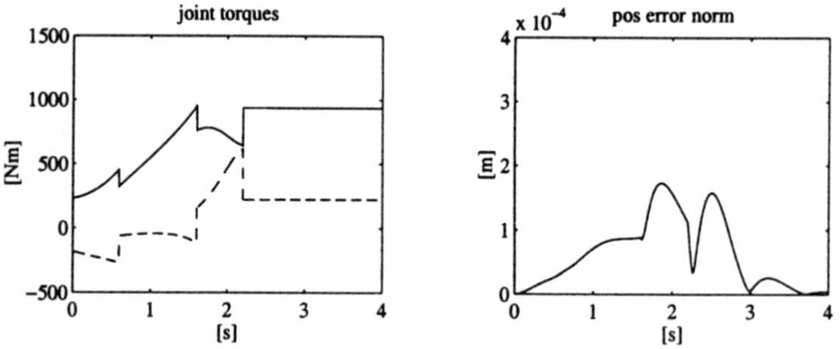

Figure 6.37 Time history of the joint torques and of the norm of tip position error for the *fast* trajectory with control scheme G.

Deviation of the actual joint trajectories from the desired ones shows that tracking performance of scheme A is quite poor (Figure 6.33). It should be noticed, however, that the largest contribution to the error is caused by a time lag of the actual trajectory behind the desired one, while the distance of the tip from the geometric path is quite contained. Similar results were obtained with scheme B, and then they have not been reported.

With schemes C and D, an appreciable tracking accuracy improvement is observed (Figure 6.34), with better performance for the second scheme, thanks to the outer ac-

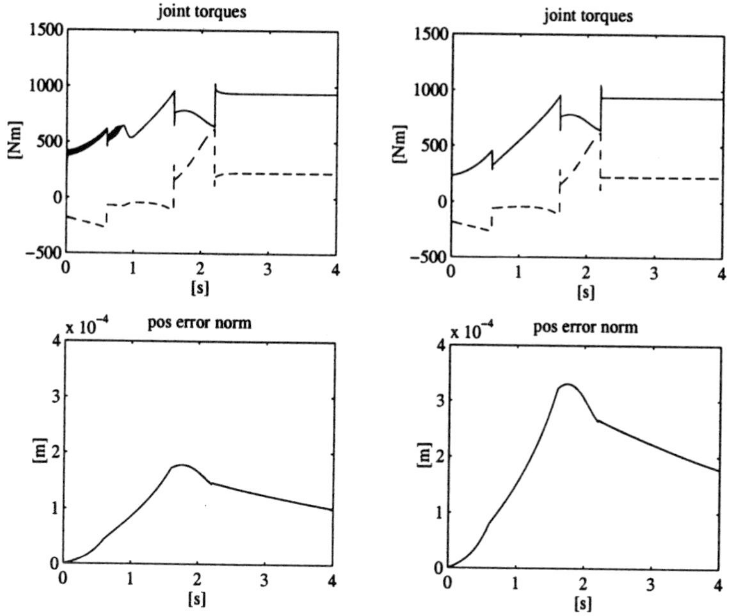

Figure 6.38 Time history of the joint torques and of the norm of tip position error for the *fast* trajectory; *left*—with control scheme H, *right*—with control scheme I.

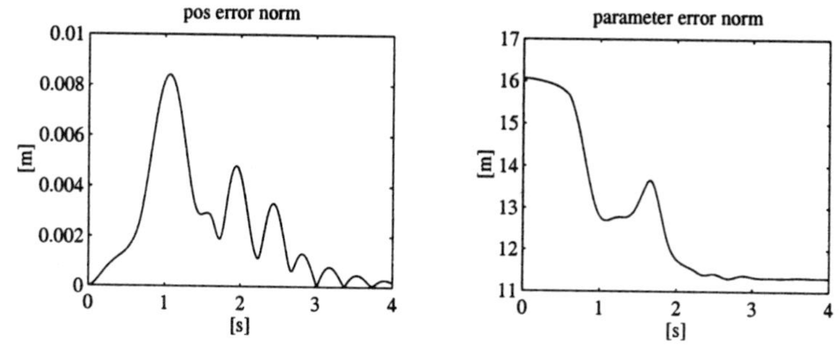

Figure 6.39 Time history of the norm of tip position error and of the norm of parameter error vector for the *fast* trajectory with control scheme J.

celeration feedback loop that allows prescribing a disturbance rejection factor twice as much as for the first scheme. Notice that the feedforward action allows obtaining a set of torques which are closer to the nominal ones required to execute the desired trajectory; the torque time history has a discontinuity in correspondence of the acceleration and deceleration fronts.

The tracking error is further decreased with scheme E (Figure 6.35), by virtue of the additional nonlinear feedforward compensation.

Scheme F guarantees stable convergence to the final arm posture with a tracking

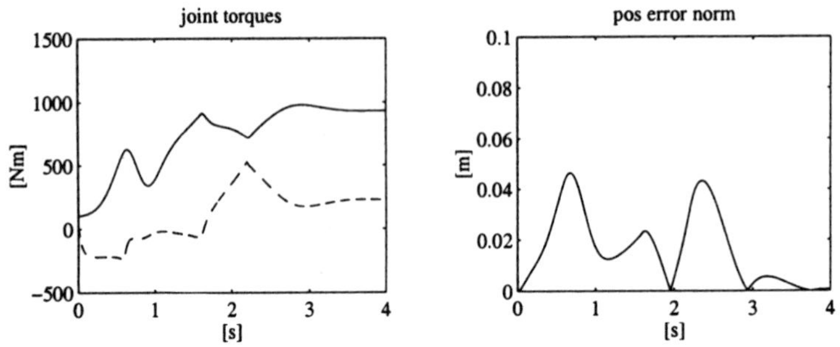

Figure 6.40 Time history of the joint torques and of the norm of tip position error for the *fast* trajectory with control scheme K.

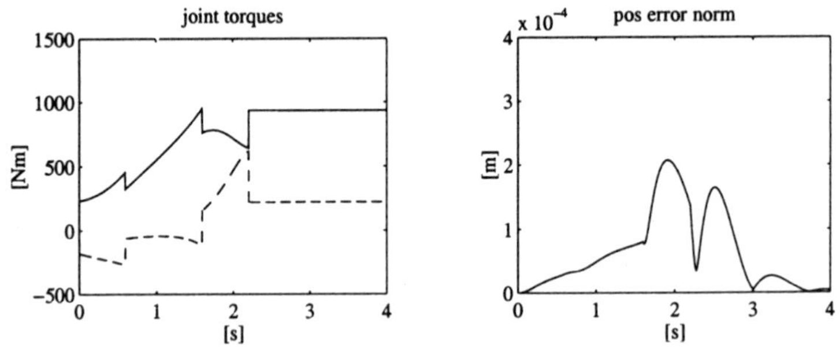

Figure 6.41 Time history of the joint torques and of the norm of tip position error for the *fast* trajectory with control scheme L.

performance which is better than that of schemes A and B, thanks to the presence of a velocity feedforward action, but worse than that of schemes C–E, in view of lack of an acceleration feedforward action (Figure 6.36).

As would be logical to expect, the best results are observed with scheme G for which the tracking error is practically zero, and it is mainly due to numerical discretization of the controller (Figure 6.37).

It is then worth comparing the performance of schemes H and I (Figure 6.38). In fact, the choice of a small threshold value for ϵ (scheme H) induces high-frequency components in Joint 1 torque (see the thick portions of the torque plot) at the advantage of a very limited tracking error. As the threshold value is increased (scheme I), the torque assumes a smoother behaviour at the expense of a doubled norm of tracking error, though.

For scheme J, a lower tracking error than that of scheme F is observed, thanks to the effectiveness of the adaptive action on the parameters of the dynamic model. Nonetheless, the parameters do not converge to their nominal values, as confirmed by the time history of the norm of the parameter error vector that reaches a nonnull steady-state value (Figure 6.39).

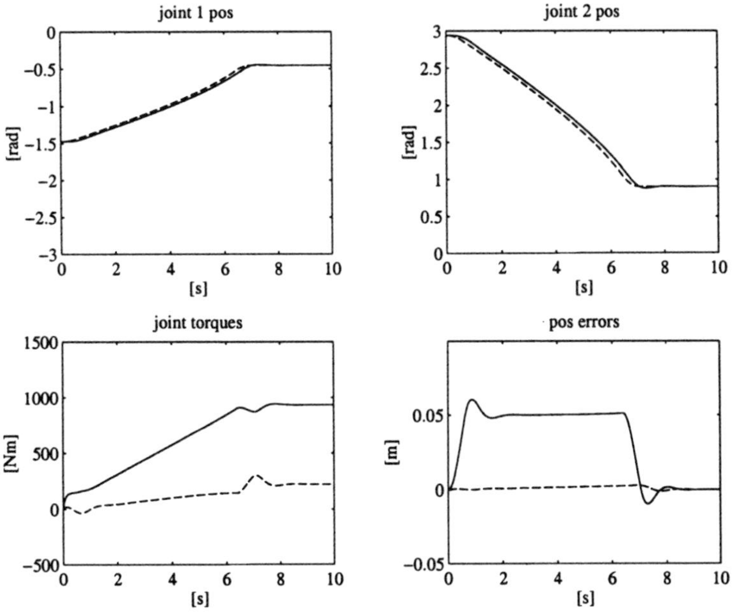

Figure 6.42 Time history of the joint positions and torques and of the tip position errors for the *slow* trajectory with control scheme A.

Finally, the performance of schemes K and L is substantially comparable to that of corresponding schemes F and G (Figures 6.40 and 6.41).

Performance of the various control schemes for the *slow* trajectory is globally better than that for the *fast* trajectory. Such improvement is particularly evident for the decentralized control schemes (Figures 6.42–6.44), whereas the tracking error reduction for the centralized control schemes is less dramatic (Figures 6.45–6.50), in view of the small order of magnitude of the errors already obtained for the *fast* trajectory. In any case, as regards performance of each single scheme, it is possible to make a number of remarks analogous to those previously made.

Problems

6.1 With reference to the block scheme with position feedback in Figure 6.6, find the transfer functions of the forward path, the return path, and the closed-loop system.

6.2 With reference to the block scheme with position and velocity feedback in Figure 6.8, find the transfer functions of the forward path, the return path, and the closed-loop system.

6.3 With reference to the block scheme with position, velocity and acceleration feedback in Figure 6.10, find the transfer functions of the forward path, the return path, and the closed-loop system.

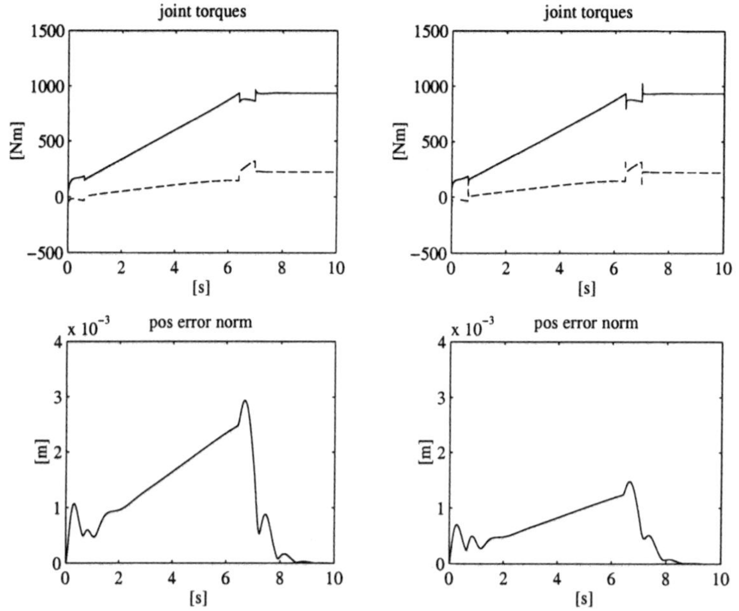

Figure 6.43 Time history of the joint torques and of the norm of tip position error for the *slow* trajectory; *left*—with control scheme C, *right*—with control scheme D.

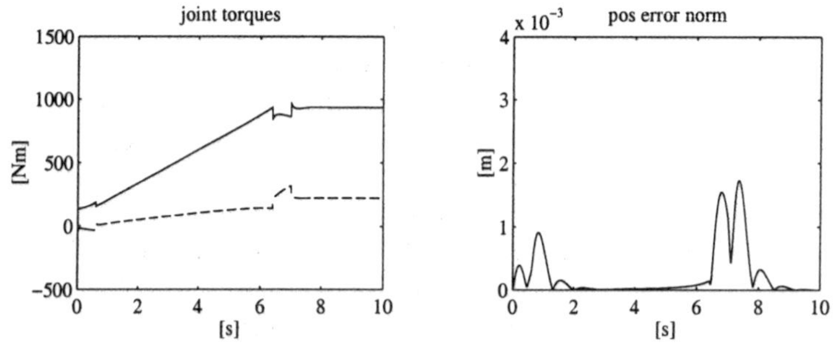

Figure 6.44 Time history of the joint torques and of the norm of tip position error for the *slow* trajectory with control scheme E.

6.4 For a single joint drive system with the data: $I = 6\,\text{kg·m}^2$, $R_a = 0.3\,\text{ohm}$, $k_t = 0.5\,\text{N·m/A}$, $k_v = 0.5\,\text{V·s/rad}$, $F_m = 0.001\,\text{N·m·s/rad}$, find the parameters of the controller with position feedback (unit transducer constant) that allows obtaining a closed-loop response with damping ratio $\zeta \geq 0.4$. Discuss disturbance rejection properties.

6.5 For the drive system of Problem 6.4, find the parameters of the controller with position and velocity feedback (unit transducer constants) that allows obtaining a closed-loop response with damping ratio $\zeta \geq 0.4$ and natural frequency $\omega_n = 20\,\text{rad/s}$. Discuss

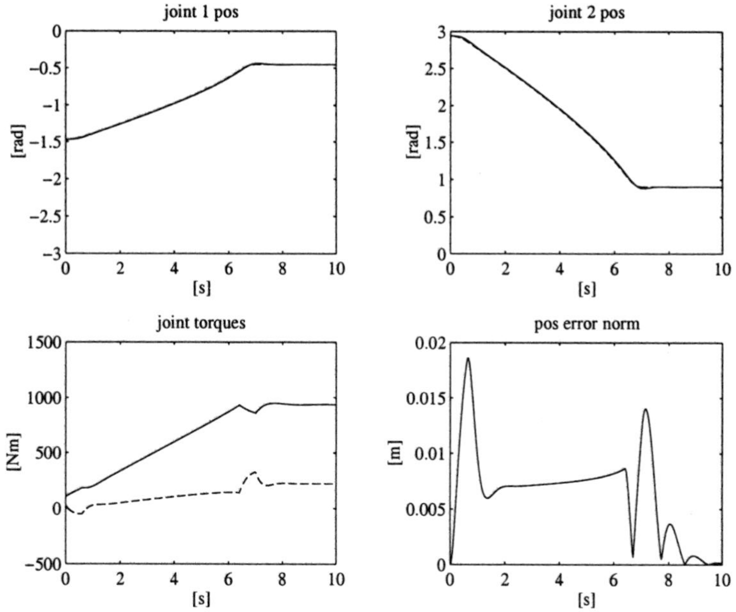

Figure 6.45 Time history of the joint positions and torques and of the norm of tip position error for the *slow* trajectory with control scheme F.

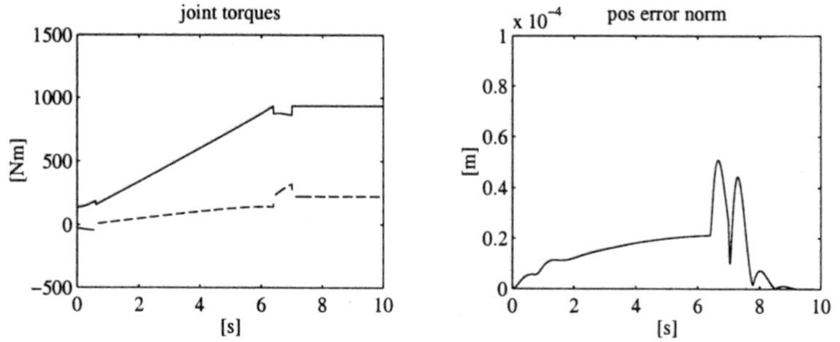

Figure 6.46 Time history of the joint torques and of the norm of tip position error for the *slow* trajectory with control scheme G.

disturbance rejection properties.

6.6 For the joint drive system of Problem 6.4, find the parameters of the controller with position, velocity and acceleration feedback (unit transducer constants) that allows obtaining a closed-loop response with damping ratio $\zeta \geq 0.4$, natural frequency $\omega_n = 20$ rad/s and disturbance rejection factor $X_R = 400$. Also, design a first-order filter that allows acceleration measurement reconstruction.

6.7 Verify that the control schemes in Figures 6.13, 6.14, 6.15 correspond to realize (6.28), (6.29), (6.30), respectively.

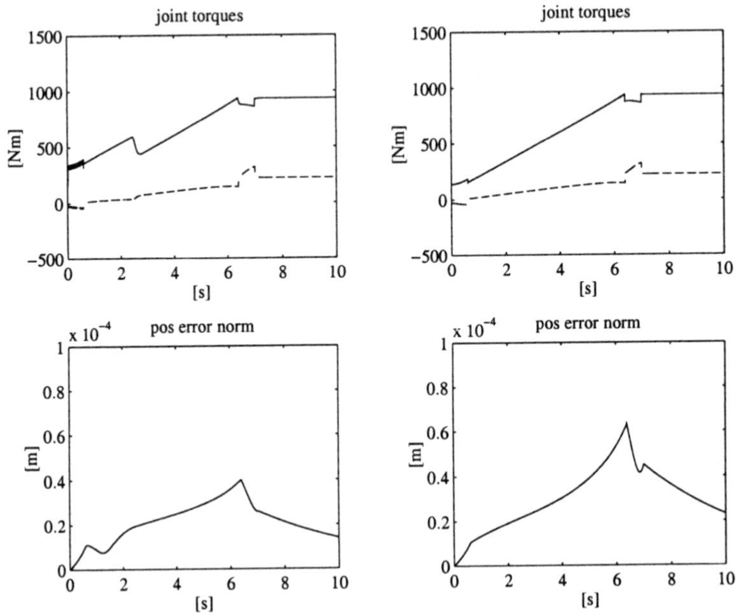

Figure 6.47 Time history of the joint torques and of the norm of tip position error for the *slow* trajectory; *left*—with control scheme H, *right*—with control scheme I.

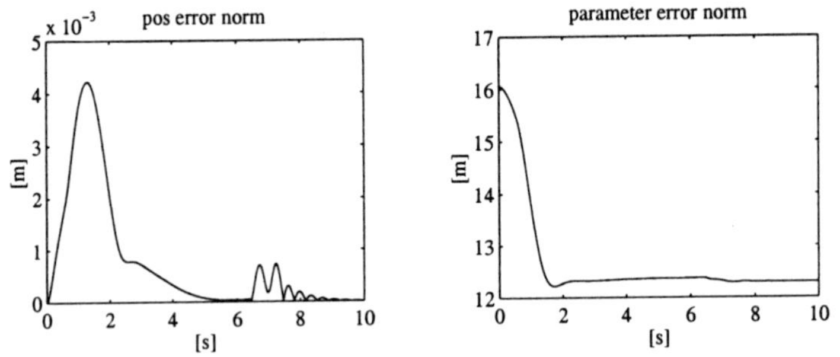

Figure 6.48 Time history of the norm of tip position error and of the norm of parameter error vector for the *slow* trajectory with control scheme J.

6.8 Verify that the standard regulation schemes in Figures 6.16, 6.17, 6.18 are equivalent to the schemes in Figures 6.13, 6.14, 6.15, respectively.

6.9 Prove inequality (6.68).

6.10 For the two-link planar arm with the same data as in Section 6.7, design a joint control of PD type with gravity compensation. By means of a computer simulation, verify stability for the following postures $\boldsymbol{q} = [\pi/4 \quad -\pi/2]^T$ and $\boldsymbol{q} = [-\pi \quad -3\pi/4]^T$, respectively. Implement the control in discrete-time with a sampling time of 1 ms.

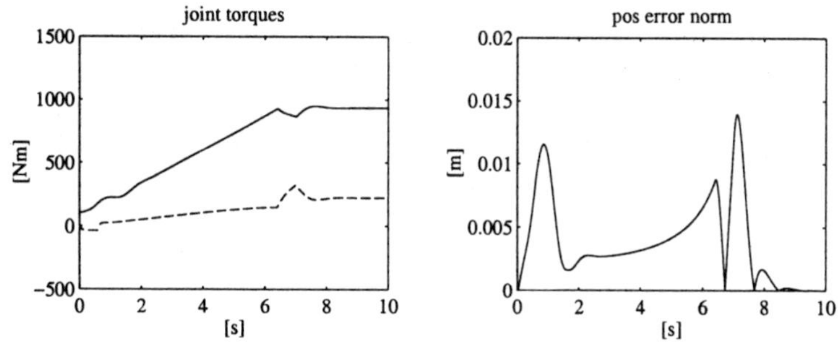

Figure 6.49 Time history of the joint torques and of the norm of tip position error for the *slow* trajectory with control scheme K.

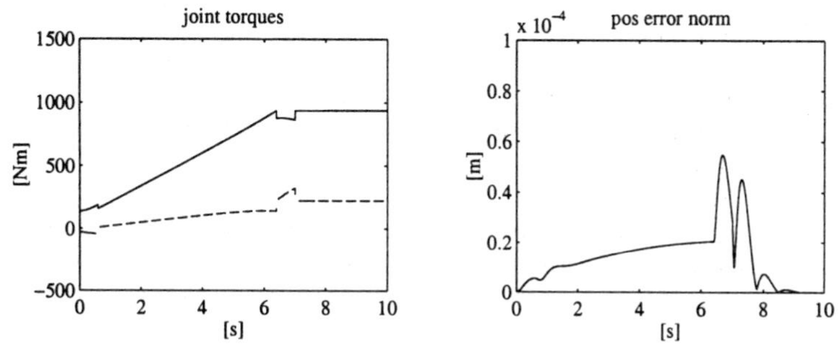

Figure 6.50 Time history of the joint torques and of the norm of tip position error for the *slow* trajectory with control scheme L.

6.11 For the two-link planar arm with the same data as in Section 6.7, on the assumption of a concentrated tip payload of mass $m_L = 10\,\text{kg}$, design an independent joint control with feedforward computed torque. Perform a computer simulation of the motion of the controlled arm along the joint space rectilinear path from $\boldsymbol{q}_i = [\,0 \quad \pi/4\,]^T$ to $\boldsymbol{q}_f = [\,\pi/2 \quad \pi/2\,]^T$ with a trapezoidal velocity profile and a trajectory duration $t_f = 1$ s. Implement the control in discrete-time with a sampling time of 1 ms.

6.12 For the two-link planar arm of Problem 6.11, design an inverse dynamics joint control. Perform a computer simulation of the motion of the controlled arm along the trajectory specified in Problem 6.11. Implement the control in discrete-time with a sampling time of 1 ms.

6.13 For the two-link planar arm of Problem 6.11, design a robust joint control. Perform a computer simulation of the motion of the controlled arm along the trajectory specified in Problem 6.11. Implement the control in discrete-time with a sampling time of 1 ms.

6.14 For the two-link planar arm of Problem 6.11, design an adaptive joint control, on the basis of a suitable parameterization of the arm dynamic model. Perform a computer simulation of the motion of the controlled arm along the trajectory specified in Problem 6.11. Implement the control in discrete-time with a sampling time of 1 ms.

6.15 For the two-link planar of Problem 6.11, design a PD control in the operational space with gravity compensation. By means of a computer simulation, verify stability for the following postures $\boldsymbol{p} = [\,0.5 \quad 0.5\,]^T$ and $\boldsymbol{p} = [\,0.6 \quad -0.2\,]^T$, respectively. Implement the control in discrete-time with a sampling time of 1 ms.

6.16 For the two-link planar arm of Problem 6.11, design an inverse dynamics control in the operational space. Perform a computer simulation of the motion of the controlled arm along the operational space rectlinear path from $\boldsymbol{p}(0) = [\,0.7 \quad 0.2\,]^T$ to $\boldsymbol{p}(1) = [\,0.1 \quad -0.6\,]^T$ with a trapezoidal velocity profile and a trajectory duration $t_f = 1$ s. Implement the control in discrete-time with a sampling time of 1 ms.

Bibliography

Abdallah C., Dawson D., Dorato P., Jamshidi M. (1991) Survey of robust control for rigid robots. *IEEE Control Systems Mag.* 11(2):24–30.

An C.H., Atkeson C.G., Hollerbach J.M. (1988) *Model-Based Control of a Robot Manipulator*. MIT Press, Cambridge, Mass.

Arimoto S., Miyazaki F. (1984) Stability and robustness of PID feedback control for robot manipulators of sensory capability. In *Robotics Research: The First International Symp.* M. Brady, R. Paul (Eds.), MIT Press, Cambridge, Mass., pp. 783–799.

Asada H., Slotine J.-J.E. (1986) *Robot Analysis and Control*. Wiley, New York.

Balestrino A., De Maria G., Sciavicco L. (1983) An adaptive model following control for robotic manipulators. *ASME J. Dynamic Systems, Measurement, and Control* 105:143–151.

Balestrino A., De Maria G., Sciavicco L. (1983) Adaptive control of manipulators in the task oriented space. In *Proc. 13th Int. Symp. Industrial Robots & Robots 7* Chicago, Ill., 13, pp. 13–28,

Bejczy A.K. (1974) *Robot Arm Dynamics and Control*. Memo. TM 33-669, Jet Propulsion Laboratory, California Institute of Technology.

Caccavale F., Natale C., Siciliano B., Villani, L. (1998) Resolved-acceleration control of robot manipulators: A critical review with experiments. *Robotica* 16:565–573.

Chiacchio P., Pierrot F., Sciavicco L., Siciliano B. (1993) Robust design of independent joint controllers with experimentation on a high-speed parallel robot. *IEEE Trans. Industrial Electronics* 40:393–403.

Chiacchio P., Sciavicco L., Siciliano B. (1990) The potential of model-based control algorithms for improving industrial robot tracking performance. In *Proc. IEEE Int. Work. Intelligent Motion Control* Istanbul, Turkey, pp. 831–836.

Corless M. (1989) Tracking controllers for uncertain systems: Application to a Manutec r3 robot. *ASME J. Dynamic Systems, Measurement, and Control* 111:609–618.

Craig J.J. (1988) *Adaptive Control of Mechanical Manipulators*. Addison-Wesley, Reading, Mass.

Craig J.J. (1989) *Introduction to Robotics: Mechanics and Control*. 2nd ed., Addison-Wesley, Reading, Mass.

Dubowsky S., DesForges D.T. (1979) The application of model referenced adaptive control to robotic manipulators. *ASME J. Dynamic Systems, Measurement, and Control* 101:193–200.

Freund E. (1982) Fast nonlinear control with arbitrary pole-placement for industrial robots and manipulators. *Int. J. Robotics Research* 1(1):65–78.

Horowitz R., Tomizuka M. (1986) An adaptive control scheme for mechanical manipulators—Compensation of nonlinearity and decoupling control. *ASME J. Dynamic Systems, Measurement, and Control* 108:127–135.

Hsia T.C.S., Lasky T.A., Guo Z. (1991) Robust independent joint controller design for industrial robot manipulators. *IEEE Trans. Industrial Electronics* 38:21–25.

Khatib O. (1987) A unified approach for motion and force control of robot manipulators: The operational space formulation. *IEEE J. Robotics and Automation* 3:43–53.

Khosla P.K., Kanade T. (1988) Experimental evaluation of nonlinear feedback and feedforward control schemes for manipulators. *Int. J. Robotics Research* 7(1):18–28.

Koivo A.J. (1989) *Fundamentals for Control of Robotic Manipulators*. Wiley, New York.

Kreutz K. (1989) On manipulator control by exact linearization. *IEEE Trans. Automatic Control* 34:763–767.

Leahy M.B., Saridis G.N. (1989) Compensation of industrial manipulator dynamics. *Int. J. Robotics Research* 8(4):73–84.

Luh J.Y.S. (1983) Conventional controller design for industrial robots: A tutorial. *IEEE Trans. Systems, Man, and Cybernetics* 13:298–316.

Luh J.Y.S., Walker M.W., Paul R.P.C. (1980) Resolved-acceleration control of mechanical manipulators. *IEEE Trans. Automatic Control* 25:468–474.

Nicolò F., Katende J. (1983) A robust MRAC for industrial robots. In *Proc. 2nd IASTED Int. Symp. Robotics and Automation* Lugano, Switzerland, pp. 162–171.

Nicosia S., Tomei P. (1984) Model reference adaptive control algorithms for industrial robots. *Automatica* 20:635–644.

Ortega R., Spong M.W. (1989) Adaptive motion control of rigid robots: a tutorial. *Automatica* 25:877–888.

Paul R.P. (1981) *Robot Manipulators: Mathematics, Programming, and Control*. MIT Press, Cambridge, Mass.

Slotine J.-J.E. (1987) Robust control of robot manipulators. *Int. J. Robotics Research* 4(2):49–64.

Slotine J.-J.E., Li W. (1987) On the adaptive control of robot manipulators. *Int. J. Robotics Research* 6(3):49–59.

Slotine J.-J.E., Li W. (1988) Adaptive manipulator control: A case study. *IEEE Trans. Automatic Control* 33:995–1003.

Slotine J.-J.E., Sastry S.S. (1983) Tracking control of nonlinear systems using sliding surfaces with application to robot manipulators. *Int. J. Control* 38:465–492.

Spong M.W. (1992) On the robust control of robot manipulators. *IEEE Trans. Automatic Control* 37:1782–1786.

Spong M.W., Ortega R., Kelly R. (1990) Comments on "Adaptive manipulator control: A case study". *IEEE Trans. Automatic Control* 35:761–762.

Spong M.W., Vidyasagar M. (1987) Robust linear compensator design for nonlinear robotic control. *IEEE J. Robotics and Automation* 3:345–351.

Spong M.W., Vidyasagar M. (1989) *Robot Dynamics and Control*. Wiley, New York.

Takegaki M., Arimoto S. (1981) A new feedback method for dynamic control of manipulators. *ASME J. Dynamic Systems, Measurement, and Control* 102:119–125.

Tarn T.-J., Bejczy A.K., Yun X., Li Z. (1991) Effect of motor dynamics on nonlinear feedback robot arm control. *IEEE Trans. Robotics and Automation* 7:114–122.

Vukobratović M., Stokić D. (1982) *Control of Manipulation Robots: Theory and Application*. Scientific Fundamentals of Robotics 2, Springer-Verlag, Berlin.

Vukobratović M., Stokić D. (1989) *Applied Control of Manipulation Robots*. Springer-Verlag, Berlin.

Vukobratović M., Stokić D., Kirćanski N. (1985) *Non-Adaptive and Adaptive Control of Manipulation Robots*. Scientific Fundamentals of Robotics 5, Springer-Verlag, Berlin.

7. Interaction Control

One of the fundamental requirements for the success of a manipulation task is the capability to handle *interaction between manipulator and environment*. The quantity that describes the state of interaction more effectively is the *contact force* at the manipulator's end effector. High values of contact force are generally undesirable since they may stress both the manipulator and the manipulated object. In this chapter, performance of operational space motion control schemes is studied first. The concepts of mechanical *compliance* and *impedance* are defined, with special regard to the problem of integrating contact force measurements into the control strategy. Then, *force control* schemes are presented which are obtained from motion control schemes suitably modified by the closure of an outer force regulation feedback loop. For the planning of control actions to perform an interaction task, *natural constraints* set by the task geometry and *artificial constraints* set by the control strategy are established; the constraints are expressed in a suitable constraint frame. The formulation is conveniently exploited to derive a *hybrid force/position control* scheme.

7.1 Manipulator Interaction with Environment

Control of interaction between a robot manipulator and the environment is crucial for successful execution of a number of practical tasks where the robot end effector has to manipulate an object or perform some operation on a surface. Typical examples include polishing, deburring, machining or assembly. A complete classification of possible robot tasks is practically infeasible in view of the large variety of cases that may occur, nor would such a classification be really useful to find a general strategy to *control interaction with environment*.

During interaction, the environment sets constraints on the geometric paths that can be followed by the end effector. This situation is generally referred to as *constrained motion*. In such a case, the use of a purely motion control strategy for controlling interaction is a candidate to fail, as explained below.

Successful execution of an interaction task with the environment by using motion control could be obtained only if the task were accurately planned. This would in turn require an accurate model of both the robot manipulator (kinematics and dynamics) and the environment (geometry and mechanical features). Manipulator modelling can be known with enough precision, but a detailed description of the environment is difficult to obtain.

To understand the importance of task planning accuracy, it is sufficient to observe that to perform a mechanical part mating with a positional approach, the relative positioning of the parts should be guaranteed with an accuracy of an order of magnitude greater than part mechanical tolerance. Once the absolute position of one part is exactly known, the manipulator should guide the motion of the other with the same accuracy.

In practice, the planning errors may give rise to a contact force causing a deviation of the end effector from the desired trajectory. On the other hand, the control system reacts to reduce such deviation. This ultimately leads to a build-up of the contact force until saturation of the joint actuators is reached or breakage of the parts in contact occurs.

The higher the environment stiffness and position control accuracy are, the easier a situation like the one just described can occur. This drawback can be overcome if a *compliant behaviour* is ensured during the interaction.

From the above discussion it should be clear that the *contact force* is the quantity describing the state of interaction in the most complete fashion; to this purpose, the availability of force measurements is expected to provide enhanced performance for controlling interaction.

Interaction control strategies can be grouped in two categories; those performing *indirect force control* and those performing *direct force control*. The main difference between the two categories is that the former achieve force control via motion control, without explicit closure of a force feedback loop; the latter, instead, offer the possibility of controlling the contact force to a desired value, thanks to the closure of a force feedback loop. To the first category belong *compliance control* and *impedance control* which are treated next. Then, *force control* schemes will follow.

7.2 Compliance Control

For a detailed analysis of interaction between the manipulator and environment it is worth considering the behaviour of the system under a position control scheme when contact forces arise. Since these are naturally described in the operational space, it is convenient to refer to *operational space control* schemes.

Consider the manipulator dynamic model (6.35). In view of (4.41), the model can be written as

$$\boldsymbol{B}(\boldsymbol{q})\ddot{\boldsymbol{q}} + \boldsymbol{C}(\boldsymbol{q},\dot{\boldsymbol{q}})\dot{\boldsymbol{q}} + \boldsymbol{F}\dot{\boldsymbol{q}} + \boldsymbol{g}(\boldsymbol{q}) = \boldsymbol{u} - \boldsymbol{J}^T(\boldsymbol{q})\boldsymbol{h}, \tag{7.1}$$

where $\boldsymbol{h}$ is the vector of contact forces exerted by the manipulator's end-effector on the environment.

It is reasonable to predict that, in the case $\boldsymbol{h} \neq \boldsymbol{0}$, the control scheme based on (6.102) no longer ensures that the end effector reaches its desired posture $\boldsymbol{x}_d$. In fact, by recalling that $\widetilde{\boldsymbol{x}} = \boldsymbol{x}_d - \boldsymbol{x}$, at the equilibrium it is

$$\boldsymbol{J}_A^T(\boldsymbol{q})\boldsymbol{K}_P\widetilde{\boldsymbol{x}} = \boldsymbol{J}^T(\boldsymbol{q})\boldsymbol{h}. \tag{7.2}$$

On the assumption of a full-rank Jacobian, one has

$$\widetilde{\boldsymbol{x}} = \boldsymbol{K}_P^{-1}\boldsymbol{T}_A^T(\boldsymbol{x})\boldsymbol{h} = \boldsymbol{K}_P^{-1}\boldsymbol{h}_A, \tag{7.3}$$

where $\boldsymbol{h}_A$ is the vector of equivalent generalized forces that can be defined according to (4.118). The expression in (7.3) shows that at the equilibrium the manipulator, under a position control action, behaves as a generalized spring in the operational space with *compliance* $\boldsymbol{K}_P^{-1}$ in respect of force $\boldsymbol{h}_A$. By recalling the expression of the transformation matrix $\boldsymbol{T}_A$ in (3.61) and assuming matrix $\boldsymbol{K}_P$ to be diagonal, it can be recognized that linear compliance (due to force components) is independent of the posture, whereas torsional compliance (due to moment components) does depend on the current manipulator configuration through the matrix $\boldsymbol{T}_A$.

On the other hand, if $\boldsymbol{h} \in \mathcal{N}(\boldsymbol{J}^T)$, one has $\widetilde{\boldsymbol{x}} = \boldsymbol{0}$ with $\boldsymbol{h} \neq \boldsymbol{0}$, *i.e.*, contact forces are completely balanced by the manipulator mechanical structure; for instance, the anthropomorphic manipulator at a shoulder singularity in Figure 3.13 does not react to any force orthogonal to the plane of the structure.

For a better understanding of interaction between manipulator and environment, it is necessary to have an analytical description of contact forces. A real contact is a naturally distributed phenomenon in which the local characteristics of both manipulator and environment are involved. In addition, friction effects between parts typically exist which greatly complicate the nature of the contact itself.

A detailed description of the contact is demanding from a modelling viewpoint. To point out the fundamental aspects of interaction control, it is convenient to resort to a simple but significant model of contact. To this purpose, a decoupled *elastically compliant environment* is considered which is described by the model

$$\boldsymbol{h} = \begin{bmatrix} \boldsymbol{f} \\ \boldsymbol{\mu} \end{bmatrix} = \begin{bmatrix} \boldsymbol{K}_f & \boldsymbol{O} \\ \boldsymbol{O} & \boldsymbol{K}_\mu \end{bmatrix} \begin{bmatrix} d\boldsymbol{p} \\ \boldsymbol{\omega} dt \end{bmatrix} = \boldsymbol{K} \begin{bmatrix} d\boldsymbol{p} \\ \boldsymbol{\omega} dt \end{bmatrix}, \tag{7.4}$$

where $d\boldsymbol{p}$ is the vector of translation along the reference frame axes and $\boldsymbol{\omega} dt$ is the vector of small rotation about the axes of such frame as in (3.94). Hence, the vector $[\, d\boldsymbol{p}^T \quad \boldsymbol{\omega}^T dt\,]^T$ describes a generalized displacement from the environment rest position. The *stiffness matrix* $\boldsymbol{K}$ is typically *positive semi-definite*. In fact, the environment does not generate reaction forces along those directions where unconstrained end-effector motion is allowed.

In view of (3.61), the expression in (7.4) can be written in terms of operational space variables as

$$\boldsymbol{h} = \boldsymbol{K}\boldsymbol{T}_A(\boldsymbol{x}) d\boldsymbol{x} \tag{7.5}$$

where $d\boldsymbol{x}$ denotes the operational space generalized displacement with respect to the undeformed environment rest position $\boldsymbol{x}_e$, *i.e.*,

$$d\boldsymbol{x} = \boldsymbol{x} - \boldsymbol{x}_e. \tag{7.6}$$

Resorting to (4.118) and (7.6) gives

$$\boldsymbol{h}_A = \boldsymbol{T}_A^T(\boldsymbol{x})\boldsymbol{K}\boldsymbol{T}_A(\boldsymbol{x}) d\boldsymbol{x} = \boldsymbol{K}_A(\boldsymbol{x})(\boldsymbol{x} - \boldsymbol{x}_e) \tag{7.7}$$

that allows relating the equivalent forces on the manipulator with the environment deformation through the matrix $\boldsymbol{K}_A$, *i.e.*, the environment stiffness matrix. The matrix $\boldsymbol{K}_A^{-1}$, if it can be defined, is the environment *compliance* matrix. It represents a *passive* compliance since it describes an inherent property of the environment in the operational space chosen to express manipulator end-effector position and orientation. By recalling that $\boldsymbol{K}_A$ is only positive semi-definite, the concept of compliance cannot be globally defined in all operational space but only along those directions, spanning $\mathcal{R}(\boldsymbol{K}_A)$, along which end-effector motion is constrained by the environment.

On the other hand, notice that the matrix $\boldsymbol{K}_P^{-1}$ in (7.3) represents an *active compliance* since it is performed on the manipulator by a suitable position control action. With the environment model (7.7), the expression in (7.3) becomes

$$\widetilde{\boldsymbol{x}} = \boldsymbol{K}_P^{-1}\boldsymbol{K}_A(\boldsymbol{x})(\boldsymbol{x} - \boldsymbol{x}_e); \tag{7.8}$$

at the equilibrium, the end-effector location is given by

$$\boldsymbol{x}_\infty = \big(\boldsymbol{I} + \boldsymbol{K}_P^{-1}\boldsymbol{K}_A(\boldsymbol{x})\big)^{-1}(\boldsymbol{x}_d + \boldsymbol{K}_P^{-1}\boldsymbol{K}_A(\boldsymbol{x})\boldsymbol{x}_e), \tag{7.9}$$

while the contact force can be shown to be

$$\boldsymbol{h}_{A\infty} = \big(\boldsymbol{I} + \boldsymbol{K}_A(\boldsymbol{x})\boldsymbol{K}_P^{-1}\big)^{-1}\boldsymbol{K}_A(\boldsymbol{x})(\boldsymbol{x}_d - \boldsymbol{x}_e). \tag{7.10}$$

Analysis of (7.9) shows that the equilibrium position depends on the environment rest position as well as on the desired position imposed by the control system to the manipulator. The interaction of the two systems (environment and manipulator) is influenced by the mutual weight of the respective compliance features. It is then possible to increase the active compliance so that the manipulator dominates the environment and vice versa. Such a dominance can be specified with reference to the single directions of the operational space. For a given environment stiffness, according to the prescribed interaction task, one may choose large values of the elements of $\boldsymbol{K}_P$ for those directions along which the environment has to comply and small values of the elements of $\boldsymbol{K}_P$ for those directions along which the manipulator has to comply.

Expression (7.10) gives the value of the contact force at the equilibrium, which reveals that it may be appropriate to tune manipulator compliance with environment compliance along certain directions of the operational space. In fact, along a direction with high environment stiffness, it is better to have a compliant manipulator so that it can taper the intensity of interaction through a suitable choice of the desired position. In this case, the end-effector equilibrium position $\boldsymbol{x}_\infty$ practically coincides with the environment undeformed position $\boldsymbol{x}_e$, and the manipulator generates an interaction force, depending on the corresponding element of $\boldsymbol{K}_P$, that is determined by the choice of the component of $(\boldsymbol{x}_d - \boldsymbol{x}_e)$ along the relative direction.

In the dual case of high environment compliance, if the manipulator is made stiff, the end-effector equilibrium position $\boldsymbol{x}_\infty$ is very close to the desired position $\boldsymbol{x}_d$, and it is the environment to generate the elastic force along the constrained directions of interest.

In certain cases, it is possible to employ mechanical devices interposed between the manipulator's end effector and the environment so as to change passive compliance along particular directions of the operational space. For instance, in a peg-in-hole insertion task, the gripper is provided with a device ensuring high stiffness along the

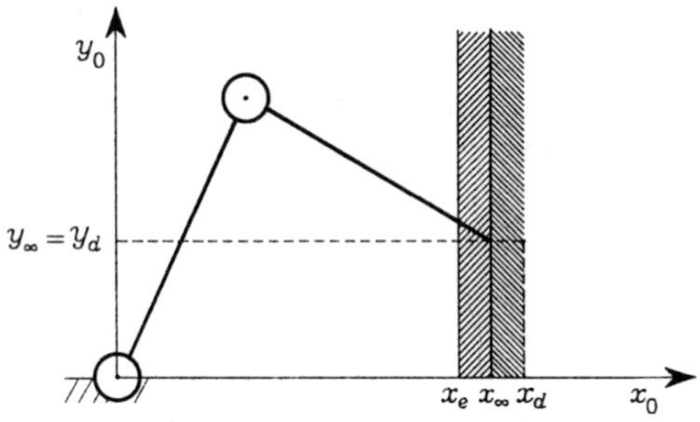

Figure 7.1 Two-link planar arm in contact with an elastically compliant plane.

insertion direction and high compliance along the other directions (*remote centre of compliance*). Therefore, in the presence of unavoidable position displacements from the planned insertion trajectory, contact forces and moments arise which modify the peg position so as to facilitate insertion.

The inconvenience of such devices is their low versatility to different operating conditions and generic interaction tasks, *i.e.*, whenever a modification of the compliant mechanical hardware is required. On the other hand, with active compliant actions the control software can be easily modified so as to satisfy the requirements of different interaction tasks.

Example 7.1

Consider the two-link planar arm whose tip is in contact with a purely frictionless elastic plane; let x_e be the equilibrium position of the plane, which is assumed to be orthogonal to axis x (Figure 7.1). The environment stiffness matrix is

$$\boldsymbol{K}_A = \boldsymbol{K}_f = \mathrm{diag}\{k_x, 0\},$$

corresponding to the absence of interaction forces along the vertical direction ($f_y = 0$). Let $\boldsymbol{p}_d = [\, x_d \quad y_d \,]^T$ be the desired tip position, which is located beyond the contact plane. The proportional control action on the arm is characterized by

$$\boldsymbol{K}_P = \mathrm{diag}\{k_{Px}, k_{Py}\}.$$

The equilibrium equations for the position in (7.9) and the force in (7.10) give

$$\boldsymbol{p}_\infty = \begin{bmatrix} \dfrac{k_{Px}x_d + k_x x_e}{k_{Px} + k_x} \\ y_d \end{bmatrix} \qquad \boldsymbol{f}_\infty = \begin{bmatrix} \dfrac{k_{Px}k_x}{k_{Px} + k_x}(x_d - x_e) \\ 0 \end{bmatrix}.$$

With reference to positioning accuracy, the arm tip reaches the vertical coordinate y_d, since the vertical motion direction is not constrained. As for the horizontal direction, the presence of the elastic plane imposes that the arm can move as far as it reaches the coordinate x_∞. The value of the horizontal contact force at the equilibrium is related to the difference between x_e and x_d by an equivalent stiffness coefficient which is given by the parallel composition of the stiffness coefficients of the two interacting systems. Hence, the arm stiffness and environment stiffness influence the resulting equilibrium configuration. In the case when

$$k_{Px}/k_x \gg 1,$$

it is

$$x_\infty \approx x_d \qquad f_{x\infty} \approx k_x(x_d - x_e)$$

and thus the arm prevails over the environment, in that the plane complies almost up to x_d and the elastic force is mainly generated by the environment (passive compliance). In the opposite case

$$k_{Px}/k_x \ll 1,$$

it is

$$x_\infty \approx x_e \qquad f_{x\infty} \approx k_{Px}(x_d - x_e)$$

and thus the environment prevails over the arm which complies up to the equilibrium x_e, and the elastic force is mainly generated by the arm (active compliance).

To complete the analysis of manipulator compliance in contact with environment, it is worth considering the effects of a joint space position control law. With reference to (6.43), in the presence of end-effector contact forces, the equilibrium posture is determined by

$$\boldsymbol{K}_P\widetilde{\boldsymbol{q}} = \boldsymbol{J}^T(\boldsymbol{q})\boldsymbol{h} \tag{7.11}$$

and then

$$\widetilde{\boldsymbol{q}} = \boldsymbol{K}_P^{-1}\boldsymbol{J}^T(\boldsymbol{q})\boldsymbol{h}. \tag{7.12}$$

On the assumption of small displacements from the equilibrium, it is reasonable to compute the resulting operational space displacement as $\widetilde{\boldsymbol{x}} \approx \boldsymbol{J}_A(\boldsymbol{q})\widetilde{\boldsymbol{q}}$ which, in view of (4.118) and (7.12), can be written as

$$\widetilde{\boldsymbol{x}} = \boldsymbol{J}_A(\boldsymbol{q})\boldsymbol{K}_P^{-1}\boldsymbol{J}_A^T(\boldsymbol{q})\boldsymbol{h}_A. \tag{7.13}$$

An active compliance in the joint space has then been obtained. Notice that, in this case, the equivalent compliance matrix $\boldsymbol{J}_A(\boldsymbol{q})\boldsymbol{K}_P^{-1}\boldsymbol{J}_A^T(\boldsymbol{q})$ is always dependent on the manipulator configuration, both for the force and moment components. Also in this case, the occurrence of manipulator Jacobian singularities is to be analyzed apart.

7.3 Impedance Control

It is now desired to analyze the interaction of manipulator with environment under the action of an inverse dynamics control in the operational space. With reference to model (7.1), consider the control law (6.49)

$$\boldsymbol{u} = \boldsymbol{B}(\boldsymbol{q})\boldsymbol{y} + \boldsymbol{n}(\boldsymbol{q}, \dot{\boldsymbol{q}}),$$

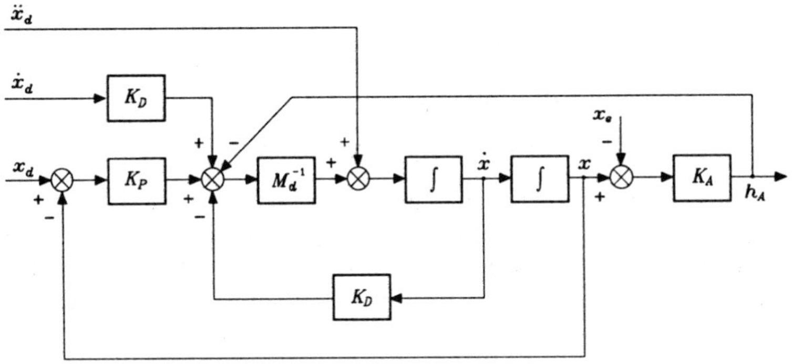

Figure 7.2 Equivalent block scheme of a manipulator in contact with an elastic environment under impedance control.

with n as in (6.48). In the presence of end-effector forces, the controlled manipulator is described by

$$\ddot{q} = y - B^{-1}(q)J^T(q)h \tag{7.14}$$

that reveals the existence of a nonlinear coupling term due to contact forces. Choose y in a way conceptually analogous to (6.106) as

$$y = J_A^{-1}(q)M_d^{-1}\left(M_d\ddot{x}_d + K_D\dot{\tilde{x}} + K_P\tilde{x} - M_d\dot{J}_A(q,\dot{q})\dot{q}\right), \tag{7.15}$$

where M_d is a positive definite diagonal matrix. Substituting (7.15) into (7.14) and accounting for second-order differential kinematics in the form (4.117) yields

$$M_d\ddot{\tilde{x}} + K_D\dot{\tilde{x}} + K_P\tilde{x} = M_dB_A^{-1}(q)h_A, \tag{7.16}$$

where

$$B_A(q) = J_A^{-T}(q)B(q)J_A^{-1}(q)$$

is the inertia matrix of the manipulator in the operational space as in (4.120′); this matrix is configuration-dependent and is positive definite if J_A has full rank.

The expression in (7.16) establishes a relationship through a generalized *mechanical impedance* between the vector of resulting forces $M_dB_A^{-1}h_A$ and the vector of displacements $\tilde{x}$ in the operational space. This impedance can be attributed to a mechanical system characterized by a mass matrix M_d, a damping matrix K_D, and a stiffness matrix K_P, which allow specifying the dynamic behaviour along the operational space directions.

The presence of B_A^{-1} makes the system coupled. If it is wished to keep linearity and decoupling during interaction with the environment, it is then necessary to *measure* the generalized contact *force*; this can be achieved by means of appropriate force sensors which are usually mounted to the manipulator wrist[1]. Choosing

$$u = B(q)y + n(q,\dot{q}) + J^T(q)h \tag{7.17}$$

[1] See the next chapter for a treatment of force sensors.

with

$$\boldsymbol{y} = \boldsymbol{J}_A^{-1}(\boldsymbol{q})\boldsymbol{M}_d^{-1}\big(\boldsymbol{M}_d\ddot{\boldsymbol{x}}_d + \boldsymbol{K}_D\dot{\widetilde{\boldsymbol{x}}} + \boldsymbol{K}_P\widetilde{\boldsymbol{x}} - \boldsymbol{M}_d\dot{\boldsymbol{J}}_A(\boldsymbol{q},\dot{\boldsymbol{q}})\dot{\boldsymbol{q}} - \boldsymbol{h}_A\big), \tag{7.18}$$

on the assumption of error-free force measurements, yields

$$\boldsymbol{M}_d\ddot{\widetilde{\boldsymbol{x}}} + \boldsymbol{K}_D\dot{\widetilde{\boldsymbol{x}}} + \boldsymbol{K}_P\widetilde{\boldsymbol{x}} = \boldsymbol{h}_A. \tag{7.19}$$

It is worth noticing that the addition of the term $\boldsymbol{J}^T\boldsymbol{h}$ in (7.17) exactly compensates the contact forces and then it renders the manipulator infinitely stiff with respect to external stress. In order to confer a compliant behaviour to the manipulator, the term $-\boldsymbol{J}_A^{-1}\boldsymbol{M}_d^{-1}\boldsymbol{h}_A$ has been introduced in (7.18) which allows characterizing the manipulator as a *linear impedance* with regard to the equivalent forces $\boldsymbol{h}_A$, as shown in (7.19). The resulting block scheme of a manipulator in contact with an elastic environment under impedance control is illustrated in Figure 7.2.

The behaviour of the system in (7.19) at the equilibrium is analogous to that described by (7.2); nonetheless, compared to a compliance control specified by $\boldsymbol{K}_P$, the equation in (7.19) allows a complete characterization of system dynamics through an *active impedance* specified by matrices $\boldsymbol{M}_d$, $\boldsymbol{K}_D$, $\boldsymbol{K}_P$. These matrices are usually taken as diagonal; also in this case, it is not difficult to recognize that impedance is configuration-independent as regards the force components, while it depends on the current manipulator configuration as regards the moment components through the matrix $\boldsymbol{T}_A$.

Furthermore, similarly to active and passive compliance, the concept of *passive impedance* can be introduced if the interaction force $\boldsymbol{h}_A$ is generated at the contact with an environment of proper mass, damping and stiffness. In this case, the system of manipulator with environment can be regarded as a mechanical system constituted by the parallel of the two impedances, and then its dynamic behaviour is conditioned by the relative weight between them. As pointed out above, one may think about constructing a mechanical device with proper passive impedance that allows the manipulator to better cope with the given interaction task.

Example 7.2

Consider the planar arm in contact with an elastically compliant plane of the previous example. Apply the impedance control with force measurement (7.17) and (7.18) characterized by:

$$\boldsymbol{M}_d = \text{diag}\{m_{dx}, m_{dy}\} \qquad \boldsymbol{K}_D = \text{diag}\{k_{Dx}, k_{Dy}\} \qquad \boldsymbol{K}_P = \text{diag}\{k_{Px}, k_{Py}\}.$$

If $\boldsymbol{x}_d$ is constant, the dynamics of the manipulator and environment system along the two directions of the operational space is described by

$$\begin{aligned} m_{dx}\ddot{x} + k_{Dx}\dot{x} + (k_{Px} + k_x)x &= k_x x_e + k_{Px}x_d \\ m_{dy}\ddot{y} + k_{Dy}\dot{y} + k_{Py}y &= k_{Py}y_d. \end{aligned}$$

Along the vertical direction, one has an unconstrained motion whose time behaviour is determined by the following natural frequency and damping factor:

$$\omega_{ny} = \sqrt{\frac{k_{Py}}{m_{dy}}} \qquad \zeta_y = \frac{k_{Dy}}{2\sqrt{m_{dy}k_{Py}}},$$

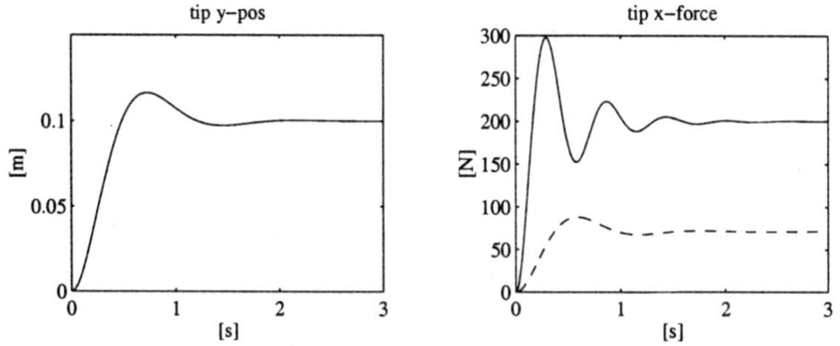

Figure 7.3 Time history of the tip position along vertical direction and of the contact force along horizontal direction with impedance control scheme for environments of different compliance.

while, along the horizontal direction, the behaviour of the contact force $f_x = k_x(x - x_e)$ is determined by

$$\omega_{nx} = \sqrt{\frac{k_{Px} + k_x}{m_{dx}}} \qquad \zeta_x = \frac{k_{Dx}}{2\sqrt{m_{dx}(k_{Px} + k_x)}}.$$

Below, the dynamic behaviour of the system is analyzed for two different values of environment compliance: $k_x = 10^3$ N/m and $k_x = 10^4$ N/m. The actual arm is the same as in Example 4.2. Apply an impedance control with force measurements of the kind (7.17) and (7.18), and PD control actions equivalent to those chosen in the simulations of Section 6.7, *i.e.*,

$$m_{dx} = m_{dy} = 100 \qquad k_{Dx} = k_{Dy} = 500 \qquad k_{Px} = k_{Py} = 2500.$$

For these values it is

$$\omega_{ny} = 5\,\text{rad/s} \qquad \zeta_y = 0.5.$$

Then, for the more compliant environment it is

$$\omega_{nx} \approx 5.9\,\text{rad/s} \qquad \zeta_x \approx 0.42,$$

whereas for the less compliant environment it is

$$\omega_{nx} \approx 11.2\,\text{rad/s} \qquad \zeta_x \approx 0.22.$$

Let the arm tip be in contact with the environment at position $\boldsymbol{p} = [\,1 \quad 0\,]^T$; it is desired to take it to position $\boldsymbol{p}_d = [\,1.1 \quad 0.1\,]^T$.

The results in Figure 7.3 show that motion dynamics along the vertical direction is the same in the two cases. As regards the contact force along the horizontal direction, for the more compliant environment (*dashed line*) a well-damped behaviour is obtained, whereas for the less compliant environment (*solid line*) the resulting behaviour is less damped. Further, at the equilibrium, in the first case a displacement of 7.14 cm with a contact force of 71.4 N are observed, whereas in the second case a displacement of 2 cm with a contact force of 200 N are observed.

7.4 Force Control

In the above schemes, the interaction force could be indirectly controlled by acting on the reference value $\boldsymbol{x}_d$ of the manipulator motion control system. Interaction between manipulator and environment is anyhow directly influenced by compliance of the environment and by either compliance or impedance of the manipulator.

If it is desired to accurately control the contact force, it is necessary to devise control schemes that allow directly specifying the desired interaction force. The development of a *force control* system, in analogy to a motion control system, would require the adoption of a stabilizing PD control action on the force error besides the usual nonlinear compensation actions. Force measurements may be corrupted by noise, and then a derivative action may not be implemented in practice. The stabilizing action is to be provided by suitable damping of velocity terms. As a consequence, a force control system typically features a control law based not only on force measurements but also on velocity measurements, and eventually position measurements, too.

The realization of a force control scheme can be entrusted to the closure of an *outer force regulation feedback loop* generating the control input for the motion control scheme the manipulator is usually endowed with. Therefore, force control schemes are presented below which are based on the use of an inverse dynamics position control. Nevertheless, notice that a force control strategy is meaningful only for those directions of the operational space along which interaction forces between manipulator and environment may arise.

7.4.1 Force Control with Inner Position Loop

With reference to the inverse dynamics law with force measurement (7.17), choose in lieu of (7.18) the control

$$\boldsymbol{y} = \boldsymbol{J}_A^{-1}(\boldsymbol{q})\boldsymbol{M}_d^{-1}\big(-\boldsymbol{K}_D\dot{\boldsymbol{x}} + \boldsymbol{K}_P(\boldsymbol{x}_F - \boldsymbol{x}) - \boldsymbol{M}_d\dot{\boldsymbol{J}}_A(\boldsymbol{q},\dot{\boldsymbol{q}})\dot{\boldsymbol{q}}\big) \tag{7.20}$$

where $\boldsymbol{x}_F$ is a suitable reference to be related to a force error. Notice that the control law (7.20) does not foresee the adoption of compensating actions relative to $\dot{\boldsymbol{x}}_F$ and $\ddot{\boldsymbol{x}}_F$. Substituting (7.20) into (7.17) leads, after similar algebraic manipulation as above, to the system described by

$$\boldsymbol{M}_d\ddot{\boldsymbol{x}} + \boldsymbol{K}_D\dot{\boldsymbol{x}} + \boldsymbol{K}_P\boldsymbol{x} = \boldsymbol{K}_P\boldsymbol{x}_F, \tag{7.21}$$

which shows how (7.17) and (7.20) perform a position control taking $\boldsymbol{x}$ to $\boldsymbol{x}_F$ with a dynamics specified by the choice of matrices $\boldsymbol{M}_d$, $\boldsymbol{K}_D$, $\boldsymbol{K}_P$.

Let $\boldsymbol{h}_{Ad}$ denote the desired *constant* force reference; the relation between $\boldsymbol{x}_F$ and the force error can be symbolically expressed as

$$\boldsymbol{x}_F = \boldsymbol{C}_F(\boldsymbol{h}_{Ad} - \boldsymbol{h}_A), \tag{7.22}$$

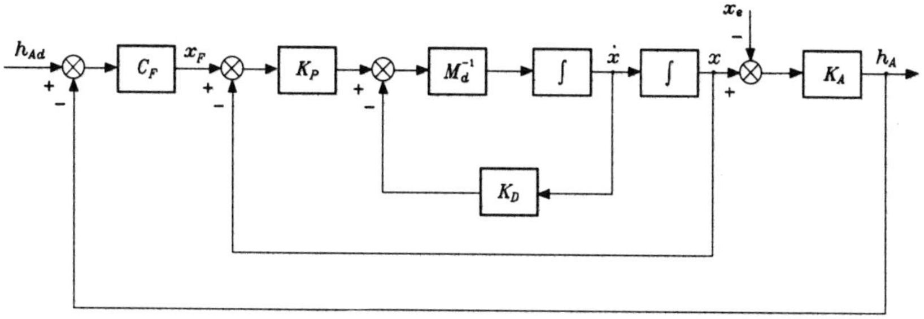

Figure 7.4 Block scheme of force control with inner position loop.

where C_F is a diagonal matrix whose elements give the control actions to perform along the operational space directions of interest. The equations in (7.21) and (7.22) reveal that force control is developed on the basis of a preexisting position control loop.

On the assumption of the elastically compliant environment described by (7.7), the equation in (7.21) with (7.22) becomes

$$\boldsymbol{M}_d\ddot{\boldsymbol{x}} + \boldsymbol{K}_D\dot{\boldsymbol{x}} + \boldsymbol{K}_P(\boldsymbol{I} + \boldsymbol{C}_F\boldsymbol{K}_A)\boldsymbol{x} = \boldsymbol{K}_P\boldsymbol{C}_F(\boldsymbol{K}_A\boldsymbol{x}_e + \boldsymbol{h}_{Ad}). \tag{7.23}$$

To decide about the kind of control action to specify with $\boldsymbol{C}_F$, it is worth representing (7.7), (7.21), (7.22) in terms of the block scheme in Figure 7.4, which is logically derived from the scheme in Figure 7.2. This scheme suggests that if $\boldsymbol{C}_F$ has a purely proportional control action, then $\boldsymbol{h}_A$ cannot reach $\boldsymbol{h}_{Ad}$, and $\boldsymbol{x}_e$ influences the interaction force also at steady state.

If $\boldsymbol{C}_F$ has also an integral control action on the components of generalized force, then it is possible to achieve $\boldsymbol{h}_A = \boldsymbol{h}_{Ad}$ at steady state and, at the same time, to reject the effect of $\boldsymbol{x}_e$ on $\boldsymbol{h}_A$. Hence, a convenient choice for $\boldsymbol{C}_F$ is a *proportional-integral* (PI) *action*

$$\boldsymbol{C}_F = \boldsymbol{K}_F + \boldsymbol{K}_I\int^t (\cdot)\, d\varsigma. \tag{7.24}$$

The dynamic system resulting from (7.23) and (7.24) is of third order, and then it is necessary to adequately choose the matrices $\boldsymbol{K}_D$, $\boldsymbol{K}_P$, $\boldsymbol{K}_F$, $\boldsymbol{K}_I$ in respect of the characteristics of the environment. Since the values of environment stiffness are typically high, the weight of the proportional and integral actions shall be contained; the choice of $\boldsymbol{K}_F$ and $\boldsymbol{K}_I$ influences the stability margins and the bandwidth of the system under force control. On the assumption that a stable equilibrium is reached, it is $\boldsymbol{h}_{A\infty} = \boldsymbol{h}_{Ad}$ and then

$$\boldsymbol{K}_A\boldsymbol{x}_\infty = \boldsymbol{K}_A\boldsymbol{x}_e + \boldsymbol{h}_{Ad}. \tag{7.25}$$

7.4.2 Force Control with Inner Velocity Loop

From the block scheme of Figure 7.4 it can be observed that, if the position feedback loop is opened, $\boldsymbol{x}_F$ represents a velocity reference, and then an integration relationship

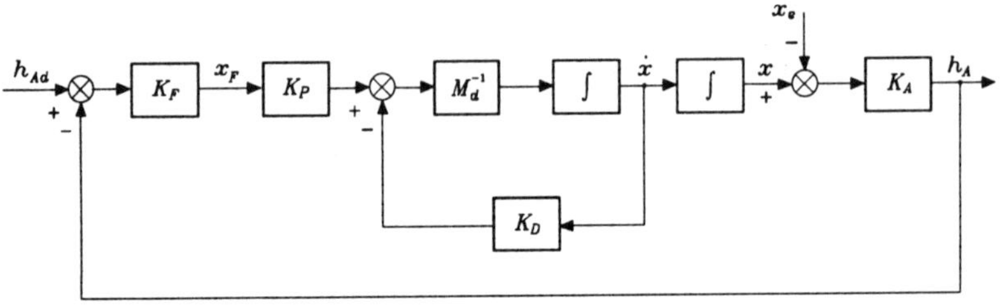

Figure 7.5 Block scheme of force control with inner velocity loop.

exists between $\boldsymbol{x}_F$ and $\boldsymbol{x}$. This leads to recognizing that, in this case, the interaction force with the environment coincides with the desired value at steady state, even with a proportional force controller $\boldsymbol{C}_F$. In fact, choosing

$$\boldsymbol{y} = \boldsymbol{J}_A^{-1}(\boldsymbol{q})\boldsymbol{M}_d^{-1}\big(-\boldsymbol{K}_D\dot{\boldsymbol{x}} + \boldsymbol{K}_P\boldsymbol{x}_F - \boldsymbol{M}_d\dot{\boldsymbol{J}}_A(\boldsymbol{q},\dot{\boldsymbol{q}})\dot{\boldsymbol{q}}\big) \tag{7.26}$$

with a purely proportional control structure ($\boldsymbol{C}_F = \boldsymbol{K}_F$) on the force error yields

$$\boldsymbol{x}_F = \boldsymbol{K}_F(\boldsymbol{h}_{Ad} - \boldsymbol{h}_A), \tag{7.27}$$

and then system dynamics is described by

$$\boldsymbol{M}_d\ddot{\boldsymbol{x}} + \boldsymbol{K}_D\dot{\boldsymbol{x}} + \boldsymbol{K}_P\boldsymbol{K}_F\boldsymbol{K}_A\boldsymbol{x} = \boldsymbol{K}_P\boldsymbol{K}_F(\boldsymbol{K}_A\boldsymbol{x}_e + \boldsymbol{h}_{Ad}). \tag{7.28}$$

The relationship between position and contact force at the equilibrium is given by (7.25). The corresponding block scheme is reported in Figure 7.5. It is worth emphasizing that control design is simplified, since the resulting system now is of second order[2]; it should be noticed, however, that the absence of an integral action in the force controller does not ensure reduction of the effects due to unmodeled dynamics.

7.4.3 Parallel Force/Position Control

The presented force control schemes require the force reference to be consistent with the geometrical features of the environment. In fact, if $\boldsymbol{h}_{Ad}$ has components outside $\mathcal{R}(\boldsymbol{K}_A)$, both (7.23) (in case of an integral action in $\boldsymbol{C}_F$) and (7.28) show that, along the corresponding operational space directions, the components of $\boldsymbol{h}_{Ad}$ are interpreted as velocity references which cause a drift of the end-effector position. If $\boldsymbol{h}_{Ad}$ is correctly planned along the directions outside $\mathcal{R}(\boldsymbol{K}_A)$, the resulting motion governed by the motion control action tends to take the end-effector position to zero in the case of (7.23), and the end-effector velocity to zero in the case of (7.28). Hence, the above

[2] The matrices $\boldsymbol{K}_P$ and $\boldsymbol{K}_F$ are not independent and one may refer to a single matrix $\boldsymbol{K}'_F = \boldsymbol{K}_P\boldsymbol{K}_F$.

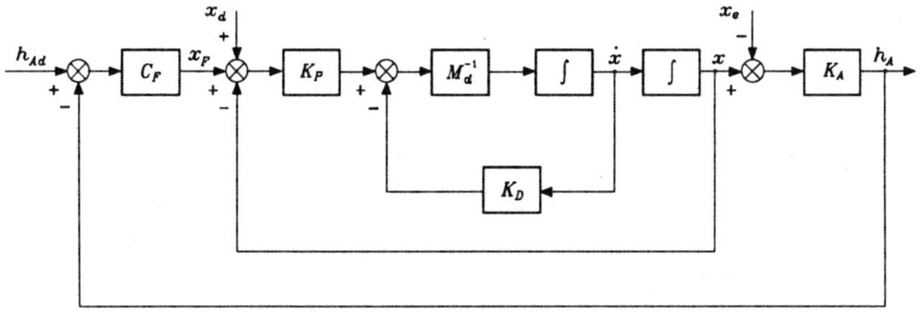

Figure 7.6 Block scheme of parallel force/position control.

control schemes do not allow motion control even along the admissible task space directions.

If it is desired to specify a desired end-effector location $\boldsymbol{x}_d$ as in pure motion control schemes, the scheme of Figure 7.4 can be modified by adding the reference $\boldsymbol{x}_d$ to the input where positions are summed. This corresponds to choosing

$$\boldsymbol{y} = \boldsymbol{J}_A^{-1}(\boldsymbol{q})\boldsymbol{M}_d^{-1}\big(-\boldsymbol{K}_D\dot{\boldsymbol{x}} + \boldsymbol{K}_P(\widetilde{\boldsymbol{x}} + \boldsymbol{x}_F) - \boldsymbol{M}_d\dot{\boldsymbol{J}}_A(\boldsymbol{q},\dot{\boldsymbol{q}})\dot{\boldsymbol{q}}\big) \qquad (7.29)$$

where $\widetilde{\boldsymbol{x}} = \boldsymbol{x}_d - \boldsymbol{x}$. The resulting scheme (Figure 7.6) is termed *parallel force/position control*, in view of the presence of a position control action $\boldsymbol{K}_P\widetilde{\boldsymbol{x}}$ in parallel to a force control action $\boldsymbol{K}_P\boldsymbol{C}_F(\boldsymbol{h}_{Ad} - \boldsymbol{h}_A)$. It is easy to verify that, in this case, the equilibrium position satisfies the equation

$$\boldsymbol{x}_\infty = \boldsymbol{x}_d + \boldsymbol{C}_F\big(\boldsymbol{K}_A(\boldsymbol{x}_e - \boldsymbol{x}_\infty) + \boldsymbol{h}_{Ad}\big). \qquad (7.30)$$

Therefore, along those directions outside $\mathcal{R}(\boldsymbol{K}_A)$ where motion is unconstrained, the position reference $\boldsymbol{x}_d$ is reached by $\boldsymbol{x}$. Vice versa, along those directions in $\mathcal{R}(\boldsymbol{K}_A)$ where motion is constrained, $\boldsymbol{x}_d$ is treated as an additional disturbance; the adoption of an integral action in $\boldsymbol{C}_F$ as for the scheme of Figure 7.4 ensures that the force reference $\boldsymbol{h}_{Ad}$ is reached at steady state, at the expense of a position error on $\boldsymbol{x}$ depending on environment compliance.

Example 7.3

Consider again the planar arm in contact with the elastically compliant plane of the above examples; let the initial contact position be the same as that of Example 7.2. Performance of the various force control schemes is analyzed; as in Example 7.2, a more compliant ($k_x = 10^3$ N/m) and a less compliant ($k_x = 10^4$ N/m) environment are considered. The position control actions $\boldsymbol{M}_d$, $\boldsymbol{K}_D$, $\boldsymbol{K}_P$ are chosen as in Example 7.2; a force control action is added along the horizontal direction, *i.e.*,

$$\boldsymbol{C}_F = \mathrm{diag}\{c_{Fx}, 0\}.$$

The reference for the contact force is chosen as $\boldsymbol{h}_{Ad} = [\,10 \quad 0\,]^T$; the position reference—meaningful only for the parallel control—is taken as $\boldsymbol{p}_d = [\,1.015 \quad 0.1\,]^T$.

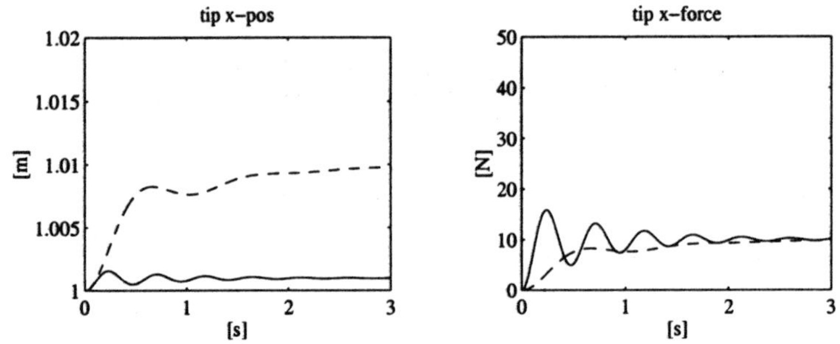

Figure 7.7 Time history of the tip position and of the contact force along horizontal direction with force control scheme with inner position loop for two environments of different compliance.

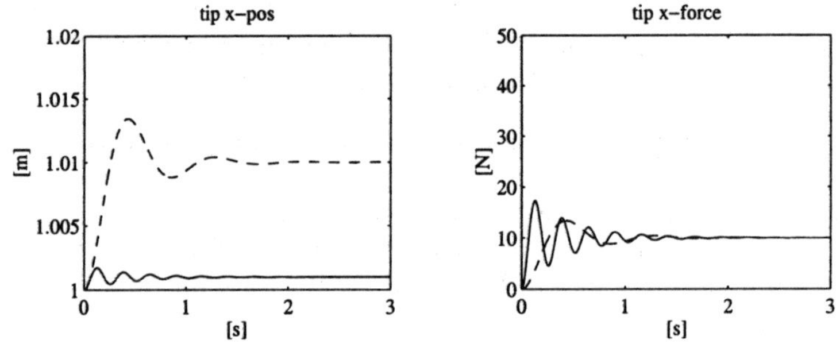

Figure 7.8 Time history of the tip position and of the contact force along horizontal direction with force control scheme with inner velocity loop for two environments of different compliance.

With regard to the scheme with inner position loop of Figure 7.4, a PI control action c_{Fx} is chosen with parameters:

$$k_{Fx} = 0.00064 \qquad k_{Ix} = 0.0016.$$

This confers two complex poles $(-1.96, \pm j5.74)$, a real pole (-1.09), and a real zero (-2.5) to the overall system, for the more compliant environment.

With regard to the scheme with inner velocity loop of Figure 7.5, the proportional control action is

$$k_{Fx} = 0.0024$$

so that the overall system, for the more compliant environment, has two complex poles $(-2.5, \pm j7.34)$.

With regard to the parallel control scheme of Figure 7.6, the PI control action c_{Fx} is chosen with the same parameters as for the first control scheme.

Figures. 7.7, 7.8, 7.9 report the time history of the tip position and contact force along axis x for the three considered schemes. A comparison between the various cases shows what follows.

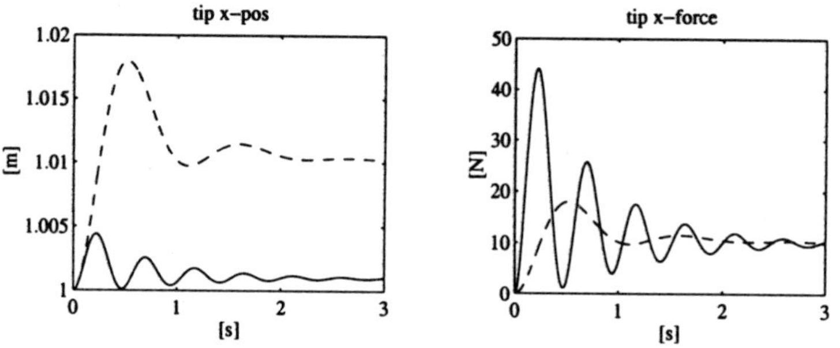

Figure 7.9 Time history of tip position and of the contact force along horizontal direction with parallel force/position control scheme for two environments of different compliance.

- All control laws guarantee a steady-state value of contact forces equal to the desired one for both the more compliant (*dashed line*) and the less compliant (*continuous line*) environment.
- For given motion control actions ($\boldsymbol{M}_d$, $\boldsymbol{K}_D$, $\boldsymbol{K}_P$), the force control with inner velocity loop presents a faster dynamics than that of the force control with inner position loop.
- The dynamic response with the parallel control shows how the addition of a position reference along the horizontal direction degrades the transient behaviour, but it does not influence the steady-state contact force. This effect can be justified by noting that a step position input is equivalent to a properly filtered impulse force input.

The reference position along axis y is obviously reached by the arm tip according to dynamics of position control; the relative time history is not reported.

7.5 Natural Constraints and Artificial Constraints

Interaction control schemes can be employed for execution of constrained motions as long as the force and position references are chosen to be compatible with environment geometry.

A real manipulation task is characterized by complex contact situations where some directions are subject to end-effector position constraints and others are subject to interaction force constraints. During task execution, the nature of constraints may vary substantially.

The need to handle complex contact situations requires the capability to specify and perform control of both end-effector position and contact force. An example is that of a surface finishing task where the tool motion is specified in the direction tangent to the piece, while along the normal direction it is desired to exert a force of given value.

A fundamental aspect to be considered is that it is not possible to simultaneously impose arbitrary values of position and force along each direction. As a consequence,

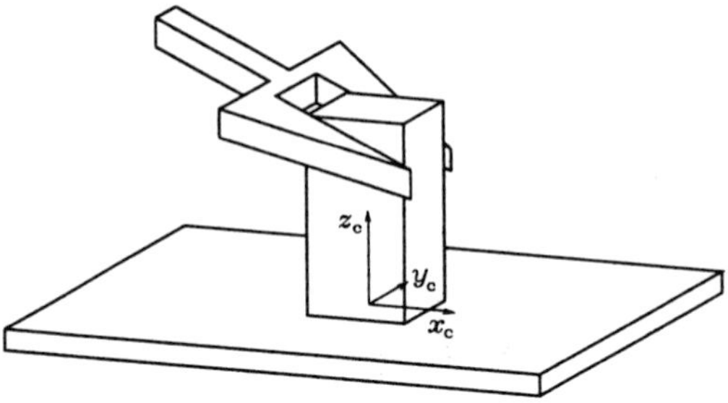

Natural Constraints	Artificial Constraints
$\dot{p}_z^c$	f_z^c
ω_x^c	μ_x^c
ω_y^c	μ_y^c
f_x^c	$\dot{p}_x^c$
f_y^c	$\dot{p}_y^c$
μ_z^c	ω_z^c

Figure 7.10 Sliding of a prismatic object on a planar surface. Variables subject to natural and artificial constraints.

one should ensure that the reference trajectories for the control system be compatible with the constraints imposed by the environment during task execution, so as to achieve a correct specification of the control problem.

A kineto-statics analysis of a situation of interaction between the manipulator and environment leads to the following considerations.

- Along each degree of freedom of the task space, the environment imposes either a position or a force constraint to the manipulator's end effector; such constraints are termed *natural constraints* since they are determined directly by task geometry.

- Along each degree of freedom of the task space, the manipulator can control only the variables which are not subject to natural constraints; the reference values for those variables are termed *artificial constraints* since they are imposed with regard to the strategy for executing the given task.

Notice that the two sets of constraints are complementary, in that they regard different variables for each degree of freedom. Also they allow a complete specification of the task, since they involve all variables. Kineto-statics variables are associated with these constraints, *i.e.*, velocities and generalized forces.

It should be pointed out that the above examples illustrating performance of motion and force control strategies have implicitly accounted for the presence of natural and artificial constraints. In view of the particular simple task geometry, the degrees of freedom were naturally described with reference to the base frame assumed to express operational space quantities.

In the general case, it is worth introducing a *constraint frame* O_c–$x_c y_c z_c$, not necessarily aligned with the base frame, in order to simplify task description and to allow determination of natural constraints and then specification of artificial constraints. The introduction of the constraint frame noticeably simplifies task planning, but the control strategy will obviously have to account for the rotation matrix—eventually time-varying during task execution—which is needed to transform quantities expressed in the base frame into the corresponding quantities in the constraint frame.

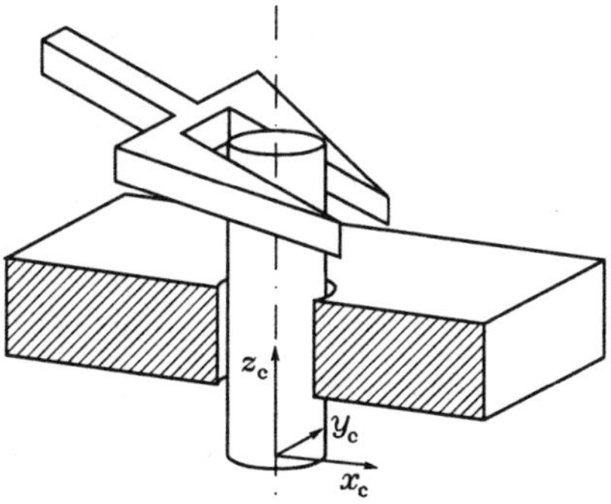

Natural Constraints	Artificial Constraints
$\dot{p}_x^c$	f_x^c
$\dot{p}_y^c$	f_y^c
ω_x^c	μ_x^c
ω_y^c	μ_y^c
f_z^c	$\dot{p}_z^c$
μ_z^c	ω_z^c

Figure 7.11 Insertion of a cylindrical peg in a hole. Variables subject to natural and artificial constraints.

7.5.1 Case Studies

To illustrate description of an interaction task in terms of natural and artificial constraints as well as to emphasize the opportunity to use a constraint frame, in the following a number of typical case studies are analyzed.

Sliding on a Planar Surface

The end-effector manipulation task is the sliding of a prismatic object on a planar surface. Task geometry suggests choosing the constraint frame as attached to the contact plane with an axis orthogonal to the plane (Figure 7.10).

Natural constraints can be determined first. Motion constraints describe the impossibility to generate arbitrary linear velocity along axis z_c and angular velocities about axes x_c and y_c; if the plane is rigid, then these velocities are null. Force constraints describe the impossibility to exert arbitrary forces along axes x_c and y_c and moment about axis z_c; if the plane is frictionless, these generalized forces are zero.

The artificial constraints regard the variables not subject to natural constraints. Hence, with reference to the natural generalized force constraints along axes x_c, y_c and about z_c, it is possible to specify artificial constraints for linear velocity along x_c, y_c and angular velocity about z_c. Similarly, with reference to natural velocity constraints along axis z_c and about axes x_c, y_c, it is possible to specify artificial constraints for force along z_c and moments about x_c, y_c. The set of constraints is summarized in the table in Figure 7.10.

Peg-in-hole

The end-effector manipulation task is the insertion of a cylindrical object (peg) in a hole. Task geometry suggests choosing the constraint frame with an axis parallel to the hole axis and the origin along this axis (Figure 7.11).

The natural constraints are determined by observing that it is not possible to generate arbitrary linear velocities along axes x_c, y_c and angular velocity about the same axes, nor is possible to exert arbitrary force along z_c and moment about z_c; all

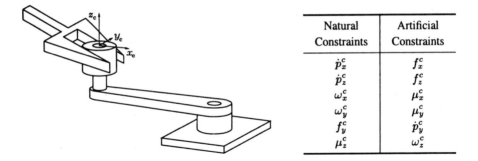

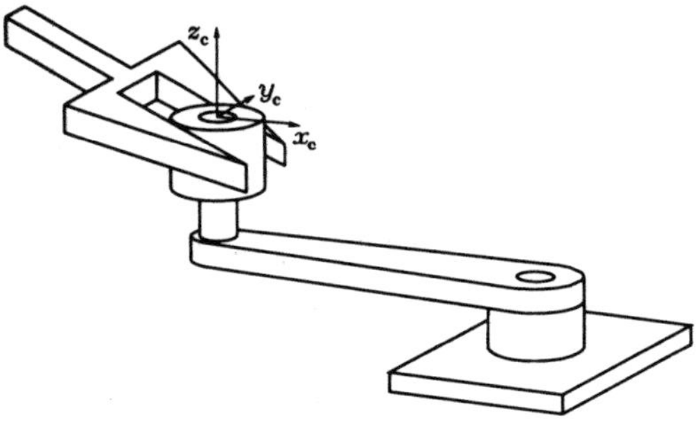

Natural Constraints	Artificial Constraints
$\dot{p}_x^c$	f_x^c
$\dot{p}_z^c$	f_z^c
ω_x^c	μ_x^c
ω_y^c	μ_y^c
f_y^c	$\dot{p}_y^c$
μ_z^c	ω_z^c

Figure 7.12 Turning a crank. Variables subject to natural and artificial constraints.

these variables are null in the case of rigid frictionless insertion. As a consequence, the artificial constraints allow specifying forces along x_c, y_c and moments about the same axes, as well as linear velocity along z_c and angular velocity about the same axis. The table in Figure 7.11 summarizes the constraints. Among the variables subject to artificial constraints, $\dot{p}_z^c \neq 0$ describes insertion while the others are typically null to effectively perform the task.

Turning a Crank

The end-effector manipulation task is the turning of a crank. Task geometry suggests choosing the constraint frame with an axis aligned with the axis of the idle handle and another axis aligned along the crank lever (Figure 7.12). Notice that in this case the constraint frame is time-varying.

The natural constraints do not allow generating arbitrary linear velocities along x_c, z_c and angular velocities about x_c, y_c, nor arbitrary force along y_c and moment about z_c. As a consequence, the artificial constraints allow specifying forces along x_c, z_c and moments about x_c, y_c, as well as a linear velocity along y_c and an angular velocity about z_c. The situation is summarized in the table in Figure 7.12. Among the variables subject to artificial constraints, forces and moments are typically null for task execution.

7.6 Hybrid Force/Position Control

Description of an interaction task between manipulator and environment in terms of natural constraints and artificial constraints, expressed with reference to the constraint frame, suggests a control structure that utilizes the artificial constraints to specify the objectives of the control system and allows controlling only those variables not subject to natural constraints. In fact, the control action shall not affect those variables constrained by the environment so as to avoid conflicts between control and interaction with environment that may lead to an improper system behaviour. Such a control structure is termed *hybrid force/position control*, since definition of artificial constraints

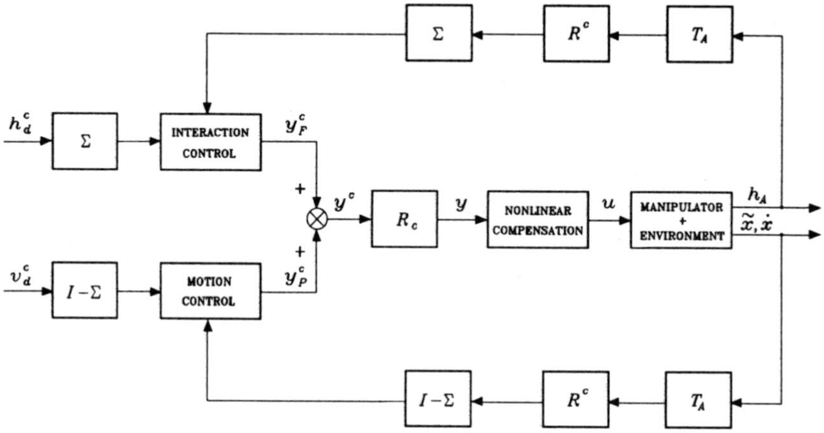

Figure 7.13 General block scheme of a hybrid force/position control structure.

involves both force and position variables.

The natural constraints are described by specifying a set of components of velocity and generalized force vectors expressed in the constraint frame. Such constraints can be written in compact form as

$$\boldsymbol{\Sigma v}^c = \boldsymbol{v}_n^c \qquad (\boldsymbol{I} - \boldsymbol{\Sigma})\boldsymbol{h}^c = \boldsymbol{h}_n^c, \tag{7.31}$$

where $\boldsymbol{\Sigma}$ is a diagonal matrix with either unit or null elements corresponding to the task components to be constrained. In (7.31), the natural constraints imposed by the environment are specified by $\boldsymbol{v}_n^c$ and $\boldsymbol{h}_n^c$; it is easy to find the matrix $\boldsymbol{\Sigma}$ for the case studies presented above. Matrices $\boldsymbol{\Sigma}$ and $(\boldsymbol{I} - \boldsymbol{\Sigma})$ are complementary, since for each degree of freedom one has a natural constraint regarding either a velocity or a force component.

Once the natural constraints have been expressed in the above form, the set of artificial constraints can be expressed in the form

$$(\boldsymbol{I} - \boldsymbol{\Sigma})\boldsymbol{v}^c = \boldsymbol{v}_a^c \qquad \boldsymbol{\Sigma h}^c = \boldsymbol{h}_a^c, \tag{7.32}$$

where $\boldsymbol{v}_a^c$ and $\boldsymbol{h}_a^c$ represent the components of velocity and generalized force that are taken to describe the artificial constraints. The matrix $\boldsymbol{\Sigma}$ is called *selection matrix*, since it allows selecting the appropriate control actions for each degree of freedom of the task.

The general block scheme of a hybrid force/position control structure is shown in Figure 7.13. It is assumed that all the output variables needed for force/position control are available.

Regarding motion control actions, in general, the end-effector position error $\tilde{\boldsymbol{x}}$ and velocity error $\dot{\boldsymbol{x}}$ are to be fed back; velocity and acceleration references are available at the input, once the artificial constraints have been specified. Regarding interaction control actions, in general the end-effector position error $\tilde{\boldsymbol{x}}$ and velocity $\dot{\boldsymbol{x}}$ are to be fed back besides the contact force $\boldsymbol{h}_A$. The force reference is available at the input, once the artificial constraints have been specified.

At first, it is necessary to use the matrix $\boldsymbol{T}_A$ to transform $\widetilde{\boldsymbol{x}}$ and $\dot{\boldsymbol{x}}$ into quantities homogeneous to those of the artificial constraints. To this regard, for limited position errors, it is reasonable to assimilate $\widetilde{\boldsymbol{x}}$ to a velocity vector and treat it in the same fashion as $\dot{\boldsymbol{x}}$. The transformation formally operates also on the force $\boldsymbol{h}_A$, although in reality the physical force $\boldsymbol{h}$ is available directly[3].

The obtained quantities are expressed with reference to the manipulator base frame; hence, it is necessary to transform them in the constraint frame where the task is naturally described. This is achieved by premultiplying them by the rotation matrix $\boldsymbol{R}^c$ transforming a vector from the base frame to the constraint frame. This transformation allows guaranteeing an effective decoupling of the task degrees of freedom through the use of selection matrices.

On the basis of the selection operated by the matrices $\boldsymbol{\Sigma}$ and $(\boldsymbol{I} - \boldsymbol{\Sigma})$ on the feedback quantities as well as on the corresponding references in view of the artificial constraints, it is possible to design interaction control and motion control actions along complementary directions of the task space. The actual control schemes to employ can be chosen among the variety of schemes presented in the above sections and in the previous chapter, with the notable exception of the parallel control which is to be regarded as an alternative solution to hybrid control.

The output vectors $\boldsymbol{y}_F^c$ and $\boldsymbol{y}_P^c$ of the control blocks can be summed to generate the control $\boldsymbol{y}^c$ expressing the actions on all task components. This vector is then to be transformed from the constraint frame back to the base frame by premultiplying its linear and angular components by the matrix $\boldsymbol{R}_c = (\boldsymbol{R}^c)^T$.

Finally, depending on the actual control scheme, it may be necessary to perform (partial or total) nonlinear compensating actions for both force and position. The resulting control law gives the joint control torques at the input of the manipulator.

According to the above analysis, the control structure is adapted to the task requirements, thanks to the use of selection matrices. Along each task component, the proper control action is activated by the selection matrix, while its dual one is ignored; this strategy avoids undesirable interference between the two controllers, since it structurally decouples force control actions from motion control actions in terms of the components of the given task.

The above considerations are founded on the assumption of perfect task planning. In those situations when hybrid control has to operate under imperfect task planning, the system behaviour may become quite critical. For instance, consider the extremal case when a hybrid controller governs manipulator motion in a situation of unplanned impact. As the selection matrix mechanism performs a model-based control logic which is not supported by any verification during task execution, it is not possible to modify the behaviour of the control scheme in respect of what actually happens in the environment. In fact, the selection matrices cancel part of the force sensor

---

[3] As will be seen in the next chapter, force measurements are typically available with reference to a frame attached to the force sensor mounted to the manipulator's wrist.

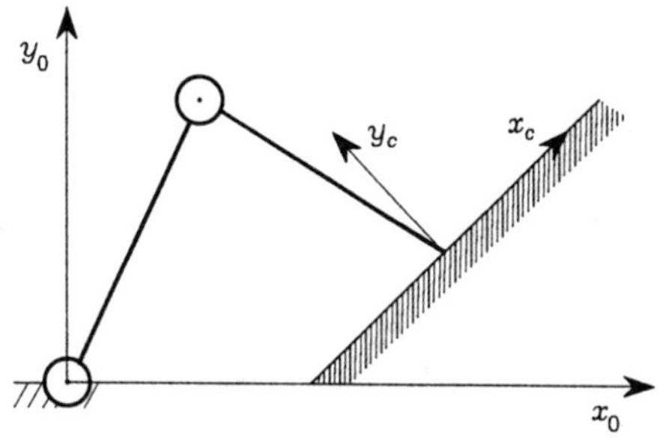

Figure 7.14 Characterization of constraint frame for a two-link planar arm in contact with an elastically compliant plane.

measurements on the assumption that this information is not useful to the controller, whereas the same information might be crucial in all such circumstances when lack of information about the environment plays a significant role.

An effective alternative to the hybrid control strategy in situations of a poorly structured environment can be provided by the parallel control strategy presented in Section 7.4.3. The absence of selection mechanisms on the force and position feedback loops confers to the parallel control a capacity of autonomous reaction whenever the natural constraints are violated, that is, when there is mismatching between planning and real geometry of the interaction.

Example 7.4

Consider a two-link planar arm in contact with a purely frictionless elastic plane; differently from the above examples, the plane is at an angle of $\pi/4$ with axis x (Figure 7.14). The natural choice of the constraint frame is that with axis x_c along the plane and axis y_c orthogonal to the plane; the task is obviously characterized by two degrees of freedom. Let $\boldsymbol{p}^c$ and $\boldsymbol{f}^c$ respectively denote the tip position and contact force, both expressed in the constraint frame; then, the selection matrix is

$$\boldsymbol{\Sigma} = \text{diag}\{0, 1\},$$

since the natural constraints regard $\dot{y}^c$ and f_x^c. As a consequence, the artificial constraints allow specifying f_y^c and $\dot{x}^c$. If the task is to slide the tip along the plane, the constraint frame orientation remains constant with respect to the base frame. The relative rotation matrix is given by

$$\boldsymbol{R}^c = \begin{bmatrix} 1/\sqrt{2} & 1/\sqrt{2} \\ -1/\sqrt{2} & 1/\sqrt{2} \end{bmatrix}. \tag{7.33}$$

According to the hybrid control structure, it is possible to perform motion control along axis x_c and interaction (compliance, impedance or force) control along axis y_c. Hence, once the measured quantities expressed in the base frame—position error $\tilde{x}$, velocity $\dot{x}$, and contact force f_y—are transformed into the constraint frame, it is possible to proceed as in the above examples. The resulting control law will have components along x_c, y_c, and then it will have to be transformed back to the base frame via the matrix $\boldsymbol{R}_c$.

Problems

7.1 Show that the equilibrium position and force for the compliance control scheme are expressed by (7.9) and (7.10), respectively.

7.2 Show that the equilibrium position for the parallel force/position control scheme satisfies (7.30).

7.3 For the manipulation task of driving a screw in a hole illustrated in Figure 7.15, find the natural constraints and artificial constraints with respect to a suitably chosen constraint frame.

7.4 Consider the planar arm in contact with the elastically compliant plane in Figure 7.14. The plane forms an angle of $\pi/4$ with axis x and its undeformed position intersects axis x in the point of coordinates $(1, 0)$; environment stiffness along axis y_c is $5 \cdot 10^3$ N/m. With the data of the arm in Section 6.7, design a hybrid control in which an inverse dynamics position control operates along axis x_c while an impedance control operates along axis y_c. Perform a computer simulation of the interaction of the controlled manipulator along the rectilinear path from position $\boldsymbol{p}_i = [\,1 + 0.1\sqrt{2} \quad 0\,]^T$ m to $\boldsymbol{p}_f = [\,1.2 + 0.1\sqrt{2} \quad 0.2\,]^T$ m with a trapezoidal velocity profile and a trajectory duration $t_f = 1$ s. Implement the control in discrete-time with a sampling time of 1 ms.

7.5 For the arm and environment of Problem 7.4, design a hybrid control in which an inverse dynamics position control operates along axis x_c, while a force control with inner position loop operates along axis y_c; let the desired contact force along axis y_c be 50 N. Perform a computer simulation of the interaction of the controlled manipulator along the trajectory of Problem 7.4. Implement the control in discrete-time with a sampling time of 1 ms.

7.6 For the arm and environment of Problem 7.4, design a hybrid control in which an inverse dynamics position control operates along axis x_c, while a force control with inner velocity loop operates along axis y_c; the force reference is the same as in Problem 7.5. Perform a computer simulation of the interaction of the controlled manipulator along the trajectory of Problem 7.4. Implement the control in discrete-time with a sampling time of 1 ms.

Bibliography

Anderson R.J., Spong M.W. (1988) Hybrid impedance control of robotic manipulators. *IEEE J. Robotics and Automation* 4:549–556.

Chiaverini S. (1990) *Controllo in Forza di Manipolatori* (in Italian). Tesi di Dottorato di Ricerca, Università degli Studi di Napoli Federico II.

Chiaverini S., Sciavicco L. (1993) The parallel approach to force/position control of robotic manipulators. *IEEE Trans. Robotics and Automation* 4:361–373.

Chiaverini S., Siciliano B., Villani L. (1994) Force/position regulation of compliant robot manipulators. *IEEE Trans. Automatic Control* 39:647–652.

De Luca A., Manes C., Nicolò F. (1988) A task space decoupling approach to hybrid control of manipulators. In *Proc. 2nd IFAC Symp. Robot Control* Karlsruhe, Germany, pp. 157–162.

De Schutter J., Van Brussel H. (1988) Compliant robot motion I. A formalism for specifying compliant motion tasks. *Int. J. Robotics Research* 7(4):3–17.

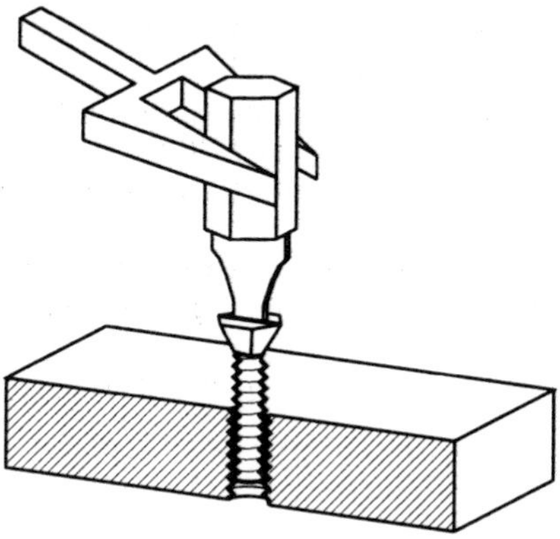

Figure 7.15 Driving a screw in a hole.

De Schutter J., Van Brussel H. (1988) Compliant robot motion II. A control approach based on external control loops. *Int. J. Robotics Research* 7(4):18–33.

Eppinger S.D., Seering W.P. (1987) Introduction to dynamic models for robot force control. *IEEE Control Systems Mag.* 7(2):48–52.

Goldenberg A.A. (1988) Implementation of force and impedance control in robot manipulators. In *Proc. 1988 IEEE Int. Conf. Robotics and Automation* Philadelphia, Penn., pp. 1626–1632.

Hogan N. (1985) Impedance control: An approach to manipulation: Part I—Theory. *ASME J. Dynamic Systems, Measurement, and Control* 107:1–7.

Hogan N. (1985) Impedance control: An approach to manipulation: Part II—Implementation. *ASME J. Dynamic Systems, Measurement, and Control* 107:8–16.

Kazerooni H., Houpt P.K., Sheridan T.B. (1986) Robust compliant motion of manipulators, part I: The fundamental concepts of compliant motion. *IEEE J. Robotics and Automation* 2:83–96.

Khatib O. (1987) A unified approach for motion and force control of robot manipulators: The operational space formulation. *IEEE J. Robotics and Automation* 3:43–53.

Lozano-Pérez T., Mason M.T., Taylor R.H. (1984) Automatic synthesis of fine-motion strategies for robots. *Int. J. Robotics Research* 3(1):3–24.

Mason M.T. (1981) Compliance and force control for computer controlled manipulators. *IEEE Trans. Systems, Man, and Cybernetics* 6:418–432.

Paul, R.P. (1981) *Robot Manipulators: Mathematics, Programming, and Control.* MIT Press, Cambridge, Mass.

Paul R.P., Shimano B. (1976) Compliance and control. In *Proc. 1976 Joint Automatic Control Conf.* San Francisco, Cal., pp. 694–699.

Raibert M.H., Craig J.J. (1981) Hybrid position/force control of manipulators. *ASME J. Dynamic Systems, Measurement, and Control* 103:126–133.

Salisbury J.K. (1980) Active stiffness control of a manipulator in Cartesian coordinates. In *Proc. 19th IEEE Conf. Decision and Control* Albuquerque, New Mex., pp. 95–100.

Siciliano B., Villani L. (1999) *Robot Force Control.* Kluwer Academic Publishers, Boston, Mass.

Volpe R., Khosla P. (1993) A theoretical and experimental investigation of explicit force control strategies for manipulators. *IEEE Trans. on Automatic Control* 38:1634–1650.

Whitney D.E. (1977) Force feedback control of manipulator fine motions. *ASME J. Dynamic Systems, Measurement, and Control* 99:91–97.

Whitney D.E. (1982) Quasi-static assembly of compliantly supported rigid parts. *ASME J. Dynamic Systems, Measurement, and Control* 104:65–77.

Whitney D.E. (1987) Historical perspective and state of the art in robot force control. *Int. J. Robotics Research* 6(1):3–14.

Yoshikawa T. (1987) Dynamic hybrid position/force control of robot manipulators—Description of hand constraints and calculation of joint driving force. *IEEE J. Robotics and Automation* 3:386–392.

Yoshikawa T. (1990) *Foundations of Robotics*. MIT Press, Cambridge, Mass.

8. Actuators and Sensors

In this chapter, two basic robot components are treated: *actuators* and *sensors*. In the first part, the features of an *actuating system* are presented in terms of the power supply, power amplifier, servomotor and transmission. In view of their control versatility, two types of servomotors are used; namely, *electric servomotors* for actuating the joints of small and medium size manipulators, and *hydraulic servomotors* for actuating the joints of large size manipulators. The models describing the input/output relationship for such servomotors are derived. In the second part, *proprioceptive sensors* are presented which allow measurements of the quantities characterizing the internal state of the manipulator; namely, *encoders* and *resolvers* for joint position measurement, *tachometers* for joint velocity measurement; further, *exteroceptive sensors* are presented including *force sensors* for end-effector force measurement and *vision sensors* for object image measurements when the manipulator interacts with the environment.

8.1 Joint Actuating System

The motion imposed to a manipulator's joint is realized by an *actuating system* which in general is constituted by:

- a *power supply*,
- a *power amplifier*,
- a *servomotor*,
- a *transmission*.

The connection between the various components is illustrated in Figure 8.1 where the exchanged powers are shown. To this purpose, recall that power can be always expressed as the product of a flow and a force quantity, whose physical context allows specifying the nature of the power (mechanical, electric, hydraulic, or pneumatic).

In terms of a global input/output relationship, P_c denotes the (usually electric) power associated with the control law signal, whereas P_u represents the mechanical power required to the joint to actuate the motion. The intermediate connections characterize the supply power P_a of the motor (of electric, hydraulic, or pneumatic type), the power provided by the primary source P_p of the same physical nature as that of P_a, and the mechanical power P_m developed by the motor. Moreover, P_{da}, P_{ds} and P_{dt} denote the powers lost for dissipation in the conversions performed respectively

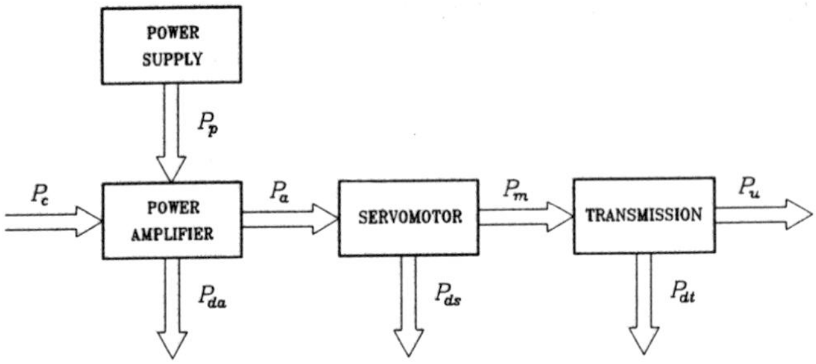

Figure 8.1 Components of a joint actuating system.

by the amplifier, motor and transmission.

To choose the components of an actuating system, it is worth starting from the requirements imposed on the mechanical power P_u by the force and velocity that describe the joint motion.

8.1.1 Transmissions

The execution of joint motions of a manipulator demands *low speeds* with *high torques*. In general, such requirements do not allow an effective use of the mechanical features of servomotors, which typically provide high speeds with low torques in optimal operating conditions. It is then necessary to interpose a *transmission* (gear train) to optimize the transfer of mechanical power from the motor (P_m) to the joint (P_u). During this transfer, the power P_{dt} is dissipated as a result of friction.

The choice of transmission depends on the power requirements, the kind of desired motion, and the allocation of the motor with respect to the joint. In fact, the transmission allows transforming the outputs of the motor both quantitatively (velocity and torque) and qualitatively (a rotational motion about the motor axis into a translational motion of the joint). Also, it allows optimizing the static and dynamic performance of a manipulator, by reducing the effective loads when the motor is located upstream of the joint; for instance, if some motors are mounted to the base of the robot, the total weight of the manipulator is decreased and the power-to-weight ratio is increased.

The following transmissions are typically used for industrial manipulators.

- *Spur gears* that modify the characteristics of the rotational motion of the motor by changing the axis of rotation and/or by translating the application point; spur gears are usually constructed with wide cross-section teeth and squat shafts.

- *Lead screws* that convert rotational motion of the motor into translational motion, as needed for actuation of prismatic joints; in order to reduce friction, ball screws are usually employed that are preloaded so as to increase stiffness and decrease backlash.

- *Timing belts* and *chains* which are equivalent from a kinematic viewpoint and are employed to locate the motor remotely from the axis of the actuated joint. The stress on timing belts may cause strain, and then these are used in applications requiring high speeds and low forces. On the other hand, chains are used in applications requiring low speeds, since their large mass may induce vibration at high speeds.

On the assumption of rigid transmissions with no backlash, the relationship between input forces (velocities) and output forces (velocities) is purely proportional.

The mechanical features of the motor used for an actuating system may sometimes allow direct connection of the motor to the joint without the use of any transmission element (*direct drive*). The drawbacks due to transmission elasticity and backlash are thus eliminated, although more sophisticated control algorithms are required, since the absence of reduction gears does not allow neglecting the nonlinear coupling terms in the dynamic model. The use of direct-drive actuating systems is not yet popular for industrial manipulators, in view of the cost and size of the motors as well as of control complexity.

8.1.2 Servomotors

Actuation of joint motions is entrusted to *motors* which allow realizing a desired motion for the mechanical system. Concerning the kind of input power P_a, motors can be classified into three groups.

- *Pneumatic motors* which utilize the pneumatic energy provided by a compressor and transform it into mechanical energy by means of pistons or turbines.
- *Hydraulic motors* which transform the hydraulic energy stored in a reservoir into mechanical energy by means of suitable pumps.
- *Electric motors* whose primary supply is the electric energy available from the electric distribution system.

A portion of the input power P_a is converted to the output as a mechanical power P_m, and the rest (P_{ds}) is dissipated because of mechanical, electric, hydraulic, or pneumatic loss.

The motors employed in robotics are the evolution of the motors employed in industrial automation having powers ranging from about ten watts to about ten kilowatts. For the typical performance required, such motors shall have the following requirements with respect to those employed in conventional applications:

- low inertia and high power-to-weight ratio,
- possibility of overload and delivery of impulse torques,
- capacity to develop high accelerations,
- wide velocity range (from 1 to 1000),
- high positioning accuracy (at least 1/1000 of a circle),
- low torque ripple so as to guarantee continuous rotation even at low speed.

These requirements are enhanced by the good trajectory tracking and positioning accuracy demanded to an actuating system for robots, and thus the motor shall play the role of a *servomotor*. In this respect, pneumatic motors are difficult to control accurately, in view of the unavoidable fluid compressibility errors. Therefore, they are not widely employed, if not for the actuation of the typical opening and closing motions of the jaws in a gripper tool, or else for the actuation of simple arms used in applications where continuous motion control is not of concern.

The most employed motors in robotic applications are the *electric servomotors*. Among them, the most popular are the permanent-magnet direct-current (dc) servomotors and the brushless dc servomotors, in view of their good control flexibility.

Not uncommon are also stepper motors. These actuators are controlled by suitable excitation sequences and their operating principle does not require measurement of motor shaft angular position. The dynamic behaviour of stepper motors is greatly influenced by payload, though. Also, they induce vibration of the mechanical structure of the manipulator. Such inconveniences confine the use of stepper motors to the field of micromanipulators, for which low-cost implementation prevails over the need for high dynamic performance.

A certain number of applications features the employment of *hydraulic servomotors*, which allow actuation of both translational motion (single piston) and rotational motion (axial or radial pistons). These servomotors offer a static and dynamic performance comparable with that offered by electric servomotors.

The differences between electric and hydraulic servomotors can be fundamentally observed from a plant viewpoint. In this respect, *electric servomotors* present the following *advantages*:

- wide-spread availability of power supply,
- low cost and wide range of products,
- high power conversion efficiency,
- easy maintenance,
- no pollution of working environment.

Instead, they present the following *limitations*:

- burnout problems at static situations caused by the effect of gravity on the manipulator; emergency brakes are then required,
- need for special protection when operating in flammable environments.

Hydraulic servomotors present the following *drawbacks*:

- need for a hydraulic power station,
- high cost, narrow range of products, and difficulty of miniaturization,
- low power conversion efficiency,
- need for operational maintenance,
- pollution of working environment due to oil leakage.

In their *favor* it is worth pointing out that they:

- do not suffer from burnout in static situations,

- are self-lubricated and the circulating fluid facilitates heat disposal,
- are inherently safe in harmful environments,
- have excellent power-to-weight ratios.

From an operational viewpoint, it can be observed what follows.

- Both types of servomotors have a good dynamic behaviour, although the electric servomotor has greater control flexibility. The dynamic behaviour of a hydraulic servomotor depends on the temperature of the compressed fluid.
- The electric servomotor is typically characterized by high speeds and low torques, and as such it requires the use of gear transmissions (causing elasticity and backlash). On the other hand, the hydraulic servomotor is capable to generate high torques at low speeds.

In view of the above remarks, hydraulic servomotors are specifically employed for manipulators that have to carry heavy payloads; in this case, not only is the hydraulic servomotor the most suitable actuator, but also the cost of the plant accounts for a reduced percentage on the total cost of the manipulation system.

8.1.3 Power Amplifiers

The *power amplifier* has the task of modulating, under the action of a control signal, the power flow which is provided by the primary supply and has to be delivered to the actuators for the execution of the desired motion. In other words, the amplifier takes a fraction of the power available at the source which is proportional to the control signal; then, it transmits this power to the motor in terms of suitable force and flow quantities.

The inputs to the amplifier are the power taken from the primary source P_p and the power associated with the control signal P_c. The total power is partly delivered to the actuator (P_a) and partly lost for dissipation (P_{da}).

Given the typical use of electric and hydraulic servomotors, the operational principles of the respective amplifiers are discussed.

To control an *electric servomotor*, it is necessary to provide it with a voltage or current of suitable form depending on the kind of servomotor employed. Voltage (or current) is direct for permanent-magnet dc servomotors, while it is alternating for brushless dc servomotors. The value of voltage for permanent-magnet dc servomotors or the values of voltage and frequency for brushless dc servomotors are determined by the control signal of the amplifier, so as to make the motor execute the desired motion.

For the power ranges typically required by joint motions (of the order of a few kilowatts), transistor amplifiers are employed which are suitably switched by using pulse-width modulation (PWM) techniques. They allow achieving a power conversion efficiency $P_a/(P_p + P_c)$ greater than 0.9 and a power gain P_a/P_c of the order of 10^6. The amplifiers employed to control permanent-magnet dc servomotors are dc-to-dc converters (*choppers*), whereas those employed to control brushless dc servomotors are dc-to-ac converters (*inverters*).

Control of a *hydraulic servomotor* is performed by varying the flow rate of the compressed fluid delivered to the motor. The task of modulating the flow rate is typically entrusted to an interface (electro-hydraulic servovalve). This allows establishing a relationship between the electric control signal and the position of a distributor which is able to vary the flow rate of the fluid transferred from the primary source to the motor. The electric control signal is usually current-amplified and feeds a solenoid which moves (directly or indirectly) the distributor, whose position is measured by a suitable transducer. In this way, a position servo on the valve stem is obtained which reduces occurrence of any stability problem that may arise on motor control. The magnitude of the control signal determines the flow rate of the compressed fluid through the distributor, according to a characteristic which is possibly made linear by means of a keen mechanical design.

8.1.4 Power Supplies

The task of the *power supply* is to supply the primary power to the amplifier which is needed for operation of the actuating system.

In the case of *electric drives*, the power supply is constituted by a transformer and a typically uncontrolled bridge rectifier. These allow converting the alternating voltage available from the distribution into a direct voltage of suitable magnitude which is required to feed the power amplifier.

In the case of *hydraulic drives*, the power supply is obviously more complex. In fact, a gear or piston pump is employed to compress the fluid which is driven by a primary motor operating at constant speed, typically a three-phase nonsynchronous motor. To reduce the unavoidable pressure oscillations provoked by a flow rate demand depending on operational conditions of the motor, a reservoir is interfaced to store hydraulic energy. Such reservoir in turn plays the same role as the filter capacitor used at the output of a bridge rectifier. The hydraulic power station is completed by the use of various components (filters, pressure valves, and check valves) that ensure proper operation of the system. Finally, it can be inferred how the presence of complex hydraulic circuits operating at high pressures (of the order of 100 atm) causes an appreciable pollution of the working environment.

8.2 Servomotors

This section is devoted to present the operation of the electric and hydraulic servomotors typically used for actuating the joints of a manipulator. Starting from the mathematical models that describe their dynamic behaviour, block schemes are derived which allow pointing out the formal analogy between the two types of servomotors.

8.2.1 Electric Servomotors

The servomotor is entrusted with the task of converting electric energy into mechanical energy, and thus it shall have a truly flexible operating characteristic. For this reason, the electric servomotors employed in robotic applications are permanent-magnet dc servomotors and brushless dc servomotors.

The *permanent-magnet dc servomotor* is constituted by:

- A stator coil that generates magnetic flux; this generator is always a permanent magnet made by ferromagnetic ceramics or rare earths (high fields in contained space).
- An armature that includes the current-carrying winding that surrounds a rotary ferromagnetic core (rotor).
- A commutator that provides an electric connection by means of brushes between the rotating armature winding and the external feed winding, according to a commutation logic determined by the rotor motion.

The *brushless dc servomotor* is constituted by:

- A rotating coil (rotor) that generates magnetic flux; this generator is a permanent magnet made by ferromagnetic ceramics or rare earths.
- A stationary armature (stator) made by a polyphase winding.
- A static commutator that, on the basis of the signals provided by a position sensor located on the motor shaft, generates the feed sequence of the armature winding phases as a function of the rotor motion.

The power amplifier for a permanent-magnet dc servomotor is typically a dc-to-dc converter (chopper), whereas for a brushless dc servomotor it is typically a dc-to-ac converter (inverter), whose frequency is dictated by the angular velocity of the rotor.

With reference to the above details of constructions, a comparison between the operating principle of a permanent-magnet dc and a brushless dc servomotor leads to the following considerations.

In the brushless dc motor, by means of the rotor position sensor, the winding orthogonal to the magnetic field of the coil is found; then, feeding the winding makes the rotor rotate. As a consequence of rotation, the electronic control module commutes the feeding on the winding of the various phases in such a way that the resulting field at the armature is always kept orthogonal to that of the coil. As regards electromagnetic interaction, such motor operates in a way similar to that of a permanent-magnet dc motor where the brushes are at an angle of $\pi/2$ with respect to the direction of the excitation flux. In fact, feeding the armature coil makes the rotor rotate, and commutation of brushes from one plate of the commutator to the other allows maintaining the rotor in rotation. The role played by the brushes and commutator in a permanent-magnet dc motor is analogous to that played by the position sensor and electronic control module in a brushless dc motor.

The main reason for using a brushless dc motor is to eliminate the problems due to mechanical commutation of the brushes in a permanent-magnet dc motor. In fact, the presence of the commutator limits the performance of a permanent-magnet dc motor, since this provokes electric loss due to voltage drops at the contact between the brushes and plates, and mechanical loss due to friction and arcing during commutation from one plate to the next one caused by the inductance of the winding. The elimination of the causes provoking such inconveniences, *i.e.*, the brushes and plates, allows an improvement of motor performance in terms of higher speeds and less material wear.

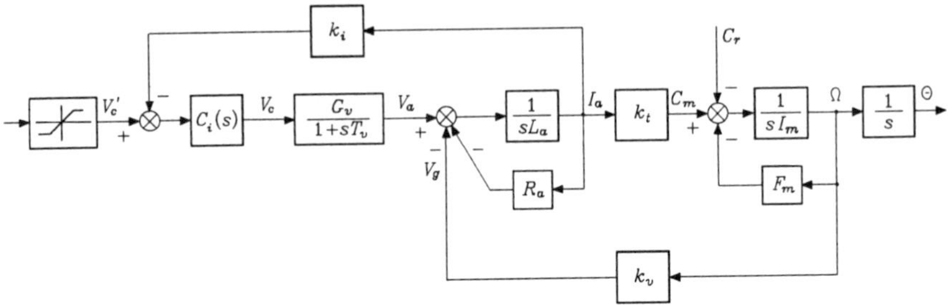

Figure 8.2 Block scheme of an electric servomotor with amplifier.

The inversion between the functions of stator and rotor leads to further advantages. The presence of a winding on the stator instead of the rotor facilitates heat disposal. The absence of a rotor winding, together with the possibility of using rare-earth permanent magnets, allows construction of more compact rotors which are in turn characterized by a low moment of inertia. Therefore, the size of a brushless dc motor is smaller than that of a permanent-magnet dc motor of the same power; an improvement of dynamic performance can also be obtained by using a brushless dc motor. For the choice of the most suitable servomotor for a specific application, the cost factor plays a relevant role, though.

From a modelling viewpoint, a permanent-magnet dc motor and a brushless dc motor provided with the commutation module and position sensor can be described by the same differential equations. In the domain of the complex variable s, the electric balance of the armature is described by the equations

$$V_a = (R_a + sL_a)I_a + V_g \tag{8.1}$$
$$V_g = k_v \Omega, \tag{8.2}$$

where V_a and I_a respectively denote armature voltage and current, R_a and L_a are respectively the armature resistance and inductance, and V_g denotes the back electromotive force which is proportional to the angular velocity Ω through the voltage constant k_v that depends on the construction details of the motor as well as on the magnetic flux of the coil.

The mechanical balance is described by the equations

$$C_m = (sI_m + F_m)\Omega + C_r \tag{8.3}$$
$$C_m = k_t I_a, \tag{8.4}$$

where C_m and C_r respectively denote the electromechanical driving torque and reaction torque, I_m and F_m are respectively the moment of inertia and viscous friction coefficient at the motor shaft, and the torque constant k_t is numerically equal to k_v in the SI unit system for a compensated motor.

Concerning the power amplifier, the input/output relationship between the control voltage V_c and the armature voltage V_a is given by the transfer function

$$\frac{V_a}{V_c} = \frac{G_v}{1 + sT_v} \tag{8.5}$$

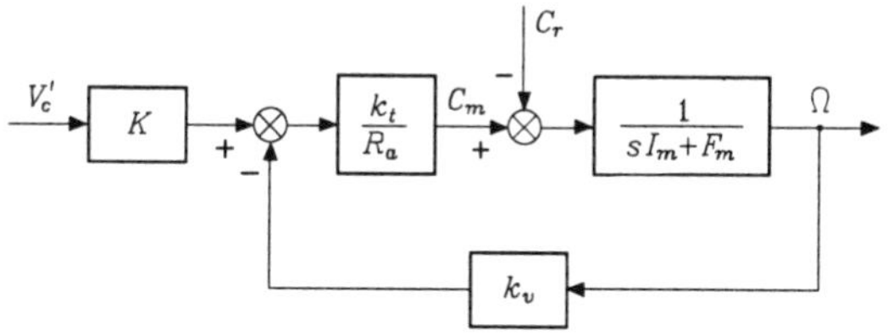

Figure 8.3 Block scheme of an electric servomotor as a velocity-controlled generator.

where G_v denotes the voltage gain and T_v is a time constant that can be neglected with respect to the other time constants of the system. In fact, by using a modulation frequency in the range of $(10, 100)$ kHz, the time constant of the amplifier is in the range of $(10^{-5}, 10^{-4})$ s.

The block scheme of the servomotor with power amplifier is illustrated in Figure 8.2. In such scheme, besides the blocks corresponding to the above relations, there is an armature *current feedback* loop where current is thought of as measured by a transducer k_i between the power amplifier and the armature winding of the motor. Further, the scheme features a current regulator $C_i(s)$ as well as an element with a nonlinear saturation characteristic. The aim of such feedback is twofold. On one hand, the voltage V_c' plays the role of a current reference and thus, by means of a suitable choice of the regulator $C_i(s)$, the lag between the current I_a and the voltage V_c' can be reduced with respect to the lag between I_a and V_c. On the other hand, the introduction of a saturation nonlinearity allows limiting the magnitude of V_c', and then it works like a current limit which ensures protection of the power amplifier whenever abnormal operating conditions occur.

The choice of the regulator $C_i(s)$ of the current loop allows obtaining a velocity-controlled or torque-controlled behaviour from the amplifier/motor complex, depending on the values attained by the loop gain. In fact, in the case of $k_i = 0$, recalling that the mechanical viscous friction coefficient is negligible with respect to the electrical friction coefficient ($F_m \ll k_v k_t / R_a$), setting $K = C_i(0)G_v$, and assuming $C_r = 0$ leads at steady state to:

$$\omega \approx \frac{K}{k_v} v_c', \tag{8.6}$$

and thus the actuating system behaves like a *velocity-controlled generator*. Instead, when $k_i \neq 0$, choosing a large loop gain for the current loop ($Kk_i \gg R_a$) leads at steady state to:

$$c_m \approx \frac{k_t}{k_i}\left(v_c' - \frac{k_v}{K}\omega\right), \tag{8.7}$$

and thus the actuating system behaves like a *torque-controlled generator*, since K is usually large and the driving torque is practically independent of the angular velocity.

As regards the dynamic behaviour, it is worth considering a *reduced-order model* which can be obtained by neglecting the electric time constant L_a/R_a with respect to the mechanical time constant I_m/F_m and assuming also $T_v \approx 0$. Choosing a purely proportional controller ($C_i G_v = K$) with $k_i = 0$ leads to the block scheme

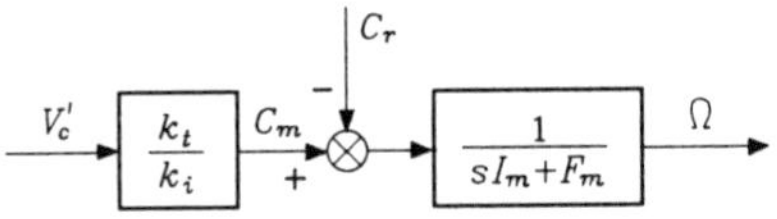

Figure 8.4 Block scheme of an electric servomotor as a torque-controlled generator.

in Figure 8.3 for the velocity-controlled generator. On the other hand, if it is assumed $Kk_i \gg R_a$ and $k_v\Omega/Kk_i \approx 0$, the resulting block scheme of the torque-controlled generator is that in Figure 8.4.

From the above schemes, the following input/output relations between control voltage, reaction torque, and angular velocity can be derived:

$$\Omega = \frac{\dfrac{K}{k_v}}{1+s\dfrac{R_aI_m}{k_vk_t}}V_c' - \frac{\dfrac{R_a}{k_vk_t}}{1+s\dfrac{R_aI_m}{k_vk_t}}C_r \tag{8.8}$$

for the velocity-controlled generator, and

$$\Omega = \frac{\dfrac{k_t}{k_iF_m}}{1+s\dfrac{I_m}{F_m}}V_c' - \frac{\dfrac{1}{F_m}}{1+s\dfrac{I_m}{F_m}}C_r \tag{8.9}$$

for the torque-controlled generator. These transfer functions show how, without current feedback, the system has a better rejection of disturbance torques in terms of both equivalent gain ($R_a/k_vk_t \ll 1/F_m$) and time response ($R_aI_m/k_vk_t \ll I_m/F_m$). This leads to drawing the following conclusions.

In all such applications where the drive system has to provide high rejection of disturbance torques, as in the case of independent joint control, it is not advisable to have a current feedback in the loop, at least when all quantities are within their nominal values. In this case, the problem of setting a protection can be solved by introducing a current limit that is not performed by a saturation on the control signal but it exploits a current feedback with a dead-zone nonlinearity on the feedback path, as shown in Figure 8.5. Therefore, an actual current limit is obtained whose precision is as high as the slope of the dead zone; it is understood that stability of the current loop is to be addressed when operating in this way.

Centralized control schemes, instead, demand the actuating system to behave as a torque-controlled generator. It is then clear that a current feedback with a suitable regulator $C_i(s)$ shall be used so as to confer a good static and dynamic behaviour to the current loop. In this case, servoing of the driving torque is achieved indirectly, since it is based on a current measurement which is related to the driving torque by means of gain $1/k_t$.

8.2.2 Hydraulic Servomotors

The hydraulic servomotors employed in robotic applications are based on the simple

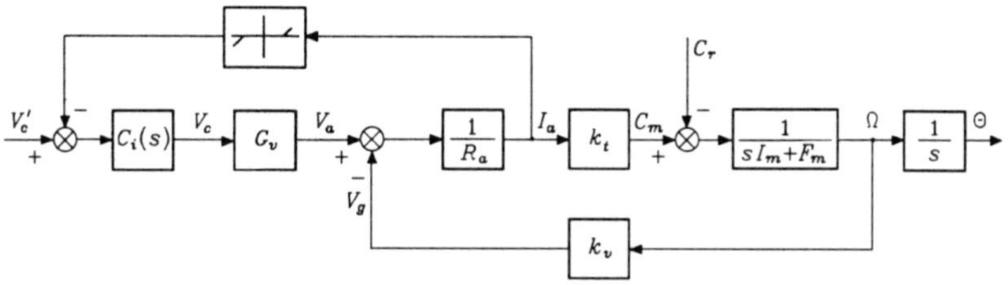

Figure 8.5 Block scheme of an electric servomotor with nonlinear current feedback.

operating principle of volume variation under the action of compressed fluid. From a construction viewpoint, they are characterized by one or more chambers made by pistons (cylinders reciprocating in tubular housings). Linear servomotors have a limited range and are constituted by a single piston. Rotary servomotors have unlimited range and are constituted by several pistons (usually an odd number) with an axial or radial disposition with respect to the motor axis of rotation.

No matter how a hydraulic servomotor is constructed, the derivation of its input/output mathematical model refers to the basic equations describing the relationship between flow rate and pressure, the relationship between the fluid and the parts in motion, and the mechanical balance of the parts in motion. Let Q represent the volume flow rate supplied by the distributor; the flow rate balance is given by the equation

$$Q = Q_m + Q_l + Q_c \tag{8.10}$$

where Q_m is the flow rate transferred to the motor, Q_l is the flow rate due to leakage, and Q_c is the flow rate related to fluid compressibility. The terms Q_l and Q_c are taken into account in view of the high operating pressures (of the order of one hundred atmospheres).

Let P denote the differential pressure of the servomotor due to the load; then, it can be assumed

$$Q_l = k_l P. \tag{8.11}$$

Regarding the loss for compressibility, if V denotes the instantaneous volume of the fluid, one has

$$Q_c = \gamma V s P \tag{8.12}$$

where γ is the uniform compressibility coefficient of the fluid. Notice that the proportional factor $k_c = \gamma V$ between the time derivative of the pressure and the flow rate due to compressibility depends on the volume of the fluid; therefore, in the case of rotary servomotors, k_c is a constant, whereas in the case of a linear servomotor, the volume of fluid varies and thus the characteristic of the response depends on the operating point.

The volume flow rate transferred to the motor is proportional to the volume variation in the chambers per time unit; with reference from now on to a rotary servomotor, such variation is proportional to the angular velocity, and then

$$Q_m = k_q \Omega. \tag{8.13}$$

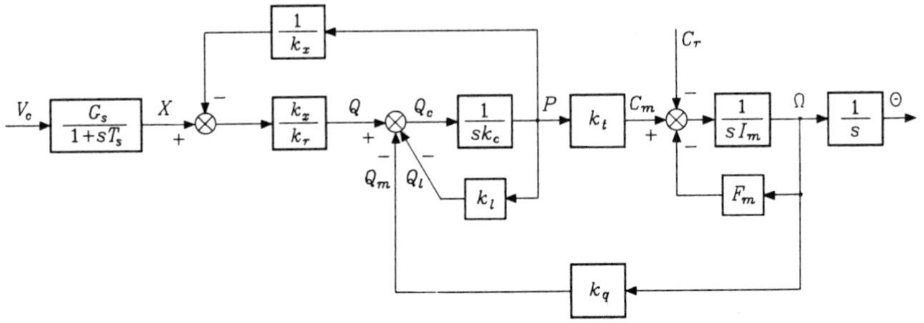

Figure 8.6 Block scheme of a hydraulic motor with servovalve and distributor.

The mechanical balance of the parts in motion is described by

$$C_m = (sI_m + F_m)\Omega + C_r \tag{8.14}$$

with obvious meaning of the symbols. Finally, the driving torque is proportional to the differential pressure of the servomotor due to the load, *i.e.*,

$$C_m = k_t P. \tag{8.15}$$

Concerning the servovalve, the transfer function between the stem position X and the control voltage V_c is expressed by

$$\frac{X}{V_c} = \frac{G_s}{1 + sT_s}, \tag{8.16}$$

thanks to the linearizing effect achieved by position feedback; G_s is the equivalent gain of the servovalve, whereas its time constant T_s is of the order of milliseconds and thus it can be neglected with respect to the other time constants of the system.

Finally, regarding the distributor, the relationship between the differential pressure, the flow rate, and the stem displacement is highly nonlinear; linearization about an operating point leads to the equation

$$P = k_x X - k_r Q. \tag{8.17}$$

By virtue of (8.10)–(8.17), the servovalve/distributor/motor complex is represented by the block scheme of Figure 8.6. A comparison between the schemes in Figs. 8.2 and 8.6 clearly shows the formal analogy in the dynamic behaviour of an electric and a hydraulic servomotor. Nevertheless, such analogy shall not induce to believe that it is possible to make a hydraulic servomotor play the role of a velocity- or torque-controlled generator, as for an electric servomotor. In this case, the pressure feedback loop (formally analogous to the current feedback loop) is indeed a structural characteristic of the system and, as such, it cannot be modified but with the introduction of suitable transducers and the realization of the relative control circuitry.

8.3 Sensors

The adoption of *sensors* is of crucial importance to achieve high-performance robotic systems. It is worth classifying sensors into *proprioceptive sensors* that measure the internal state of the manipulator, and *exteroceptive sensors* that provide the robot with knowledge of the surrounding environment.

In order to guarantee that a coordinated motion of the mechanical structure is obtained in correspondence of the task planning, algorithms are used which have been widely discussed in the preceding chapters, e.g., as in kinematic calibration, dynamic parameter identification and manipulator control. Such algorithms require the on-line measurement, by means of proprioceptive sensors, of the quantities characterizing the internal state of the manipulator, *i.e.*,

- joint position,
- joint velocity,
- joint torque.

On the other hand, typical exteroceptive sensors include:

- force sensors,
- tactile sensors,
- proximity sensors,
- range sensors,
- vision sensors.

The goal of such sensors is to extract the features characterizing the interaction of the robot with the objects in the environment, so as to enhance the degree of autonomy of the system. To this class also belong those sensors which are specific for the robotic application, such as sound, humidity, smoke, pressure, and temperature sensors. Fusion of the available sensory data can be used for (high-level) task planning, which in turn characterizes a *robot* as the *intelligent connection of perception to action*.

In the following, the main features of the sensors used to measure position, velocity, force and visual information are illustrated.

8.3.1 Position Transducers

The aim of *position transducers* is to provide an electric signal proportional to the linear or angular displacement of a mechanical apparatus with respect to a given reference position. They are mostly utilized for control of machine tools, and thus their range is wide. Potentiometers, linear variable-differential transformers (LVDT), and inductosyns may be used to measure linear displacements. Potentiometers, encoders, resolvers and synchros may be used to measure angular displacements.

Angular displacement transducers are typically employed in robotic applications since, also for prismatic joints, the servomotor is of rotary type. In view of their precision, robustness and reliability, the most common transducers are the *encoders* and *resolvers*, whose operating principles are detailed in what follows.

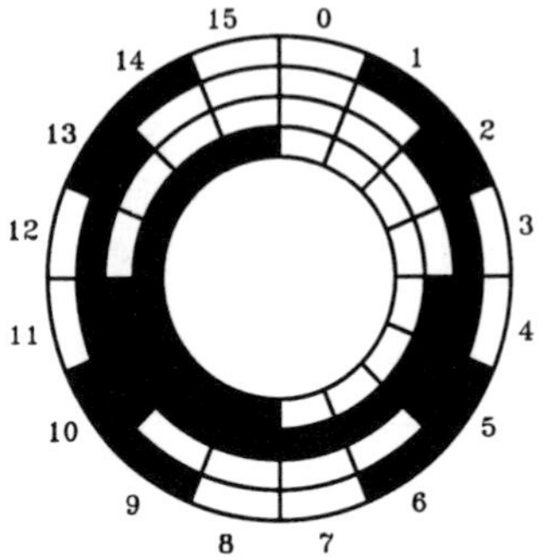

#	Code	#	Code
0	0000	8	1100
1	0001	9	1101
2	0011	10	1111
3	0010	11	1110
4	0110	12	1010
5	0111	13	1011
6	0101	14	1001
7	0100	15	1000

Figure 8.7 Schematic representation of an absolute encoder with Gray-code table.

On the other hand, linear displacement transducers (LVDT's and inductosyns) are mainly employed in measuring robots.

Encoder

There are two types of encoder: absolute and incremental. The *absolute encoder* is constituted by an optical-glass disk on which concentric circles (tracks) are disposed; each track has an alternating sequence of transparent sectors and matte sectors obtained by deposit of a metallic film. A light beam is emitted in correspondence of each track which is intercepted by a photodiode or a phototransistor located on the opposite side of the disk. By a suitable arrangement of the transparent and matte sectors, it is possible to convert a finite number of angular positions into corresponding digital data. The number of tracks determines the length of the word, and thus the resolution of the encoder.

To avoid problems of incorrect measurement in correspondence of a simultaneous multiple transition between matte and transparent sectors, it is worth utilizing a Gray-code encoder whose schematic representation is given in Figure 8.7 with reference to the implementation of 4 tracks that allow discriminating 16 angular positions. It can be noticed that measurement ambiguity is eliminated, since only one change of contrast occurs at each transition. For the typical resolution required for joint control, absolute encoders with a minimum number of 12 tracks (bits) are employed (resolution of 1/4096 per circle).

Incremental encoders have a wider use than absolute encoders, since they are simpler from a construction viewpoint and thus cheaper. Like the absolute one, the incremental encoder is constituted by an optical disk on which two tracks are disposed, whose transparent and matte sectors (in equal number on the two tracks) are mutually in quadrature. The presence of two tracks allows detecting, besides the number of transitions associated with any angular rotation, also the sign of rotation. Often a third track is present with one single matte sector which allows defining an absolute mechanical zero as a reference for angular position. A schematic representation is illustrated in Figure 8.8.

The use of an incremental encoder for a joint actuating system clearly demands

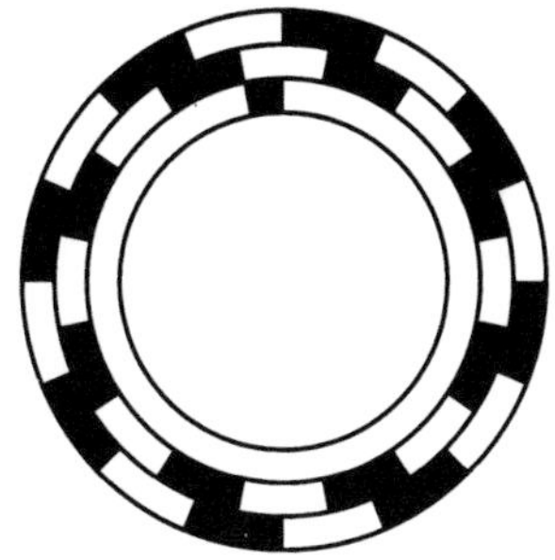

Figure 8.8 Schematic representation of an incremental encoder.

the evaluation of absolute positions. This is performed by means of suitable counting and storing electronic circuits. To this purpose, it is worth noticing that the position information is available on volatile memories, and thus it can be corrupted due to the effect of disturbances acting on the electronic circuit, or else fluctuations in the supply voltage. Such limitation does obviously not occur for absolute encoders, since the angular position information is coded directly on the optical disk.

The optical encoder has its own signal processing electronics inside the case, which provides direct digital position measurements to be interfaced with the control computer. If an external circuitry is employed, velocity measurements can be reconstructed from position measurements. In fact, if a pulse is generated at each transition, a velocity measurement can be obtained in three possible ways; namely, by using a voltage-to-frequency converter (with analog output), by (digitally) measuring the frequency of the pulse train, or by (digitally) measuring the sampling time of the pulse train. Between these last two techniques, the former is suitable for high-speed measurements while the latter is suitable for low-speed measurements.

Resolver

The resolver is an electromechanical position transducer which is compact and robust. Its operating principle is based on the mutual induction between two electric circuits which allow continuous transmission of angular position without mechanical limits. The information on the angular position is associated with the magnitude of two sinusoidal voltages, which are treated by a suitable resolver-to-digital converter (RDC) to obtain the digital data corresponding to the position measurement. The electric scheme of a resolver with the functional diagram of a tracking-type RDC is illustrated in Figure 8.9.

From a construction viewpoint, the resolver is a small electric machine with a rotor and a stator; the inductance coil is on the rotor while the stator has two windings at 90 electrical degrees one from the other. By feeding the rotor with a sinusoidal voltage $V\sin\omega t$ (with typical frequencies in the range of (0.4,10) kHz), a voltage is induced on the stator windings whose magnitude depends on the rotation angle θ. The two voltages are fed to two digital multipliers, whose input is α and whose outputs are algebraically summed to achieve $V\sin\omega t\sin(\theta-\alpha)$; this signal is then amplified and sent to the

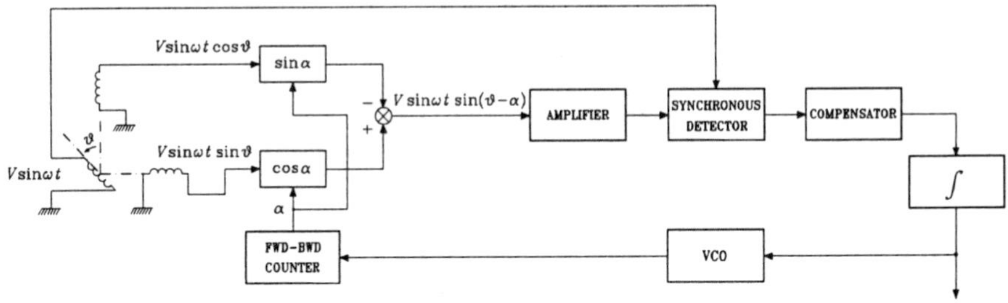

Figure 8.9 Electric scheme of a resolver with functional diagram of a tracking-type RDC.

input of a synchronous detector, whose filtered output is proportional to the quantity $\sin(\theta - \alpha)$. The resulting signal, after a suitable compensating action, is integrated and then sent to the input of a voltage-controlled oscillator (VCO) (a voltage-to-frequency converter) whose output pulses are input to a forward-backward counter. Digital data of the quantity α are available on the output register of the counter, which represent a measurement of the angle θ.

It can be recognized that the converter works according to a feedback principle. The presence of two integrators (one is represented by the forward-backward counter) in the loop ensures that the (digital) position and (analog) velocity measurements are error-free as long as the rotor rotates at constant speed; actually, a round-off error occurs on the word α and thus affects the position measurement. The compensating action is needed to confer suitable stability properties and bandwidth to the system. Whenever digital data are wished also for velocity measurements, it is necessary to use an analog-to-digital converter. Since the resolver is a very precise transducer, a resolution of 1 bit out of 16 can be obtained at the output of the RDC.

8.3.2 Velocity Transducers

Even though velocity measurements can be reconstructed from position transducers, often it is preferred to resort to direct measurements of velocity, by means of suitable transducers. *Velocity transducers* are employed in a wide number of applications and are termed *tachometers*. The most common devices of this kind are based on the operating principles of electric machines. The two basic types of tachometers are the *direct-current (dc) tachometer* and the *alternating-current (ac) tachometer*.

DC Tachometer

The direct-current tachometer is the most used transducer in the applications. It is a small dc generator whose magnetic field is provided by a permanent magnet. Special care is paid in its construction, so as to achieve a linear input/output relationship and to reduce the effects of magnetic hysteresis and temperature. Since the field flux is constant, when the rotor is set in rotation, its output voltage is proportional to angular speed according to the constant characteristic of the machine.

Because of the presence of a commutator, the output voltage has a residual ripple which cannot be eliminated by proper filtering, since its frequency depends on angular speed. A linearity range of 0.1 to 1% can be obtained, whereas the residual ripple coefficient is of 2 to 5% of the mean value of the output signal.

AC Tachometer

In order to avoid the drawbacks caused by the presence of a residual ripple in the output of a dc tachometer, one may resort to an ac tachometer. While the dc tachometer is a true dc generator, the ac tachometer differs from a generator. In fact, if a synchronous generator would be used, the frequency of the output signal would be proportional to the angular speed.

To obtain an alternating voltage whose magnitude is proportional to speed, one may resort to an electric machine that is structurally different from the synchronous generator. The ac tachometer has two windings on the stator mutually in quadrature and a cup rotor. If one of the windings is fed by a constant-magnitude sinusoidal voltage, a sinusoidal voltage is induced on the other winding which has the same frequency, a magnitude proportional to angular speed, and a phase equal or opposite to that of the input voltage according to the sign of rotation; the exciting frequency is usually set to 400 Hz. The use of a synchronous detector then allows obtaining an analog measurement of angular velocity. In this case, the output ripple can be eliminated by a proper filter, since its fundamental frequency is twice as much as the supply frequency.

The performance of ac tachometers is comparable to that of dc tachometers. Two further advantages of ac tachometers are the lack of wiping contacts and the presence of a low moment of inertia, thanks to the use of a lightweight cup rotor. However a residual voltage occurs, even when the rotor is still, because of the unavoidable parasitic couplings between the stator coil and the measurement circuitry.

8.3.3 Force Sensors

Measurement of a force or torque is usually reduced to measurement of the strain induced by the force (torque) applied to an extensible element of suitable features. Therefore, an indirect measure of force is obtained by means of measurements of small displacements. The basic component of a force sensor is the *strain gauge* which uses the change of electric resistance of a wire under strain.

Strain Gauge

The strain gauge is constituted by a wire of low temperature coefficient. The wire is disposed on an insulated support (Figure 8.10a) which is glued to the element subject to strain under the action of a stress. Dimensions of the wire change and then they cause a change of electric resistance.

The strain gauge is chosen in such a way that the resistance R_s changes linearly in the range of admissible strain for the extensible element. To transform changes of resistance into an electric signal, the strain gauge is inserted in one arm of a Wheatstone bridge which is balanced in the absence of stress on the strain gauge itself. From Figure 8.10b it can be understood that the voltage balance in the bridge is

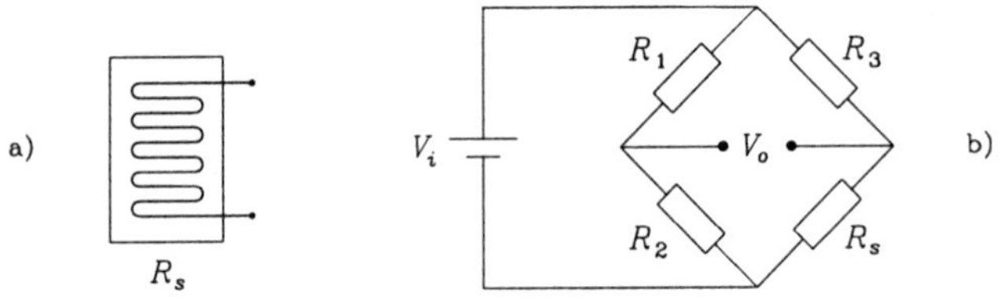

Figure 8.10 Schematic representation of a strain gauge (a), and its insertion in a Wheatstone bridge (b).

described by

$$V_o = \left(\frac{R_2}{R_1 + R_2} - \frac{R_s}{R_3 + R_s}\right) V_i. \tag{8.18}$$

If temperature variations occur, the wire changes its dimension without application of any external stress. To reduce the effect of temperature variations on the measurement output, it is worth inserting another strain gauge in an adjacent arm of the bridge, which is glued on a portion of the extensible element not subject to strain.

Finally, to increase bridge sensitivity, two strain gauges may be used which shall be glued on the extensible element in a way that one strain gauge is subject to traction and the other to compression; the two strain gauges then have to be inserted in two adjacent arms of the bridge.

Shaft Torque Sensor

In order to employ a servomotor as a torque-controlled generator, an indirect measure of the driving torque is typically used, e.g., through the measure of armature current in a permanent-magnet dc servomotor. If it is desired to guarantee insensitivity to change of parameters relating torque to the measured physical quantities, it is necessary to resort to a direct torque measurement.

The torque delivered by the servomotor to the joint can be measured by strain gauges mounted on an extensible apparatus interposed between the motor and the joint, e.g., a hollow shafting. Such apparatus shall have low torsional stiffness and high bending stiffness, and it shall ensure a proportional relationship between the applied torque and the induced strain.

By connecting the strain gauges mounted on the hollow shafting (in a Wheatstone bridge configuration) to a slip ring by means of graphite brushes, it is possible to feed the bridge and measure the resulting unbalanced signal which is proportional to the applied torque.

The measured torque is that delivered by the servomotor to the joint, and thus it does not coincide with the driving torque C_m in the block schemes of the actuating systems in Figure 8.2 and in Figure 8.6. In fact, such measurement does not account for the inertial and friction torque contributions as well as for the transmission located upstream of the measurement point. It follows that the implementation of a centralized control scheme in which the servomotor behaves as a torque-controlled generator shall be made on the basis of a manipulator dynamic model which does not account for the inertial and friction effects of the servomotor itself.

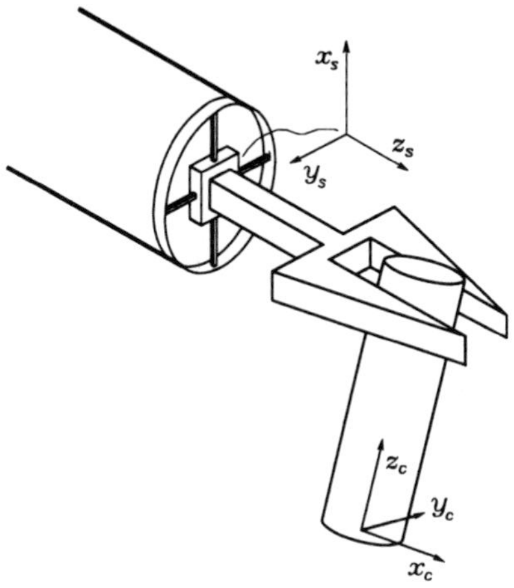

Figure 8.11 Use of a force sensor on the outer link of a manipulator.

Wrist Force Sensor

When the manipulator's end effector is in contact with the working environment, the control algorithms presented in the previous chapter require contact force measurements. The *force sensor* allows measuring the three components of a force and the three components of a moment with respect to a frame attached to it.

As illustrated in Figure 8.11, the sensor is employed as a connecting apparatus at the wrist between the outer link of the manipulator and the end effector. The connection is made by means of a suitable number of extensible elements subject to strain under the action of a force and a moment. Strain gauges are glued on each element which provide strain measurements. The elements have to be disposed in a keen way so that at least one element is appreciably deformed for any possible orientation of forces and moments.

Furthermore, the single force component with respect to the frame attached to the sensor should induce the least possible number of deformations, so as to obtain good structural decoupling of force components. Since a complete decoupling cannot be achieved, the number of significant deformations to reconstruct the six components of the force and moment vector is greater than six.

A typical force sensor is that where the extensible elements are disposed as in a Maltese cross; this is schematically indicated in Figure 8.12. The elements connecting the outer link with the end effector are four bars with a rectangular parallelepiped shape. On the opposite sides of each bar, a pair of strain gauges is glued that constitute two arms of a Wheatstone bridge; there is a total of eight bridges and thus the possibility of measuring eight strains.

The matrix relating strain measurements to the force components expressed in a Frame s attached to the sensor is termed sensor *calibration matrix*. Let w_i, for $i = 1, \dots, 8$, denote the outputs of the eight bridges providing measurement of the strains induced by the applied forces on the bars according to the directions specified

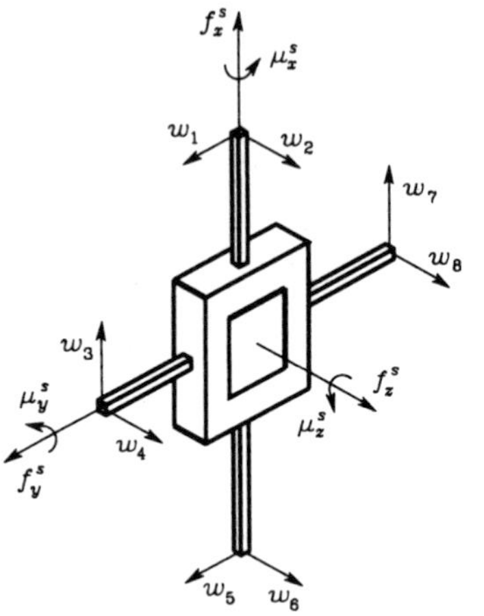

Figure 8.12 Schematic representation of a Maltese-cross force sensor.

in Figure 8.12. Then, the calibration matrix is given by the transformation

$$\begin{bmatrix} f_x^s \\ f_y^s \\ f_z^s \\ \mu_x^s \\ \mu_y^s \\ \mu_z^s \end{bmatrix} = \begin{bmatrix} 0 & 0 & c_{13} & 0 & 0 & 0 & c_{17} & 0 \\ c_{21} & 0 & 0 & 0 & c_{25} & 0 & 0 & 0 \\ 0 & c_{32} & 0 & c_{34} & 0 & c_{36} & 0 & c_{38} \\ 0 & 0 & 0 & c_{44} & 0 & 0 & 0 & c_{48} \\ 0 & c_{52} & 0 & 0 & 0 & c_{56} & 0 & 0 \\ c_{61} & 0 & c_{63} & 0 & c_{65} & 0 & c_{67} & 0 \end{bmatrix} \begin{bmatrix} w_1 \\ w_2 \\ w_3 \\ w_4 \\ w_5 \\ w_6 \\ w_7 \\ w_8 \end{bmatrix}. \tag{8.19}$$

Reconstruction of force measures through the calibration matrix is entrusted to suitable signal processing circuitry available in the sensor.

Typical sensors have a diameter of about 10 cm and a height of about 5 cm, with a measurement range of (50,500) N for the forces and of (5,70) N·m for the torques, and a resolution of the order of 0.1% of the maximum force and of 0.05% of the maximum torque, respectively; the sampling frequency at the output of the processing circuitry is of the order of 1 kHz.

Finally, it is worth noticing that force sensor measurements cannot be directly used by a force/motion control algorithm, since they describe the equivalent forces acting on the sensors which differ from the forces applied to the manipulator's end effector (Figure 8.11). It is therefore necessary to transform those forces from the sensor Frame s into the constraint Frame c introduced in Section 7.5; in view of the transformation in (3.104), one has

$$\begin{bmatrix} \boldsymbol{f}_c^c \\ \boldsymbol{\mu}_c^c \end{bmatrix} = \begin{bmatrix} \boldsymbol{R}_s^c & \boldsymbol{O} \\ \boldsymbol{S}(\boldsymbol{r}_{cs}^c)\boldsymbol{R}_s^c & \boldsymbol{R}_s^c \end{bmatrix} \begin{bmatrix} \boldsymbol{f}_s^s \\ \boldsymbol{\mu}_s^s \end{bmatrix}, \tag{8.20}$$

which requires knowledge of the position $\boldsymbol{r}_{cs}^c$ of the origin of Frame s with respect to Frame c as well as of the orientation $\boldsymbol{R}_s^c$ of Frame s with respect to Frame c. Both such

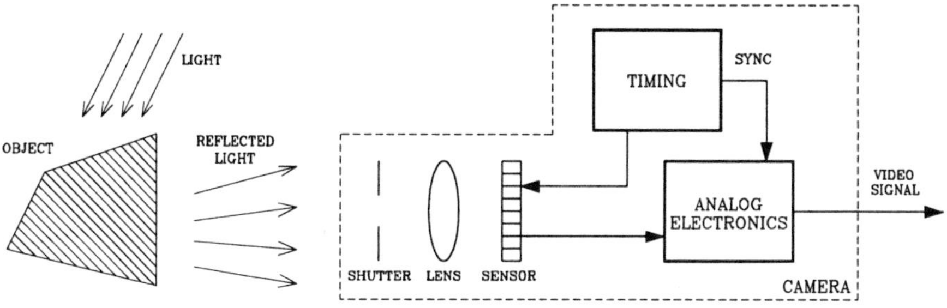

Figure 8.13 Schematic representation of vision system.

quantities are expressed in Frame c, and thus they are constant only if the end effector is still, once contact has been achieved.

8.3.4 Vision Sensors

The task of a camera as a *vision sensor* is to measure the intensity of the light reflected by an object. To this purpose, a photosensitive element, termed *pixel* (or *photosite*), is employed, which is capable to transform light energy into electric energy. Different types of sensors are available depending on the physical principle exploited to realize the energy transformation. The most widely used devices are the CCD and the CMOS sensors based on the photoelectric effect of semiconductors.

CCD

A CCD (Charge Coupled Device) sensor is constituted by a rectangular array of photosites. Due to the photoelectric effect, when a photon hits the semiconductor surface, a number of free electrons are created, so that each element accumulates a charge depending on the time integral of the incident illumination over the photosensitive element. This charge is then passed by a transport mechanism (similar to an analog shift register) to the output amplifier, while at the same time the photosite is discharged. The electric signal is to be further processed in order to produce the real *video signal*.

CMOS

A CMOS (Complementary Metal Oxide Semiconductor) sensor is constituted by a rectangular array of photodiodes. The junction of each photodiode is precharged and it is discharged when hit by photons. An amplifier integrated in each pixel can transform this charge into a voltage or current level. The main difference with the CCD sensor is that the pixels of a CMOS sensor are non-integrating devices; after being activated they measure throughput, not volume. In this manner, a saturated pixel will never overflow and influence a neighboring pixel. This prevents the effect of *blooming*, which indeed affects the CCD sensors.

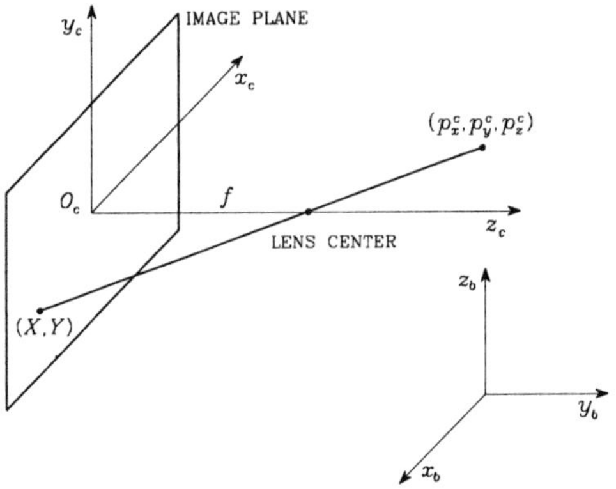

Figure 8.14 Perspective transformation.

Camera

As sketched in Figure 8.13, a camera is a complex system comprising several devices other than the photosensitive sensor, *i.e.*, a *shutter*, a *lens* and *analog preprocessing electronics*. The lens is in charge of focusing the light reflected by the object on the plane where the photosensitive sensor lies, called the *image plane*.

With reference to Figure 8.14, consider a frame O_c–$x_c y_c z_c$ attached to the camera, whose location with respect to the base frame is identified by the homogeneous transformation matrix $\boldsymbol{T}_c^b$. Take a point of the object of coordinates $\boldsymbol{p}^c = [\,p_x^c \quad p_y^c \quad p_z^c\,]^T$; typically, the centroid of the object is chosen. Then, the coordinate transformation from the base frame to the camera frame is described as

$$\tilde{\boldsymbol{p}}^c = \boldsymbol{T}_b^c \tilde{\boldsymbol{p}}, \tag{8.22}$$

where $\boldsymbol{p}$ denotes the object position with respect to the base frame and homogeneous representations of vectors have been used.

Due to the refraction phenomenon, the point in the camera frame is transformed into a point in the image plane via the *perspective transformation*, *i.e.*,

$$\begin{aligned} X &= \frac{f p_x^c}{f - p_z^c} \\ Y &= \frac{f p_y^c}{f - p_z^c}, \end{aligned} \tag{8.23}$$

where (X, Y) are the new coordinates and f is the *focal length* of the lens. Notice that these coordinates are expressed in metric units and the above transformation is singular at $p_z^c = f$. Nonetheless, these relationships hold only in theory, since the real lenses are always affected by imperfections, which cause image quality degradation. Two types of distortions can be recognized; namely, *aberrations* and *geometric distortion*. The former can be reduced by restricting the light rays to a small central region of the lens; the effects of the latter can be compensated on the basis of a suitable model whose parameters are to be identified.

A visual information is typically elaborated by a digital processor, and thus the measurement principle is to transform the light intensity $I(X,Y)$ of each point in the image plane into a number. It is clear that a *spatial sampling* is needed since an infinite number of points in the image plane exist, as well as a *temporal sampling* since the image can change during time. The CCD or CMOS sensors play the role of spatial samplers, while the shutter in front of the lens plays the role of the temporal sampler.

The spatial sampling unit is the pixel, and thus the coordinates (X,Y) of a point in the image plane are to be expressed in pixels, *i.e.*, (X_I, Y_I). Due to the photosite finite dimensions, the pixel coordinates of the point are related to the coordinates in metric units through two scale factors α_x and α_y; namely,

$$\begin{aligned} X_I &= \frac{\alpha_x f p_x^c}{f - p_z^c} + X_0 \\ Y_I &= \frac{\alpha_y f p_y^c}{f - p_z^c} + Y_0, \end{aligned} \tag{8.24}$$

where X_0 and Y_0 are the offsets which take into account the position of the origin of the pixel coordinate system with respect to the optical axis. This nonlinear transformation can be written in a linear form by resorting to the homogeneous representation of the point (x_I, y_I, z_I) via the relationships:

$$\begin{aligned} X_I &= \frac{x_I}{z_I} \\ Y_I &= \frac{y_I}{z_I}. \end{aligned}$$

As a consequence, (8.24) can be rewritten as

$$\begin{bmatrix} x_I \\ y_I \\ z_I \end{bmatrix} = \boldsymbol{\Omega} \begin{bmatrix} p_x^c \\ p_y^c \\ p_z^c \\ 1 \end{bmatrix} \tag{8.25}$$

where

$$\boldsymbol{\Omega} = \begin{bmatrix} \alpha_x & 0 & X_0 & 0 \\ 0 & \alpha_y & Y_0 & 0 \\ 0 & 0 & 1 & 0 \end{bmatrix} \begin{bmatrix} 1 & 0 & 0 & 0 \\ 0 & 1 & 0 & 0 \\ 0 & 0 & -1/f & 1 \\ 0 & 0 & 0 & 1 \end{bmatrix}. \tag{8.26}$$

At this point, the overall transformation from the Cartesian space of the observed object to the *image space* of its image in pixels is characterized by composing the transformations in (8.22) and (8.25) as

$$\boldsymbol{\Xi} = \boldsymbol{\Omega} \boldsymbol{T}_b^c \tag{8.27}$$

which represents the so-called *camera calibration* matrix. It is worth pointing out that such a matrix contains *intrinsic parameters* $(\alpha_x, \alpha_y, X_0, Y_0, f)$ in $\boldsymbol{\Omega}$ depending on the sensor and lens characteristics as well as *extrinsic parameters* in T_c^b depending on the relative position and orientation of the camera with respect to the base frame. Several calibration techniques exist to identify these parameters in order to compute the transformation between the Cartesian space and the image space as accurately as possible.

If a monochrome CCD camera[1] is of concern, the output amplifier of the sensor produces a signal which is processed by a timing analog electronics in order to generate an electric signal according to one of the existing *video standards*, *i.e.*, the CCIR European and Australian standard, or the RS170 American and Japanese standard. In any case, the video signal is a voltage of 1 V peak-to-peak whose amplitude represents sequentially the image intensity.

The entire image is divided into a number of lines (625 for the CCIR standard and 525 for the RS170 standard) to be sequentially scanned. The raster scan proceeds horizontally across each line and each line from top to bottom, but first all the even lines, forming the first *field*, and then all the odd lines, forming the second *field*, so that a *frame* is composed by two successive fields. This technique, called *interlacing* allows updating the image either at frame rate or at field rate; in the former case the update frequency is that of the entire frame (25 Hz for the CCIR standard and 30 Hz for the RS170 standard), while in the latter case the update frequency can be doubled as long as half the vertical resolution can be tolerated.

The last step of the measurement process is to digitize the analog video signal. The special analog-to-digital converters adopted for video signal acquisition are called *frame grabbers*. By connecting the output of the camera to the frame grabber, the video waveform is sampled and quantized and the values stored in a two-dimensional memory array representing the spatial sample of the image, known as *framestore*; this array is then updated at field or frame rate.

In the case of CMOS cameras (currently available only for monochrome images), thanks to the CMOS technology which allows the integration of the analog-to-digital converter in each pixel, the output of the camera is directly a two-dimensional array, whose elements can be accessed randomly. Such advantage, with respect to CCD cameras, leads to the possibility of higher frame rates if only parts of the entire frame are accessed.

The above described sequence of steps from image formation to image acquisition can be classified as a process of *low-level vision*; this includes the extraction of elementary image features, e.g., centroid and intensity discontinuities. On the other hand, a robotic system can be considered really autonomous only if procedures for emulating cognition are available, e.g., recognizing an observed object among a set of CAD models stored into a data base. In this case, the artificial vision process can be referred to as *high-level vision*.

---

[1] Colour cameras are equipped with special CCD's sensitive to three basic colours (RGB); the most sophisticated cameras have three separate sensors, one per each basic colour.

Problems

8.1 Prove (8.6)–(8.9).

8.2 Consider the dc servomotor with the data: $I_m = 0.0014\,\text{kg}\cdot\text{m}^2$, $F_m = 0.01\,\text{N}\cdot\text{m}\cdot\text{s/rad}$, $L_a = 2\,\text{mH}$, $R_a = 0.2\,\text{ohm}$, $k_t = 0.2\,\text{N}\cdot\text{m/A}$, $k_v = 0.2\,\text{V}\cdot\text{s/rad}$, $C_iG_v = 1$, $T_v = 0.1\,\text{ms}$, $k_i = 0$. Perform a computer simulation of the current and velocity response to a unit step voltage input V_c'. Adopt a sampling time of 1 ms.

8.3 For the servomotor of Problem 8.2, design the controller of the current loop $C_i(s)$ so that the current response to a unit step voltage input V_c' is characterized by a settling time of 2 ms. Compare the velocity response with that obtained in Problem 8.2.

8.4 Find the control voltage/output position and reaction torque/output position transfer functions for the scheme of Figure 8.6.

8.5 For a Gray-code optical encoder, find the interconversion logic circuit which allows obtaining a binary-coded output word.

8.6 With reference to a contact situation of the kind illustrated in Figure 8.11, let:

$$\boldsymbol{r}_{cs}^{c} = [\,-0.3 \quad 0 \quad 0.2\,]^T \text{ m} \qquad \boldsymbol{R}_s^c = \begin{bmatrix} 0 & 0 & 1 \\ 0 & -1 & 0 \\ 1 & 0 & 0 \end{bmatrix}$$

$$\boldsymbol{f}_s^s = [\,20 \quad 0 \quad 0\,]^T \text{ N} \qquad \boldsymbol{\mu}_s^s = [\,0 \quad 6 \quad 0\,]^T \text{ N}\cdot\text{m}.$$

Compute the equivalent force and moment in the contact frame.

8.7 Consider the SCARA manipulator in Figure 2.34 with link lengths $a_1 = a_2 = 0.5$ m. Let the base frame be located at the intersection between the first link and the base link with axis z pointing downwards and axis x in the direction of the first link when $\vartheta_1 = 0$. Assume that a CCD camera is mounted on the wrist so that the camera frame is aligned with the end-effector frame. The camera parameters are: $f = 8$ mm, $\alpha_x = 79.2$ pixel/mm, $\alpha_y = 120.5$ pixel/mm, $X_0 = 250$, $Y_0 = 250$. An object is observed by the camera and is described by the point of coordinates $\boldsymbol{p} = [\,0.8 \quad 0.5 \quad 0.9\,]^T$ m. Compute the pixel coordinates of the point when the manipulator is at the configuration $\boldsymbol{q} = [\,0 \quad \pi/4 \quad 0.1 \quad 0\,]$.

Bibliography

Asada H., Youcef-Toumi K. (1987) *Direct-Drive Robots*. MIT Press, Cambridge, Mass.

Blackburn J.F., Reethof G., Shearer J.L. (1960) *Fluid Power Control*. Technology Press & Wiley, New York.

Corke, P.I. (1996) *Visual Control of Robots*. Research Studies Press, Taunton, England.

Critchlow A.J. (1985) *Introduction to Robotics*. Macmillan, New York.

Electro-Craft Corporation (1980) *DC Motors, Speed Controls, Servo Systems*. 5th ed., Engineering Handbook, Hopkins, Minn.

Good M.C., Sweet L.M., Strobel K.L. (1985) Dynamic models for control system design of integrated robot and drive systems. *ASME J. Dynamic Systems, Measurement, and Control* 107:53–59.

Kenjo T., Nagamori S. (1985) *Permanent-Magnet and Brushless DC Motors*. Clarendon Press, Oxford, England.

Lhote F., Kauffmann J.-M., André P., Taillard J.-P. (1984) *Robot Components and Systems*. Robot Technology 4, Kogan Page, London.

McKerrow, P.J. (1991) *Introduction to Robotics*. Addison-Wesley, Sydney.

Rivin, E.I. (1988) *Mechanical Design of Robots*. McGraw-Hill, New York.

Rosen, C.A., and D. Nitzan (1977) Use of sensors in programmable automation. *IEEE Computer* 10(12):12–23.

Snyder, W.E. (1985) *Industrial Robots: Computer Interfacing and Control*. Prentice-Hall, Englewood Cliffs, N.J.

SRI International (1988) *Robot Design Handbook*. G.B. Andeen (Ed.), McGraw-Hill, New York.

Tsai, R. (1987) A versatile camera calibration technique for high accuracy 3-D machine vision metrology using off-the-shelf TV cameras and lenses. *IEEE Trans. Robotics and Automation* 3:323–344.

9. Control Architecture

The final chapter is devoted to set the concepts presented in the previous chapters on modelling, planning, control, actuators and sensors in a wide and distinct framework. A reference model for the *functional architecture* of an industrial robot's *control system* is presented. The *hierarchical structure* and its articulation into *functional modules* allows finding the requirements and characteristics of the *programming environment* and *hardware architecture*.

9.1 Functional Architecture

The *control system* to supervise the activities of a robotic system shall be endowed with a number of tools providing the following functions:

- capability of moving physical objects in the working environment, *i.e.*, *manipulation* ability;
- capability of obtaining information on the state of the system and working environment, *i.e.*, *sensory* ability;
- capability of exploiting information to modify system behaviour in a preprogrammed manner, *i.e.*, *intelligent* behaviour ability;
- capability of storing, elaborating and providing data on system activity, *i.e.*, *data processing* ability.

An effective implementation of these functions can be obtained by means of a *functional architecture* which is thought of as the superposition of several *activity levels* arranged in a *hierarchical structure*. The lower levels of the structure are oriented to physical motion execution, whereas the higher levels are oriented to logical action planning. The levels are connected by data flows; those directed towards the higher levels regard measurements and/or results of actions, while those directed towards the lower levels regard transmission of directions.

With reference to the control system functions implementing management of the above listed system activities, in general it is worth allocating three *functional models* at each level. A first module is devoted to sensory data management (sensory module). A second module is devoted to provide knowledge of the relevant world (modelling module). A third module is devoted to decide the policy of the action (decision module).

More specifically, the *sensory modules* acquire, elaborate, correlate and integrate

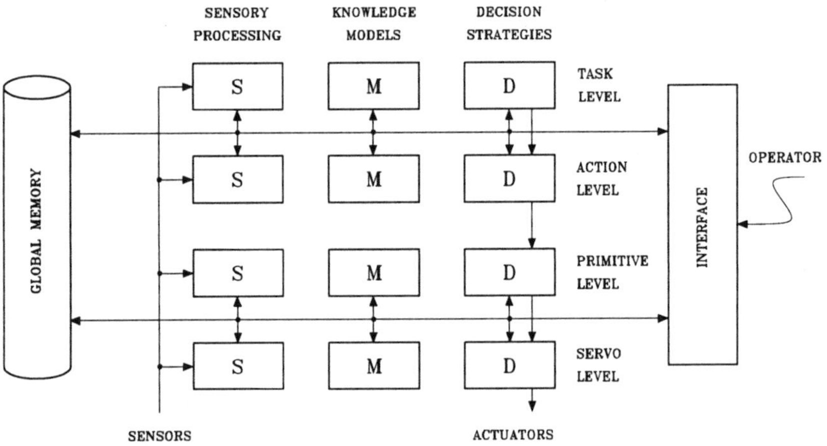

Figure 9.1 Reference model for a control system functional architecture.

sensory data in time and space, in order to recognize and measure the system state and environment characteristic; clearly, the functions of each module are oriented to the management of the relevant sensory data for that level.

On the other hand, the *modelling modules* contain models derived on the basis of a priori knowledge of system and environment; these models are updated by the information coming from the sensory modules, while the activation of the required functions is entrusted to the decision modules.

Finally, the *decision modules* perform decomposition of high-level tasks into low-level actions; such task decomposition concerns both decomposition in time of sequential actions and decomposition in space of concurrent actions. Each decision module is entrusted with the functions concerning management of elementary action assignments, task planning and execution.

The functions of a decision module characterize the level of the hierarchy and determine the functions required to the modelling and sensory modules operating at the same level. This implies that the contents of these two modules do not allow uniquely determining the hierarchical level, since the same function may be present at more levels depending on the needs of the decision modules at the relative levels.

The functional architecture needs an *operator interface* at each level of the hierarchy, so as to allow an operator to perform supervisory and intervention functions on the robotic system.

The instructions imparted to the decision module at a certain level may be provided either by the decision module at the next higher level or by the operator interface, or else by a combination of the two. Moreover, the operator, by means of suitable communication tools, can be informed on the system state and thus can contribute his/her own knowledge and decisions to the modelling and sensory modules.

In view of the high data flow concerning the exchange of information between the various levels and modules of the functional architecture, it is worth allocating a shared *global memory* which contains the updated estimates on the state of the whole system and environment.

The structure of the reference model for the functional architecture is represented in Figure 9.1, where the *four hierarchical levels* potentially relevant for robotic systems in industrial applications are illustrated. Such levels regard definition of the *task*, its decomposition into elementary *actions*, assignment of *primitives* to the actions, and implementation of control actions on the *servo*manipulator. In the following, the general functions of the three modules at each level are described.

At the *task level*, the user specifies the task which the robotic system shall execute; this specification is performed at a high level of abstraction. The goal of the desired task is analyzed and decomposed into a sequence of actions which are coordinated in space and time and allow implementation of the task. The choice of actions is performed on the basis of knowledge models as well as of the scene of interest for the task. For instance, consider the application of a robot installed in an assembly line which is required to perform a specific assembly task. To define the elementary actions that have to be transmitted to the decision module at the next lower level, the decision module shall consult its knowledge base available in the modelling module, e.g., type of assembly, components of the object to assembly, assembly sequence, and choice of tools. This knowledge base shall be continuously updated by the information provided by the sensory module concerning location of the parts to assembly; such information is available thanks to a high-level vision system operating in a scarcely structured environment, or else thanks to simple sensors detecting the presence of an object in a structured environment.

At the *action level*, the symbolic commands coming from the task level are translated into sequences of intermediate configurations which characterize a motion path for each elementary action. The choice of the sequences is performed on the basis of models of the manipulator and environment where the action is to take place. With reference to one of the actions generated by the above assembly task, the decision module chooses the most appropriate coordinate system to compute manipulator's end-effector locations, by separating translation from rotation if needed, it decides whether to operate in the joint or operational space, it computes the path or via points, and for the latter it defines the interpolation functions. By doing so, the decision module shall compare the sequence of configurations with a model of the manipulator as well as with a geometric description of the environment, which are both available in the modelling model. In this way, action feasibility is ascertained in terms of obstacle-collision avoidance, motion in the neighbourhood of kinematically singular configurations, occurrence of mechanical joint limits, and eventually utilization of available redundant degrees of mobility. The knowledge base is updated by the information on the portion of scene where the single action takes place which is provided by the sensory module, e.g., by means of a low-level vision system or range sensors.

At the *primitive level*, on the basis of the sequence of configurations received by the action level, admissible motion trajectories are computed and the control strategy is decided. The motion trajectory is interpolated so as to generate the references for the servo level. The choice of motion and control primitives is conditioned by the features of the mechanical structure and its degree of interaction with the environment. Still with reference to the above case study, the decision module computes the geometric path and the relative trajectory on the basis of the knowledge of the manipulator dynamic model available in the modelling module. Moreover, it defines the type of control algorithm, e.g., decentralized control, centralized control, or interaction control; it specifies the relative gains; and it performs proper coordinate transformations, e.g., kinematic inversion if needed. The sensory module provides information on the occurrence of conflicts between motion planning and execution, by means of, e.g., force sensors, low-level vision systems and proximity sensors.

At the *servo level*, on the basis of the motion trajectories and control strategies imparted by the primitive level, control algorithms are implemented which provide the driving signals to the joint servomotors. The control algorithm operates on error signals between the reference and the actual values of the controlled quantities, by utilizing knowledge of manipulator dynamic model, and of kinematics if needed. In particular, the decision module performs a microinterpolation on the reference trajectory to fully exploit the dynamic characteristic of the drives, it computes the control law, and it generates the (voltage or current) signals for controlling the specific drives. The modelling module elaborates the terms of the control law depending on the manipulator current configuration and pass them to the decision module; such terms are computed on the basis of knowledge of manipulator dynamic model. Finally, the sensory module provides measurements of the proprioceptive sensors (position, velocity and contact force if needed); these measurements are used by the decision module to compute the servo errors and, if required, by the modelling module to update the configuration-dependent terms in the model.

The specification of the functions associated with each level points out that the implementation of such functions shall be performed at different time rates, in view of their complexity and requirements. On one hand, the functions associated with the higher levels are not subject to demanding real-time constraints, since they regard planning activities. On the other hand, their complexity is notable, since scheduling, optimization, resource management and high-level sensory system data processing are required to update complex models.

At the lowest level, demanding real-time operation prevails in order to obtain high dynamic performance of the mechanical structure. The above remarks allow concluding that, at the servo level, it is necessary to provide the driving commands to the motors and to detect the proprioceptive sensory data at sampling rates of the order of the millisecond, while sampling rates of the order of the minute are admissible at the task level.

With respect to this reference model of functional architecture, current industrial robot's control systems are not endowed with all the functions illustrated, because of both technology and cost limitations. In this regard, the task level is not implemented at all since there do not yet exist effective and reliable application software packages allowing support of the complex functions required at this level.

It is worth characterizing those functional levels of the reference models which are typically implemented in *advanced industrial robot's control systems*. The details of Figure 9.2 show what follows.

- The modelling and sensory modules are always present at the lowest level, because of demanding requirements at the servo level for high dynamic performance robots to be employed even in relatively simple applications.

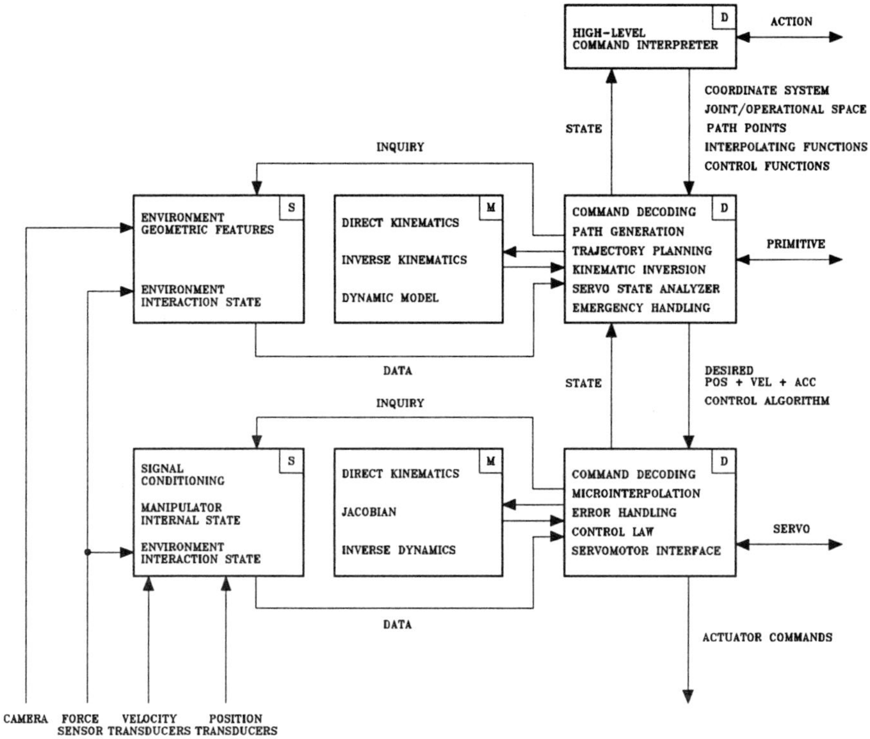

Figure 9.2 Hierarchical levels of a functional architecture for industrial robots.

- At the primitive level, the modelling module is usually present while the sensory module is present only in a reduced number of applications that require robot interaction with a less structured environment.
- At the action level, the decision module is present only as an interpreter of the high-level commands imparted by the operator. All the task decomposition functions are entrusted to the operator, and thus the modelling and sensory module are absent at this level. Possible checking of action feasibility is moved down to the primitive level where a modelling module exists.

In view of the highly-structured reference model of functional architecture illustrated above, evolution of the control system towards more and more powerful capabilities is possible. In fact, one may foresee that information technology progress may allow adding hierarchically higher levels than the task level. These should functionally characterize complex tasks to be decomposed into elementary tasks and yet, at an even higher level, missions to be decomposed into complex tasks. A six-level hierarchical structure of the above kind has been proposed as the reference model for the functional architecture of the control system of a service robotic system for space applications (NASREM). In this framework, one may allocate the functions required to *advanced robotics* systems devoted to applications in, e.g., agriculture, archaeology, assistance to the disabled, domestic use, operation in contaminated environments, medicine, surveillance and undersea.

9.2 Programming Environment

Programming a robotic system requires definition of a *programming environment* supported by suitable *languages*, which allows the operator imparting the task directions the robot shall execute. The programming environment is entrusted not only with the function of translating statements by means of a suitable language, but also with the function of checking correct execution of a task being executed by the robot. Therefore, robot programming environments, besides having some features in common with computer programming environments, present a number of issues related to the observation that program execution produces effects on the physical world. In other words, even if a very accurate description of physical reality is available in the programming environment, a number of situations will unavoidably occur which have not been or cannot be predicted.

As a consequence, a robot programming environment shall be endowed with the following features:

- real-time operating system,
- world modelling,
- motion control,
- sensory data reading,
- interaction with physical system,
- error detection capability,
- recovery of correct operational functions,
- specific language structure.

Therefore, the requirements on a programming environment may naturally stem from the articulation into models of the preceding reference model of functional architecture. Such environment will be clearly conditioned by the level of the architecture at which operator access is allowed. In the following, the requirements imposed on the programming environment by the functions respectively characterizing the sensory, modelling and decision modules are presented with reference to the hierarchical levels of the functional architecture.

Sensory data handling is the determining factor which qualifies a programming environment. At the servo level, real-time proprioceptive sensory data conditioning is required. At the primitive level, sensory data have to be expressed in the relevant reference frames. At the action level, geometric features of the objects interested to the action have to be extracted by high-level sensory data. At the task level, tools allowing recognition of the objects present in the scene are required.

The ability of *consulting knowledge models* is a support for a programming environment. At the servo level, on-line numerical computation of the models utilized by control algorithms is to be performed on the basis of sensory data. At the primitive level, coordinate transformations have to be operated. At the action level, it is crucial to have tools allowing system simulation and CAD modelling of elementary objects. At the task level, the programming environment shall assume the functions of an expert system.

Decision functions play a fundamental role in a programming environment, since they allow defining the flow charts. At the servo level, on-line computation ability is required to generate the driving signals for the mechanical system. At the primitive level, logic conditioning is to be present. At the action level, process synchronization options shall be available in order to implement nested loops, parallel computation and interrupt system. At the task level, the programming environment shall allow management of concurrent processes, and it shall be endowed with tools to test for, locate and remove mistakes from a program (debuggers) at a high-interactive level.

The evolution of programming environments has been conditioned by technology development of computer science. An analysis of this evolution leads to finding three generations of environments with respect to their functional characteristics; namely, *teaching-by-showing programming*, *robot-oriented programming*, and *object-oriented programming*. In the evolution of the environments, the next generation usually incorporates the functional characteristics of the previous generation.

This classification regards those features of the programming environment relative to the operator interface, and thus it has a direct correspondence with the hierarchical levels of the reference model of functional architecture. The functions associated with the servo level lead to understanding that a programming environment problem does not really exist for the operator. In fact, low-level programming concerns the use of traditional programming languages (Assembly, C) for development of real-time systems. The operator is only left with the possibility of intervening by means of simple command actuation (point-to-point, reset), reading of proprioceptive sensory data, and limited editing capability.

9.2.1 Teaching-by-showing

The first generation has been characterized by programming techniques of *teaching-by-showing* type. The operator guides the manipulator manually or by means of a teach pendant along the desired motion path. During motion execution, the data read by joint position transducers are stored and thus they can be utilized later as references for the joint drive servos; in this way, the mechanical structure is capable of executing (playing back) the motion taught by a direct acquisition on the spot.

The programming environment does not allow implementation of logic conditioning and queuing, and thus the associated computational hardware plays elementary functions. The operator is not required to have special programming skill, and thus he/she can be a plant technician. The setup of a working program obviously requires the robot to be available to the operator at the time of teaching, and thus the robot itself has to be taken off production. Typical applications that can be solved by this programming technique regard spot welding, spray painting and, in general, simple palletizing.

With regard to the reference model of functional architecture, a programming environment based on the teaching-by-showing technique allows operator access at the primitive level.

The drawbacks of such an environment may be partially overcome by the adoption of simple programming languages which allow:

- acquiring meaningful posture by teaching,
- computing end-effector location with respect to a reference frame, by means of a direct kinematics transformation,
- assigning a motion primitive and the trajectory parameters (usually, velocity as a percentage of the maximum velocity),
- computing servo references, by means of an inverse kinematics transformation,
- conditioning teaching sequences to the use of simple external sensors (presence of an object at the gripper),
- correcting motion sequences by using simple text editors,
- performing simple connections between subsets of elementary sequences.

Examples of teaching-by-showing languages provided with some of these functions are T3 by Milacron and FUNKY by IBM. Providing a teaching-by-showing environment with the the above-listed functions can be framed as an attempt to develop a structured programming environment.

9.2.2 Robot-oriented Programming

Thanks to the advent of efficient low-cost computational means, *robot-oriented* programming environments have been developed. The need for interaction of the environment with physical reality has imposed integration of several functions, typical of high-level programming languages (BASIC, PASCAL), with those specifically required by robotic applications. In fact, many robot-oriented languages have retained the teaching-by-showing programming mode, in view of its natural characteristic of accurate interface with the physical world.

Since the general framework is that of a computer programming environment, two alternatives have been considered:

- to develop *ad hoc languages* for robotic applications,
- to develop robot *program libraries* supporting standard programming languages.

The current situation features the existence of numerous new proprietary languages, whereas it would be desirable to develop either robotic libraries to be used in the context of consolidated standards or new general-purpose languages for industrial automation applications.

Robot-oriented languages are *structured programming* languages which incorporate high-level statements and have the characteristic of an interpreted language, in order to obtain an interactive environment allowing the programmer to check the execution of each source program statement before proceeding to the next one. Common features of such languages are:

- text editor,
- complex data representation structures,
- extensive use of predefined state variable,
- execution of matrix algebra operations,
- extensive use of symbolic representations for coordinate frames,
- possibility to specify the coordinated motion of more frames rigidly attached to objects by means of a single frame,
- inclusion of subroutines with data and parameter exchange,
- use of logic conditioning and queuing by means of flags,
- capability of parallel computing,
- functions of programmable logic controller (PLC).

With respect to the reference model of functional architecture, it can be recognized that a robot-oriented programming environment allows operator access at the action level.

In view of the structured language characteristic, the operator in this case shall be an expert language programmer. Editing an application program may be performed off line, *i.e.*, without physical availability of the robot to the operator; off-line programming demands a perfectly structured environment, though. A robotic system endowed with a robot-oriented programming language allows execution of complex applications where the robot is inserted in a work cell and interacts with other machines and devices to perform complex tasks, such as part assembly.

Examples of the most common robot-oriented programming languages are VAL II by Unimation, AML by IBM, and PDL 2 by Comau.

Finally, a programming environment that allows accessing at the task level of a reference model of functional architecture is characterized by an *object-oriented* language. Such environment shall have the capability of specifying a task by means of high-level statements allowing automatic execution of a number of actions on the objects present in the scene. Robot programming languages belonging to this generation are currently under development and thus they are not yet available on the market. They can be framed in the field of expert systems and artificial intelligence.

9.3 Hardware Architecture

The hierarchical structure of the functional architecture adopted as a reference model for an industrial robot's control system, together with its articulation into different functional modules, suggests a hardware implementation which exploits distributed computational resources interconnected by means of suitable communication channels. To this purpose, it is worth recalling that the functions implemented in current

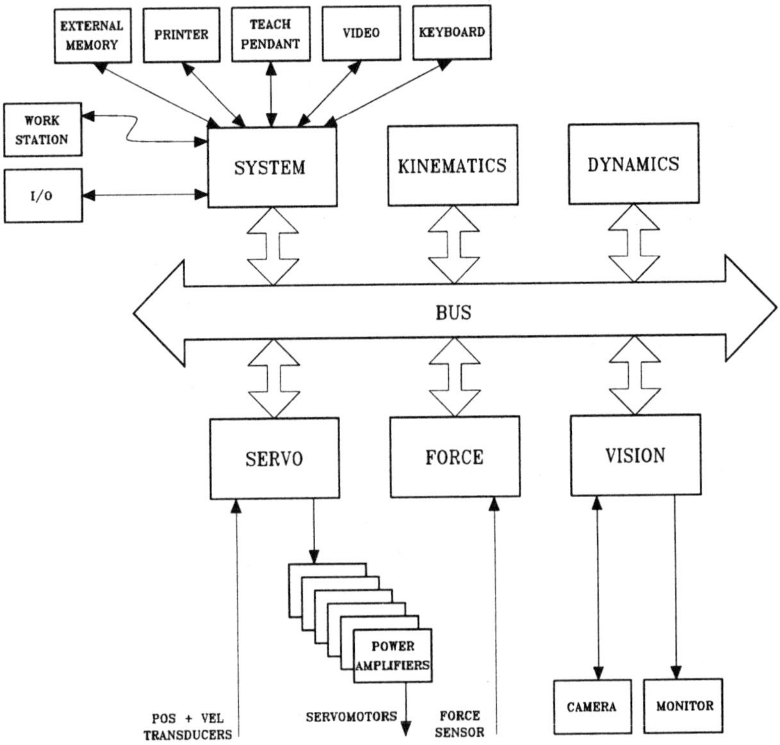

Figure 9.3 General model of the hardware architecture of an industrial robot's control system.

control systems regard the three levels from servo to action, with a typically limited development of the functions implemented at the action level. At the servo and primitive levels, computational capabilities are required with demanding real-time constraints.

A general model of the *hardware architecture* for the control system of an industrial robot is illustrated in Figure 9.3. In this figure, proper *boards* with autonomous computational capabilities have been associated with the functions indicated in the reference model of functional architecture of Figure 9.2. The boards are connected to a *bus*, e.g., a VME bus, which allows supporting communication data flow; the bus bandwidth shall be wide enough so as to satisfy the requirements imposed by real-time constraints.

The *system* board is typically a CPU endowed with:

- a microprocessor with mathematical coprocessor,
- a bootstrap EPROM memory,
- a local RAM memory,
- a RAM memory shared with the other boards through the bus,
- a number of serial and parallel ports interfacing the bus and the external world,

- counters, registers and timers,
- an interrupt system.

The following functions are to be implemented in the system board:

- operator interface through teach pendant, keyboard, video and printer,
- interface with an external memory (hard disk) used to store data and application programs,
- interface with workstations and other control systems by means of a local communication network, e.g., Ethernet,
- I/O interface with peripheral devices in the working area, e.g., feeders, conveyors and ON/OFF sensors,
- system bootstrap,
- programming language interpreter,
- bus arbiter.

The other boards facing the bus may be endowed, besides the basic components of the system board, with a supplementary or alternative processor (DSP, Transputer) for implementation of computationally demanding or dedicated functions. With reference to the architecture in Figure 9.3, the following functions are implemented in the *kinematics* board:

- computation of motion primitives,
- computation of direct kinematics, inverse kinematics and Jacobian,
- test for trajectory feasibility,
- handling of kinematic redundancy.

The *dynamics* board is devoted to:

- computation of inverse dynamics.

The *servo* board has the functions of:

- microinterpolation of references,
- computation of control algorithm,
- digital-to-analog conversion and interface with power amplifiers,
- handling of position and velocity transducer data,
- motion interruption in case of malfunction.

The remaining boards in the figure have been considered for the sake of an example to illustrate how the use of sensors may require local processing capabilities to retrieve significant information from the given data which can be effectively used in the sensory system. The *force* board performs the following operations:

- conditioning of data provided by the force sensor,
- representation of forces in a given coordinate frame.

The *vision* board is in charge of:

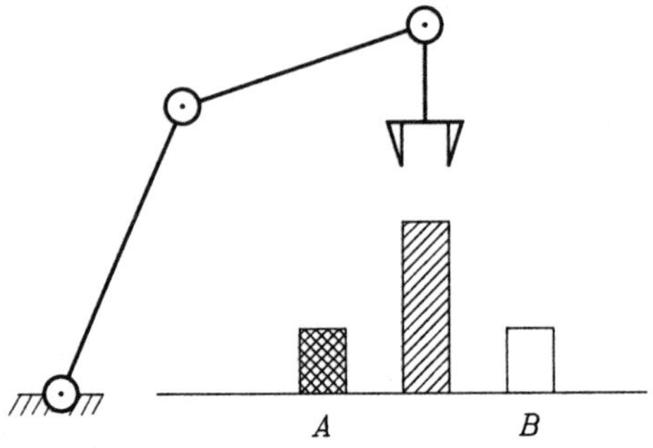

Figure 9.4 Object pick-and-place task.

- processing data provided by the camera,
- extracting geometric features of the scene,
- localizing objects in given coordinate frames.

Although the boards face the same bus, the frequency at which data are exchanged needs not to be the same for each board. Those board connected to the proprioceptive sensors indeed have the necessity to exchange date with the robot at the highest possible frequency (from 100 to 1000 Hz) to ensure high dynamic performance to motion control as well as to reveal end-effector contact in a very short time.

On the other hand, the kinematics and dynamics boards implement modelling functions and, as such, they do not require data update at a rate as high as that required by the servo board. In fact, manipulator configuration does not appreciably vary in a very short time, at least with respect to typical operational velocities and/or accelerations of current industrial robots. Common sampling frequencies are in the range of (10,100) Hz.

Also the vision board does not require a high update rate, both because the scene is generally quasi-static, and because processing of interpretive functions are typically complex. Typical frequencies are in the range of (1,10) Hz.

In sum, the board access to the communication bus of a hardware control architecture may be performed according to a multirate logic which allows solving bus data overflow problems.

Problems

9.1 With reference to the situation illustrated in Figure 9.4, describe the sequence of actions required to the manipulator to pick up an object at location A and place it at location B.

9.2 For the situation of Problem 9.1, find the motion primitives in the cases of given via points and given path points.

9.3 The planar arm indicated in Figure 9.5 is endowed with a wrist force sensor which

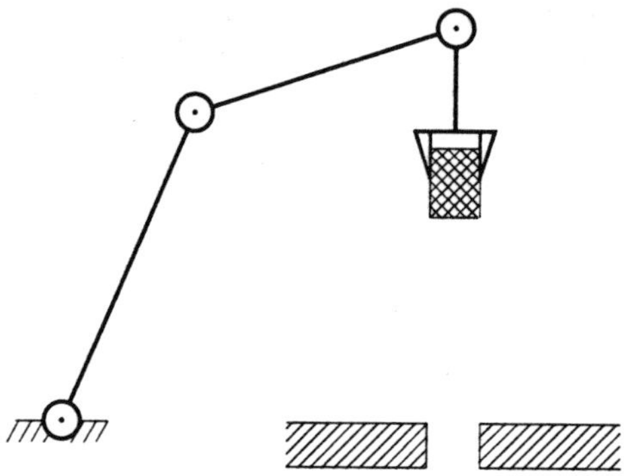

Figure 9.5 Peg-in-hole task.

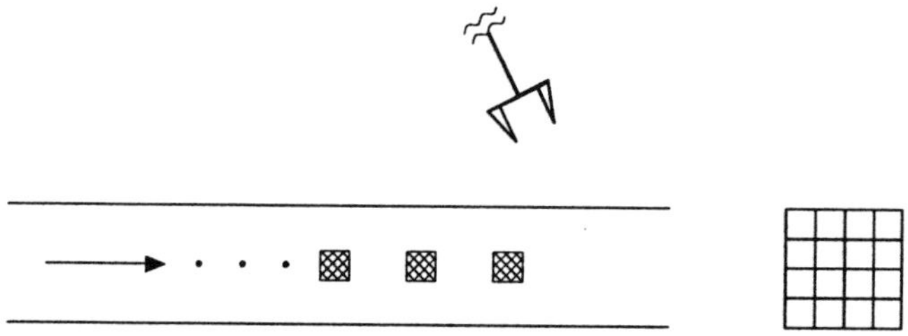

Figure 9.6 Palletizing task of objects available on a conveyor.

allows measuring the relevant force and moment components for the execution of a peg-in-hole task. Draw the flow chart for writing a program to execute the described task.

9.4 A palletizing problem is represented in Figure 9.6. Sixteen equal objects have to be loaded on the pallet. The manipulator's end effector has to pick up the objects from a conveyor, whose feeding is commanded by the robot in such a way that the objects are found always in the same location to be picked. Write a PASCAL program to execute the task.

Bibliography

Albus J.S., McCain H.G., Lumia R. (1987) NASA/NBS Standard Reference Model for Telerobot Control System Architecture (NASREM). NBS tech. note 1235, Gaithersburg, Md.

Brady M. (1985) Artificial intelligence and robotics. *Artificial Intelligence* 26:79–121.

Comau (1990) C3G 9000 Control System Documentation: Electronic Configuration, Options and Accessories. Handbook, Beinasco, Italy.

Gruver W.A., Soroka B.I., Craig J.J., Turner T.L. (1984) Industrial robot programming languages: A comparative evaluation. *IEEE Trans. Systems, Man, and Cybernetics* 14:565–570.

Lozano-Pérez T. (1983) Robot programming. *Proc. IEEE* 71:821–841.

McKerrow P.J. (1991) *Introduction to Robotics*. Addison-Wesley, Sydney.

Takase K., Paul R., Berg E. (1981) A structured approach to robot programming and teaching. *IEEE Trans. Systems, Man, and Cybernetics* 11:274–289.

Taylor R.H., Grossman D.D. (1983) An integrated robot system architecture. *Proc. IEEE* 71:842–856.

Appendix A. Linear Algebra

Since modelling and control of robot manipulators requires an extensive use of *matrices* and *vectors* as well as of matrix and vector *operations*, the goal of this appendix is to provide a brush-up of *linear algebra*.

A.1 Definitions

A *matrix* of dimensions $(m \times n)$, with m and n positive integers, is an array of elements a_{ij} arranged into m *rows* and n *columns*:

$$\boldsymbol{A} = [a_{ij}]_{\substack{i=1,\ldots,m\\ j=1,\ldots,n}} = \begin{bmatrix} a_{11} & a_{12} & \ldots & a_{1n} \\ a_{21} & a_{22} & \ldots & a_{2n} \\ \vdots & \vdots & \ddots & \vdots \\ a_{m1} & a_{m2} & \ldots & a_{mn} \end{bmatrix}. \tag{A.1}$$

If $m = n$, the matrix is said to be *square*; if $m < n$, the matrix has more columns than rows; if $m > n$ the matrix has more rows than columns. Further, if $n = 1$, the notation (A.1) is used to represent a (column) vector $\boldsymbol{a}$ of dimensions $(m \times 1)$[1]; the elements a_i are said to be vector components. A square matrix $\boldsymbol{A}$ of dimensions $(n \times n)$ is said to be *upper triangular* if $a_{ij} = 0$ for $i > j$:

$$\boldsymbol{A} = \begin{bmatrix} a_{11} & a_{12} & \ldots & a_{1n} \\ 0 & a_{22} & \ldots & a_{2n} \\ \vdots & \vdots & \ddots & \vdots \\ 0 & 0 & \ldots & a_{nn} \end{bmatrix};$$

the matrix is said to be *lower triangular* if $a_{ij} = 0$ for $i < j$.

An $(n \times n)$ square matrix $\boldsymbol{A}$ is said to be *diagonal* if $a_{ij} = 0$ for $i \neq j$, *i.e.*,

$$\boldsymbol{A} = \begin{bmatrix} a_{11} & 0 & \ldots & 0 \\ 0 & a_{22} & \ldots & 0 \\ \vdots & \vdots & \ddots & \vdots \\ 0 & 0 & \ldots & a_{nn} \end{bmatrix} = \mathrm{diag}\{a_{11}, a_{22}, \ldots, a_{nn}\}.$$

[1] According to standard mathematical notation, small boldface is used to denote vectors while capital boldface is used to denote matrices. Scalars are denoted by roman characters.

If an $(n \times n)$ diagonal matrix has all unit elements on the diagonal ($a_{ii} = 1$), the matrix is said to be *identity* and is denoted by $\boldsymbol{I}_n$[2]. A matrix is said to be *null* if all its elements are null and is denoted by $\boldsymbol{O}$. The null column vector is denoted by $\boldsymbol{0}$.

The *transpose* $\boldsymbol{A}^T$ of a matrix $\boldsymbol{A}$ of dimensions $(m \times n)$ is the matrix of dimensions $(n \times m)$ which is obtained from the original matrix by interchanging its rows and columns:

$$\boldsymbol{A}^T = \begin{bmatrix} a_{11} & a_{21} & \dots & a_{m1} \\ a_{12} & a_{22} & \dots & a_{m2} \\ \vdots & \vdots & \ddots & \vdots \\ a_{1n} & a_{2n} & \dots & a_{mn} \end{bmatrix}. \tag{A.2}$$

The transpose of a column vector $\boldsymbol{a}$ is the row vector $\boldsymbol{a}^T$.

An $(n \times n)$ square matrix $\boldsymbol{A}$ is said to be *symmetric* if $\boldsymbol{A}^T = \boldsymbol{A}$, and thus $a_{ij} = a_{ji}$:

$$\boldsymbol{A} = \begin{bmatrix} a_{11} & a_{12} & \dots & a_{1n} \\ a_{12} & a_{22} & \dots & a_{2n} \\ \vdots & \vdots & \ddots & \vdots \\ a_{1n} & a_{2n} & \dots & a_{nn} \end{bmatrix}.$$

An $(n \times n)$ square matrix $\boldsymbol{A}$ is said to be *skew-symmetric* if $\boldsymbol{A}^T = -\boldsymbol{A}$, and thus $a_{ij} = -a_{ji}$ for $i \neq j$ and $a_{ii} = 0$, leading to

$$\boldsymbol{A} = \begin{bmatrix} 0 & a_{12} & \dots & a_{1n} \\ -a_{12} & 0 & \dots & a_{2n} \\ \vdots & \vdots & \ddots & \vdots \\ -a_{1n} & -a_{2n} & \dots & 0 \end{bmatrix}.$$

A *partitioned* matrix is a matrix whose elements are matrices (*blocks*) of proper dimensions:

$$\boldsymbol{A} = \begin{bmatrix} \boldsymbol{A}_{11} & \boldsymbol{A}_{12} & \dots & \boldsymbol{A}_{1n} \\ \boldsymbol{A}_{21} & \boldsymbol{A}_{22} & \dots & \boldsymbol{A}_{2n} \\ \vdots & \vdots & \ddots & \vdots \\ \boldsymbol{A}_{m1} & \boldsymbol{A}_{m2} & \dots & \boldsymbol{A}_{mn} \end{bmatrix}.$$

A partitioned matrix may be block-triangular or block-diagonal. Special partitions of a matrix are that by columns

$$\boldsymbol{A} = [\,\boldsymbol{a}_1 \quad \boldsymbol{a}_2 \quad \dots \quad \boldsymbol{a}_n\,]$$

and that by rows

$$\boldsymbol{A} = \begin{bmatrix} \boldsymbol{a}_1^T \\ \boldsymbol{a}_2^T \\ \vdots \\ \boldsymbol{a}_m^T \end{bmatrix}.$$

[2] Subscript n is usually omitted if the dimensions are clear from the context.

Given a square matrix $\boldsymbol{A}$ of dimensions $(n \times n)$, the *algebraic complement* $\boldsymbol{A}_{(ij)}$ of element a_{ij} is the matrix of dimensions $((n-1) \times (n-1))$ which is obtained by eliminating row i and column j of matrix $\boldsymbol{A}$.

A.2 Matrix Operations

The *trace* of an $(n \times n)$ square matrix $\boldsymbol{A}$ is the sum of the elements on the diagonal:

$$\operatorname{Tr}(\boldsymbol{A}) = \sum_{i=1}^{n} a_{ii}. \tag{A.3}$$

Two matrices $\boldsymbol{A}$ and $\boldsymbol{B}$ of the same dimensions $(m \times n)$ are equal if $a_{ij} = b_{ij}$. If $\boldsymbol{A}$ and $\boldsymbol{B}$ are two matrices of the same dimensions, their *sum* is the matrix

$$\boldsymbol{C} = \boldsymbol{A} + \boldsymbol{B} \tag{A.4}$$

whose elements are given by $c_{ij} = a_{ij} + b_{ij}$. The following properties hold:

$$\begin{aligned}
&\boldsymbol{A} + \boldsymbol{O} = \boldsymbol{A} \\
&\boldsymbol{A} + \boldsymbol{B} = \boldsymbol{B} + \boldsymbol{A} \\
&(\boldsymbol{A} + \boldsymbol{B}) + \boldsymbol{C} = \boldsymbol{A} + (\boldsymbol{B} + \boldsymbol{C}).
\end{aligned}$$

Notice that two matrices of the same dimensions and partitioned in the same way can be summed formally by operating on the blocks in the same position and treating them like elements.

The product of a scalar α by an $(m \times n)$ matrix $\boldsymbol{A}$ is the matrix $\alpha\boldsymbol{A}$ whose elements are given by αa_{ij}. If $\boldsymbol{A}$ is an $(n \times n)$ diagonal matrix with all equal elements on the diagonal $(a_{ii} = a)$, it follows that $\boldsymbol{A} = a\boldsymbol{I}_n$.

If $\boldsymbol{A}$ is a square matrix, one may write

$$\boldsymbol{A} = \boldsymbol{A}_s + \boldsymbol{A}_a \tag{A.5}$$

where

$$\boldsymbol{A}_s = \frac{1}{2}(\boldsymbol{A} + \boldsymbol{A}^T) \tag{A.6}$$

is a symmetric matrix representing the *symmetric* part of $\boldsymbol{A}$, and

$$\boldsymbol{A}_a = \frac{1}{2}(\boldsymbol{A} - \boldsymbol{A}^T) \tag{A.7}$$

is a skew-symmetric matrix representing the *skew-symmetric* part of $\boldsymbol{A}$.

The row-by-column *product* of a matrix $\boldsymbol{A}$ of dimensions $(m \times p)$ by a matrix $\boldsymbol{B}$ of dimensions $(p \times n)$ is the matrix of dimensions $(m \times n)$

$$\boldsymbol{C} = \boldsymbol{A}\boldsymbol{B} \tag{A.8}$$

whose elements are given by $c_{ij} = \sum_{k=1}^{p} a_{ik} b_{kj}$. The following properties hold:

$$\begin{aligned}
\boldsymbol{A} &= \boldsymbol{A}\boldsymbol{I}_p = \boldsymbol{I}_m\boldsymbol{A}\\
\boldsymbol{A}(\boldsymbol{B}\boldsymbol{C}) &= (\boldsymbol{A}\boldsymbol{B})\boldsymbol{C}\\
\boldsymbol{A}(\boldsymbol{B}+\boldsymbol{C}) &= \boldsymbol{A}\boldsymbol{B}+\boldsymbol{A}\boldsymbol{C}\\
(\boldsymbol{A}+\boldsymbol{B})\boldsymbol{C} &= \boldsymbol{A}\boldsymbol{C}+\boldsymbol{B}\boldsymbol{C}\\
(\boldsymbol{A}\boldsymbol{B})^T &= \boldsymbol{B}^T\boldsymbol{A}^T.
\end{aligned}$$

Notice that, in general, $\boldsymbol{A}\boldsymbol{B} \neq \boldsymbol{B}\boldsymbol{A}$, and $\boldsymbol{A}\boldsymbol{B} = \boldsymbol{O}$ does not imply that $\boldsymbol{A} = \boldsymbol{O}$ or $\boldsymbol{B} = \boldsymbol{O}$; further, notice that $\boldsymbol{A}\boldsymbol{C} = \boldsymbol{B}\boldsymbol{C}$ does not imply that $\boldsymbol{A} = \boldsymbol{B}$.

If an $(m \times p)$ matrix $\boldsymbol{A}$ and a $(p \times n)$ matrix $\boldsymbol{B}$ are partitioned in such a way that the number of blocks for each row of $\boldsymbol{A}$ is equal to the number of blocks for each column of $\boldsymbol{B}$, and the blocks $\boldsymbol{A}_{ik}$ and $\boldsymbol{B}_{kj}$ have dimensions compatible with product, the matrix product $\boldsymbol{A}\boldsymbol{B}$ can be formally obtained by operating by rows and columns on the blocks of proper position and treating them like elements.

For an $(n \times n)$ *square* matrix $\boldsymbol{A}$, the *determinant* of $\boldsymbol{A}$ is the scalar given by the following expression, which holds $\forall i = 1, \ldots, n$:

$$\det(\boldsymbol{A}) = \sum_{j=1}^{n} a_{ij}(-1)^{i+j}\det\big(\boldsymbol{A}_{(ij)}\big). \tag{A.9}$$

The determinant can be computed according to any row i as in (A.9); the same result is obtained by computing it according to any column j. If $n = 1$, then $\det(a_{11}) = a_{11}$. The following property holds:

$$\det(\boldsymbol{A}) = \det(\boldsymbol{A}^T).$$

Moreover, interchanging two generic columns p and q of a matrix $\boldsymbol{A}$ yields

$$\det\big([\,\boldsymbol{a}_1 \ldots \boldsymbol{a}_p \ldots \boldsymbol{a}_q \ldots \boldsymbol{a}_n\,]\big) = -\det\big([\,\boldsymbol{a}_1 \ldots \boldsymbol{a}_q \ldots \boldsymbol{a}_p \ldots \boldsymbol{a}_n\,]\big).$$

As a consequence, if a matrix has two equal columns (rows), then its determinant is null. Also, it is $\det(\alpha\boldsymbol{A}) = \alpha^n\det(\boldsymbol{A})$.

Given an $(m \times n)$ matrix $\boldsymbol{A}$, the determinant of the square block obtained by selecting an equal number k of rows and columns is said to be k-order *minor* of matrix $\boldsymbol{A}$. The minors obtained by taking the *first* k rows and columns of $\boldsymbol{A}$ are said to be *principal* minors.

If $\boldsymbol{A}$ and $\boldsymbol{B}$ are square matrices, then

$$\det(\boldsymbol{A}\boldsymbol{B}) = \det(\boldsymbol{A})\det(\boldsymbol{B}). \tag{A.10}$$

If $\boldsymbol{A}$ is an $(n \times n)$ triangular matrix (in particular diagonal), then

$$\det(\boldsymbol{A}) = \prod_{i=1}^{n} a_{ii}.$$

More generally, if A is block-triangular with m blocks A_{ii} on the diagonal, then

$$\det(A) = \prod_{i=1}^{m} \det(A_{ii}).$$

A square matrix A is said to be *singular* when $\det(A) = 0$.

The *rank* $\varrho(A)$ of a matrix A of dimensions $(m \times n)$ is the maximum integer r so that at least a nonnull minor of order r exists. The following properties hold:

$$\begin{aligned}
&\varrho(A) \le \min\{m, n\} \\
&\varrho(A) = \varrho(A^T) \\
&\varrho(A^T A) = \varrho(A) \\
&\varrho(AB) \le \min\{\varrho(A), \varrho(B)\}.
\end{aligned}$$

A matrix so that $\varrho(A) = \min\{m, n\}$ is said to be *full-rank*.

The *adjoint* of a square matrix A is the matrix

$$\mathbf{Adj}\, A = [(-1)^{i+j} \det(A_{(ij)})]^T_{\substack{i = 1,\dots,n \\ j = 1,\dots,n}}. \tag{A.11}$$

An $(n \times n)$ square matrix A is said to be *invertible* if a matrix A^{-1} exists, termed *inverse* of A, so that

$$A^{-1}A = AA^{-1} = I_n.$$

Since $\varrho(I_n) = n$, an $(n \times n)$ square matrix A is invertible if and only if $\varrho(A) = n$, *i.e.*, $\det(A) \neq 0$ (nonsingular matrix). The inverse of A can be computed as

$$A^{-1} = \frac{1}{\det(A)} \mathbf{Adj}\, A. \tag{A.12}$$

The following properties hold:

$$\begin{aligned}
&(A^{-1})^{-1} = A \\
&(A^T)^{-1} = (A^{-1})^T.
\end{aligned}$$

If the inverse of a square matrix is equal to its transpose

$$A^T = A^{-1}, \tag{A.13}$$

then the matrix is said to be *orthogonal*; in this case it is

$$AA^T = A^T A = I. \tag{A.14}$$

If A and B are invertible square matrices of the same dimensions, then

$$(AB)^{-1} = B^{-1}A^{-1}. \tag{A.15}$$

Given n square matrices $\boldsymbol{A}_{ii}$ all invertible, the following expression holds:

$$\left(\operatorname{diag}\{\boldsymbol{A}_{11},\ldots,\boldsymbol{A}_{nn}\}\right)^{-1} = \operatorname{diag}\{\boldsymbol{A}_{11}^{-1},\ldots,\boldsymbol{A}_{nn}^{-1}\}.$$

where $\operatorname{diag}\{\boldsymbol{A}_{11},\ldots,\boldsymbol{A}_{nn}\}$ denotes the block-diagonal matrix.

If $\boldsymbol{A}$ and $\boldsymbol{C}$ are invertible square matrices of proper dimensions, the following expression holds:

$$(\boldsymbol{A}+\boldsymbol{BCD})^{-1} = \boldsymbol{A}^{-1} - \boldsymbol{A}^{-1}\boldsymbol{B}(\boldsymbol{DA}^{-1}\boldsymbol{B}+\boldsymbol{C}^{-1})^{-1}\boldsymbol{DA}^{-1},$$

where the matrix $\boldsymbol{DA}^{-1}\boldsymbol{B}+\boldsymbol{C}^{-1}$ must be invertible.

If a block-partitioned matrix is invertible, then its inverse is given by the general expression

$$\begin{bmatrix} \boldsymbol{A} & \boldsymbol{D} \\ \boldsymbol{C} & \boldsymbol{B} \end{bmatrix}^{-1} = \begin{bmatrix} \boldsymbol{A}^{-1}+\boldsymbol{E\Delta}^{-1}\boldsymbol{F} & -\boldsymbol{E\Delta}^{-1} \\ -\boldsymbol{\Delta}^{-1}\boldsymbol{F} & \boldsymbol{\Delta}^{-1} \end{bmatrix} \tag{A.16}$$

where $\boldsymbol{\Delta} = \boldsymbol{B} - \boldsymbol{CA}^{-1}\boldsymbol{D}$, $\boldsymbol{E} = \boldsymbol{A}^{-1}\boldsymbol{D}$ and $\boldsymbol{F} = \boldsymbol{CA}^{-1}$, on the assumption that the inverses of matrices $\boldsymbol{A}$ and $\boldsymbol{\Delta}$ exist. In the case of a block-triangular matrix, invertibility of the matrix requires invertibility of the blocks on the diagonal. The following expressions hold:

$$\begin{bmatrix} \boldsymbol{A} & \boldsymbol{O} \\ \boldsymbol{C} & \boldsymbol{B} \end{bmatrix}^{-1} = \begin{bmatrix} \boldsymbol{A}^{-1} & \boldsymbol{O} \\ -\boldsymbol{B}^{-1}\boldsymbol{CA}^{-1} & \boldsymbol{B}^{-1} \end{bmatrix}$$

$$\begin{bmatrix} \boldsymbol{A} & \boldsymbol{D} \\ \boldsymbol{O} & \boldsymbol{B} \end{bmatrix}^{-1} = \begin{bmatrix} \boldsymbol{A}^{-1} & -\boldsymbol{A}^{-1}\boldsymbol{DB}^{-1} \\ \boldsymbol{O} & \boldsymbol{B}^{-1} \end{bmatrix}.$$

The *derivative* of an $(m \times n)$ matrix $\boldsymbol{A}(t)$, whose elements $a_{ij}(t)$ are differentiable functions, is the matrix

$$\dot{\boldsymbol{A}}(t) = \frac{d}{dt}\boldsymbol{A}(t) = \left[\frac{d}{dt}a_{ij}(t)\right]_{\substack{i=1,\ldots,m \\ j=1,\ldots,n}}. \tag{A.17}$$

If an $(n \times n)$ square matrix $\boldsymbol{A}(t)$ is so that $\varrho(\boldsymbol{A}(t)) = n \; \forall t$ and its elements $a_{ij}(t)$ are differentiable functions, then the derivative of the *inverse* of $\boldsymbol{A}(t)$ is given by

$$\frac{d}{dt}\boldsymbol{A}^{-1}(t) = -\boldsymbol{A}^{-1}(t)\dot{\boldsymbol{A}}(t)\boldsymbol{A}^{-1}(t). \tag{A.18}$$

Given a scalar function $f(\boldsymbol{x})$, endowed with partial derivatives with respect to the elements x_i of the $(n \times 1)$ vector $\boldsymbol{x}$, the *gradient* of function f with respect to vector $\boldsymbol{x}$ is the $(n \times 1)$ column vector

$$\mathbf{grad}_{\boldsymbol{x}} f(\boldsymbol{x}) = \left(\frac{\partial f(\boldsymbol{x})}{\partial \boldsymbol{x}}\right)^T = \begin{bmatrix} \dfrac{\partial f(\boldsymbol{x})}{\partial x_1} & \dfrac{\partial f(\boldsymbol{x})}{\partial x_2} & \ldots & \dfrac{\partial f(\boldsymbol{x})}{\partial x_n} \end{bmatrix}^T. \tag{A.19}$$

Further, if $\boldsymbol{x}(t)$ is a differentiable function with respect to t, then

$$\dot{f}(\boldsymbol{x}) = \frac{d}{dt}f(\boldsymbol{x}(t)) = \frac{\partial f}{\partial \boldsymbol{x}}\dot{\boldsymbol{x}} = \mathbf{grad}_{\boldsymbol{x}}^T f(\boldsymbol{x})\dot{\boldsymbol{x}}. \tag{A.20}$$

Given a vector function $\boldsymbol{g}(\boldsymbol{x})$ of dimensions $(m \times 1)$, whose elements g_i are differentiable with respect to the vector $\boldsymbol{x}$ of dimensions $(n \times 1)$, the Jacobian matrix (or simply *Jacobian*) of the function is defined as the $(m \times n)$ matrix

$$\boldsymbol{J}_g(\boldsymbol{x}) = \frac{\partial \boldsymbol{g}(\boldsymbol{x})}{\partial \boldsymbol{x}} = \begin{bmatrix} \dfrac{\partial g_1(\boldsymbol{x})}{\partial \boldsymbol{x}} \\ \dfrac{\partial g_2(\boldsymbol{x})}{\partial \boldsymbol{x}} \\ \vdots \\ \dfrac{\partial g_m(\boldsymbol{x})}{\partial \boldsymbol{x}} \end{bmatrix}. \tag{A.21}$$

If $\boldsymbol{x}(t)$ is a differentiable function with respect to t, then

$$\dot{\boldsymbol{g}}(\boldsymbol{x}) = \frac{d}{dt}\boldsymbol{g}(\boldsymbol{x}(t)) = \frac{\partial \boldsymbol{g}}{\partial \boldsymbol{x}}\dot{\boldsymbol{x}} = \boldsymbol{J}_g(\boldsymbol{x})\dot{\boldsymbol{x}}. \tag{A.22}$$

A.3 Vector Operations

Given n vectors $\boldsymbol{x}_i$ of dimensions $(m \times 1)$, they are said to be *linearly independent* if the expression

$$k_1\boldsymbol{x}_1 + k_2\boldsymbol{x}_2 + \ldots + k_n\boldsymbol{x}_n = \mathbf{0}$$

holds only when all the constants k_i vanish. A necessary and sufficient condition for the vectors $\boldsymbol{x}_1, \boldsymbol{x}_2 \ldots, \boldsymbol{x}_n$ to be linearly independent is that the matrix

$$\boldsymbol{A} = [\,\boldsymbol{x}_1 \quad \boldsymbol{x}_2 \quad \ldots \quad \boldsymbol{x}_n\,]$$

has rank n; this implies that a necessary condition for linear independence is that $n \leq m$. If instead $\varrho(\boldsymbol{A}) = r < n$, then only r vectors are linearly independent and the remaining $n - r$ vectors can be expressed as a linear combination of the previous ones.

A system of vectors $\mathcal{X}$ is a *vector space* on the field of real numbers $\mathbb{R}$ if the operations of *sum of two vectors* of $\mathcal{X}$ and *product of a scalar by a vector* of $\mathcal{X}$ have values in $\mathcal{X}$ and the following properties hold:

$$
\begin{aligned}
&\boldsymbol{x}+\boldsymbol{y}=\boldsymbol{y}+\boldsymbol{x} && \forall \boldsymbol{x},\boldsymbol{y}\in\mathcal{X}\\
&(\boldsymbol{x}+\boldsymbol{y})+\boldsymbol{z}=\boldsymbol{x}+(\boldsymbol{y}+\boldsymbol{z}) && \forall \boldsymbol{x},\boldsymbol{y},\boldsymbol{z}\in\mathcal{X}\\
&\exists \boldsymbol{0}\in\mathcal{X}:\boldsymbol{x}+\boldsymbol{0}=\boldsymbol{x} && \forall \boldsymbol{x}\in\mathcal{X}\\
&\forall \boldsymbol{x}\in\mathcal{X},\quad \exists(-\boldsymbol{x})\in\mathcal{X}:\boldsymbol{x}+(-\boldsymbol{x})=\boldsymbol{0} &&\\
&1\boldsymbol{x}=\boldsymbol{x} && \forall \boldsymbol{x}\in\mathcal{X}\\
&\alpha(\beta\boldsymbol{x})=(\alpha\beta)\boldsymbol{x} && \forall\alpha,\beta\in\mathbb{R}\quad \forall \boldsymbol{x}\in\mathcal{X}\\
&(\alpha+\beta)\boldsymbol{x}=\alpha\boldsymbol{x}+\beta\boldsymbol{x} && \forall\alpha,\beta\in\mathbb{R}\quad \forall \boldsymbol{x}\in\mathcal{X}\\
&\alpha(\boldsymbol{x}+\boldsymbol{y})=\alpha\boldsymbol{x}+\alpha\boldsymbol{y} && \forall\alpha\in\mathbb{R}\quad \forall \boldsymbol{x},\boldsymbol{y}\in\mathcal{X}.
\end{aligned}
$$

The *dimension* of the space $\dim(\mathcal{X})$ is the maximum number of linearly independent vectors $\boldsymbol{x}$ in the space. A set $\{\boldsymbol{x}_1,\boldsymbol{x}_2,\ldots,\boldsymbol{x}_n\}$ of linearly independent vectors is a *basis* of vector space $\mathcal{X}$, and each vector $\boldsymbol{y}$ in the space can be uniquely expressed as a linear combination of vectors from the basis:

$$\boldsymbol{y}=c_1\boldsymbol{x}_1+c_2\boldsymbol{x}_2+\ldots+c_n\boldsymbol{x}_n \tag{A.23}$$

where the constants $c_1,c_2,\ldots,c_n$ are said to be the *components* of the vector $\boldsymbol{y}$ in the basis $\{\boldsymbol{x}_1,\boldsymbol{x}_2,\ldots,\boldsymbol{x}_n\}$.

A subset $\mathcal{Y}$ of a vector space $\mathcal{X}$ is a *subspace* $\mathcal{Y}\subseteq\mathcal{X}$ if it is a vector space with the operations of vector sum and product of a scalar by a vector, *i.e.*,

$$\alpha\boldsymbol{x}+\beta\boldsymbol{y}\in\mathcal{Y}\qquad \forall\alpha,\beta\in\mathbb{R}\quad\forall\boldsymbol{x},\boldsymbol{y}\in\mathcal{Y}.$$

According to a geometric interpretation, a subspace is a *hyperplane* passing by the origin (null element) of $\mathcal{X}$.

The *scalar product* $<\boldsymbol{x},\boldsymbol{y}>$ of two vectors $\boldsymbol{x}$ and $\boldsymbol{y}$ of dimensions $(m\times 1)$ is the scalar that is obtained by summing the products of the respective components in a given basis:

$$<\boldsymbol{x},\boldsymbol{y}>=x_1y_1+x_2y_2+\ldots+x_my_m=\boldsymbol{x}^T\boldsymbol{y}=\boldsymbol{y}^T\boldsymbol{x}. \tag{A.24}$$

Two vectors are said to be *orthogonal* when their scalar product is null:

$$\boldsymbol{x}^T\boldsymbol{y}=0. \tag{A.25}$$

The *norm* of a vector can be defined as

$$\|\boldsymbol{x}\|=\sqrt{\boldsymbol{x}^T\boldsymbol{x}}. \tag{A.26}$$

It is possible to show that both the *triangle inequality*

$$\|\boldsymbol{x}+\boldsymbol{y}\|\le\|\boldsymbol{x}\|+\|\boldsymbol{y}\| \tag{A.27}$$

and the *Schwarz' inequality*

$$|\boldsymbol{x}^T\boldsymbol{y}|\le\|\boldsymbol{x}\|\,\|\boldsymbol{y}\| \tag{A.28}$$

hold. A *unit vector* $\hat{\boldsymbol{x}}$ is a vector whose *norm* is unity, *i.e.*, $\hat{\boldsymbol{x}}^T\hat{\boldsymbol{x}} = 1$. Given a vector $\boldsymbol{x}$, its unit vector is obtained by dividing each component by its norm:

$$\hat{\boldsymbol{x}} = \frac{1}{\|\boldsymbol{x}\|}\boldsymbol{x}. \tag{A.29}$$

A typical example of vector space is the *Euclidean space* whose dimension is 3; in this case a basis is constituted by the unit vectors of a coordinate frame.

The *vector product* of two vectors $\boldsymbol{x}$ and $\boldsymbol{y}$ in the Euclidean space is the vector

$$\boldsymbol{x} \times \boldsymbol{y} = \begin{bmatrix} x_2y_3 - x_3y_2 \\ x_3y_1 - x_1y_3 \\ x_1y_2 - x_2y_1 \end{bmatrix}. \tag{A.30}$$

The following properties hold:

$$\begin{aligned} &\boldsymbol{x} \times \boldsymbol{x} = \boldsymbol{0} \\ &\boldsymbol{x} \times \boldsymbol{y} = -\boldsymbol{y} \times \boldsymbol{x} \\ &\boldsymbol{x} \times (\boldsymbol{y} + \boldsymbol{z}) = \boldsymbol{x} \times \boldsymbol{y} + \boldsymbol{x} \times \boldsymbol{z}. \end{aligned}$$

The vector product of two vectors $\boldsymbol{x}$ and $\boldsymbol{y}$ can be expressed also as the product of a matrix operator $\boldsymbol{S}(\boldsymbol{x})$ by the vector $\boldsymbol{y}$. In fact, by introducing the *skew-symmetric* matrix

$$\boldsymbol{S}(\boldsymbol{x}) = \begin{bmatrix} 0 & -x_3 & x_2 \\ x_3 & 0 & -x_1 \\ -x_2 & x_1 & 0 \end{bmatrix} \tag{A.31}$$

obtained with the components of vector $\boldsymbol{x}$, the vector product $\boldsymbol{x} \times \boldsymbol{y}$ is given by

$$\boldsymbol{x} \times \boldsymbol{y} = \boldsymbol{S}(\boldsymbol{x})\boldsymbol{y} = -\boldsymbol{S}(\boldsymbol{y})\boldsymbol{x} \tag{A.32}$$

as can be easily verified. Moreover, the following properties hold:

$$\begin{aligned} &\boldsymbol{S}(\boldsymbol{x})\boldsymbol{x} = \boldsymbol{S}^T(\boldsymbol{x})\boldsymbol{x} = \boldsymbol{0} \\ &\boldsymbol{S}(\alpha\boldsymbol{x} + \beta\boldsymbol{y}) = \alpha\boldsymbol{S}(\boldsymbol{x}) + \beta\boldsymbol{S}(\boldsymbol{y}) \\ &\boldsymbol{S}(\boldsymbol{x})\boldsymbol{S}(\boldsymbol{y}) = \boldsymbol{y}\boldsymbol{x}^T - \boldsymbol{x}^T\boldsymbol{y}\boldsymbol{I} \end{aligned}$$

$$\boldsymbol{S}(\boldsymbol{S}(\boldsymbol{x})\boldsymbol{y}) = \boldsymbol{y}\boldsymbol{x}^T - \boldsymbol{x}\boldsymbol{y}^T.$$

Given three vector $\boldsymbol{x}$, $\boldsymbol{y}$, $\boldsymbol{z}$ in the Euclidean space, the following expressions hold for the *scalar triple products*:

$$\boldsymbol{x}^T(\boldsymbol{y} \times \boldsymbol{z}) = \boldsymbol{y}^T(\boldsymbol{z} \times \boldsymbol{x}) = \boldsymbol{z}^T(\boldsymbol{x} \times \boldsymbol{y}). \tag{A.33}$$

If any two vectors of three are equal, then the scalar triple product is null; e.g.,

$$\boldsymbol{x}^T(\boldsymbol{x} \times \boldsymbol{y}) = 0.$$

A.4 Linear Transformations

Consider a vector space $\mathcal{X}$ of dimension n and a vector space $\mathcal{Y}$ of dimension m with $m \leq n$. The *linear transformation* between the vectors $\boldsymbol{x} \in \mathcal{X}$ and $\boldsymbol{y} \in \mathcal{Y}$ can be defined as

$$\boldsymbol{y} = \boldsymbol{A}\boldsymbol{x} \tag{A.34}$$

in terms of the matrix $\boldsymbol{A}$ of dimensions $(m \times n)$. The *range space* (or simply range) of the transformation is the subspace

$$\mathcal{R}(\boldsymbol{A}) = \{\boldsymbol{y} : \boldsymbol{y} = \boldsymbol{A}\boldsymbol{x},\ \boldsymbol{x} \in \mathcal{X}\} \subseteq \mathcal{Y}, \tag{A.35}$$

which is the subspace generated by the linearly independent columns of matrix $\boldsymbol{A}$ taken as a basis of $\mathcal{Y}$. It is easy to recognize that

$$\varrho(\boldsymbol{A}) = \dim(\mathcal{R}(\boldsymbol{A})). \tag{A.36}$$

On the other hand, the *null space* (or simply null) of the transformation is the subspace

$$\mathcal{N}(\boldsymbol{A}) = \{\boldsymbol{x} : \boldsymbol{A}\boldsymbol{x} = \boldsymbol{0},\ \boldsymbol{x} \in \mathcal{X}\} \subseteq \mathcal{X}. \tag{A.37}$$

Given a matrix $\boldsymbol{A}$ of dimensions $(m \times n)$, the notable result holds:

$$\varrho(\boldsymbol{A}) + \dim(\mathcal{N}(\boldsymbol{A})) = n. \tag{A.38}$$

Therefore, if $\varrho(\boldsymbol{A}) = r \leq \min\{m, n\}$, then $\dim(\mathcal{R}(\boldsymbol{A})) = r$ and $\dim(\mathcal{N}(\boldsymbol{A})) = n - r$. It follows that if $m < n$, then $\mathcal{N}(\boldsymbol{A}) \neq \emptyset$ independently of the rank of $\boldsymbol{A}$; if $m = n$, then $\mathcal{N}(\boldsymbol{A}) \neq \emptyset$ only in the case of $\varrho(\boldsymbol{A}) = r < m$.

If $\boldsymbol{x} \in \mathcal{N}(\boldsymbol{A})$ and $\boldsymbol{y} \in \mathcal{R}(\boldsymbol{A}^T)$, then $\boldsymbol{y}^T\boldsymbol{x} = 0$, *i.e.*, the vectors in the null space of $\boldsymbol{A}$ are orthogonal to each vector in the range space of the transpose of $\boldsymbol{A}$. It can be shown that the set of vectors orthogonal to each vector of the range space of $\boldsymbol{A}^T$ coincides with the null space of $\boldsymbol{A}$, whereas the set of vectors orthogonal to each vector in the null space of $\boldsymbol{A}^T$ coincides with the range space of $\boldsymbol{A}$. In symbols:

$$\mathcal{N}(\boldsymbol{A}) \equiv \mathcal{R}^{\perp}(\boldsymbol{A}^T) \qquad \mathcal{R}(\boldsymbol{A}) \equiv \mathcal{N}^{\perp}(\boldsymbol{A}^T) \tag{A.39}$$

where $\perp$ denotes the *orthogonal complement* of a subspace.

A linear transformation allows defining the *norm* of a matrix $\boldsymbol{A}$ induced by the norm defined for a vector $\boldsymbol{x}$ as follows. In view of the property

$$\|\boldsymbol{A}\boldsymbol{x}\| \leq \|\boldsymbol{A}\|\,\|\boldsymbol{x}\|, \tag{A.40}$$

the norm of $\boldsymbol{A}$ can be defined as

$$\|\boldsymbol{A}\| = \sup_{\boldsymbol{x} \neq \boldsymbol{0}} \frac{\|\boldsymbol{A}\boldsymbol{x}\|}{\|\boldsymbol{x}\|} \tag{A.41}$$

which can be computed also as

$$\max_{||\boldsymbol{x}||=1} ||\boldsymbol{A}\boldsymbol{x}||.$$

A direct consequence of (A.40) is the property

$$||\boldsymbol{A}\boldsymbol{B}|| \leq ||\boldsymbol{A}|| \, ||\boldsymbol{B}||. \tag{A.42}$$

A.5 Eigenvalues and Eigenvectors

Consider the linear transformation on a vector $\boldsymbol{u}$ established by an $(n \times n)$ square matrix $\boldsymbol{A}$. If the vector resulting from the transformation has the same direction of $\boldsymbol{u}$ (with $\boldsymbol{u} \neq \boldsymbol{0}$), then

$$\boldsymbol{A}\boldsymbol{u} = \lambda\boldsymbol{u}. \tag{A.43}$$

The equation in (A.43) can be rewritten in matrix form as

$$(\lambda\boldsymbol{I} - \boldsymbol{A})\boldsymbol{u} = \boldsymbol{0}. \tag{A.44}$$

For the homogeneous system of equations in (A.44) to have a solution different from the trivial one $\boldsymbol{u} = \boldsymbol{0}$, it must be

$$\det(\lambda\boldsymbol{I} - \boldsymbol{A}) = 0 \tag{A.45}$$

which is termed *characteristic equation*. Its solutions $\lambda_1, \ldots, \lambda_n$ are the *eigenvalues* of matrix $\boldsymbol{A}$; they coincide with the eigenvalues of matrix $\boldsymbol{A}^T$. On the assumption of distinct eigenvalues, the n vectors $\boldsymbol{u}_i$ satisfying the equation

$$(\lambda_i\boldsymbol{I} - \boldsymbol{A})\boldsymbol{u}_i = \boldsymbol{0} \qquad i = 1, \ldots, n \tag{A.46}$$

are said to be the *eigenvectors* associated with the eigenvalues λ_i.

The matrix $\boldsymbol{U}$ formed by the column vectors $\boldsymbol{u}_i$ is invertible and constitutes a basis in the space of dimension n. Further, the *similarity transformation* established by $\boldsymbol{U}$:

$$\boldsymbol{\Lambda} = \boldsymbol{U}^{-1}\boldsymbol{A}\boldsymbol{U} \tag{A.47}$$

is so that $\boldsymbol{\Lambda} = \text{diag}\{\lambda_1, \ldots, \lambda_n\}$. It follows that $\det(\boldsymbol{A}) = \prod_{i=1}^{n} \lambda_i$.

If the matrix $\boldsymbol{A}$ is *symmetric*, its eigenvalues are real and $\boldsymbol{\Lambda}$ can be written as

$$\boldsymbol{\Lambda} = \boldsymbol{U}^T\boldsymbol{A}\boldsymbol{U}; \tag{A.48}$$

hence, the eigenvector matrix $\boldsymbol{U}$ is orthogonal.

A.6 Bilinear Forms and Quadratic Forms

A *bilinear form* in the variables x_i and y_j is the scalar

$$B = \sum_{i=1}^{m}\sum_{j=1}^{n} a_{ij}x_iy_j$$

which can be written in matrix form

$$B(\boldsymbol{x},\boldsymbol{y}) = \boldsymbol{x}^T\boldsymbol{A}\boldsymbol{y} = \boldsymbol{y}^T\boldsymbol{A}^T\boldsymbol{x} \tag{A.49}$$

where $\boldsymbol{x} = [\,x_1 \quad x_2 \quad \dots \quad x_m\,]^T$, $\boldsymbol{y} = [\,y_1 \quad y_2 \quad \dots \quad y_n\,]^T$, and $\boldsymbol{A}$ is the $(m \times n)$ matrix of the coefficients a_{ij} representing the core of the form.

A special case of bilinear form is the *quadratic form*

$$Q(\boldsymbol{x}) = \boldsymbol{x}^T\boldsymbol{A}\boldsymbol{x} \tag{A.50}$$

where $\boldsymbol{A}$ is an $(n \times n)$ square matrix. Hence, for computation of (A.50), the matrix $\boldsymbol{A}$ can be replaced with its symmetric part $\boldsymbol{A}_s$ given by (A.6). It follows that if $\boldsymbol{A}$ is a *skew-symmetric* matrix, then

$$\boldsymbol{x}^T\boldsymbol{A}\boldsymbol{x} = \boldsymbol{0} \qquad \forall \boldsymbol{x}.$$

The quadratic form (A.50) is said to be *positive definite* if

$$\begin{aligned} \boldsymbol{x}^T\boldsymbol{A}\boldsymbol{x} &> 0 \qquad \forall \boldsymbol{x} \neq \boldsymbol{0} \\ \boldsymbol{x}^T\boldsymbol{A}\boldsymbol{x} &= 0 \qquad \boldsymbol{x} = \boldsymbol{0}. \end{aligned} \tag{A.51}$$

The matrix $\boldsymbol{A}$ core of the form is also said to be *positive definite*. Analogously, a quadratic form is said to be *negative definite* if it can be written as $-Q(\boldsymbol{x}) = -\boldsymbol{x}^T\boldsymbol{A}\boldsymbol{x}$ where $Q(\boldsymbol{x})$ is positive definite.

A necessary condition for a square matrix to be positive definite is that its elements on the diagonal are strictly positive. Further, in view of (A.48), the eigenvalues of a positive definite matrix are all positive. If the eigenvalues are not known, a necessary and sufficient condition for a symmetric matrix to be positive definite is that its principal minors are strictly positive (*Sylvester's criterion*). It follows that a positive definite matrix is full-rank and thus it is always invertible.

A symmetric positive definite matrix $\boldsymbol{A}$ can always be decomposed as

$$\boldsymbol{A} = \boldsymbol{U}^T\boldsymbol{\Lambda}\boldsymbol{U}, \tag{A.52}$$

where $\boldsymbol{U}$ is an orthogonal matrix of eigenvectors ($\boldsymbol{U}^T\boldsymbol{U} = \boldsymbol{I}$) and $\boldsymbol{\Lambda}$ is the diagonal matrix of the eigenvalues of $\boldsymbol{A}$.

Let $\lambda_{\min}(\boldsymbol{A})$ and $\lambda_{\max}(\boldsymbol{A})$ respectively denote the smallest and largest eigenvalues of a positive definite matrix $\boldsymbol{A}$ ($\lambda_{\min}, \lambda_{\max} > 0$). Then, the quadratic form in (A.50) satisfies the following inequality:

$$\lambda_{\min}(\boldsymbol{A})\|\boldsymbol{x}\|^2 \le \boldsymbol{x}^T\boldsymbol{A}\boldsymbol{x} \le \lambda_{\max}(\boldsymbol{A})\|\boldsymbol{x}\|^2. \tag{A.53}$$

An $(n \times n)$ square matrix $\boldsymbol{A}$ is said to be *positive semi-definite* if

$$\boldsymbol{x}^T\boldsymbol{A}\boldsymbol{x} \ge 0 \qquad \forall \boldsymbol{x}. \tag{A.54}$$

This definition implies that $\varrho(\boldsymbol{A}) = r < n$, and thus r eigenvalues of $\boldsymbol{A}$ are positive and $n-r$ are null. Therefore, a positive semi-definite matrix $\boldsymbol{A}$ has a null space of finite dimension, and specifically the form vanishes when $\boldsymbol{x} \in \mathcal{N}(\boldsymbol{A})$. A typical example of a positive semi-definite matrix is the matrix $\boldsymbol{A} = \boldsymbol{H}^T\boldsymbol{H}$ where $\boldsymbol{H}$ is an $(m \times n)$ matrix with $m < n$. In an analogous way, a *negative semi-definite* matrix can be defined.

Given the *bilinear form* in (A.49), the *gradient* of the form with respect to $\boldsymbol{x}$ is given by

$$\mathbf{grad}_{\boldsymbol{x}} B(\boldsymbol{x}, \boldsymbol{y}) = \left(\frac{\partial B(\boldsymbol{x}, \boldsymbol{y})}{\partial \boldsymbol{x}}\right)^T = \boldsymbol{A}\boldsymbol{y}, \tag{A.55}$$

whereas the gradient of B with respect to $\boldsymbol{y}$ is given by

$$\mathbf{grad}_{\boldsymbol{y}} B(\boldsymbol{x}, \boldsymbol{y}) = \left(\frac{\partial B(\boldsymbol{x}, \boldsymbol{y})}{\partial \boldsymbol{y}}\right)^T = \boldsymbol{A}^T\boldsymbol{x}. \tag{A.56}$$

Given the *quadratic form* in (A.50) with $\boldsymbol{A}$ *symmetric*, the *gradient* of the form with respect to $\boldsymbol{x}$ is given by

$$\mathbf{grad}_{\boldsymbol{x}} Q(\boldsymbol{x}) = \left(\frac{\partial Q(\boldsymbol{x})}{\partial \boldsymbol{x}}\right)^T = 2\boldsymbol{A}\boldsymbol{x}. \tag{A.57}$$

Further, if $\boldsymbol{x}$ and $\boldsymbol{A}$ are differentiable functions of t, then

$$\dot{Q}(\boldsymbol{x}) = \frac{d}{dt} Q(\boldsymbol{x}(t)) = 2\boldsymbol{x}^T\boldsymbol{A}\dot{\boldsymbol{x}} + \boldsymbol{x}^T\dot{\boldsymbol{A}}\boldsymbol{x}; \tag{A.58}$$

if $\boldsymbol{A}$ is constant, then the second term obviously vanishes.

A.7 Pseudo-inverse

The inverse of a matrix can be defined only when the matrix is square and nonsingular. The inverse operation can be extended to the case of non-square matrices. Given a matrix $\boldsymbol{A}$ of dimensions $(m \times n)$ with $\varrho(\boldsymbol{A}) = \min\{m, n\}$, if $n < m$, a *left inverse* of $\boldsymbol{A}$ can be defined as the matrix $\boldsymbol{A}_l$ of dimensions $(n \times m)$ so that

$$\boldsymbol{A}_l\boldsymbol{A} = \boldsymbol{I}_n.$$

If instead $n > m$, a *right inverse* of $\boldsymbol{A}$ can be defined as the matrix $\boldsymbol{A}_r$ of dimensions $(n \times m)$ so that

$$\boldsymbol{A}\boldsymbol{A}_r = \boldsymbol{I}_m.$$

If $\boldsymbol{A}$ has more rows than columns $(m > n)$ and has rank n, a special left inverse is the matrix

$$\boldsymbol{A}_l^{\dagger} = (\boldsymbol{A}^T\boldsymbol{A})^{-1}\boldsymbol{A}^T \tag{A.59}$$

which is termed *left pseudo-inverse*, since $\boldsymbol{A}_l^{\dagger}\boldsymbol{A} = \boldsymbol{I}_n$. If $\boldsymbol{W}_l$ is an $(m \times m)$ *positive definite* matrix, a *weighted* left pseudo-inverse is given by

$$\boldsymbol{A}_l^{\dagger} = (\boldsymbol{A}^T\boldsymbol{W}_l^{-1}\boldsymbol{A})^{-1}\boldsymbol{A}^T\boldsymbol{W}_l^{-1}. \tag{A.60}$$

If A has more columns than rows ($m < n$) and has rank m, a special right inverse is the matrix

$$A_r^\dagger = A^T(AA^T)^{-1} \tag{A.61}$$

which is termed *right pseudo-inverse*, since $AA_r^\dagger = I_m$[3]. If W_r is an $(n \times n)$ *positive definite* matrix, a *weighted* right pseudo-inverse is given by

$$A_r^\dagger = W_r^{-1}A^T(AW_r^{-1}A^T)^{-1}. \tag{A.62}$$

The pseudo-inverse is very useful to invert a linear transformation $y = Ax$ with A a full-rank matrix. If A is a square nonsingular matrix, then obviously $x = A^{-1}y$ and then $A_l^\dagger = A_r^\dagger = A^{-1}$.

If A has more columns than rows ($m < n$) and has rank m, then the solution x for a given y is not unique; it can be shown that the expression

$$x = A^\dagger y + (I - A^\dagger A)k, \tag{A.63}$$

with k an arbitrary $(n \times 1)$ vector and $A^\dagger$ as in (A.61), is a solution to the system of linear equations established by (A.34). The term $A^\dagger y \in \mathcal{N}^\perp(A) \equiv \mathcal{R}(A^T)$ minimizes the norm of the solution $\|x\|$, while the term $(I - A^\dagger A)k$ is the projection of k in $\mathcal{N}(A)$ and is termed *homogeneous solution*.

On the other hand, if A has more rows than columns ($m > n$), the equation in (A.34) has no solution; it can be shown that an *approximate* solution is given by

$$x = A^\dagger y \tag{A.64}$$

where $A^\dagger$ as in (A.59) minimizes $\|y - Ax\|$.

A.8 Singular Value Decomposition

For a nonsquare matrix it is not possible to define eigenvalues. An extension of the eigenvalue concept can be obtained by singular values. Given a matrix A of dimensions $(m \times n)$, the matrix A^TA has n nonnegative eigenvalues $\lambda_1 \geq \lambda_2 \geq \ldots \geq \lambda_n \geq 0$ (ordered from the largest to the smallest) which can be expressed in the form

$$\lambda_i = \sigma_i^2 \qquad \sigma_i \geq 0.$$

The scalars $\sigma_1 \geq \sigma_2 \geq \ldots \geq \sigma_n \geq 0$ are said to be the *singular values* of matrix A. The *singular value decomposition* (SVD) of matrix A is given by

$$A = U\Sigma V^T \tag{A.65}$$

---

[3] Subscripts l and r are usually omitted whenever the use of a left or right pseudo-inverse is clear from the context.

where $\boldsymbol{U}$ is an $(m \times m)$ orthogonal matrix

$$\boldsymbol{U} = [\,\boldsymbol{u}_1 \quad \boldsymbol{u}_2 \quad \dots \quad \boldsymbol{u}_m\,], \tag{A.66}$$

$\boldsymbol{V}$ is an $(n \times n)$ orthogonal matrix

$$\boldsymbol{V} = [\,\boldsymbol{v}_1 \quad \boldsymbol{v}_2 \quad \dots \quad \boldsymbol{v}_n\,] \tag{A.67}$$

and $\boldsymbol{\Sigma}$ is an $(m \times n)$ matrix

$$\boldsymbol{\Sigma} = \begin{bmatrix} \boldsymbol{D} & \boldsymbol{O} \\ \boldsymbol{O} & \boldsymbol{O} \end{bmatrix} \qquad \boldsymbol{D} = \mathrm{diag}\{\sigma_1, \sigma_2, \dots, \sigma_r\} \tag{A.68}$$

where $\sigma_1 \geq \sigma_2 \geq \ldots \geq \sigma_r > 0$. The number of nonnull singular values is equal to the rank r of matrix $\boldsymbol{A}$.

The columns of $\boldsymbol{U}$ are the eigenvectors of the matrix $\boldsymbol{A}\boldsymbol{A}^T$, whereas the columns of $\boldsymbol{V}$ are the eigenvectors of the matrix $\boldsymbol{A}^T\boldsymbol{A}$. In view of the partitions of $\boldsymbol{U}$ and $\boldsymbol{V}$ in (A.66) and (A.67), it is: $\boldsymbol{A}\boldsymbol{v}_i = \sigma_i \boldsymbol{u}_i$, for $i = 1, \dots, r$ and $\boldsymbol{A}\boldsymbol{v}_i = \boldsymbol{0}$, for $i = r+1, \dots, n$.

Singular value decomposition is useful for analysis of the linear transformation $\boldsymbol{y} = \boldsymbol{A}\boldsymbol{x}$ established in (A.34). According to a geometric interpretation, the matrix $\boldsymbol{A}$ transforms the unit sphere in $\mathbb{R}^n$ defined by $\|\boldsymbol{x}\| = 1$ into the set of vectors $\boldsymbol{y} = \boldsymbol{A}\boldsymbol{x}$ which define an *ellipsoid* of dimension r in $\mathbb{R}^m$. The singular values are the lengths of the various axes of the ellipsoid. The *condition number* of the matrix

$$\kappa = \frac{\sigma_1}{\sigma_r}$$

is related to the eccentricity of the ellipsoid and provides a measure of ill-conditioning ($\kappa \gg 1$) for numerical solution of the system established by (A.34).

Bibliography

Boullion T.L., Odell P.L. (1971) *Generalized Inverse Matrices*. Wiley, New York.

DeRusso P.M., Roy R.J., Close C.M., A.A. Desrochers (1998) *State Variables for Engineers*. 2nd ed., Wiley, New York.

Gantmacher F.R. (1959) *Theory of Matrices*. Vols. I & II, Chelsea Publishing Company, New York.

Golub G.H., Van Loan C.F. (1989) *Matrix Computations*. 2nd ed., The Johns Hopkins University Press, Baltimore, Md.

Noble B. (1977) *Applied Linear Algebra*. 2nd ed., Prentice-Hall, Englewood Cliffs, N.J.

Appendix B. Rigid Body Mechanics

The goal of this appendix is to recall some fundamental concepts of *rigid body mechanics* which are preliminary to the study of manipulator *kinematics*, *statics* and *dynamics*.

B.1 Kinematics

A *rigid body* is a system characterized by the constraint that the distance between any two points is always constant.

Consider a rigid body $\mathcal{B}$ moving with respect to an orthonormal reference frame O–xyz of unit vectors $\boldsymbol{x}$, $\boldsymbol{y}$, $\boldsymbol{z}$, called *fixed frame*. The rigidity assumption allows introducing an orthonormal frame O'–$x'y'z'$ attached to the body, called *moving frame*, with respect to which the position of any point of $\mathcal{B}$ is independent of time. Let $\boldsymbol{x}'(t)$, $\boldsymbol{y}'(t)$, $\boldsymbol{z}'(t)$ be the unit vectors of the moving frame expressed in the fixed frame at time t.

The orientation of the moving frame O'–$x'y'z'$ at time t with respect to the fixed frame O–xyz can be expressed by means of the *orthogonal* (3×3) matrix

$$\boldsymbol{R}(t) = \begin{bmatrix} \boldsymbol{x}'^T(t)\boldsymbol{x} & \boldsymbol{y}'^T(t)\boldsymbol{x} & \boldsymbol{z}'^T(t)\boldsymbol{x} \\ \boldsymbol{x}'^T(t)\boldsymbol{y} & \boldsymbol{y}'^T(t)\boldsymbol{y} & \boldsymbol{z}'^T(t)\boldsymbol{y} \\ \boldsymbol{x}'^T(t)\boldsymbol{z} & \boldsymbol{y}'^T(t)\boldsymbol{z} & \boldsymbol{z}'^T(t)\boldsymbol{z} \end{bmatrix}, \tag{B.1}$$

which is termed *rotation matrix*. The columns of the matrix in (B.1) represent the components of the unit vectors of the moving frame when expressed in the fixed frame, whereas the rows represent the components of the unit vectors of the fixed frame when expressed in the moving frame.

Let $\boldsymbol{p}'$ be the *constant* position vector of a generic point P of $\mathcal{B}$ in the moving frame O'–$x'y'z'$. The motion of P with respect to the fixed frame O–xyz is described by the equation

$$\boldsymbol{p}(t) = \boldsymbol{p}_{O'}(t) + \boldsymbol{R}(t)\boldsymbol{p}', \tag{B.2}$$

where $\boldsymbol{p}_{O'}(t)$ is the position vector of origin O' of the moving frame with respect to the fixed frame.

Notice that a position vector is a *bound vector* since its line of application and point of application are both prescribed, in addition to its direction; the point of application typically coincides with the origin of a reference frame. Therefore, to transform a bound vector from a frame to another, both translation and rotation between the two frames must be taken into account.

If the positions of the points of $\mathcal{B}$ in the moving frame are known, it follows from (B.2) that the motion of each point of $\mathcal{B}$ with respect to the fixed frame is uniquely determined once the position of the origin and the orientation of the moving frame with respect to the fixed frame are specified in time. The origin of the moving frame is determined by *three* scalar functions of time. Since the orthonormality conditions impose six constraints on the nine elements of matrix $\boldsymbol{R}(t)$, the *orientation* of the moving frame depends only on *three* independent scalar functions. Therefore, a rigid body motion is described by arbitrarily specifying *six* scalar functions of time.

The expression in (B.2) continues to hold if the position vector $\boldsymbol{p}_{O'}(t)$ of the origin of the moving frame is replaced with the position vector of any other point of $\mathcal{B}$, *i.e.*,

$$\boldsymbol{p}(t) = \boldsymbol{p}_Q(t) + \boldsymbol{R}(t)(\boldsymbol{p}' - \boldsymbol{p}'_Q), \tag{B.3}$$

where $\boldsymbol{p}_Q(t)$ and $\boldsymbol{p}'_Q$ are the position vectors of a point Q of $\mathcal{B}$ in the fixed and moving frames, respectively.

In the following, for simplicity of notation, the dependence on the time variable t will be dropped.

Differentiating (B.3) with respect to time gives the known velocity composition rule

$$\dot{\boldsymbol{p}} = \dot{\boldsymbol{p}}_Q + \boldsymbol{\omega} \times (\boldsymbol{p} - \boldsymbol{p}_Q), \tag{B.4}$$

where $\boldsymbol{\omega}$ is the *angular velocity* of rigid body $\mathcal{B}$.

Notice that $\boldsymbol{\omega}$ is a *free vector* since its point of application is not prescribed. To transform a free vector from a frame to another, only rotation between the two frames must be taken into account.

By recalling the definition of the skew-symmetric operator $\boldsymbol{S}(\cdot)$ in (A.31), the expression in (B.4) can be rewritten as

$$\begin{aligned}\dot{\boldsymbol{p}} &= \dot{\boldsymbol{p}}_Q + \boldsymbol{S}(\boldsymbol{\omega})(\boldsymbol{p} - \boldsymbol{p}_Q)\\ &= \dot{\boldsymbol{p}}_Q + \boldsymbol{S}(\boldsymbol{\omega})\boldsymbol{R}(\boldsymbol{p}' - \boldsymbol{p}'_Q).\end{aligned}$$

Comparing this equation with the formal time derivative of (B.3) leads to the result

$$\dot{\boldsymbol{R}} = \boldsymbol{S}(\boldsymbol{\omega})\boldsymbol{R}. \tag{B.5}$$

In view of (B.4), the *elementary displacement* of a point P of the rigid body $\mathcal{B}$ in the time interval $(t, t + dt)$ is

$$\begin{aligned}d\boldsymbol{p} = \dot{\boldsymbol{p}}dt &= \big(\dot{\boldsymbol{p}}_Q + \boldsymbol{\omega} \times (\boldsymbol{p} - \boldsymbol{p}_Q)\big)dt\\ &= d\boldsymbol{p}_Q + \boldsymbol{\omega}dt \times (\boldsymbol{p} - \boldsymbol{p}_Q).\end{aligned} \tag{B.6}$$

Differentiating (B.4) with respect to time yields the following expression for acceleration:

$$\ddot{\boldsymbol{p}} = \ddot{\boldsymbol{p}}_Q + \dot{\boldsymbol{\omega}} \times (\boldsymbol{p} - \boldsymbol{p}_Q) + \boldsymbol{\omega} \times \big(\boldsymbol{\omega} \times (\boldsymbol{p} - \boldsymbol{p}_Q)\big). \tag{B.7}$$

B.2 Dynamics

Let ρdV be the mass of an elementary particle of a rigid body $\mathcal{B}$, where ρ denotes the density of the particle of volume dV. Let also $V_{\mathcal{B}}$ be the body volume and $m = \int_{V_{\mathcal{B}}} \rho dV$ its *total mass* assumed to be constant. If $\boldsymbol{p}$ denotes the position vector of the particle of mass ρdV in the frame O–xyz, the *centre of mass* of $\mathcal{B}$ is defined as the point C whose position vector is

$$\boldsymbol{p}_C = \frac{1}{m}\int_{V_{\mathcal{B}}} \boldsymbol{p}\rho dV. \tag{B.8}$$

In the case when $\mathcal{B}$ is the union of n distinct parts of mass $m_1, \ldots, m_n$ and centres of mass $\boldsymbol{p}_{C1} \ldots \boldsymbol{p}_{Cn}$, the centre of mass of $\mathcal{B}$ can be computed as

$$\boldsymbol{p}_C = \frac{1}{m}\sum_{i=1}^{n} m_i \boldsymbol{p}_{Ci}$$

with $m = \sum_{i=1}^{n} m_i$.

Let r be a line passing by O and $d(\boldsymbol{p})$ the distance from r of the particle of $\mathcal{B}$ of mass ρdV and position vector $\boldsymbol{p}$. The *moment of inertia* of body $\mathcal{B}$ with respect to line r is defined as the positive scalar

$$I_r = \int_{V_{\mathcal{B}}} d^2(\boldsymbol{p})\rho dV.$$

Let $\boldsymbol{r}$ denote the unit vector of line r; then, the moment of inertia of $\mathcal{B}$ with respect to line r can be expressed as

$$I_r = \boldsymbol{r}^T\left(\int_{V_{\mathcal{B}}} \boldsymbol{S}^T(\boldsymbol{p})\boldsymbol{S}(\boldsymbol{p})\rho dV\right)\boldsymbol{r} = \boldsymbol{r}^T \boldsymbol{I}_O \boldsymbol{r}, \tag{B.9}$$

where $\boldsymbol{S}(\cdot)$ is the skew-symmetric operator in (A.31), and the *symmetric positive definite* matrix

$$\boldsymbol{I}_O = \begin{bmatrix} \int_{V_{\mathcal{B}}}(p_y^2+p_z^2)\rho dV & -\int_{V_{\mathcal{B}}} p_x p_y \rho dV & -\int_{V_{\mathcal{B}}} p_x p_z \rho dV \\ * & \int_{V_{\mathcal{B}}}(p_x^2+p_z^2)\rho dV & -\int_{V_{\mathcal{B}}} p_y p_z \rho dV \\ * & * & \int_{V_{\mathcal{B}}}(p_x^2+p_y^2)\rho dV \end{bmatrix}$$

$$= \begin{bmatrix} I_{Oxx} & -I_{Oxy} & -I_{Oxz} \\ * & I_{Oyy} & -I_{Oyz} \\ * & * & I_{Ozz} \end{bmatrix} \tag{B.10}$$

is termed *inertia tensor* of body $\mathcal{B}$ relative to pole O[1]. The (positive) elements I_{Oxx}, I_{Oyy}, I_{Ozz} are the *inertia moments* with respect to three coordinate axes of the reference

[1] The symbol '$*$' has been used to avoid rewriting the symmetric elements.

frame, whereas the elements $I_{Oxy}, I_{Oxz}, I_{Oyz}$ (of any sign) are said to be *products of inertia*.

The expression of the inertia tensor of a rigid body $\mathcal{B}$ depends both on the pole and the reference frame. If orientation of the reference frame with origin at O is changed according to a rotation matrix $\boldsymbol{R}$, the inertia tensor $\boldsymbol{I}'_O$ in the new frame is related to $\boldsymbol{I}_O$ by the relationship

$$\boldsymbol{I}_O = \boldsymbol{R}\boldsymbol{I}'_O\boldsymbol{R}^T. \tag{B.11}$$

The way an inertia tensor is transformed when the pole is changed can be inferred by the following equation, also known as *Steiner's theorem* or parallel axis theorem:

$$\boldsymbol{I}_O = \boldsymbol{I}_C + m\boldsymbol{S}^T(\boldsymbol{p}_C)\boldsymbol{S}(\boldsymbol{p}_C), \tag{B.12}$$

where $\boldsymbol{I}_C$ is the inertia tensor relative to the centre of mass of $\mathcal{B}$, when expressed in a frame parallel to the frame with origin at O and with origin at the centre of mass C.

Since the inertia tensor is a symmetric positive definite matrix, there always exists a reference frame in which the inertia tensor attains a diagonal form; such frame is said to be a *principal frame* (relative to pole O) and its coordinate axes are said to be *principal axes*. In the case when pole O coincides with the centre of mass, the frame is said to be a *central frame* and its axes are said to be *central axes*.

Notice that if the rigid body is moving with respect to the reference frame with origin at O, then the elements of the inertia tensor $\boldsymbol{I}_O$ become a function of time. With respect to a pole and a reference frame attached to the body (moving frame), instead, the elements of the inertia tensor represent six structural constants of the body which are known once the pole and reference frame have been specified.

Let $\dot{\boldsymbol{p}}$ be the velocity of a particle of $\mathcal{B}$ of elementary mass ρdV in frame O–xyz. The *linear momentum* of body $\mathcal{B}$ is defined as the vector

$$\boldsymbol{l} = \int_{V_\mathcal{B}} \dot{\boldsymbol{p}}\rho dV = m\dot{\boldsymbol{p}}_C. \tag{B.13}$$

Let Ω be any point in space and $\boldsymbol{p}_\Omega$ its position vector in frame O–xyz; then, the *angular momentum* of body $\mathcal{B}$ relative to pole Ω is defined as the vector

$$\boldsymbol{k}_\Omega = \int_{V_\mathcal{B}} \dot{\boldsymbol{p}} \times (\boldsymbol{p}_\Omega - \boldsymbol{p})\rho dV.$$

The pole can be either fixed or moving with respect to the reference frame. The angular momentum of a rigid body has the following notable expression:

$$\boldsymbol{k}_\Omega = \boldsymbol{I}_C\boldsymbol{\omega} + m\dot{\boldsymbol{p}}_C \times (\boldsymbol{p}_\Omega - \boldsymbol{p}_C), \tag{B.14}$$

where $\boldsymbol{I}_C$ is the inertia tensor relative to the centre of mass, when expressed in a frame parallel to the reference frame with origin at the centre of mass.

Forces acting on a generic system of material particles can be distinguished into *internal* forces and *external* forces.

Internal forces, exerted by one part of the system on another, have null linear and angular momentum and thus they do not influence rigid body motion.

External forces, exerted on the system by an agency outside the system, in the case of a rigid body $\mathcal{B}$ are distinguished into *active* forces and *reaction* forces.

Active forces can be either *concentrated* forces or *body* forces. The former are applied to specific points of $\mathcal{B}$, whereas the latter act on all elementary particles of the body. An example of body force is the *gravitational force* which, for any elementary particle of mass ρdV, is equal to $\boldsymbol{g}_0 \rho dV$ where $\boldsymbol{g}_0$ is the gravity acceleration vector.

Reaction forces are those exerted because of surface contact between two or more bodies. Such forces can be distributed on the contact surfaces or they can be assumed to be concentrated.

For a rigid body $\mathcal{B}$ subject to gravitational force, as well as to active and or reaction forces $\boldsymbol{f}_1 \dots \boldsymbol{f}_n$ concentrated at points $\boldsymbol{p}_1 \dots \boldsymbol{p}_n$, the *resultant* of the external forces $\boldsymbol{f}$ and the *resultant moment* $\boldsymbol{\mu}_\Omega$ with respect to a pole Ω are respectively:

$$\boldsymbol{f} = \int_{V_\mathcal{B}} \boldsymbol{g}_0 \rho dV + \sum_{i=1}^{n} \boldsymbol{f}_i = m\boldsymbol{g}_0 + \sum_{i=1}^{n} \boldsymbol{f}_i \tag{B.15}$$

$$\begin{aligned} \boldsymbol{\mu}_\Omega &= \int_{V_\mathcal{B}} \boldsymbol{g}_0 \times (\boldsymbol{p}_\Omega - \boldsymbol{p})\rho dV + \sum_{i=1}^{n} \boldsymbol{f}_i \times (\boldsymbol{p}_\Omega - \boldsymbol{p}_i) \\ &= m\boldsymbol{g}_0 \times (\boldsymbol{p}_\Omega - \boldsymbol{p}_C) + \sum_{i=1}^{n} \boldsymbol{f}_i \times (\boldsymbol{p}_\Omega - \boldsymbol{p}_i). \end{aligned} \tag{B.16}$$

In the case when $\boldsymbol{f}$ and $\boldsymbol{\mu}_\Omega$ are known and it is desired to compute the resultant moment with respect to a point Ω' other than Ω, the following relation holds:

$$\boldsymbol{\mu}_{\Omega'} = \boldsymbol{\mu}_\Omega + \boldsymbol{f} \times (\boldsymbol{p}_{\Omega'} - \boldsymbol{p}_\Omega). \tag{B.17}$$

Consider now a generic system of material particles subject to *external forces* of resultant $\boldsymbol{f}$ and resultant moment $\boldsymbol{\mu}_\Omega$. The motion of the system in a frame O–xyz is established by the following *fundamental principles of dynamics* (Newton's laws of motion):

$$\boldsymbol{f} = \dot{\boldsymbol{l}} \tag{B.18}$$

$$\boldsymbol{\mu}_\Omega = \dot{\boldsymbol{k}}_\Omega \tag{B.19}$$

where Ω is a pole fixed or coincident with the centre of mass C of the system. These equations hold for any mechanical system and can be used even in the case of variable mass. For a system with constant mass, computing the time derivative of the momentum in (B.18) gives *Newton's equations of motion* in the form

$$\boldsymbol{f} = m\ddot{\boldsymbol{p}}_C \tag{B.20}$$

where the quantity on the right-hand side represents the *resultant of inertia forces*.

If, besides the assumption of constant mass, the assumption of rigid system holds too, the expression in (B.14) of the angular momentum with (B.19) yield *Euler equations of motion* in the form

$$\boldsymbol{\mu}_\Omega = \boldsymbol{I}_\Omega \dot{\boldsymbol{\omega}} + \boldsymbol{\omega} \times (\boldsymbol{I}_\Omega \boldsymbol{\omega}) \tag{B.21}$$

where the quantity on the right-hand side represents the *resultant moment of inertia forces*.

For a system constituted by a set of rigid bodies, external forces obviously do not include the reaction forces exerted between the bodies belonging to the same system.

B.3 Work and Energy

Given a force $\boldsymbol{f}_i$ applied at a point of position $\boldsymbol{p}_i$ with respect to frame O–xyz, the *elementary work* of the force $\boldsymbol{f}_i$ on the displacement $d\boldsymbol{p}_i = \dot{\boldsymbol{p}}_i dt$ is defined as the scalar

$$dW_i = \boldsymbol{f}_i^T d\boldsymbol{p}_i.$$

For a rigid body $\mathcal{B}$ subject to a system of forces of resultant $\boldsymbol{f}$ and resultant moment $\boldsymbol{\mu}_Q$ with respect to any point Q of $\mathcal{B}$, the elementary work on the rigid displacement (B.6) is given by

$$dW = (\boldsymbol{f}^T \dot{\boldsymbol{p}}_Q + \boldsymbol{\mu}_Q^T \boldsymbol{\omega})dt = \boldsymbol{f}^T d\boldsymbol{p}_Q + \boldsymbol{\mu}_Q^T \boldsymbol{\omega} dt. \tag{B.22}$$

The *kinetic energy* of a body $\mathcal{B}$ is defined as the scalar quantity

$$\mathcal{T} = \frac{1}{2}\int_{V_\mathcal{B}} \dot{\boldsymbol{p}}^T \dot{\boldsymbol{p}} \rho dV$$

which, for a rigid body, takes on the notable expression

$$\mathcal{T} = \frac{1}{2} m \dot{\boldsymbol{p}}_C^T \dot{\boldsymbol{p}}_C + \frac{1}{2}\boldsymbol{\omega}^T \boldsymbol{I}_C \boldsymbol{\omega}, \tag{B.23}$$

where $\boldsymbol{I}_C$ is the inertia tensor relative to the centre of mass expressed in a frame parallel to the reference frame with origin at the centre of mass.

A system of position forces, *i.e.*, forces depending only on the positions of the points of application, is said to be *conservative* if the work done by each force is independent of the trajectory described by the point of application of the force but it depends only on the initial and final positions of the point of application. In this case, the elementary work of the system of forces is equal to minus the total differential of a scalar function termed *potential energy*, *i.e.*,

$$dW = -d\mathcal{U}. \tag{B.24}$$

An example of a conservative system of forces on a rigid body is gravitational force, with which is associated the potential energy

$$\mathcal{U} = -\int_{V_\mathcal{B}} \boldsymbol{g}_0^T \boldsymbol{p} \rho dV = -m\boldsymbol{g}_0^T \boldsymbol{p}_C. \tag{B.25}$$

B.4 Constrained Systems

Consider a system $\mathcal{B}_r$ of r rigid bodies and assume that all the elements of $\mathcal{B}_r$ can reach any position in space. In order to uniquely find a generic configuration of the system, it is necessary to assign a vector $\boldsymbol{x} = [\, x_1 \quad \dots \quad x_m \,]^T$ of $6r = m$ parameters. These parameters are termed *Lagrangian* or *generalized coordinates* and m determines the number of *degrees of freedom* of the *unconstrained* system $\mathcal{B}_r$.

Any limitation on the mobility of the system $\mathcal{B}_r$ is termed *constraint*. A constraint acting on $\mathcal{B}_r$ is said to be *holonomic* if it is expressed by a system of equations

$$\boldsymbol{h}(\boldsymbol{x}, t) = \mathbf{0}, \tag{B.26}$$

where $\boldsymbol{h}$ is a vector of dimensions $(s \times 1)$, with $s < m$[2]. For simplicity, only equality constraints are considered. If the equations in (B.26) do not explicitly depend on time, the constraint is said to be *time-invariant*.

On the assumption that $\boldsymbol{h}$ has continuous and continuously differentiable components, and its Jacobian $\partial\boldsymbol{h}/\partial\boldsymbol{x}$ has full rank, the equations in (B.26) allow eliminating s out of m coordinates of the system $\mathcal{B}_r$. With the remaining $n = m - s$ coordinates it is possible to uniquely determine the configurations of $\mathcal{B}_r$ satisfying the constraints (B.26). Such coordinates are the *Lagrangian* or *generalized coordinates* and n is the number of *degrees of freedom* of the *constrained* system $\mathcal{B}_r$.

The motion of a system $\mathcal{B}_r$ with n degrees of freedom and holonomic equality constraints can be described by equations of the form

$$\boldsymbol{x} = \boldsymbol{x}(\boldsymbol{\lambda}(t), t), \tag{B.27}$$

where $\boldsymbol{\lambda}(t) = [\, \lambda_1(t) \quad \dots \quad \lambda_n(t) \,]^T$ is a vector of Lagrangian coordinates.

The *elementary displacement* of system (B.27) relative to the interval $(t, t + dt)$ is defined as

$$d\boldsymbol{x} = \frac{\partial \boldsymbol{x}(\boldsymbol{\lambda}, t)}{\partial \boldsymbol{\lambda}} \dot{\boldsymbol{\lambda}} dt + \frac{\partial \boldsymbol{x}(\boldsymbol{\lambda}, t)}{\partial t} dt. \tag{B.28}$$

The *virtual displacement* of system (B.27) at time t, relative to an increment $\delta\boldsymbol{\lambda}$, is defined as the quantity

$$\delta\boldsymbol{x} = \frac{\partial \boldsymbol{x}(\boldsymbol{\lambda}, t)}{\partial \boldsymbol{\lambda}} \delta\boldsymbol{\lambda}. \tag{B.29}$$

The difference between the elementary displacement and the virtual displacement is that the former is relative to an actual motion of the system in an interval $(t, t + dt)$ which is consistent with the constraints, while the latter is relative to an imaginary motion of the system when the constraints are made invariant and equal to those at time t.

For a system with time-invariant constraints, the equations of motion (B.27) become

$$\boldsymbol{x} = \boldsymbol{x}(\boldsymbol{\lambda}(t)), \tag{B.30}$$

and then, by setting $\delta\boldsymbol{\lambda} = d\boldsymbol{\lambda} = \dot{\boldsymbol{\lambda}} dt$, virtual displacements (B.29) coincide with elementary displacements (B.28).

---

[2] On the other hand, a constraint in the form $\boldsymbol{h}(\boldsymbol{x}, \dot{\boldsymbol{x}}, t) = \mathbf{0}$ which is nonintegrable is said to be *nonholonomic*.

With the concept of virtual displacement it can be associated that of *virtual work* of a system of forces, by considering a virtual displacement instead of an elementary displacement.

If external forces are distinguished into *active forces* and *reaction forces*, a direct consequence of the principles of dynamics (B.18) and (B.19) applied to the system of rigid bodies $\mathcal{B}_r$ is that, for each virtual displacement, the following relation holds:

$$\delta W_m + \delta W_a + \delta W_h = 0, \tag{B.31}$$

where δW_m, δW_a, δW_h are the total virtual works done by the inertia, active, reaction forces, respectively.

In the case of *frictionless* equality constraints, reaction forces are exerted orthogonally to the contact surfaces and the virtual work is always null. Hence, (B.31) reduces to

$$\delta W_m + \delta W_a = 0. \tag{B.32}$$

For a steady system, inertia forces are identically null. Then, the condition for the equilibrium of system $\mathcal{B}_r$ is that the virtual work of the active forces is identically null on any virtual displacement, which gives the fundamental equation of *statics* of a constrained system

$$\delta W_a = 0, \tag{B.33}$$

known as *principle of virtual work*. Expressing (B.33) in terms of the increment $\delta\boldsymbol{\lambda}$ of generalized coordinates leads to

$$\delta W_a = \boldsymbol{\zeta}^T \delta\boldsymbol{\lambda} = 0 \tag{B.34}$$

where $\boldsymbol{\zeta}$ denotes the $(n \times 1)$ vector of active *generalized* forces.

In the dynamic case, it is worth distinguishing active forces into *conservative* (that can be derived from a potential) and *nonconservative*. The virtual work of conservative forces is given by

$$\delta W_c = -\frac{\partial \mathcal{U}}{\partial \boldsymbol{\lambda}} \delta\boldsymbol{\lambda} \tag{B.35}$$

where $\mathcal{U}(\boldsymbol{\lambda})$ is the total potential energy of the system. The work of nonconservative forces can be expressed in the form

$$\delta W_{nc} = \boldsymbol{\xi}^T \delta\boldsymbol{\lambda} \tag{B.36}$$

where $\boldsymbol{\xi}$ denotes the vector of nonconservative generalized forces. It follows that the vector of active generalized forces is

$$\boldsymbol{\zeta} = \boldsymbol{\xi} - \left(\frac{\partial \mathcal{U}}{\partial \boldsymbol{\lambda}}\right)^T. \tag{B.37}$$

Moreover, the work of inertia forces can be computed from the total kinetic energy of system $\mathcal{T}$ as

$$\delta W_m = \left(\frac{\partial \mathcal{T}}{\partial \boldsymbol{\lambda}} - \frac{d}{dt}\frac{\partial \mathcal{T}}{\partial \dot{\boldsymbol{\lambda}}}\right) \delta\boldsymbol{\lambda}. \tag{B.38}$$

Substituting (B.35), (B.36), (B.38) into (B.32) and observing that (B.32) is true for any increment $\delta\boldsymbol{\lambda}$ leads to *Lagrange's equations*

$$\frac{d}{dt}\frac{\partial\mathcal{L}}{\partial\dot{\lambda}_i} - \frac{\partial\mathcal{L}}{\partial\lambda_i} = \xi_i \qquad i = 1,\ldots,n \tag{B.39}$$

where

$$\mathcal{L} = \mathcal{T} - \mathcal{U} \tag{B.40}$$

is the *Lagrangian* of the system. The equations in (B.39) completely describe the dynamic behaviour of an n-degree-of-freedom system with holonomic equality constraints.

The sum of kinetic and potential energy of a system with time-invariant constraints is termed *Hamiltonian* function

$$\mathcal{H} = \mathcal{T} + \mathcal{U}. \tag{B.41}$$

Conservation of energy dictates that the time derivative of the Hamiltonian must balance the power generated by the nonconservative forces acting on the system, *i.e.*,

$$\frac{d\mathcal{H}}{dt} = \boldsymbol{\xi}^T\dot{\boldsymbol{\lambda}}. \tag{B.42}$$

In view of (B.37) and (B.41), the equation in (B.42) becomes

$$\frac{d\mathcal{T}}{dt} = \boldsymbol{\zeta}^T\dot{\boldsymbol{\lambda}}. \tag{B.43}$$

Bibliography

Goldstein H. (1980) *Classical Mechanics*. 2nd ed., Addison-Wesley, Reading, Mass.
Meirovitch L. (1990) *Dynamics and Control of Structures*. Wiley, New York.
Symon K.R. (1971) *Mechanics*. 3rd ed., Addison-Wesley, Reading, Mass.

Appendix C. Feedback Control

As a premise to the study of manipulator decentralized control and centralized control, the fundamental principles of *feedback control* of *linear systems* are recalled, and an approach to the determination of control laws for *nonlinear systems* based on the use of *Lyapunov functions* is presented.

C.1 Control of Single-input/Single-output Linear Systems

Classical *automatic control* theory of *linear time-invariant single-input/single-output systems* demonstrates that, in order to servo the output $y(t)$ of a system to a reference $r(t)$, it is worth adopting a *negative feedback control* structure. This structure indeed allows use of approximate mathematical models to describe the input/output relationship of the system to control, since negative feedback has a potential for reducing the effects of system parameter variations and nonmeasurable disturbance inputs $d(t)$ on the output.

This structure can be represented in the *domain of complex variable s* as in the block scheme of Figure C.1, where $G(s)$, $H(s)$ and $C(s)$ are the transfer functions of the system to control, the transducer and the controller, respectively. From this scheme it is easy to derive

$$Y(s) = W(s)R(s) + W_D(s)D(s), \tag{C.1}$$

where

$$W(s) = \frac{C(s)G(s)}{1 + C(s)G(s)H(s)} \tag{C.2}$$

is the *closed-loop input/output transfer function* and

$$W_D(s) = \frac{G(s)}{1 + C(s)G(s)H(s)} \tag{C.3}$$

is the *disturbance/output transfer function*.

The goal of controller design is to find a control structure $C(s)$ ensuring that the output variable $Y(s)$ tracks a reference input $R(s)$. Further, the controller shall guarantee that the effects of the disturbance input $D(s)$ on the output variable are

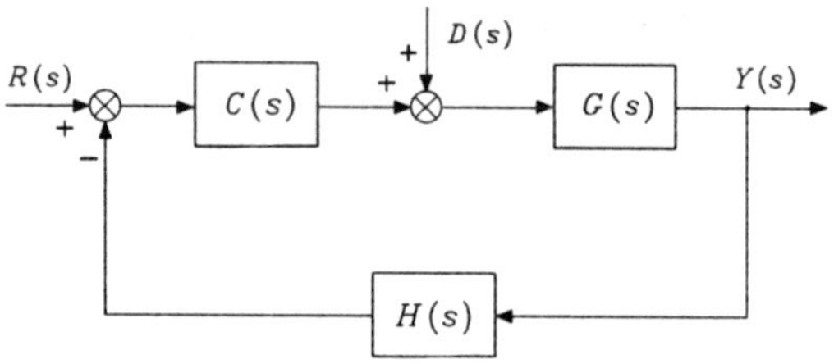

Figure C.1 Feedback control structure.

suitably reduced. The goal is then twofold; namely, *reference tracking* and *disturbance rejection*.

The basic problem for controller design consists of the determination of an action $C(s)$ which can make the system *asymptotically stable*. In the absence of positive or null real part pole/zero and zero/pole cancellation in the *open-loop* function $F(s) = C(s)G(s)H(s)$, a necessary and sufficient condition for asymptotic stability is that the *poles* of $W(s)$ and $W_D(s)$ have all *negative real parts*; such poles coincide with the zeros of the rational transfer function $1 + F(s)$. Testing for this condition can be performed by resorting to stability criteria, thus avoiding computation of the function zeros.

Routh's criterion allows determining the sign of the real parts of the zeros of the function $1 + F(s)$ by constructing a table with the coefficients of the polynomial at the numerator of $1 + F(s)$ (*characteristic polynomial*).

Routh's criterion is easy to apply for testing stability of a feedback system, but it does not provide a direct relationship between the open-loop function and stability of the closed-loop system. It is then worth resorting to *Nyquist criterion* which is based on the representation, in the complex plane, of the open-loop transfer function $F(s)$ evaluated in the *domain of real angular frequency* ($s = j\omega, -\infty < \omega < +\infty$).

Drawing of Nyquist plot and computation of the number of circles made by the vector representing the complex number $1 + F(j\omega)$ when ω continuously varies from $-\infty$ to $+\infty$ allows testing whether or not the *closed-loop* system is asymptotically stable. It is also possible to determine the number of positive, null and negative real part roots of the characteristic polynomial, similarly to application of Routh's criterion. Nonetheless, Nyquist criterion is based on the plot of the open-loop transfer function, and thus it allows establishing a direct relationship between this function and closed-loop system stability. It is then possible from an examination of the Nyquist plot to draw suggestions on the controller structure $C(s)$ which ensures closed-loop system asymptotic stability.

If the closed-loop system is asymptotically stable, the *steady-state response* to a sinusoidal input $r(t)$, with $d(t) = 0$, is sinusoidal, too. In this case, the function $W(s)$, evaluated for $s = j\omega$, is termed *frequency response function*; the frequency response function of a feedback system can be assimilated to that of a low-pass filter with the possible occurrence of a *resonance peak* inside its *bandwidth*.

As regards the transducer, this shall be chosen so that its bandwidth is much greater than the feedback system bandwidth, in order to ensure a nearly instantaneous response for any value of ω inside the bandwidth of $W(j\omega)$. Therefore, setting $H(j\omega) \approx H_0$ and assuming that the *loop gain* $|C(j\omega)G(j\omega)H_0| \gg 1$ in the same bandwidth, the expression in (C.1) for $s = j\omega$ can be approximated as

$$Y(j\omega) \approx \frac{R(j\omega)}{H_0} + \frac{D(j\omega)}{C(j\omega)H_0}.$$

Taking $R(j\omega) = H_0 Y_d(j\omega)$ leads to

$$Y(j\omega) \approx Y_d(j\omega) + \frac{D(j\omega)}{C(j\omega)H_0}, \tag{C.4}$$

i.e., the output tracks the desired output $Y_d(j\omega)$ and the frequency components of the disturbance in the bandwidth of $W(j\omega)$ produce an effect on the output which can be reduced by increasing $|C(j\omega)H_0|$. Furthermore, if the disturbance input is a constant, the steady-state output is not influenced by the disturbance as long as $C(s)$ has at least a pole at the origin.

Therefore, a feedback control system is capable of establishing a proportional relationship between the desired output and the actual output, as evidenced by (C.4). This equation, however, requires that the frequency content of the input (desired output) be inside the frequency range for which the loop gain is much greater than unity.

The previous considerations show the advantage of including a *proportional action* and an *integral action* in the controller $C(s)$, leading to the transfer function

$$C(s) = K_I \frac{1 + sT_I}{s} \tag{C.5}$$

of a *proportional-integral controller* (PI); T_I is the time constant of the integral action and the quantity $K_I T_I$ is called proportional sensitivity.

The adoption of a PI controller is effective for low-frequency response of the system, but it may involve a reduction of *stability margins* and/or a reduction of closed-loop system bandwidth. To avoid these drawbacks, a *derivative action* can be added to the proportional and integral actions, leading to the transfer function

$$C(s) = K_I \frac{1 + sT_I + s^2 T_D T_I}{s} \tag{C.6}$$

of a *proportional-integral-derivative controller* (PID); T_D denotes the time constant of the derivative action. Notice that physical realizability of (C.6) demands the introduction of a high-frequency pole which little influences the input/output relationship in the system bandwidth. The transfer function in (C.6) is characterized by the presence of two zeros which provide a stabilizing action and an enlargement of the closed-loop system bandwidth. Bandwidth enlargement implies shorter *response time* of the system, in terms of both variations of the reference signal and recovery action of the feedback system to output variations induced by the disturbance input.

The parameters of the adopted control structure shall be chosen so as to satisfy requirements on the system behaviour at *steady state* and during the *transient*. Classical tools to determine such parameters are *root locus* in the domain of the complex variable s or *Nichols chart* in the domain of the real angular frequency ω. The two tools are conceptually equivalent. Their potential is different in that root locus allows finding the control law that assigns the exact parameters of the closed-loop system time response, whereas Nichols chart allows specifying a controller that confers good transient and steady-state behaviour to the system response.

A feedback system with strict requirements on the steady-state and transient behaviour, typically, has a response that can be assimilated to that of a *second-order system*. In fact, even for closed-loop functions of greater order, it is possible to identify a pair of complex conjugate poles whose real part absolute value is smaller than the real part absolute values of the other poles. Such pair of poles is *dominant* in that its contribution to the transient response prevails over that of the other poles. It is then possible to approximate the input/output relationship with the transfer function

$$W(s) = \frac{k_W}{1 + \dfrac{2\zeta s}{\omega_n} + \dfrac{s^2}{\omega_n^2}} \tag{C.7}$$

which has to be realized by a proper choice of the controller. Regarding the values to assign to the parameters characterizing the transfer function in (C.7), the following remarks are in order. The constant k_W represents the input/output *steady-state gain*, which is equal to $1/H_0$ if $C(s)G(s)H_0$ has at least a pole at the origin. The *natural frequency* ω_n is the modulus of the complex conjugate poles, whose real part is given by $-\zeta\omega_n$ where ζ is the *damping ratio* of the pair of poles.

The influence of parameters ζ and ω_n on the closed-loop frequency response can be evaluated in terms of the resonance peak magnitude

$$M_r = \frac{1}{2\zeta\sqrt{1-\zeta^2}}$$

occurring at the resonant frequency

$$\omega_r = \omega_n\sqrt{1-2\zeta^2},$$

and of the 3 dB bandwidth

$$\omega_3 = \omega_n\sqrt{1-2\zeta^2+\sqrt{2-4\zeta^2+4\zeta^4}}.$$

A step input is typically used to characterize the transient response in the time domain. The influence of parameters ζ and ω_n on the *step response* can be evaluated in terms of the percentage of *overshoot*

$$s\% = 100\exp(-\pi\zeta/\sqrt{1-\zeta^2}),$$

of the *rise time*

$$t_r \approx \frac{1.8}{\omega_n},$$

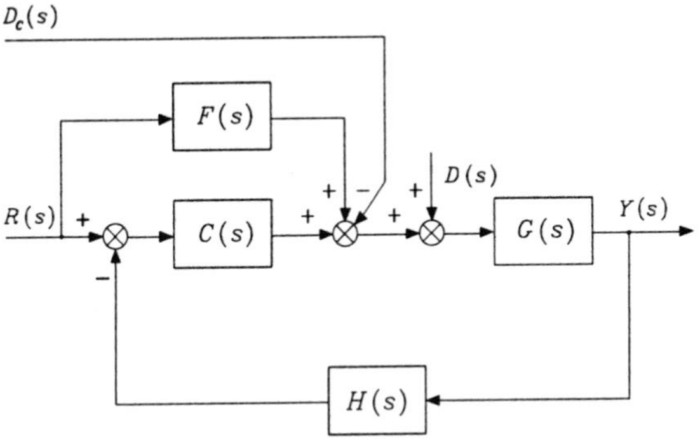

Figure C.2 Feedback control structure with feedforward compensation.

and of the *settling time* within 1%

$$t_s = \frac{4.6}{\zeta\omega_n}.$$

The adoption of a *feedforward compensation* action represents a feasible solution both for tracking a time-varying reference input and for enhancing rejection of the effects of a disturbance on the output. Consider the general scheme in Figure C.2. Let $R(s)$ denote a given input reference and $D_c(s)$ denote a computed estimate of the disturbance $D(s)$; the introduction of the feedforward action allows obtaining the input/output relationship

$$\begin{aligned} Y(s) = &\left(\frac{C(s)G(s)}{1+C(s)G(s)H(s)} + \frac{F(s)G(s)}{1+C(s)G(s)H(s)}\right)R(s) \\ &+ \frac{G(s)}{1+C(s)G(s)H(s)}\big(D(s)-D_c(s)\big). \end{aligned} \tag{C.8}$$

By assuming that the desired output is related to the reference input by a constant factor K_d and regarding the transducer as an instantaneous system ($H(s) \approx H_0 = 1/K_d$) for the current operating conditions, the choice

$$F(s) = \frac{K_d}{G(s)} \tag{C.9}$$

allows obtaining the input/output relationship

$$Y(s) = Y_d(s) + \frac{G(s)}{1+C(s)G(s)H_0}\big(D(s)-D_c(s)\big). \tag{C.10}$$

If $|C(j\omega)G(j\omega)H_0| \gg 1$, the effect of the disturbance on the output is further reduced by means of an accurate estimate of the disturbance.

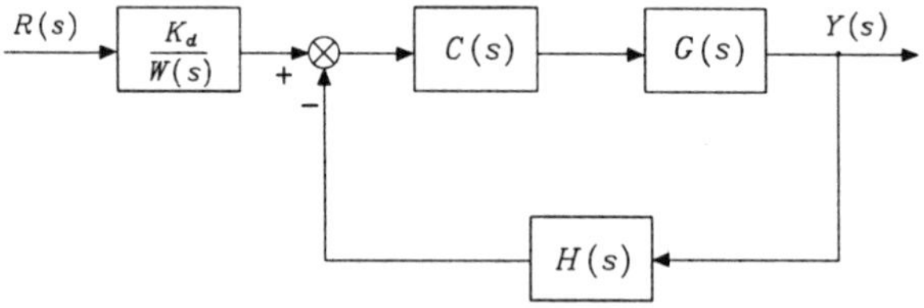

Figure C.3 Feedback control structure with inverse model technique.

Feedforward compensation technique may lead to a solution, termed *inverse model control*, illustrated in the scheme of Figure C.3. It should be remarked, however, that such a solution is based on dynamics cancellation, and thus it can be employed only for a minimum-phase system, *i.e.*, a system whose poles and zeros have all strictly negative real parts. Further, one should consider physical realizability issues as well as effects of parameter variations which prevent perfect cancellation.

C.2 Control of Nonlinear Mechanical Systems

If the system to control does not satisfy the linearity property, the control design problem becomes more complex. The fact that a *system* is qualified as *nonlinear*, whenever linearity does not hold, leads to understanding how it is not possible to resort to general techniques for control design, but it is necessary to face the problem for each class of nonlinear systems which can be defined through imposition of special properties.

On the above premise, the control design problem of nonlinear systems described by the dynamic model

$$\boldsymbol{H}(\boldsymbol{x})\ddot{\boldsymbol{x}} + \boldsymbol{h}(\boldsymbol{x}, \dot{\boldsymbol{x}}) = \boldsymbol{u} \tag{C.11}$$

is considered, where $[\,\boldsymbol{x}^T \quad \dot{\boldsymbol{x}}^T\,]^T$ denotes the $(2n \times 1)$ *state* vector of the system, $\boldsymbol{u}$ is the $(n \times 1)$ *input* vector, $\boldsymbol{H}(\boldsymbol{x})$ is an $(n \times n)$ *positive definite* (and thus invertible) matrix depending on $\boldsymbol{x}$, and $\boldsymbol{h}(\boldsymbol{x}, \dot{\boldsymbol{x}})$ is an $(n \times 1)$ vector depending on state. Several *mechanical systems* can be reduced to this class, including manipulators with rigid links and joints.

The *control* law can be found through a nonlinear compensating action obtained by choosing the following *nonlinear state feedback* law (*inverse dynamics* control):

$$\boldsymbol{u} = \widehat{\boldsymbol{H}}(\boldsymbol{x})\boldsymbol{v} + \widehat{\boldsymbol{h}}(\boldsymbol{x}, \dot{\boldsymbol{x}}), \tag{C.12}$$

where $\widehat{\boldsymbol{H}}(\boldsymbol{x})$ and $\widehat{\boldsymbol{h}}(\boldsymbol{x})$ respectively denote the *estimates* of the terms $\boldsymbol{H}(\boldsymbol{x})$ and $\boldsymbol{h}(\boldsymbol{x})$, computed on the basis of measures on the system state, and $\boldsymbol{v}$ is a new control input to be defined later. In general, it is

$$\begin{aligned} \widehat{\boldsymbol{H}}(\boldsymbol{x}) &= \boldsymbol{H}(\boldsymbol{x}) + \Delta\boldsymbol{H}(\boldsymbol{x}) \\ \widehat{\boldsymbol{h}}(\boldsymbol{x}, \dot{\boldsymbol{x}}) &= \boldsymbol{h}(\boldsymbol{x}, \dot{\boldsymbol{x}}) + \Delta\boldsymbol{h}(\boldsymbol{x}, \dot{\boldsymbol{x}}), \end{aligned} \tag{C.13}$$

because of the unavoidable modelling approximations or as a consequence of an intentional simplification in the compensating action. Substituting (C.12) into (C.11) and accounting for (C.13) yields

$$\ddot{\boldsymbol{x}} = \boldsymbol{v} + \boldsymbol{z}(\boldsymbol{x}, \dot{\boldsymbol{x}}, \boldsymbol{v}) \tag{C.14}$$

where

$$\boldsymbol{z}(\boldsymbol{x}, \dot{\boldsymbol{x}}, \boldsymbol{v}) = \boldsymbol{H}^{-1}(\boldsymbol{x})\big(\Delta\boldsymbol{H}(\boldsymbol{x})\boldsymbol{v} + \Delta\boldsymbol{h}(\boldsymbol{x}, \dot{\boldsymbol{x}})\big).$$

If *tracking* of a trajectory $(\boldsymbol{x}_d(t), \dot{\boldsymbol{x}}_d(t), \ddot{\boldsymbol{x}}_d(t))$ is desired, the tracking error can be defined as

$$\boldsymbol{e} = \begin{bmatrix} \boldsymbol{x}_d - \boldsymbol{x} \\ \dot{\boldsymbol{x}}_d - \dot{\boldsymbol{x}} \end{bmatrix} \tag{C.15}$$

and it is necessary to derive the error dynamics equation to study convergence of the actual state to the desired one. To this purpose, the choice

$$\boldsymbol{v} = \ddot{\boldsymbol{x}}_d + \boldsymbol{w}(\boldsymbol{e}), \tag{C.16}$$

substituted into (C.14), leads to the error equation

$$\dot{\boldsymbol{e}} = \boldsymbol{F}\boldsymbol{e} - \boldsymbol{G}\boldsymbol{w}(\boldsymbol{e}) - \boldsymbol{G}\boldsymbol{z}(\boldsymbol{e}, \boldsymbol{x}_d, \dot{\boldsymbol{x}}_d, \ddot{\boldsymbol{x}}_d), \tag{C.17}$$

where the $(2n \times 2n)$ and $(2n \times n)$ matrices, respectively,

$$\boldsymbol{F} = \begin{bmatrix} \boldsymbol{O} & \boldsymbol{I} \\ \boldsymbol{O} & \boldsymbol{O} \end{bmatrix} \qquad \boldsymbol{G} = \begin{bmatrix} \boldsymbol{O} \\ \boldsymbol{I} \end{bmatrix}$$

follow from the error definition in (C.15). Control law design consists of finding the error function $\boldsymbol{w}(\boldsymbol{e})$ which makes (C.17) *globally asymptotically stable*[1], *i.e.*,

$$\lim_{t\to\infty} \boldsymbol{e}(t) = \boldsymbol{0}.$$

In the case of *perfect* nonlinear compensation ($\boldsymbol{z}(\cdot) = \boldsymbol{0}$), the simplest choice of the control action is the *linear* one

$$\begin{aligned} \boldsymbol{w}(\boldsymbol{e}) &= -\boldsymbol{K}_P(\boldsymbol{x}_d - \boldsymbol{x}) - \boldsymbol{K}_D(\dot{\boldsymbol{x}}_d - \dot{\boldsymbol{x}}) \\ &= [\,-\boldsymbol{K}_P \quad -\boldsymbol{K}_D\,]\,\boldsymbol{e}, \end{aligned} \tag{C.18}$$

where asymptotic stability of the error equation is ensured by choosing *positive definite* matrices $\boldsymbol{K}_P$ and $\boldsymbol{K}_D$. The error transient behaviour is determined by the eigenvalues of the matrix

$$\boldsymbol{A} = \begin{bmatrix} \boldsymbol{O} & \boldsymbol{I} \\ -\boldsymbol{K}_P & -\boldsymbol{K}_D \end{bmatrix} \tag{C.19}$$

[1] *Global* asymptotic stability is invoked to remark that the equilibrium state is asymptotically stable for any perturbation.

characterizing the error dynamics

$$\dot{e} = Ae. \tag{C.20}$$

If compensation is *imperfect*, then $z(\cdot)$ cannot be neglected and the error equation in (C.17) takes on the general form

$$\dot{e} = f(e). \tag{C.21}$$

It may be worth choosing the control law $w(e)$ as the sum of a nonlinear term and a linear term of the kind in (C.18); in this case, the error equation can be written as

$$\dot{e} = Ae + k(e), \tag{C.22}$$

where A is given by (C.19) and $k(e)$ is available to make the system globally asymptotically stable. The equations in (C.21) and (C.22) express nonlinear differential equations of the error. To test for stability and obtain advise on the choice of suitable control actions, one may resort to *Lyapunov direct method* illustrated below.

C.3 Lyapunov Direct Method

The philosophy of *Lyapunov direct method* is the same as that of most methods used in control engineering to study stability; namely, testing for stability without solving the differential equations describing the dynamic system.

This method can be presented in short on the basis of the following reasoning. If it is possible to associate an energy-based description with a (linear or nonlinear) autonomous dynamic system and, for each system state with the exception of the equilibrium state, the time rate of such energy is negative, then energy decreases along any system trajectory until it attains its minimum at the equilibrium state; this argument justifies an intuitive concept of stability.

With reference to (C.21), by setting $f(0) = 0$, the *equilibrium state* is $e = 0$. A scalar function $V(e)$ of the system state, continuous together with its first derivative, is defined a *Lyapunov function* if the following properties hold:

$$\begin{aligned} V(e) &> 0 & \forall e &\neq 0 \\ V(e) &= 0 & e &= 0 \\ \dot{V}(e) &< 0 & \forall e &\neq 0 \\ V(e) &\to \infty & \|e\| &\to \infty. \end{aligned}$$

The existence of such a function ensures *global asymptotic stability* of the equilibrium $e = 0$. In practice, the equilibrium $e = 0$ is globally asymptotically stable if a positive definite, radially unbounded function $V(e)$ is found so that its time derivative along the system trajectories is negative definite.

If positive definiteness of $V(e)$ is realized by the adoption of a *quadratic form*, *i.e.*,

$$V(e) = e^T Q e \tag{C.23}$$

with Q a symmetric positive definite matrix, then in view of (C.21) it follows

$$\dot{V}(e) = 2e^T Q f(e). \tag{C.24}$$

If $f(e)$ is so as to render the function $\dot{V}(e)$ negative definite, the function $V(e)$ is a *Lyapunov function*, since the choice (C.23) allows proving system global asymptotic stability. If $\dot{V}(e)$ in (C.24) is not negative definite for the given $V(e)$, nothing can be inferred on the stability of the system, since Lyapunov method gives only a *sufficient* condition. In such cases one should resort to different choices of $V(e)$ in order to find, if possible, a negative definite $\dot{V}(e)$.

In the case when the property of negative definiteness does not hold, but $\dot{V}(e)$ is only *negative semi-definite*

$$\dot{V}(e) \leq 0,$$

global asymptotic stability of the equilibrium state is ensured if the only system trajectory for which $\dot{V}(e)$ is *identically* null ($\dot{V}(e) \equiv 0$) is the equilibrium trajectory $e \equiv 0$ (a consequence of *La Salle's theorem*).

Finally, consider the stability problem of the nonlinear system in the form (C.22); on the assumption that $k(0) = 0$, it is easy to verify that $e = 0$ is an equilibrium state for the system. The choice of a Lyapunov function candidate as in (C.23) leads to the following expression for its derivative:

$$\dot{V}(e) = e^T(A^T Q + QA)e + 2e^T Q k(e). \tag{C.25}$$

By setting

$$A^T Q + QA = -P, \tag{C.26}$$

the expression in (C.25) becomes

$$\dot{V}(e) = -e^T P e + 2e^T Q k(e). \tag{C.27}$$

The matrix equation in (C.26) is said to be a *Lyapunov equation*; for any choice of a symmetric positive definite matrix P, the solution matrix Q exists and is symmetric positive definite if and only if the eigenvalues of A have all negative real parts. Since matrix A in (C.19) verifies such condition, it is always possible to assign a positive definite matrix P and find a positive definite matrix solution Q to (C.26). It follows that the first term on the right-hand side of (C.27) is negative definite and the stability problem is reduced to searching a control law so that $k(e)$ renders the total $\dot{V}(e)$ negative (semi-)definite.

Bibliography

Franklin G.F., Powell J.D., Emami-Naeini A. (1994) *Feedback Control of Dynamic Systems*. 3rd ed., Addison-Wesley, Reading, Mass.

Ogata K. (1997) *Modern Control Engineering*. 3rd ed., Prentice-Hall, Upper Saddle River, N.J.

La Salle J., Lefschetz S. (1961) *Stability by Lyapunov's Direct Method*. Academic Press, New York.

Slotine J.-J.E., Li W. (1991) *Applied Nonlinear Control*. Prentice-Hall, Englewood Cliffs, N.J.

Index

W